The Cosmic Microwave Background

NATO ASI Series

Advanced Science Institutes Series

A Series presenting the results of activities sponsored by the NATO Science Committee, which aims at the dissemination of advanced scientific and technological knowledge, with a view to strengthening links between scientific communities.

The Series is published by an international board of publishers in conjunction with the NATO Scientific Affairs Division

A Life Sciences	Plenum Publishing Corporation
B Physics	London and New York
C Mathematical and Physical Sciences	Kluwer Academic Publishers
D Behavioural and Social Sciences	Dordrecht, Boston and London
E Applied Sciences	
F Computer and Systems Sciences	Springer-Verlag
G Ecological Sciences	Berlin, Heidelberg, New York, London,
H Cell Biology	Paris and Tokyo
I Global Environmental Change	

PARTNERSHIP SUB-SERIES

1. **Disarmament Technologies**	Kluwer Academic Publishers
2. **Environment**	Springer-Verlag / Kluwer Academic Publishers
3. **High Technology**	Kluwer Academic Publishers
4. **Science and Technology Policy**	Kluwer Academic Publishers
5. **Computer Networking**	Kluwer Academic Publishers

The Partnership Sub-Series incorporates activities undertaken in collaboration with NATO's Cooperation Partners, the countries of the CIS and Central and Eastern Europe, in Priority Areas of concern to those countries.

NATO-PCO-DATA BASE

The electronic index to the NATO ASI Series provides full bibliographical references (with keywords and/or abstracts) to more than 50000 contributions from international scientists published in all sections of the NATO ASI Series.
Access to the NATO-PCO-DATA BASE is possible in two ways:

– via online FILE 128 (NATO-PCO-DATA BASE) hosted by ESRIN,
Via Galileo Galilei, I-00044 Frascati, Italy.

– via CD-ROM "NATO-PCO-DATA BASE" with user-friendly retrieval software in English, French and German (© WTV GmbH and DATAWARE Technologies Inc. 1989).

The CD-ROM can be ordered through any member of the Board of Publishers or through NATO-PCO, Overijse, Belgium.

Series C: Mathematical and Physical Sciences – Vol. 502

The Cosmic Microwave Background

edited by

C. H. Lineweaver, J. G. Bartlett, A. Blanchard
Observatoire Astronomique de Strasbourg,
Strasbourg, France

M. Signore
Ecole normale supérieure,
Paris, France

and

J. Silk
Departments of Astronomy and Physics,
University of California at Berkeley,
Berkeley, CA, U.S.A.

Kluwer Academic Publishers

Dordrecht / Boston / London

Published in cooperation with NATO Scientific Affairs Division

Proceedings of the NATO Advanced Study Institute on
The Cosmological Background Radiation
Strasbourg, France
May 27–June 7, 1996

A C.I.P. Catalogue record for this book is available from the Library of Congress.

ISBN-13:978-94-010-6512-2   e-ISBN-13:978-94-009-0051-6
DOI:10.1007/978-94-009-0051-6

Published by Kluwer Academic Publishers,
P.O. Box 17, 3300 AA Dordrecht, The Netherlands.

Sold and distributed in the U.S.A. and Canada
by Kluwer Academic Publishers,
101 Philip Drive, Norwell, MA 02061, U.S.A.

In all other countries, sold and distributed
by Kluwer Academic Publishers,
P.O. Box 322, 3300 AH Dordrecht, The Netherlands.

Dedication

We dedicate these proceedings to the people who paid for it: taxpayers of the NATO alliance.

Table of Contents

Preface

This volume consists of invited lectures presented at the NATO Advanced Study Institute "The Cosmic Background Radiation", held in Strasbourg, France from May 27 to June 7, 1996. The aim of the school was to provide students and young researchers with an overview of the cosmic microwave background (CMB) radiation and an understanding of the latest research in this field which seems to be expanding faster than the Universe. The lectures cover the anisotropy, spectrum and polarization of the CMB from both the observational and theoretical viewpoints. The importance of the CMB is discussed in lectures on general relativity, structure formation, inflation, nucleosynthesis and primordial molecules. Statistics and data analysis techniques useful for CMB analysis are covered in detail.

We would like to thank the NATO Science Committee for generous support. The lectures took place at Louis Pasteur University with logistical support from the Strasbourg Astronomical Observatory. We are grateful to both of these institutions and their personnel. We are very thankful to Jim Bartlett and Charley Lineweaver who have strongly contributed to various aspects of the preparation of this Advanced Study Institute. Special thanks to Domingos Barbosa, Erwan Lastennet, and Houri Ziaeepour for making sure the school ran smoothly. The organization of this school would not have been possible without the active support of Michèle Michel.

Finally, we thank the lecturers and contributors to this volume for the high quality of their presentations.

Alain Blanchard, Monique Signore & Joseph Silk
Directors of the School

International Scientific Committee:
Bartlett, J.G. (Strasbourg, France), Blanchard, A. (Strasbourg, France), Martinez E. (Santander, Spain), Peacock J.(Edinburgh, Scotland), Puget J.L. (Orsay, France), Peebles J. (Princeton, USA), Signore, M. (Paris, France), Silk, J. (Berkeley, USA), Smoot, G.F. (Berkeley, USA)

Local Organizing Committee:
Jim Bartlett, Alain Blanchard, Charley Lineweaver, Michèle Michel, David Valls-Gabaud

Editors:
C.H. Lineweaver, J.G. Bartlett, A. Blanchard, M. Signore, J. Silk

List of Participants

Abel, Tom, Max Planck Institute, Garching, Germany
Aghanim, Nabila, IAS, Orsay, France
Barbosa, Domingos, Strasbourg Observatory, France
Barreiro, Rita Belen, Santander, Spain
Bayrakceken, Fuat, Ankara University of Sciences, Turkey
Bernadeau, Francis, SPT, CEA, Saclay, France
Bonanno, Alfio, Louis Pasteur University, Strasbourg, France
Bracco, Christian, Observatoire de Haute Provence, France
Caldwell, Robert, DAMTP, Cambridge, England
Cheung, Charlotte, Imperial College, London, England
Escoubes, Bruno, CRN, Strasbourg, England
Fresneau, Alain, Strasbourg Observatory, France
Gardini, Alessandro, University of Milan, Italy
Gawiser, Eric, University of California, Berkeley, U.S.A.
Giardino, Giovanna, University of Milan, Italy
Heck, André, Strasbourg Observatory, France
Hivon, Eric, T.A.C., Copenhagen, Denmark
Hwang, Jai-chan, Kyungpook National University, Taegu, Korea
Jeppsson, Thorbjoern List, T.A.C., Copenhagen, Denmark
Korn, Dyveke Raisa, T.A.C., Copenhagen, Denmark
Lastennet, Erwan, Strasbourg Observatory, France
Lehner, Nicolas, Strasbourg Observatory, France
Lesgourgues, Julien, Univ. of Tours, France
Maiani, Tito, University of Rome "La Sapienza", Italy
Maoli, Roberto, DEMIRM, Paris, France
Mei, Simona, Strasbourg Observatory, France
Melchiorri, Alessandro, IFA-CNR, University of Rome "La Sapienza", Italy
Mendes, Luis, Instituto Superior Tecnico, Lisboa, Portugal
Metcalf, Ben, University of California, Berkeley, U.S.A.
Montecchio, Chiara, University of Rome "La Sapienza", Italy

Noh, Mrs. Hyerim, Korea Astronomy Observatory, Taejeon, Korea
Olivo Melchiorri, Bianca, IFA-CNR, Roma, Italy
Pandozy, Giovanna, International Space University, Strasbourg, France
Pierpaoli, Elena, SISSA, Trieste, Italy
Pisano, Giampaolo, University of Rome "La Sapienza", Italy
Prunet, Simon, IAS, Orsay, France
Retzlaff, Joerg, Astrophysical Institute, Potsdam, Germany
Roukema, Boudewijn, National Astronomical Observatory, Mitaka, Japan
Sadat, Rachida, Strasbourg Observatory, France
Schneider, Raffaella, University of Rome "La Sapienza", Italy
Stirling, Alison, Institute for Astronomy, Edinburgh, Scotland
Wagner, Stefan, Landessternwarte Heidelberg-Koenigstuhl, Germany
Wandelt, Benjamin, Imperial College, London, England
Wang, Yun, Fermilab, Batavia, U.S.A.
Ziaeepour, Houri, ESO, Garching, Germany

List of Contributors

Emory F. Bunn
Astronomy Department, Campbell Hall
University of California at Berkeley
Berkeley, CA 94720, U.S.A.

Bernard J.T. Jones
NORDITA, Nordisk Institut for Teoretisk Fysik
Blegdamsvej 17, DK-2100
Copenhagen, Denmark

Alessandro Melchiorri
Dept. of Physics, University of Rome 1
Piazzale Aldo Moro 2
Rome, 00141, Italy

Francesco Melchiorri
Dept. of Physics, University of Rome 1
Piazzale Aldo Moro 2
Rome, 00141, Italy

Bruce Partridge
Department of Astronomy
Haverford College, PA 19041, U.S.A.

Pierre Salati
Laboratoire de Physique Théorique
ENSLAPP, LAPP, B.P. 110
74941 Annecy-Le-Vieux cedex, France

Jose L. Sanz
Instituto de Fisica de Cantabria
Facultad de Ciencias, Universidad de Cantabria
Av. de los Castros S.A. 39005 Santander, Spain

Monique Signore
Radioastronomie, Ecole Normale Supérieure
DEMIRM/ENS, 24, rue Lhomond
75231 Paris cedex 05, France

Joseph Silk
Astronomy Department
601 Campbell Hall, University of California
Berkeley, CA 94720-3411, U.S.A.

George Smoot
Lawrence Berkeley Laboratories
Berkeley, CA 94720, U.S.A.

Albert Stebbins
Nasa/Fermilab Astrophysics Center
Fermi National Accelerator Laboratory
Batavia, IL 60510-0500, U.S.A.

Michael S. Turner
Nasa/Fermilab Astrophysics Center
Fermi National Accelerator Laboratory
Batavia, IL 60510-0500, U.S.A.

Nicola Vittorio
University of Rome II "Tor Vergata"
Rome, Italy

AN INTRODUCTION TO CBR STUDIES: SPECTRUM, DEGREE-SCALE FLUCTUATIONS, FOREGROUNDS AND INTEROMETRY

R.B. PARTRIDGE
HAVERFORD COLLEGE
HAVERFORD, PA 19041 USA

1. Abstract

From the viewpoint of an observer, I first review briefly the current status of CBR measurements. Then I review the observational techniques used in measurements of the temperature and (very briefly) of the degree-scale anisotropy of the CBR. Potential observational problems and sources of noise and systematic errors in such measurements are considered next.

I then try to project the status of observational programs planned for the next few years. One such program is the measurement of T_0 at long wavelengths: Is a $\pm 1\%$ measurement of T_0 possible at $\lambda = 10 - 20\ cm$?

The final part of this presentation is an introduction to the use of interferometry in CBR observations, in searches for both fine scale ($\lesssim 1'$) and degree-scale anisotropies.

2. Introduction

It is an honor to be asked to open this workshop. As it happens, it was almost exactly 30 years ago when I gave my first talk on the cosmic microwave background (in Miami). I have no hesitation in saying that the present is the most exciting time of those past 30 years. We have come to appreciate the power of measurements of the spectrum of fluctuations in the CBR, and, with the new satellite programs now approved, we should soon be able to make full use of the information encoded in the CBR.

In a sense my remarks serve as an overture to this workshop. In the case of an opera, the overture serves two purposes. The first is to allow late comers to settle in. The second is to reveal and give a preview of all the good melodies to follow. Giving you a preview of all the good things to

1

C. H. Lineweaver et al. (eds.), The Cosmic Microwave Background, 1–31.
© 1997 *Kluwer Academic Publishers.*

follow is what I have been asked to do. As I go along, however, I will try also to give some sense of the history of the field. It is my belief that CBR studies are now a mature enough field to begin to deserve a history. Since I am an observer, not a theorist, I will emphasize primarily observational techniques and difficulties, though I may pause from time to time to present some very simple, pedagogical ideas from theory to provide a context in which to understand the observations.

3. The 8.5 Observable Parameters of the CBR

I claim that eight and a half parameters encode most of the astrophysical and cosmological information provided by the CBR. These are listed in Table 1, which also provides the most recent observational values and appropriate references. (The "1/2 parameter" is evidence for or limits on distortions in the CBR spectrum.)

Theorists may object that the division of the angular spectrum of fluctuations in the CBR into four bins in angular scale is quite arbitrary. I agree; that division is determined primarily by observational considerations, particularly the techniques used to make measurements at the various angular scales listed, *not* by the underlying physics, which will be described by Ted Bunn among others here.

In my lectures, I want to pick three of these 8 1/2 parameters for special emphasis because they provide a good introduction to the observational difficulties, successes and prospects in the field. The three are the temperature of the microwave background (T_0) and possible spectral distortions; anisotropies on degree scales (in part because terrestrial and astronomical foregrounds present a particular problem for such observations and in part because it is an area of great current interest); and observations on scales ($\lesssim 1'$), because they require a new technology and are relevant to questions of reionization.

4. T_0 and Spectral Distortions

Let me begin with a bit of history by reminding you that in the early years of CBR studies, there were real doubts that the signal detected by Penzias and Wilson (1965) was in fact either cosmic or blackbody. The aim of most early spectral measurements was to decide that question. If measurements made at a variety of different wavelengths all produced the same value for T_0, then the radiation was indeed blackbody (and the argument for cosmic origin was strengthened). By the late 1960's, both ground-based observations and CN measurements had confirmed the approximately blackbody nature of the spectrum. Even now, however, the detection of spectral distortions, rather than more and more precise determinations of T_0 remains the main goal of temperature measurements.

TABLE 1. NEWEST RESULTS ON 8.5 FUNDAMENTAL PARAMETERS
OF THE COSMIC MICROWAVE BACKGROUND RADIATION (CBR)

	Quantity	Value or Upper Limit (generally at 95% confidence)	Reference
1.	Temperature T_0, COBE	2.728 ± 0.004 K	Fixsen et al., in press; Mather et al., Ap. J. **420**, 439
	T_0 from dipole, COBE	2.714 ± 0.022 K	Fixsen et al., Ap. J., **420**, 445
	T_0, rocket	2.736 ± 0.017 K	Gush et al., PRL, **65**, 537
	T_0, non-COBE	2.732 ± 0.028 K	Partridge, 3 K
1.5	Distortions of spectrum	$\|y\| \leq 1.5 \times 10^{-5}$ $\|\mu\| \leq 0.9 \times 10^{-4}$	Fixsen et al., in press; Mather et al., Ap. J. **420**, 439
2.	Dipole, T_1	$3.358 \pm .002 \pm .023$ $\alpha = 11^h\,12^m \pm 0\stackrel{m}{.}4, \delta = -7\stackrel{\circ}{.}22 \pm 0\stackrel{\circ}{.}08$ 3.343 ± 0.016 mK	Lineweaver et al., Ap. J. **470**, 38 Fixsen et al., Ap. J., **420**, 445
3.	Formal quadrupole, T_2	4–28 μK	Kogut et al., Ap. J. Lett., **464**, L5
4.	Large-scale polarization	<200 μK	Lubin et al., Ap. J. **273**, L51
5.	$\Delta T/T \sim 10°$ scale, COBE	29 ± 2 μK rms	Banday et al., Ap. J. Lett., submitted
6.	$\Delta T/T \sim 0°5$-$2°$ scale	2–3.5×10^{-5}	Netterfield et al., Ap. J., **445**, L69 Ruhl et al., Ap. J., **453**, L1 Cheng et al., Ap. J., **456**, L71
7.	$\Delta T/T \sim 10'$ scale	$\leq 1.9 \times 10^{-5}$	Readhead et al., Ap. J., **346**, 566
8.	$\Delta T/T \leq 1'$ scale	$\leq 2 \times 10^{-5}$	Fomalont et al., Ap. J., **404**, 8 Subrahmanyan et al., MN, **263**, 416

4.1. MEASURING T_0

The early ground-based measurements, particularly those of the Princeton group (Wilkinson, 1967; Stokes, Partridge and Wilkinson, 1967), illustrate well the observational techniques used and problems encountered in making measurements of T_0.

First, measurements of T_0 must obviously be *absolute* measurements. For that reason, both exact calibration of the detector and an exact zero point are needed. I sketch below in Fig. 1a an *ideal* experiment designed to measure T_0. In passing, let me mention that the COBE FIRAS instrument came very close to meeting this ideal. Measurements made from the ground, however, encounter a variety of observational problems. Since I believe the next wave of spectral measurements will be made from the ground, let me describe some of the problems that can be encountered. Fig. 1b shows a cartoon of the actual, not the ideal, situation. On the right, we see an antenna pointed down into a cold–load calibrator which provides a zero point for the measurements. In virtually all ground-based experiments, liquid he-

4

lium is used to provide the zero point; since liquid helium is not emissive at microwave frequencies, one employs a (nominally) perfect absorber immersed in, and in thermal equilibrium with, the liquid helium, as sketched in Fig. 1b. The signal picked up by the antenna is then given by Eq. 1,

$$S_c = g(T_{zp} + T_{wall} + T_{window}) \qquad (1)$$

where g is the conversion between antenna temperature and the measured signal in volts and T_{zp} is the absorber temperature. Here and throughout, T will be used to express *antenna* temperature , which is by definition directly proportional to the power received. It is related to thermodynamic temperature T_θ as follows (e.g., Kraus, 1966): $T = T_\theta(\frac{x}{e^x-1})$ with $x = \frac{h\upsilon}{kT}$. Evidently, $T \to T_\theta$ for low x or $\lambda > 1$ cm. In addition to the zero point temperature, radiation may enter the antenna from the walls or a window, (needed to prevent condensation of atmospheric gasses on the cold wall of the calibrator).

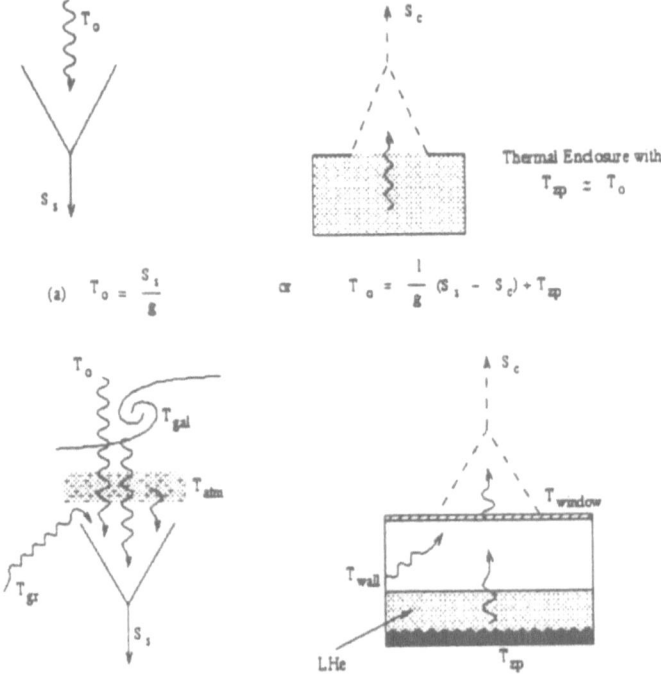

Figure1.(a) An ideal measurement of T_0, in which no foregrounds appear and T_0 can be determined by direct comparison to the temperature of a zero point load, as $T_0 = \frac{1}{g}(S_s - S_c) + T_{zp}$ (b) The real (and more complicated) situation discussed in the text.

Now consider the same antenna pointing towards the zenith. The signal entering the antenna is then given by Eq. 2,

$$S_s = g(T_0 + T_{gr} + T_{atm} + T_{Gal} + T_{offset})$$ (2)

where T_{atm} and T_{Gal} are the emission from the earth's atmosphere and the Galaxy, respectively, and T_{gr} is radiation that may be diffracted into the antenna from nearby hot surroundings (the ground). Finally, T_{offset} is a term which collects all the changes in the measured zero point level that may occur when the antenna is moved from its downward pointing to its upward pointing position. Measurements of T_0 are made by starting with the difference signal $S_s - S_c$ and measuring the eight other variables that occur in Eqs. 1 and 2. In the next few pages, I will list each of the eight, and describe the techniques used to measure it or minimize it by both ground-based observers and the COBE FIRAS satellite mission.

g Measure by using calibration loads of different temperature in front of the antenna. To minimize the effect of uncertainties in g, particularly nonlinearities, keep T_c and T_s close to each other (Figure 1). In the FIRAS experiment, the effect of gain variations was essentially eliminated by seeking a null signal by adjusting T_c to match T_s.

T_{wall} Use smooth, non–emissive walls, and keep them as cold as possible. The FIRAS cold load enclosure was essentially isothermal at $\sim$ $3K$.

T_{window} Keep thin and non–emissive. COBE needed no window.

T_{zp} As noted above, generally an absorber in liquid helium is used. Small corrections are needed for the reflectivity of the helium-air interface and for possible imperfect emissivity of the absorber. FIRAS used an isothermal enclosure which dropped into the antenna mouth and whose temperature could be varied to match the sky temperature closely.

T_{gr} All measurements of T_0, including COBE's, used ground screens to reflect away radiation that might be received from the ground (Earth). Fig. 2 shows the ground screens used in the 1967 experiment on White Mountain (Stokes *et al.* 1967). Careful antenna design can also be used to minimize the sidelobes of the antenna employed, thus reducing the sensitivity to ground emission.

Clearly, as a simple consequence of diffraction, minimizing ground radiation becomes more and more difficult as the wavelength of the observation increases.

6

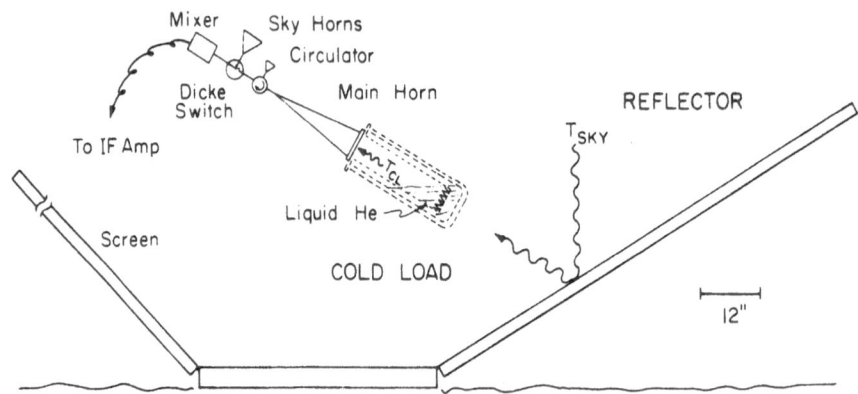

RADIOMETER

Figure 2. Schematic of the radiometers used by Wilkinson (1967)and by Stokes *et al.* (1967) to measure the CBR spectrum at 3.2, 1.58 and 0.856 cm wavelength. Note the use of a reflector: 'T_{sky},' here includes foreground emission. One ground screen is shown.

T_{offset} Often a major source of systematic error in such experiments. In the 1967 experiments (Wilkinson, 1967; Stokes, Partridge and Wilkinson, 1967), we sought to minimize this by not moving the horn antenna. It was positioned to look down (at an angle) into the cold load Dewar. Zenith observations were made by removing the cold load and placing a properly oriented (Fig. 2) plane mirror in the beam. Other experimenters, who move the apparatus (e.g., Smoot *et al.* 1987; Bersanelli *et al.* 1994), have sought to reduce T_{offset} by using carefully designed microwave switches. The COBE satellite got around this problem by essentially putting the cold load calibrator on top of the antenna (easier to do in zero gravity than on the ground!). That "cold load above" technique was pioneered by a Russian group (Puzanov *et al.* 1968; Kislyakov *et al.* 1971) but has the obvious disadvantage of needing a much stronger "window" to separate the cold load from the antenna and also to hold the cryogen in, in the case of ground-based observations.

4.2. CONTROLLING SOURCES OF ERROR

The six parameters listed above can all be controlled to some degree by the observer. In an ideal experiment, g would be known exactly, as would T_{zp}, and all the other terms would be as close to zero as possible. In a sense, T_{atm} can also be controlled by the observer. COBE reduced this signal to zero. Ground–based observers seek high altitude and dry sites (since water vapor produces a major and frequency dependent amount of atmospheric emission) and choose their frequencies wisely to avoid atmospheric molecular lines. After thus minimizing T_{atm}, ground-based observers then generally measure the remaining contribution. In most cases, the technique used is one introduced by Robert Dicke during the Second World War—tipping the antenna to various zenith angles, and measuring the increase in received radiation as the slant height through the atmosphere increases with increasing zenith angle ($\propto \sec z$). To avoid moving the antenna, which might change T_{offset}, the zenith angle can be varied by using a flat mirror, a technique introduced by the Princeton group (Wilkinson, 1967; Fig.2). One then makes (small) corrections for angle–dependent emission by the mirror.

If we make the assumption that T_{gr} and T_{Gal} as well as the offset remain unchanged as the beam is switched from the zenith to a zenith angle z where

$$S_s(z) = g(T_0 + T_{gr} + T_{Gal} + T_{offset}) + gT_{atm} \sec z,$$

we find

$$T_{atm} = \frac{S_s(z) - S_s(0)}{g(\sec z - 1)}. \tag{3}$$

However, as we will soon see, T_{Gal} will *not* necessarily be constant as z changes except under special circumstances.[1]

So let us turn now to T_{Gal}, an unavoidable source of systematic error in all measurements of T_0. Galactic emission is a minimum near the Galactic poles, where the *approximate* formula

$$T_{Gal} \approx 2 \left(\frac{\nu}{1\,GHz} \right)^{-2.8} K \tag{4}$$

holds. It may be seen that Galactic emission is a negligible problem for T_0 measurements at wavelengths $\lambda \lesssim$ 4-5 cm, i.e., $\nu \gtrsim$ 6-8 GHz. In addition, as large–scale maps of Galactic radio emission are improved and pushed to shorter wavelengths, our knowledge of Galactic emission is improving, though slowly, allowing us to correct our sky measurements for T_{Gal}.

Let me also mention explicitly that the variation of Galactic emission across the sky complicates, especially at long wavelengths, the technique described above for measuring atmospheric emission. As the antenna is tipped to change the zenith angle, not only the atmospheric emission but Galactic emission can change. Thus there is a kind of "cross-talk" between Galactic and atmospheric emission that makes separating the two terms difficult. I will have more to say about this point in §7.

---

[1]Care must also be taken to ensure that ground pickup does not change; see e.g., Partridge *et al.* 1984.

4.3. DETECTORS

A few remarks now about detectors. First, sensitivity is not at all a problem in measuring T_0. In the case of relatively narrow band ($\Delta\nu/\nu \sim 0.1$) radiometers, the rms statistical noise expressed in antenna temperature is

$$\Delta T_{rms} = \frac{T_{sys}}{\sqrt{\Delta\nu\Delta t}} \tag{5}$$

where Δt is the duration of the observation, and T_{sys} is the system temperature. At any frequency of interest to CBR studies, T_{sys} can easily be kept below 1000 K, and $\Delta\nu$ above 100 MHz. Using these figures, we see that 1% measurements of T_0 can be made in a minute or less. While sensitivity is no problem, gain stability and linearity of the receiver are; hence the need to keep $S_s - S_c$ small.

Receivers used for observations of the temperature of the CBR fall into two general classes. The first are relatively narrow bandwidth superheterodyne or direct amplification HEMT receivers in which the bandwidth is typically fixed by one or more mechanical or electronic elements in the receiver, such as the wave guide switch, or first amplifier. For these, $\Delta\nu/\nu \lesssim 0.1$. The second are bolometric detectors where the bandwidth is determined by external filters. For the latter (see Partridge, 1995, Chap. 4), care has to be taken to avoid high or low frequency leaks through the filter, or emission by the filters themselves (which could be included approximately in the "T_{window}" term in Eq. 1). For work on the Earth's surface, radiometric receivers, with their relatively narrow bandwidth, offer one obvious advantage–they can easily be tuned to work in the atmospheric windows between molecular lines. On the other hand, in space, the wide bandwidth available from bolometric receivers is an advantage, and COBE employed bolometric receivers.

Let me also mention an alternative to the use of filters to define the bandpass in bolometric receivers. It is to use a Michelson interferometer in front of the bolometer to scan a wide range of frequencies. This technique was used in the COBE FIRAS instrument (Mather, 1982) and also by Gush, Halpern and their Canadian colleagues (1990) in a clever, low-budget, rocket experiment that produced a measurement of T_0 with an accuracy close to that achieved by COBE.

4.4. BEST VALUES OF T_0

The best current measurements of T_0 are summarized in Table 1 (see also George Smoot's paper on the CBR spectrum in this volume). All four measurements or averages listed agree to better than 1%, and the precision of the COBE measurement is at a level of 0.2% of T_0–would that we knew other cosmological parameters as well!

I will have little to say about the constraints that these measurements set on distortions of the CBR spectrum (see Mather *et al.* 1994; Fixsen *et al.* 1996; and the articles by Smoot, Stebbins and Signore in this volume). It is worth remarking, however, that the COBE mission was ideally suited to establishing tight limits on the y parameter, where the distortion is largest at high frequencies. It was less well suited to determining the μ parameter, which is best seen by *comparing* short wavelength measurements to those at wavelengths $\gtrsim 10$ cm. I will return to that point below in §7. It is worth mentioning, however, that even though the longest wavelength measured by COBE was $\lambda = 1$ cm, it is these extraordinarily precise results that provide the best current limit on μ.

5. ΔT/T Measurements on Degree Scales

Extending the COBE measurements of anisotropies to angular scales 10-20 times smaller than the 7° resolution of the DMR instruments is now the major observational activity in our field. Such observations have been made from balloons (Clapp *et al.* 1994; Devlin *et al.* 1994; de Bernardis *et al.* 1994; Cheng *et al.* 1996) and at antarctic or arctic sites with low and stable atmospheric emission (Gunderson *et al.* 1995; Ruhl *et al.* 1995; Netterfield *et al.* 1995 and 1997). Since most mechanisms responsible for CBR fluctuations produce fluctuations that are independent of the wavelength, the choice of the wavelength is made primarily to avoid foreground emission (see §5.3 below). Most of the results listed above were obtained at wavelengths of 1.1-11 mm (though shorter wavelength channels were used in some of the balloon experiments, mainly to monitor Galactic dust emission). Since George Smoot will discuss the results of these experiments further, I'll treat briefly only the basic techniques employed and major problems encountered.

5.1. EXPERIMENTAL TECHNIQUES

In searching for fluctuations, only comparative measurements are needed, not the absolute measurements discussed above. That has consequences for both observational methods and for sources of noise. For instance, the amplitude of CBR fluctuations is small enough to create real demands on the sensitivity and gain stability of the receivers used. As we have seen, these fall into two general classes, relatively narrow bandwidth radiometric detectors and bolometric detectors. At present, radiometric detectors are normally used below 100 GHz and bolometric detectors are normally used at higher frequencies. In the case of a radiometric detector with absolutely stable gain, the rms noise expressed in antenna temperature is given by Eq. 5. Good radiometric detectors can have system noise as low as 30 K, and, at the frequencies typically employed for CBR anisotropy searches, bandwidths $\Delta \nu$ of several GHz. If $\Delta \nu = 1$ GHz, we see $\Delta T_{rms} \approx 1$ mK for a one second integration.

In an effort to reduce over–all system noise, experimenters are now turning to one stage of amplification before the crucial mixing stage in a superheterodyne receiver. Typically, cooled HEMT amplifiers are used. These low noise devices boost the incoming signal so that it is much larger in comparison to the noise generated in the mixer. Unfortunately, particularly at high operating frequencies (e.g., 90 GHz), the gain stability of HEMTs is not good. Short term gain fluctuations naturally affect the measured signal, and must be taken into account when deriving an overall rms system sensitivity, which becomes (see Partridge, 1995, Chap. 3, for a pedagogical "derivation"):

$$\Delta T_{rms} = T_{sys} \left[\frac{1}{\Delta \nu \Delta t} + \left(\frac{\Delta g}{g} \right)^2 \right]^{1/2} \qquad (6)$$

In HEMT receivers, gain fluctuations impose additional noise on the output which varies approximately as $1/f$, where f is the sampling frequency. There is a characteristic frequency f_0 where the two terms in square brackets in Eq. 6 are equal; this is the "1/f knee." Much of the current effort in detector development is to reduce gain fluctuations in HEMT amplifiers to lower the frequency of the 1/f knee. If, for instance, f_0 is below 0.1 Hz, then beam switching (see below) at a few Hz will effectively eliminate the effect of gain fluctuations in differential measurements of $\Delta T/T$.

I will defer to others to discuss bolometric detectors, except to note that they are generally intrinsically noisier (that is have higher "system temperature") than good radiometric detectors; at higher operating frequencies, they more than make up for it by having much larger bandwidths.

The newest development in this field is the use of arrays of detector elements. To date, only the South Pole experiment of Ruhl et al. (1995) has produced measurements from arrays, but other groups are moving in this direction (e.g., the MAXIMA effort at CFPA, Berkeley). Also, the two satellites planned for CBR fluctuation searches both employ detector arrays.

Since $\Delta T/T$ measurements are *comparative*, and are made by comparing the antenna temperature of one small region of the sky with another. To do so, all recent experiments have used some form of beam switching. The resulting patterns for simple beam switching ("two-point") and for the so-called "on-off" technique ("three-point") are shown in Fig. 3. The latter was introduced into CBR studies by Epstein (1967). When comparing the limits on $\Delta T/T$ produced by beam switching experiments, it is important to bear in mind that these two beam patterns sample the correlation function $C(\theta)$ differently. This sampling is generally included in the window function (see e.g., Bond, 1990) for a given experimental arrangement. In the case of a simple beam switching experiment, however, it is obvious that the results will contain useful information only in an approximate range $\theta_{1/2}$ to θ_s, where $\theta_{1/2}$ is the full width at half power of the beam.

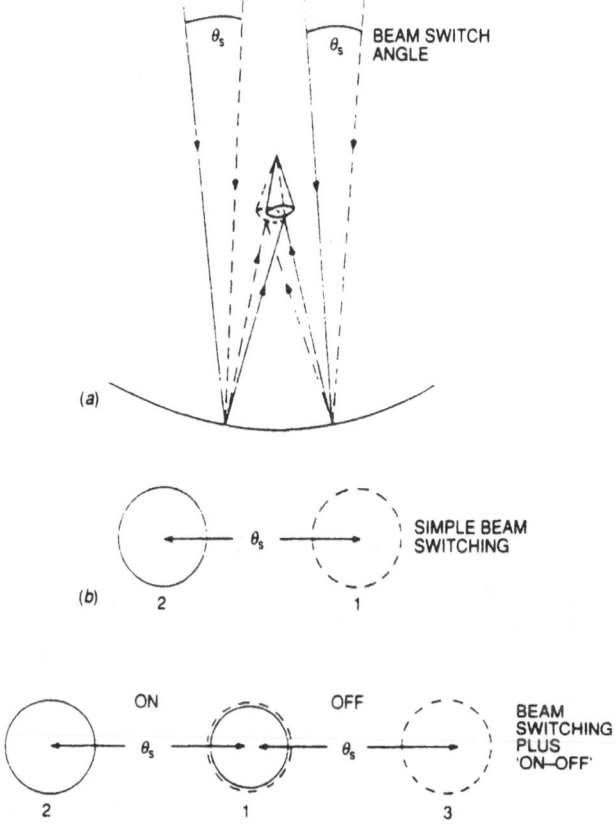

Figure 3.(a) Beam pattern resulting from beam switching through an angle θ_s. The source is centered at location 1; 2 is the reference position. (b) Beam pattern resulting from 'on-off' observations; again the source is at 1, while 2 and 3 are reference positions.

5.2. PROBLEMS ENCOUNTERED, ESPECIALLY FOREGROUND SOURCES

Providing one is willing to integrate long enough, reaching a sensitivity of $\Delta T/T = 10^{-6}$ is in principle possible if gain fluctuations can be controlled; for instance, if $T_{sys} = 30$ K and $\Delta\nu = 10$ GHz, an integrating time Δt of a few hours will suffice. None of the degree scale experiments has come close to that sensitivity, and the major reasons are foreground sources of noise.

I will touch on three sources of foreground emission here (see also George Smoot's anisotropy paper in this volume).

The first is emission from foreground radio sources. For a radio telescope with a beam solid angle Ω making observations at wavelength λ in the Rayleigh Jeans region, the connection between antenna temperature and flux is

$$T = \frac{S\lambda^2}{2k\Omega} \cong 10^{-4} S\lambda^2, \tag{7}$$

where Ω is the solid angle of the antenna beam and k is Boltzmann's constant. The numerical expression given above assumes a beam of $1°$ half width at full maximum, with λ expressed in centimeters, and S, the flux density, in Jy. As I've noted, most degree-scale searches are conducted at wavelengths < 1 cm; at these wavelengths there are relatively few sources brighter than 1 Jy. Thus extragalactic radio sources will not be large contributors to fluctuations in the microwave sky. In addition, since the counts of extragalactic radio sources as a function of their flux density are well known, at least at long wavelengths ($\lambda \geq 3.6$ cm), we can calculate the fluctuation level produced by foreground radio sources as a function of angular scale (e.g., Fig. 4, taken from Franceschini et al. 1989). Sources of different flux densities dominate the sky variance at different angular scales. It is easy to see that the variance will be a maximum when there is ~ 1 source per beam. That occurs for sources of flux density $N(>S) \sim 1/\Omega \sim 1/\theta^2$ where $N(>S)$ is the integral source count. At the highest frequency (8.4 GHz) for which reliable source counts exist (Windhorst et al. 1993; Kellermann et al. 1996), $N(>S) = (18 \pm 1)S^{-1.2\pm0.2}$ arcmin^{-2}, in convenient units where S is in μJy. Note that if sources above a certain threshold can be excluded (as is true for the interferometric observations to be discussed in §8 below), $\Delta T/T$ can be sharply reduced on large angular scales. For instance, if all sources with $S>10$ μJy can be identified and removed, the remaining fluctuations produced by weaker ($S < 10$ μJy) sources fall off rapidly at $\theta \gtrsim 1/2'$, as shown by the dashed line in Fig. 4.

Next, Galactic microwave and far infrared emission contributes to the variance of the sky. This problem has been extensively studied (e.g., Banday et al. 1991; Partridge, 1995, Chap. 7; Kogut et al. 1995; Bouchet et al. 1995; and Smoot's anisotropy paper in this volume). I will simply reproduce a figure (Fig. 5 here) taken from Bouchet et al. which displays the essential features of Galactic foreground emission, in this case calculated for a $1°$ beam. Patchy synchrotron (at $\nu \gtrsim 10$ GHz) and free-free (at $\nu \lesssim 10$ GHz) emission produces the dominant fluctuations at long wavelengths; patchy thermal emission at $\lambda \lesssim 3$ mm ($\nu \gtrsim 100$ GHz).

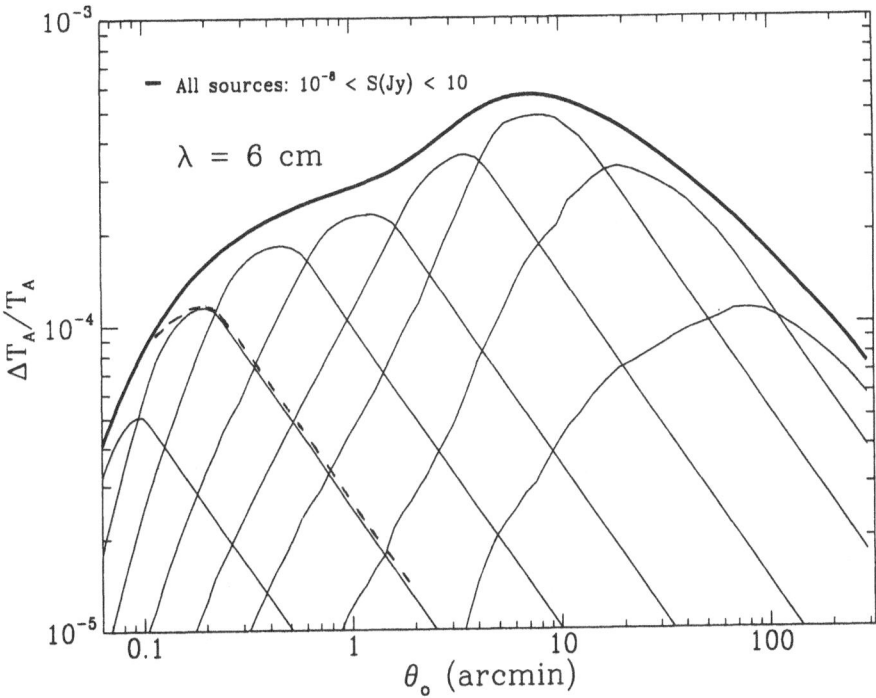

Figure 4. Fluctuations resulting from foreground extragalactic radio sources (from Franceschini *et al.* 1989, with permission). At $\lambda = 6$ cm, $\Delta T/T \gtrsim 10^{-4}$ for all angular scales from $1'$ to $2°$. The eight lower (lighter) curves indicate the contribution to $\Delta T/T$ from sources in eight decades of flux. Starting at the left is the curve for contributions by sources from 0.1-1.0 μJy; ending at the right is the curve for contributions by sources with $S = 1$-10 Jy.

From the figure, we see that observers are restricted to a frequency range of $\sim$30-200 GHz. It is no accident that most observations, whether balloon-borne or ground-based, have been made in this frequency range.

Unfortunately, that entire frequency range is not accessible from the ground because of strong emission lines or bands of atmospheric gasses. The two major culprits are oxygen and water vapor. The opacity of these two constituents is shown in Fig. 6. Since O_2 is well mixed in the atmosphere, its contribution to atmospheric opacity is impossible to avoid; on the other hand, also because it is well mixed, the point-to-point variations in oxygen emission are relatively small. With H_2O, the opposite is true-it is not well mixed, so fluctuation levels are larger; but its scale height is much lower than that for O_2, so moving to high altitude sites helps more.

Not surprisingly, most ground-based observations are made in the "valleys" between O_2 and H_2O lines. In §7, we will consider why the opacity does not go to zero between lines.

14

For comparative measurements, it is the fluctuations in T_{atm} that matter. Unfortunately, the time–spectrum of atmospheric fluctuations is not terribly well understood. One frequently used approximation is the Kolmogorov, for which the rms noise scales as $\theta^{5/6}$, or $t^{5/6}$ for constant wind velocities. Smoot *et al.* (1986) find the exponent $\sim 0.7 \pm 0.2$ in acceptable agreement with this model. While these results are not definitive, they do indicate that fluctuation levels increase with both the beam switch angle and time. Thus, to reduce atmospheric fluctuations, a small beam switch angle and a rapid switching frequency are required. An alternative, to which we turn in §8, is to use interferometric techniques, in which atmospheric emission essentially cancels out.

All three sources of foreground fluctuations–radio sources, the Galaxy and the atmosphere–have singly or in concert impeded each of the searches for anisotropies on degree scales. It has taken much hard work and observational ingenuity to produce the results cited in this section and summarized in Table 1.

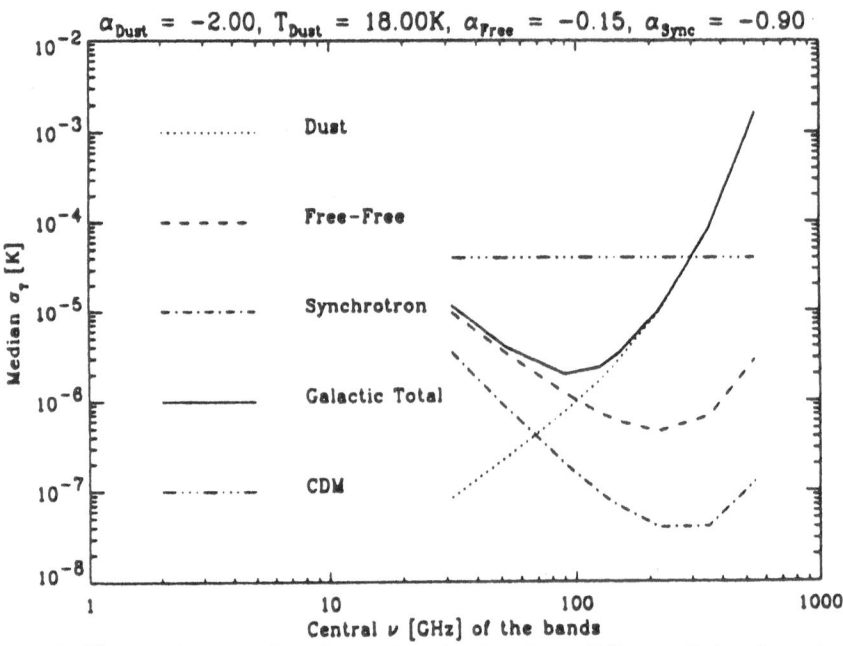

Figure 5. Fluctuations produced on 1° scales by three different Galactic emission mechanisms, compared with expected level of $\Delta T/T$ fluctuations in a standard CDM model (horizontal line).

6. The Next Few Years

Let me turn now to some more general thoughts on where the field will be going over the next four or five years. Many of us in the field will be concentrating on the planning for and construction of two major satellites

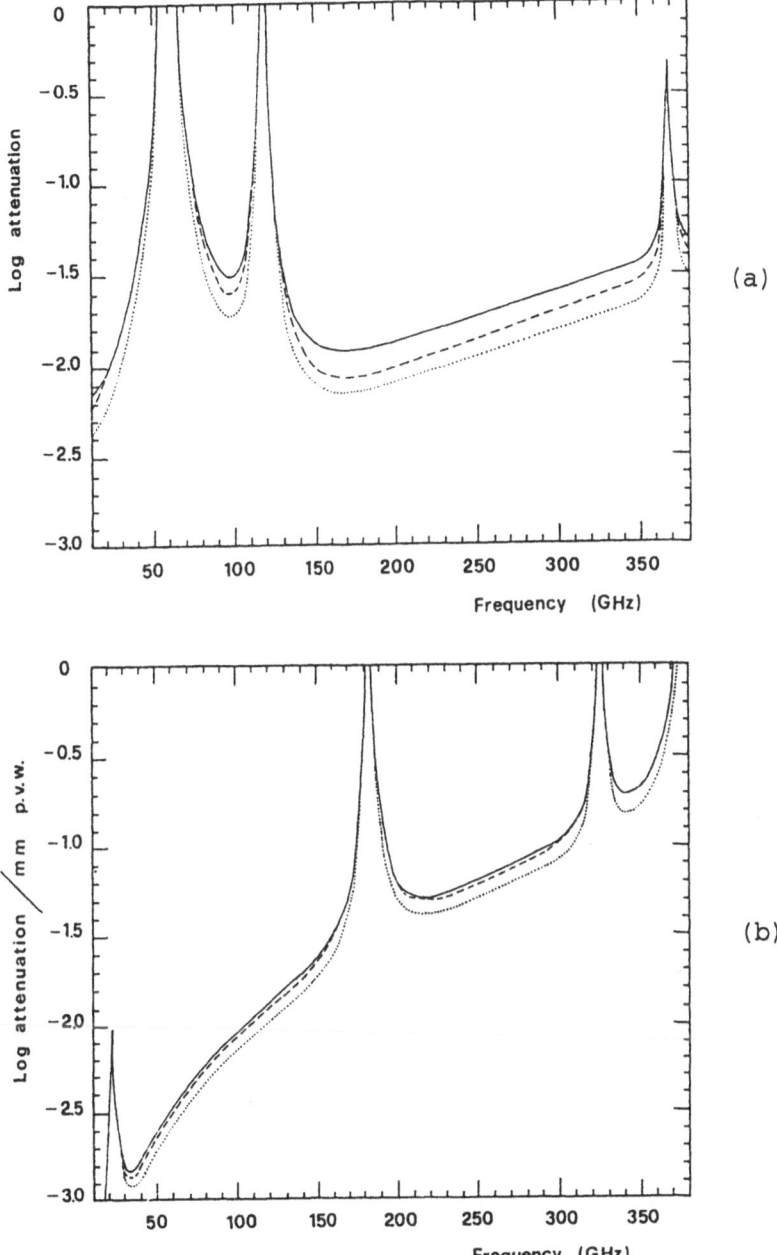

Figure 6.(a) Atmospheric attenuation k due to O_2 for Kitt Peak (solid line), the South pole (dashed line) and Mauna Kea (dotted line). $T_{atm} \sim 250$ K. (b) Attenuation for H_2O per mm of precipitable water vapor.

dedicated to CBR studies: MAP, a US satellite to be launched in the year 2000, and Planck, an ESA satellite for launch about four years later. I believe it is also safe to predict that considerable effort will be devoted to improving the measurements of CBR anisotropies on scales of 0.5°-5°, both from the ground and from balloons. From their first tentative beginnings, studies of the Sunyaev-Zel'dovich (SZ) effect are now becoming "mature" in the sense that they are providing important data on clusters of galaxies and Hubble's constant. Such work will undoubtedly continue, even after the launch of Planck which has measurements of the Sunyaev–Zel'dovich effect as one of its scientific goals. I suspect searches for both $\Delta T/T$ and SZ fluctuations are increasingly likely to be carried out using interferometric techniques, to which I'll turn later. Finally, there are two other of the 8.5 parameters that I believe are ripe for further work. The first is the large-scale, linear polarization of the CBR. I know that work is underway at Brown University by Peter Timbie and his colleagues, and in Italy as well; (see Melchiorri *et al.* this volume); but at the moment, the best published values date back 13 years (Lubin *et al.* 1983). And finally, there is the question of long wavelength measurements of the temperature of the CBR, measurements needed to confirm its blackbody nature and to constrain the μ parameter. As an example of possible future measurements, let me turn to that question first.

7. Measuring T_0 at $\lambda = 10\text{-}20$ cm

The FIRAS instrument aboard COBE provided an extraordinarily accurate measurement of the CBR spectrum at wavelengths less than 1 cm (Mather *et al.* 1994). Although they are in general agreement with the COBE measurement, individual ground- and balloon-based measurements have much larger error bars. In Table 2, I summarize some of the most recent results. As Drs. Stebbins, Smoot and Salati will explain more fully, measurements at $\lambda = 10$ cm and above are particularly important to constraining the μ parameter, which in turn is our most significant clue to energy-emitting processes at high redshifts. A precise measurement of T_0 at $\lambda = 10\text{-}20$ cm, combined with the very accurate COBE measurements at the peak of the 2.73 K curve, would go a long way to improving our knowledge of μ. Is it reasonable to hope, for instance, for a 1% measurement of T_0 in this wavelength range?

A satellite experiment (Kogut, 1995; Sironi *et al.* 1995) could presumably extend COBE–type measurements to longer wavelengths, though it would be difficult because of the need for large antennas (these are needed to keep the beam to a reasonable size, so that the Galactic contribution can be properly estimated; see below).

7.1. FROM THE GROUND!?

Let us instead investigate the possibility of making a 1% measurement in the 10-20 cm range from the ground. As I've already noted, detector sensitivity is not a problem, though the design of the zero–point calibrator would be nontrivial. It might, for instance, be necessary to cool the antenna, and return to the technique in which the cold-load calibrator is placed above the antenna. Radiation from the ground can surely be controlled to much less than 1% of T_0 by careful design of ground screens (I note that ground contributions were 0.2 – 0.4% of T_0 at comparable wavelengths in the 1980 measurements of the White Mountain group; see Smoot *et al.* 1987 or Partridge, 1995 for summaries). That leaves us with the two main sources of systematic error, atmospheric emission and emission from our Galaxy. As I have already noted, there is "cross talk" between these two.

TABLE 2. RECENT AND LONG-WAVELENGTH MEASUREMENTS OF T_0

Reference	λ, cm	Measured T_0, K	Comment
Howell and Shakeshaft (1967a)	74 + 49	3.7 ± 1.2	T_{atm} calculated, not measured.
Smith (1996)	50	$T_0 + T_{Gal} =$ 6.98 ± 0.32	(T_{Gal} ~ 5 K ??; if so T_0 ~ 2 K.)
Bensadoun (1993)	20	2.26 ± 0.19	
Staggs *et al.* (1996)	20	2.65 ± 0.33	
Bersanelli *et al.* (1994)	15	2.55 ± 0.14	T_{atm} measured.
Otoshi and Stelzried (1975)	13	2.66 ± 0.26	Closer to 2.76 ± 0.30 when T_{atm} corrected.
De Amici *et al.* (1990)	8	2.64 ± 0.07	
Mandolesi *et al.* (1986)	6.3	2.70 ± 0.07	
Fixsen *et al.* (1996; COBE FIRAS)	≤1	2.728 ± 0.004	

Note: no recent measurement exceeds the FIRAS result for T_0, though several are consistent with it. The quoted errors are 1σ.

The maps of Haslam *et al.* (1982), when corrected for both the CBR background and emission from discrete, extragalactic radio sources, produce an antenna temperature at the Galactic pole of 14.6 ± 2.2 K at his observing frequency of 408 MHz, or $\lambda = 75$ cm. At these long wavelengths, and extending down to wavelengths well below 10-20 cm, the dominant emission mechanism is synchrotron, for which a typical frequency spectral index of -0.8 ± 0.1 may be assumed, so that $T(\nu) \approx \nu^{-2.8}$. As a sample calculation, let us extrapolate the value given above down to $\lambda = 15$ cm (2 GHz), taking into account both the error in the Haslam figure above and a 0.1 uncertainty in the synchrotron spectral index. We find $T_{Gal} = 0.171 \pm 0.038$ K with the uncertainty arising roughly equally from uncertainties in the spectral index and in the absolute value from Haslam *et al.* (I've combined the two errors in quadrature.) Thus the *uncertainty* in T_0 introduced by T_{Gal} is only 1-2% of T_0 *if* the Haslam flux scale is accurate. For a limited portion of the sky near the Galactic pole, I believe better and better calibrated low frequency measurements of the Galactic emission could be hoped for (e.g., De Amici *et al.* 1995; Sironi *et al.* 1995). If these measurements were made at several frequencies, both the zero point and the spectral index could be pinned down more precisely. Thus I believe it is not out of line to hope for estimates of T_{Gal} accurate to ~1% of T_0 in the wavelength range 10-20 cm.

On the other hand, I suspect that measuring atmospheric emission to an accuracy of 0.02-0.03 K will prove very difficult. In order to obtain a tight enough beam so that Galactic emission can be accurately estimated and allowed for when finding T_{atm}, a large antenna will be needed. Moving a large antenna to do zenith scans is likely to produce mechanical stresses contributing to T_{offset}. Small offset terms are then magnified by $1/(sec\ z - 1)$ when calculating T_{atm} (see Eq. 3). There is hope, however. It may prove possible to *calculate* T_{atm} to sufficient precision so that it becomes unnecessary to actually measure it. The hope lies in the fact that atmospheric emission at the frequencies of interest here is entirely dominated by nonresonant absorption of O_2. At $\lambda = 10$-20 cm, T_{atm} is 1.5-2.5 K at sea level, and $\lesssim 1$ K at mountain sites. Thus to obtain a 1% measurement of T_0, we need a 1-2% accurate calculation of both O_2 absorption and the temperature profile of oxygen through the atmosphere.

I'd like to examine this proposition in a little more detail. First, let's confirm that H_2O emission will not present problems, provided that measurements are made from a dry, high altitude site in good weather. At frequencies much below the 22 GHz water vapor line, non-resonant emission dominates, and the absorption (or emission) coefficient of water vapor, k_{WVC}, is $\approx \nu^2$ (Liebe, 1984). The semi–empirical model of Danese and Partridge (1989) gives the coefficient of proportionality for the H_2O continuum emission, expressed in antenna temperature:

$$T_{WVC} = 2\ mK \left(\frac{\nu}{2.5\ GHz}\right)^2 \left(\frac{p}{1\ mm}\right), \tag{8}$$

where p is the precipitable water vapor in a column above the observing site, expressed in mm. This coefficient is given for a mid–latitude site, and for typical clear summer weather; it includes a small correction for emission from liquid water droplets. At good, high altitude sites in good weather, p $\lesssim 5$ mm; we see that $T_{WVC} \lesssim 15$ mK in the wavelength range 10-20 cm of interest, and could be somewhat less at a polar site or a very dry one. Furthermore, T_{WVC} can be calculated to something like 20% accuracy, so the residual uncertainty in the H_2O contribution to T_{atm} will be only ≈ 3 mK, that is of order 0.1% of T_0.

The O_2 *line* contribution at $\lambda = 10$-20 cm can be estimated to comparable overall accuracy ($\approx 0.1\%$ of T_0). It does, however, depend on the altitude h of the site used, because both the amount of O_2 overhead and the pressure broadening of the 60 GHz band depend on h. For a high altitude site, $T_{0\ell} \lesssim 10$ mK.

The frequency dependence of the O_2 nonresonant emission is more complicated and less steep than for H_2O : from the work of Liebe (1984, as modified by Danese and Partridge, 1989), we take for the absorption co-efficient in the continuum:

$$k_{oc} = \left(APT^{-2} \frac{\tau_0}{\tau_0^2 + \nu^2} + BP^2 T^{-2.5} \right) \nu^2, \qquad (9)$$

where T and P are the atmospheric temperature and pressure, respectively, and A and B are constants. If the second term in parentheses in Eq. (9) dominates, $k \propto \nu^2$, as for H_2O. If the first term dominates, the frequency dependence of k_{oc} changes from ν^0 to ν^2 as ν drops below τ_0, as sketched in Fig. 7. Complicating matters further, τ_0 itself depends on both P and T (Danese and Partridge, 1989).

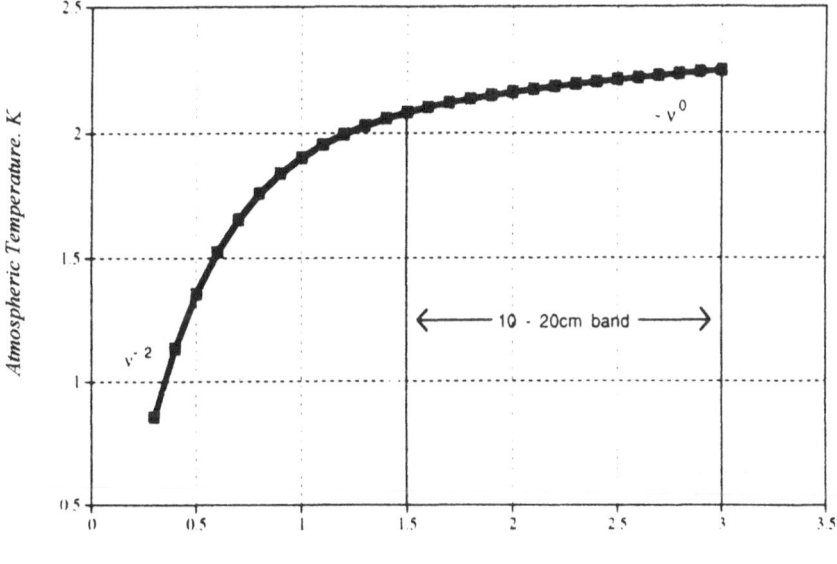

Figure 7. Atmospheric temperature at sea level, from the model used in Danese and Partridge (1989).

Measurements of atmospheric opacity at 0.4-10 GHz (e.g., Howell and Shakeshaft, 1967b; Encrenaz et al. 1970; Partridge et al. 1984) have allowed us to refine our values of τ_0 (and the coefficients A and B in Eq. 9), and hence to calculate T_{oc} to modest accuracy, as shown in Fig. 7 (in this case, for a sea-level site). At present, the crucial term τ_0 is known to only $\approx 10\%$ accuracy, so that neither the amplitude of T_{oc} nor its exact frequency dependence is known well enough to give values of T_{atm} precise to ± 20-30 mK, or $\pm 1\%$ of T_0. We need an improvement by a factor of 3-5 in accuracy. It would help to have a couple of very precise measurements of T_{oc} or k_{oc} at 10-20 cm wavelength.

8. Interferometric (Aperture Synthesis) Measurements of the CBR

The use of radio–frequency interferometers to observe the CBR offers several advantages. The first of these is increased angular resolution, but interferometric observations are also free of some of the sources of error which plague other methods, as described earlier. Interferometric techniques were first used in CBR studies (Martin, Partridge and Rood, 1980) to achieve high angular resolution. And until recently, most results were obtained at arcminute scales and below using telescopes like the Australia Telescope (Subrahmanyan *et al.* 1993) IRAM (Radford, 1993) and the Very Large Array (VLA) of the National Radio Astronomy Observatory in New Mexico (e.g., Knoke *et al.* 1984; Fomalont *et al.* 1984; Fomalont *et al.* 1993). But interferometric techniques also hold great promise for observations on degree scales, as we shall see. In keeping with the introductory spirit of these lectures, let me now outline this technique and some of its special features.

8.1. A SIMPLE INTERFEROMETER

We begin with the simplest case, two detecting elements separated by a distance B. We'll assume that each element is a circular antenna of diameter D_p, and we will introduce the quantity $b_\lambda \equiv B/\lambda$ (Fig. 8). All of you have calculated the two slit interference pattern; here let me use the reciprocity theorem [2] to argue that the beam pattern of this simple two–element interferometer will resemble the one sketched in Fig. 8. The separation of the individual maxima is given to a good approximation by $\theta = 1/b_\lambda$. The angular width of the envelope of the interference pattern is determined by the diameter of the individual elements of the array; it is approximately $\theta_p = \lambda/D_p$.

Given this beam pattern, let us analyze the response of the interferometer to two different classes of radio sources. First, consider a point source. If it is stationary with respect to the beam, the amplitude and sign of the output signal will depend on its location relative to one of the maxima. If it moves across the beam, the output of the interferometer will swing from positive to negative values as the source passes through the interference pattern. A similar argument can be used to show that the response of

---

[2] In rough language, the reciprocity theorem states that the angular response of a certain geometrical array of antennas is the same as the radiation pattern that would be produced by a similar array when broadcasting waves. Thus, in this case, two circular antennas receiving radio waves from the sky will have a beam pattern equivalent to the interference pattern produced by two illuminated circular apertures separated by the same distance. For details, see Rohlfs, (1986).

the interferometer to a uniform background will be essentially zero. (The positive and negative contributions cancel.)

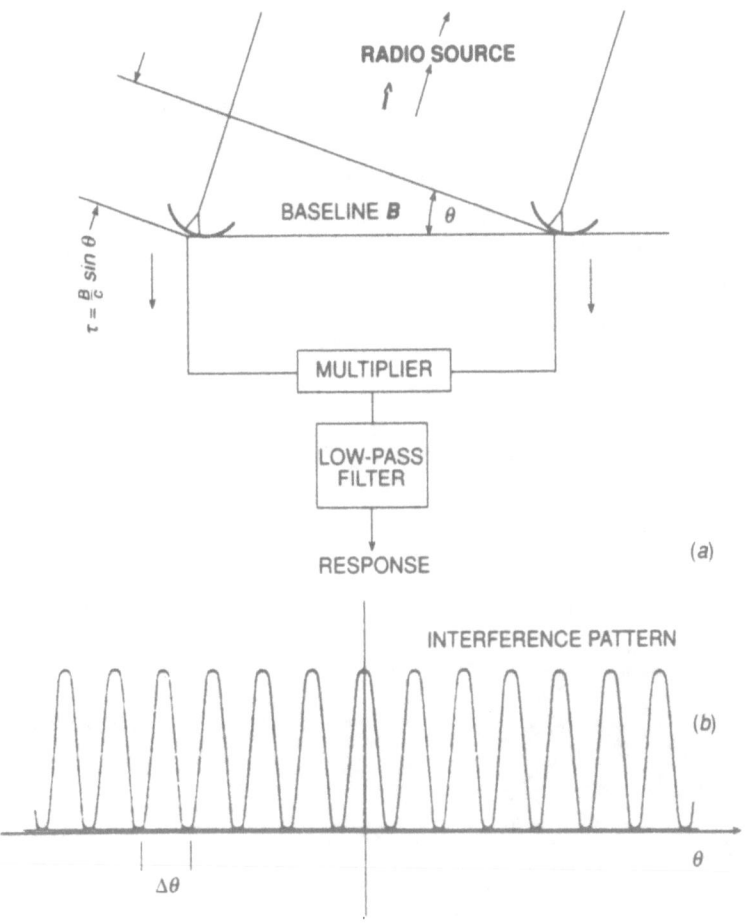

Figure 8.(a) Schematic of a two-element interferometer. The resulting interference pattern for an instantaneous observation is sketched in (b).

Next, let us ask what happens if the line between the two elements of the array rotates with respect to the vector ℓ directed towards the sky? It is easy to see that the fringe pattern will also rotate. In the simplest case where B is perpendicular to ℓ, and we allow rotation through a full 180°, a circular beam pattern will be produced. The central maximum is referred to as the synthesized main beam; the circular, secondary maxima as side lobes. Allowing the baseline between antennas to rotate reduces the amplitude of the side lobes relative to the main beam. Note that the synthesized beam width θ_s is $\sim 1/b_\lambda$ in this simple case.

22

8.2. MULTIPLE-ELEMENT ARRAYS

Another way of increasing the amplitude of the central maximum compared to the side lobes is to add elements to the array. For instance, a simple 3–element, linear array with equal spacings will produce a beam pattern with alternate maxima suppressed (see any optics text).

In the case of the VLA, there are 27 elements in the array, and the array is not linear but (approximately) an equal-angle Y. Since it is a two-dimensional, rather than a linear array, the interference pattern, or beam pattern, for the VLA is more complicated than a simple set of fringes. It is sketched schematically in Fig. 9a for an instantaneous observation, and in more detail in Fig. 9b for a longer observation in which the rotation of the earth plays an important role.

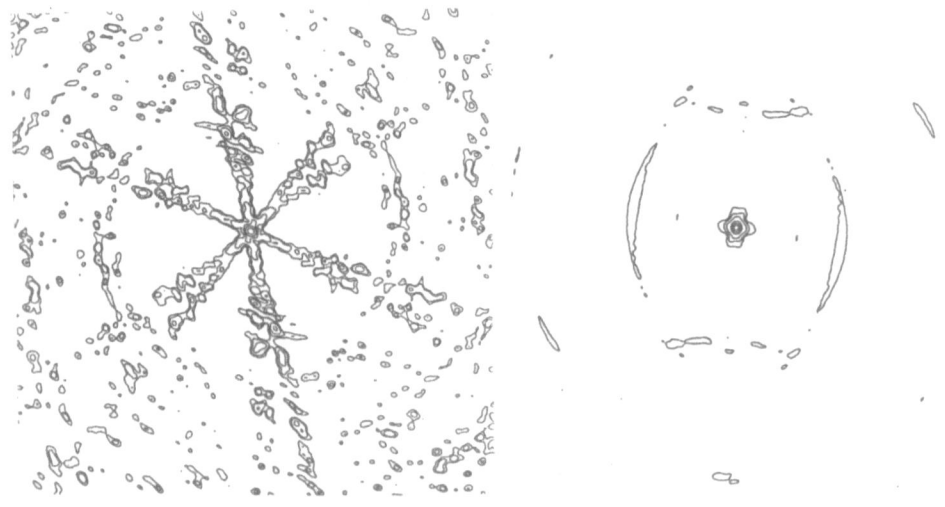

Figure 9. Beam pattern of the 27–element VLA. (a) For a pair of short observations of a position near declination 30°; (b) for a 12 hour observation of a source at roughly the same declination. The contour levels are at 2%, 4%,...64% of the peak amplitude.

These complexities should not obscure several basic results. The first is that the response of the VLA, like the response of the simple 2–element interferometer considered above, is zero to uniform backgrounds. That has two obvious consequences: the first is that the interferometric technique is sensitive only to fluctuations in the CBR, not its overall temperature; and the second is that interferometers are essentially insensitive to emission from the earth's atmosphere. In addition, the angular resolution of an array like the VLA is approximately equivalent to some angle $\theta_s \sim \lambda/D_a$, where D_a is the characteristic scale of the entire array. In addition, the instrument will not be sensitive to fluctuations in the sky for angular scales larger than the primary beam angle, θ_p, which is given by $\theta_p \sim \lambda/D_p$. The number of resolution elements within the primary beam is thus given approximately by $(\theta_p/\theta_s)^2$, which in turn is approximately $(D_a/D_p)^2$. In the case of the most close–packed configuration of the VLA, for instance, the quantity D_a is slightly less than 1 km, and D_p is 25 m; thus, each VLA image contains $\sim$1000 resolution elements. In principle, the range of angular scales to which the instrument is sensitive ranges from $\theta_p - \theta_s$; in reality, however, because of a number of effects, the sensitivity is greatly reduced on angular scales much larger than 5-10 θ_s.

8.3. INTERFEROMETRIC RECEIVERS

So far, we have said nothing about how the various portions of the wave front reaching the elements of the array are combined to produce an interferometric signal. The simplest kind of receiver brings the signals from the elements together and multiplies them (Fig.10). Note that it is helpful to introduce a time delay in one arm of the interferometer (to reduce beamwidth smearing; see Partridge, 1995, or below). It is easy to show that adding an appropriate time delay in effect tilts the baseline to an orientation perpendicular to the direction to the source.

If a proper interference signal is to be produced, the amplifiers, multipliers, and so on, need to preserve the phase of the incoming signals. It is not easy to do this over a large frequency bandwidth. For that reason, most conventional interferometric arrays operate with relatively narrow bandwidths (100 MHz in each of two circular polarizations at the VLA). As we will see in a moment, however, the large number of correlated signals in a multi-element array helps make up for the narrow bandwidth.

The rms sensitivity per resolution element of an N-element array with $N(N-1)/2$ antenna pairs array may be written as

$$\Delta T_{rms} = \frac{T_{sys}}{\sqrt{\Delta\nu \Delta t N(N-1)2}} \left(\frac{\theta_p}{\theta_s}\right)^2 \qquad (10)$$

which may be compared with Eq. 5. In the case of the 27-element VLA, $N(N-1)/2 = 351$, so the raw sensitivity of the VLA detectors corresponds to a total–power receiver of bandwidth $\Delta\nu \sim 3.5$ GHz. However, as we had noted above, each VLA image contains $\sim$1000 resolution elements. In some cases, that may be an unnecessarily large number, so that observers are in effect "wasting time" observing too many individual pieces of the sky.

24

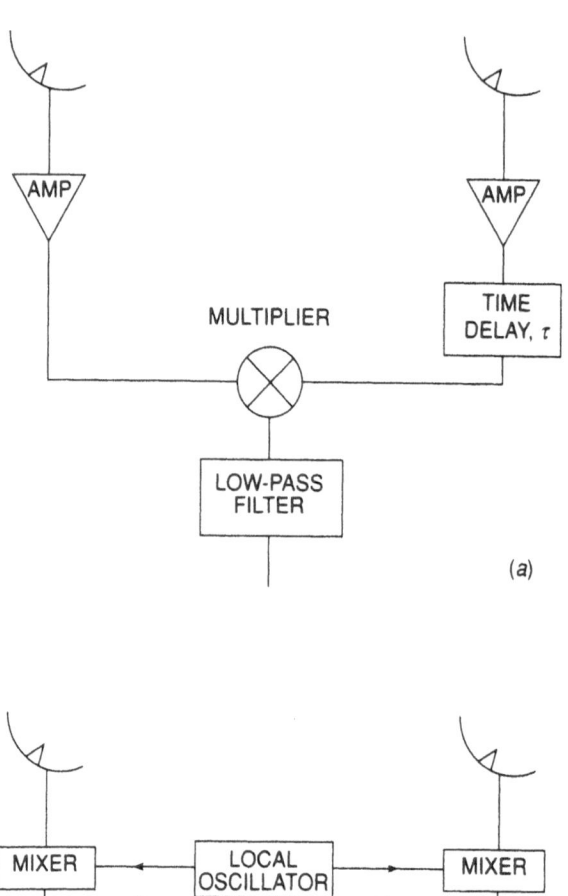

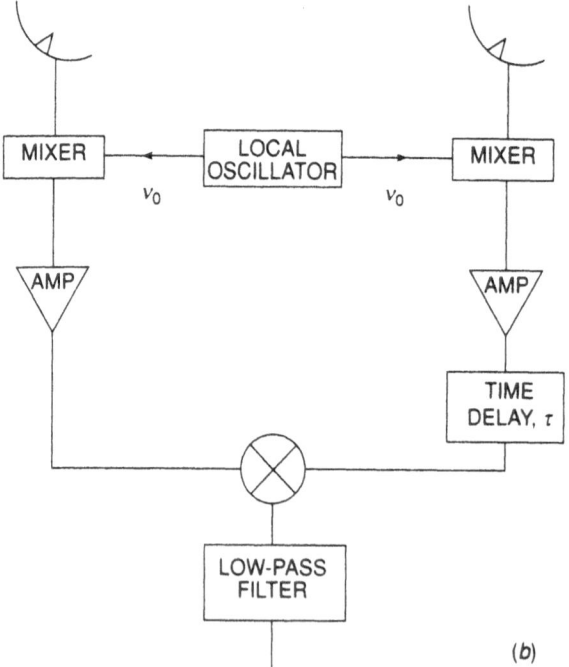

Figure 10 (a) The simplest interferometric receiver. The two signals are combined in the multiplier. For sources not on the axis of symmetry of the antenna pair, a time delay is introduced in one of the signals. (b) A heterodyne receiver, as at the VLA. Signals from the same local oscillator are fed to both mixers.

8.4. SPECIALLY DESIGNED INTERFEROMETERS FOR LARGER ANGULAR SCALES

The two points just raised above have helped determine the design of interferometers specially constructed for CBR observations. First, for most of these instruments (e.g., Timbie and Wilkinson, 1988; Scott *et al.* 1996), the size of the array, D_a, is much closer to the size of the individual elements of the array, D_p, than is true for the VLA. Thus each image made by these specially–designed instruments has fewer resolution elements, increasing the efficiency of the observations. In addition, substantial efforts are being made to increase the bandwidth $\Delta\nu$ for these instruments. For instance, the Cambridge CAT instrument has correlators with a bandwidth $\Delta\nu = 500$ MHz (Scott *et al.* 1996).

Finally, these specially designed instruments are consciously built to work on angular scales larger than those available at the VLA, with θ_s typically in the range of $0.1° - 1°$. These are the angular scales of most direct interest to theories of large–scale structure of the Universe.

8.5. PROBLEMS OF THE INTERFEROMETRIC TECHNIQUE

Scattered throughout the material above are a number of arguments in favor of the interferometric technique–the potential for higher angular resolution, the freedom from atmospheric emission and the ability to make a two–dimensional map of the sky with a substantial number of resolution elements. But there are also some problems unique to this technique. I will move fairly rapidly through these; for more details see Chapter 3 of my 1995 book, 3 K.

First, there is a purely instrumental problem, cross–talk from one antenna to the other. Signals generated in the receiver attached to one antenna can be broadcast through that antenna and be picked up by another antenna in the array. In arrays where the elements are widely spaced, this is not a particularly severe problem. But in close–packed arrays, with D_a not much larger than D_p, careful attention needs to be paid to reducing cross–talk.

Next, there is the issue of side lobes. In arrays with a small number of elements, as we've seen in §8.1 above, the side lobes can have nearly the same amplitude as the main beam. This makes for a rather awkward beam pattern on the sky, and a corresponding complex window function when the results are compared with theory. Clearly, side lobes are a bigger problem for arrays with a small number of elements (e.g., Timbie and Wilkinson, 1988 and 1990). At the VLA, the side lobes are typically $\lesssim$1-2% in relative amplitude.

Finally, there is the problem of bandwidth smearing. As we saw much earlier in this section, the separation of interference maxima depends on the wavelength. Thus the separation will be slightly different at one extreme of the detected wavelength band from the other. It is easy enough to work out that the fringe pattern will begin to wash out and average to zero at an angular distance $\sim (\lambda/\Delta\lambda b_\lambda)$ from the image center, where $\Delta\lambda$ is the bandwidth. Thus the desire to increase the bandwidth has the effect of limiting the angular response of an interferometer. This is one reason, incidentally, why the VLA is sensitive to fluctuations only in the relatively restricted angular scale from θ_s to 5-10 θ_s, not up to $\sim$30 θ_s. Bandwidth smearing can be reduced by subdividing the bandwidth $\Delta\nu$ into a number of frequency channels, then scaling the resulting images appropriately, and adding them. This technique is now in routine use at the VLA and other interferometers.

Now let us consider the nature of artifacts in an interferometric image produced by various sources of instrumental noise. First (see Knoke *et al.* 1984), random Gaussian noise from either the instrument or the atmosphere will be spread uniformly over the image. A noise blip momentarily affecting a single correlator will produce a set of parallel stripes in the image (bear Fig. 8 in mind). A correlator that remains "hot" for an entire, extended observation will introduce a set of concentric rings or ellipses in an image. All of these effects can be searched for and controlled, at least to some level.

Since random noise processes are spread smoothly over an image, but true fluctuations in the sky will be recorded only in the central θ_p of an image, we may separate true sky signals from instrument noise by comparing the variances measured from the center and the edge of an interferometric image, providing that image is substantially larger than θ_p. This is the technique that has been used by observers to separate out and set limits on sky variance (beginning with Martin *et al.* 1980; see also Knoke *et al.* 1984; Fomalont *et al.* 1984; Fomalont *et al.* 1993). Observers can thus measure or set limits on the variance of the microwave sky (including CBR fluctuations) at levels below the noise in the image.

8.6. SOME RESULTS

Let me turn now to one set of results; VLA searches for CBR fluctuations. The best VLA work has been conducted at $\lambda = 3.6$ cm with the instrument in its most compact D or C configurations. With my colleagues, Fomalont, Kellermann, Richards and Windhorst, I've now observed four separate regions of the microwave sky, most recently the Hubble Deep Field. Results on three of these areas are now available. In some cases, because we have C-configuration observations, our resolution extends down to 6", and we still have some usable data on angular scales as large as 80".

Because we operate at a relatively long wavelength, foreground radio sources are very evident in all our fields (even though we have always selected fields free of very bright foreground sources). We are trying to reach sensitivity limits of a few μJy, and typical 3.6 cm VLA fields will contain sources as bright as several hundred μJy. These bright sources, and the side lobes they generate, need to be accurately measured and subtracted before we have any hope of finding background fluctuations. The techniques we use for this subtraction are described, for instance, in Fomalont *et al.* (1993). But radio sources fainter than a 5σ threshold are almost certainly present as well, and will contribute to the sky variance at 3.6 cm. Thus we model the faint source contribution and subtract it from the observed sky variance. We have done this by extrapolating the counts of radio sources (see §5.2) to values below the 5σ threshold on our maps, then using Monte Carlo calculations to estimate the variance produced in our images by a random distribution of sources obeying that source count law (Fomalont *et al.* 1993; Partridge *et al.* 1996). This "weak source" variance is then subtracted from the variance of the microwave sky observed at the center of the image to produce a "residual variance," which includes any fluctuations in the CBR.

TABLE 3. TENTATIVE VLA LIMITS ON CBR FLUCTUATIONS

AT λ = 3.6 cm (Partridge *et al.*, 1996)

Angular Scale, arcsec²	Image Variance μJy²	Variance due to Weak Sources, μJy²	95% Confidence Limit on ΔT/T × 10⁻⁵	Limit with S-Z Region Removed × 10⁻⁵
6	0.77 ± 0.17	0.60 ± 0.11	12	10
10	1.75 ± 0.42	1.29 ± 0.23	8	6
18	5.74 ± 1.76	3.65 ± 0.63	5	3.2
30	12.36 ± 3.37	6.11 ± 1.11	3.6	3.3
60	31.9 ± 18.0	16.3 ± 7.8	2.2	2.3
80	119 ± 72	34 ± 21	2.5	2.1

In Table 3, I summarize some tentative results from our most recent set of VLA observations (Richards *et al.* 1995; Partridge *et al.* 1996). It is interesting to note that, even after the "weak source" variance has been subtracted, there remains barely significant variance in the VLA image. Some of the variance may be residual systematic effects, or caused by radio sources below our 7 μJy threshold if our correction for "weak sources" is underestimated. Some may be subarcminute fluctuations in the CBR. In addition, however, some of the excess variance is due to a single, extended, negative feature in the image (Richards *et al.* 1995). When convolved with a synthesized beam of 30", roughly matching its minor axis, the negative feature is found to have an amplitude of -200 to -300 μK, i.e., close to 10^{-4} of T_0. The last column of Table 3 shows the effect of excising this small region from the map—in general, the "residual" variance is decreased, leading to somewhat lower limits on $\Delta T/T$. What are we to make of this negative feature? We have performed a variety of tests (Richards *et al.* 1996) to make certain it is not an instrumental artifact.

It appears in all of the data and at all resolutions. We thus conclude it is a feature of the microwave sky, and have tentatively suggested that it may be a Sunyaev-Zel'dovich signal from a distant cluster of galaxies. If it were an SZ signal from a *nearby* cluster, then we should see the individual cluster galaxies (and they are not apparent on an HST image of this region of the sky, one of the areas of the Medium Deep Survey). In addition, with reasonable assumptions about the properties of the intracluster gas producing the SZ signal, we would also expect to see X-rays if the cluster redshift were of order 0.5; an examination of a deep ROSAT image shows no such evidence. Bolstering the interpretation that the cluster is at substantial redshift is the detection of two quasars near this feature, both having redshifts 2.56 (Windhorst *et al.* 1995). Finally, the amplitude of this negative signal, and its angular extent, are consistent with an SZ signal from a high redshift, fairly massive cluster.

Over the past few months, my colleagues and I have been making very similar observations of the Hubble Deep Field. These data have yet to be analyzed, but we can already say that no strong, negative feature like the one discussed above has been found in the HDF (see Fomalont *et al.* 1997).

Finally, in Table 4, I've assembled results from the literature on searches for CBR anisotropies on scales of arcminutes and below. These do not include the tentative results of Table 3, but it may be seen that they are consistent with these new results.

TABLE 4. PREVIOUS OBSERVATIONAL RESULTS, $\theta \lesssim 2'$

Resolution	λ (cm)	Stokes Parameter	$\Delta T/T$ (a) $\times (10^5)$	Reference
6″	6	I	<320	Knoke et al. (1984)
5.3″	2	I	<63	Hogan and Partridge (1989)
30″ (b)	0.13	I	<26	Kreysa and Chini (1989)
10″	0.34	I	<9	Radford (1993)
10″	3.6	I	<7.2	Fomalont et al. (1993)
18″	"	I	<5.8	"
30″	"	I	<4.0	"
60″	"	I	<2.3	"
80″	"	I	<1.9	"
10″	"	V	<5.9	"
30″	"	V	<3.6	"
80″	"	V	<2.2	"
10″	"	$\sqrt{U^2+Q^2}$	<6.9	"
30″	"	$\sqrt{U^2+Q^2}$	<3.3	"
80″	"	$\sqrt{U^2+Q^2}$	<2.1	"
~ 120″	3.5	I	<0.9	Subrahmanyan et al. (1993)

(a) At 95% confidence.
(b) Beam-switch angle.

References

1. Banday, A. J., Giler, M., Szabelska, B., Szabelski, J., and Wolfendale, A. W. 1991, *Ap. J.*, **375**, 432, and references therein.
2. Bensadoun, M., Bersanelli, M., De Amici, G., Kogut, A., Levin, S. M., Limon, M., Smoot, G. F., and Witebsky, C. 1993, *Ap. J.*, **409**, 1.
3. Bersanelli, M., Bensadoun, M., De Amici, G., Levin, S., Limon, M., Smoot, G. F. and Vinje, W. 1994, *Ap. J.*, **424**, 517.
4. Bond, J.R. 1990, in *Frontiers in Physics–From Colliders to Cosmology*, eds. B. Campbell and F.Khana, World Scientific, Singapore.
5. Bouchet, F. R., Gispert, R., and Puget, J. L. 1995, as given in *Cobras/Samba: Report on Phase A Study*.
6. Cheng, E. S. et al. 1996, *Ap. J. (Letters)*, **456**, L71.
7. Clapp, A. C. et al. 1994, *Ap. J. (Letters)*, **433**, L57.
8. Danese, L., and Partridge, R. B. 1989, *Ap. J.*, **342**, 604.
9. De Amici, G., Bensadoun, M., Bersanelli, M., Kogut, A., Levin, S., Smoot, G. F., and Witebsky, C. 1990, *Ap. J.*, **359**, 219.
10. De Amici, G., Bensadoun, M., Limon, M., Smoot, G. F., Witebsky, C. and Bersanelli, M. 1995, *Astrophys. Lett. and Comm.*, **32**, 153.

30

11. de Bernardis, P. et al. 1994, *Ap. J. (Letters)* , **422**, L33.
12. Devlin, M. J. et al. 1994, *Ap. J. (Letters)* , **430**, L1.
13. Encrenaz, P. J., Penzias, A. A., and Wilson, R. W. 1970, *Astron. and Astrophys.*,**9**, 51.
14. Epstein, E. E. 1967, *Ap. J.(Letters)* , **148**, L157.
15. Fixsen, D. J., Cheng, E. S., Gales, J. M., Mather, J. C., Shafer, R. A., and Wright, E. L. 1996, *Ap. J.*, **473**, 576
16. Fomalont, E. B., Kellermann, K. I., and Wall, J. V. 1984, *Ap. J.(Letters)* , **277**, L23.
17. Fomalont, E. B., Partridge, R. B., Lowenthal, J. D., and Windhorst, R. A. 1993, *Ap. J.*, **404**, 8.
18. Fomalont, E. B., Kellermann, K.I., Richards, E.A., Windhorst, R.A. and Partridge, R.B. 1997, *Ap. J. (Letters)*, **475**, L5.
19. Franceschini, A., Toffolatti, L., Danese, L., and De Zotti, G. 1989, *Ap. J.*, **344**, 35.
20. Gunderson, J. O., et al. 1995, *Ap. J.(Letters)* , **443**, L57.
21. Gush, H. P., Halpern, M., and Wishnow, E. H. 1990, *Phys. Rev. Lett.*, **65**, 537.
22. Haslam, C. G. T., Salter, C. J., Stoffel, H., and Wilson, W. E. 1982, *Astron. and Astrophys. Suppl.*, **47**, 1.
23. Hogan, C. J., and Partridge, R. B. 1989, *Ap. J. (Letters)* , **341**, L29.
24. Howell, T. F. and Shakeshaft, J. R. 1967a, *Nature*, **216**, 753.
25. Howell, T. F. and Shakeshaft, J. R. 1967b, *J. Atm. Terres. Phys.*, **29**, 1559.
26. Kellermann, K.I., Fomalont, E. B., Partridge, R. B., Richards, E., and Windhorst, R. A. 1996, in preparation for *Ap.J.*
27. Kislyakov, A. G., Chernyshev, V. I., Lebskii, Yu. V., Mal'tsev, V. A., and Serov, N. V. 1971, *Sov. A. J.*, **15**, 29.
28. Knoke, J. E., Partridge, R. B., Ratner, M. I., and Shapiro, I. I. 1984, *Ap. J.*, **284**, 479.
29. Kogut, A. 1995, *Astrophys. Lett. and Comm.*, **32**, 37.
30. Kraus, J. D. 1966, *Radio Astronomy*, McGraw-Hill, New York.
31. Kreysa, E., and Chini, R. 1989, in *Third ESO/CERN Symposium, Astronomy, Cosmology and Fundamental Physics*, eds. M. Caffo et al., Kluwer Academic Publishers, Dordrecht, Netherlands.
32. Liebe, H. J. 1984, *Internat. J. IR and Millimeter Waves*, **5**, 207.
33. Liebe, H. J. 1985, *Radio Sci.*, **20**, 1069.
34. Lubin, P., Melese, P. and Smoot, G. 1983, *Ap. J.(Letters)* , **273**, L51.
35. Mandolesi, N. et al. 1996, *Ap. J.* **310**, 561.
36. Martin, H. M., Partridge, R. B., and Rood, R. T. 1980, *Ap. J.(Letters)* , **240**, L79.
37. Mather, J. C. 1982, *Optical Engineering*, **21**, 769.
38. Mather, J. C. et al. 1994, *Ap. J.*, **420**, 439.
39. Netterfield, C. B., Jarosik, N., Page, L., Wilkinson, D., and Wollack, E. 1995, *Ap. J.(Letters)* ,**445**, L69.
40. Netterfield, C. B., Devlin, M. J., Jarosik, N., Page, L., and Wollack, E. J. 1997, *Ap. J.* **474**, 47, preprint, Astro-ph/9601197.
41. Otoshi, T. Y., Stelzreid, C. T. 1975, *IEEE Trans. on Instrumentation and Measurements*, **24**, 174.
42. Partridge, R. B. 1995, *3 K: The Cosmic Microwave Background Radiation*, Cambridge Univ. Press, Cambridge.
43. Partridge, R. B., Cannon, J., Foster, R., Johnson, C., Rubinstein, E., Rudolph, A., Danese, L., and DeZotti, G. 1984, *Phys. Rev.* **D29**, 2683.
44. Partridge, R. B., Richards, E., Fomalont, E. B., Kellermann, K. I., and Windhorst, R. A. 1996, *Ap. J.* (submitted).
45. Penzias, A. A., and Wilson, R. W. 1965, *Ap. J.*, **142**, 419.
46. Puzanov, V. I., Salomonovich, A. E., and Stankevich, K. S. 1968, Sov. *Ap. J.*, **11**, 905.
47. Radford, S. J. E. 1993, *Ap. J.(Letters)* , **404**, L33.

48. Richards, E. et al. 1996, in *IAU Symposium 175, Extragalactic Radio Sources*, Kluwer Acad. Publ. Co., Dordrecht, the Netherlands.
49. Rohlfs, K. 1986, *Tools of Radio Astronomy*, Springer-Verlag, Heidelberg and New York.
50. Ruhl, J. E., Dragovan, M., Platt, S. R., Kovac, J., and Novak, G. 1995, *Ap. J.(Letters)*, **453**, L1.
51. Scott, P. F. et al. 1996, *Ap. J.(Letters)*, **461**, L1.
52. Sironi, G., Bonelli, G., Dall'Oglio, G., Pagana, E., De Angeli, S., and Perelli, M. 1995, *Astrophys. Lett. and Comm.*, **32**, 31.
53. Smith, C., and Wilkinson, D. T. 1995, *Bull. A.A.S.*, **27**, 1377.
54. Smoot, G., Bensadoun, M., Bersanelli, M., De Amici, G., Kogut, A., Levin, S., Witebsky, C. 1987, *Ap. J.(Letters)*, **317**, L45.
55. Smoot, G. F., Levin, S. M., De Amici, G., and Witebsky, C. 1986, *Radio Science*.
56. Staggs, S. T., Jarosik, N. C., Wilkinson, D. T., and Wollack, E. J. 1996, *Ap. J.*, **458**, 407.
57. Stokes, R. A., Partridge, R. B., and Wilkinson, D. T. 1967, *Phys. Rev. Letters*, **19**, 1199.
58. Subrahmanyan, R., Ekers, R. D., Sinclair, M., and Silk, J. 1993, *Mon. Not. Roy. Astr. Soc.*, **263**, 416.
59. Timbie, P. T., and Wilkinson, D. T. 1988, *Rev. Sci. Instr.*, **59**, 914.
60. Timbie, P. T., and Wilkinson, D. T. 1990, *Ap. J.*, **353**, 140.
61. Wilkinson, D. T. 1967, *Phys. Rev. Letters*, **19**, 1195.
62. Windhorst, R. A., Fomalont, E. B., Kellermann, K. I., Partridge, R. B., Richards, E., Franklin, B. E., Pascarelle, S. M., and Griffiths, R. E. 1995, *Nature*, **375**, 471.
63. Windhorst, R. A., Fomalont, E. B., Partridge, R. B., and Lowenthal, J. D. 1993, *Ap. J.*, **405**, 498.

ELEMENTS OF GENERAL RELATIVITY, COSMOLOGY AND THE COSMIC MICROWAVE BACKGROUND

JOSE L. SANZ

IFCA

Facultad de Ciencias, Av. de los Castros s.n., 39005-Santander, Spain

1. Abstract

We describe basic tools of General Relativity and cosmology necessary to understand the cosmic microwave background radiation. We emphasize the gauge-invariant technique and the fluid-flow approach. A basic study of the anisotropies of such a radiation is undertaken.

2. Basics of General Relativity

The cosmic microwave background radiation (CMB) is, without any doubt, the most powerful test for cosmology. The interpretation of such a radiation as a relic of the hot and dense past in the history of the Universe supports the Big Bang model on two sides: on the one hand the perfect black-body spectrum of the radiation supports a hot past and the fact that the history of the Universe has been very quiet. On the other hand, the anisotropy detected at the level of 10^{-5} on large angular–scales, is consistent with our ideas about the formation of structure in the Universe via gravitational instability.

One interesting question is: why do we need General Relativity (GR) to study either cosmology or the CMB? There are several reasons for this: on the one hand, if we are interested in the study of anisotropies (using either the Boltzmann equation or free propagation after recombination) or in the influence of gravitational lensing on the CMB, we need to take into account the propagation of light along geodesics within the framework of GR. On the other hand, the study of the perturbations of Friedmann-Robertson-Walker (FRW) models (perturbations with scales of the order of

C. H. Lineweaver et al. (eds.), The Cosmic Microwave Background, 33–65.
© 1997 *Kluwer Academic Publishers.*

the horizon size, perturbations in the radiation era, gravitational waves) or the ideas about inflation must be also done within the framework of GR.

GR is a theory, its foundations due basically to A. Einstein (1879-1955), of both *gravitation* and *spacetime* itself. The *equivalence* and the *covariance* principles are the basic rules that allow one to construct GR. In particular, the basic equations of the theory: the geodesic equation that is the law governing the motion of test particles and the Einstein field equations that give the law for the gravitational field (or spacetime).

We will employ the following notation: the indices a, b, c, d, e, f run from $0, 1, 2, 3$ and will apply to spacetime components whereas the indices i, j, k, l run from $1, 2, 3$ and apply to space components. We follow the standard Einstein rule of addition over repeated indices which appear on contravariant and covariant positions. A "*comma*" denotes a partial derivative with respect to a coordinate whereas a "semicolon" denotes a covariant derivative. We choose units such that $c = 8\pi G = 1$; c is the velocity of light in vacuum and G is the gravitational constant.

2.1. THE EQUIVALENCE PRINCIPLE

The equivalence principle states that "at every spacetime point it is possible to choose a *local inertial system* such that the laws of nature take the same form as in *Lorentztian systems* in the absence of gravitation". This allows us to introduce *the metric*, i.e., using a global coordinate system (x^a) $(x^0 \equiv t$ is the time coordinate and x^i are the space coordinates), one can obtain the spacetime interval between two nearby events as

$$ds^2 = g_{ab}dx^a dx^b. \tag{1}$$

The metric $g_{ab}(x^c)$ satisfies the following properties: i) $g_{ab} = g_{ba}$, ii) $det(g_{ab}) \neq 0$, iii) g_{ab} transform like a 2-tensor and iv) there exist *local inertial systems* (ξ^a) such that at any fixed but arbitrary point $g_{ab} = \eta_{ab}$, the latter being the Minkowskian metric $(\eta_{ab} = diag(-1, 1, 1, 1))$. All of these properties characterize g as the metric tensor, so the spacetime is a 4D Riemannian manifold.

On the other hand, the gravitational field is identified with g. This tight relation can be understood if one tries to ask the following question: what are the equations that describe the motion of test particles under gravity? The equivalence principle states that test particles move along straight lines in a *local inertial system*. So, if we rewrite the local 4-acceleration using a global coordinate system, we find the *geodesic equation*

$$\frac{d^2x^a}{dp^2} + \Gamma^a_{bc}\frac{dx^b}{dp}\frac{dx^c}{dp} = 0, \tag{2}$$

where $\Gamma^a_{bc} = \frac{1}{2}g^{ad}[g_{bd,c} + g_{dc,b} - g_{bc,d}]$ are the Christoffel symbols and p is an afin parameter. Null geodesics, describing the free propagation of photons, are characterized by a null tangent vector k^a

$$g_{ab}k^ak^b = 0, \quad k^a \equiv \frac{dx^a}{dp}, \tag{3}$$

whereas massive particles satisfy: $g_{ab}p^ap^b = -m^2$, $p^a = m\,u^a$, $u^a \equiv \frac{dx^a}{ds}$, m being the mass of the particle.

2.2. THE COVARIANCE PRINCIPLE

This principle assumes the group (G) of general (regular) transformations on the 4D manifold as the basic one. In this sense, the objects of the theory must transform as *tensor* fields, i.e., for an r-contravariant, s-covariant tensor

$$t^{a'_1...a'_r}_{b'_1...b'_s}(x^{c'}) = \frac{\partial x^{a'_1}}{\partial x^{a_1}} \cdots \frac{\partial x^{a'_r}}{\partial x^{a_r}} \frac{\partial x^{b_1}}{\partial x^{b'_1}} \cdots \frac{\partial x^{b_s}}{\partial x^{b'_s}} t^{a_1...a_r}_{b_1...b_s}(x^c), \tag{4}$$

when the coordinate change $x^{c'} = x^{c'}(x^c)$ is considered. Also the dynamical laws in GR must be *generally covariant*, i.e., they must preserve their form under G. All of these constitute a *principle of relativity*, putting on an equal footing all the coordinate systems to describe the physics related to gravitation. The principle can be used, from the practical point of view, to find the generalization to GR of some laws that apply in Special Relativity and also to find the equations that govern the gravitational field itself. In this context, a useful generalization of the usual derivative is introduced: the covariant derivative of the tensor $t^{a_1...a_r}_{b_1...b_s}$ with respect to the coordinate x^c is defined by

$$t^{a_1...a_r}_{b_1...b_s;c} \equiv t^{a_1...a_r}_{b_1...b_s,c} + \Gamma^{a_1}_{dc}\,t^{da_2...a_r}_{b_1...b_s} + ... + \Gamma^{a_r}_{dc}\,t^{a_1...a_{r-1}d}_{b_1....b_s} -$$

$$\Gamma^d_{b_1c}\,t^{a_1...a_r}_{db_2...b_s} - ... - \Gamma^d_{b_sc}\,t^{a_1...a_r}_{b_1...b_{s-1}d}. \tag{5}$$

The covariant derivative of a tensor produces another tensor, the metric has zero covariant derivative and standard algebraic rules apply but one: covariant derivatives do not commute, i.e., $t_{;ab} \neq t_{;ba}$ except if t is a scalar.

2.3. CURVATURE

Non-local observations are able to detect a gravitational field with the Riemmann tensor (or curvature tensor) $R^a{}_{bcd}$

$$-R^a{}_{bcd} = \Gamma^a_{bc,d} - \Gamma^a_{bd,c} + \Gamma^e_{bc}\Gamma^a_{de} - \Gamma^e_{bd}\Gamma^a_{ce}. \tag{6}$$

Moreover, the geodesic equation gives the evolution of the separation vector between two neighboring particles $x^a(p)$ and $x^a(p) + \eta^a$ in the form

$$\frac{d^2\eta^a}{dp^2} = -R^a{}_{bcd}\eta^c \frac{dx^b}{dp} \frac{dx^d}{dp}. \tag{7}$$

So, gravitation is not pure fiction; rather, it is the second order derivatives of g_{ab} combined to form the Riemann tensor that characterize gravitation. For a static field described by a small gravitational potential, some components of that tensor reduce to the "tidal" field of standard Newtonian mechanics.

On the other hand, from the geometrical point of view, the Riemann tensor defines the Gaussian curvature at any point of the spacetime. This is the natural generalization to 4D space of basic ideas – elaborated by Gauss– related to the intrinsic curvature of 2D surfaces (i.e., the product of the principal curvatures). The Gaussian curvature (K) at point x^a in the directions $v_1^a \neq v_2^a$ is the 2D Gaussian curvature at that point of the 2D *geodesic* surface generated by v_1^a and v_2^a. One can prove that

$$K(x^a, v_1^b, v_2^c) = A^{-1} R_{abcd} v_1^a v_1^c v_2^b v_2^d, \quad A \equiv (g_{ac}g_{bd} - g_{ad}g_{bc})v_1^a v_1^c v_2^b v_2^d. \tag{8}$$

The previous formula applies to nD Riemannian spaces. An interesting case, from the cosmological point of view, corresponds to 3D space with constant curvature, i. e. $K = constant$. In such a case, the Riemann tensor has a simple form: $R_{ijkl} = K(g_{ik}g_{jl} - g_{il}g_{jk})$.

A last comment concerns the characterization of flatness: the spacetime is flat *iff* (if and only if) the Riemann tensor vanishes at any point (or equivalently, $K = 0$). In this case, there exist global coordinates that allow one to rewrite the metric tensor in the Minkowski form.

2.4. RELEVANT GEOMETRICAL TENSORS

One and two contractions over the Riemann tensor give the Ricci tensor ($R_{bd} \equiv R^e{}_{bed}$) and the curvature scalar ($R \equiv g^{ab}R_{ab}$), respectively, whereas the following combination

$$S_{ab} \equiv R_{ab} - \frac{1}{2}R\, g_{ab}, \tag{9}$$

called the Einstein tensor, will appear as the foremost element of the gravitational field equations. The Weyl tensor (C_{abcd}) is defined as the traceless part of the Riemann tensor

$$C_{abcd} \equiv R_{abcd} - \frac{1}{2}(g_{ac}R_{bd} - g_{ad}R_{bc} - g_{bc}R_{ad} + g_{bd}R_{ac}) + \frac{1}{6}R(g_{ac}g_{bd} - g_{ad}g_{bc}). \tag{10}$$

The Weyl tensor is not directly related to the energy-momentum content and describes the free gravitational field.

Finally, another interesting tensor from the viewpoint of the geometry of a 3D space is the Cotton-York tensor Y_{ij} (York 1971)

$$Y_{ij} = \epsilon_i{}^{kl}({}^3R_{jk} - \frac{1}{4}{}^3R\,{}^3g_{jk})_{;l} \quad , \tag{11}$$

where ${}^3R_{ij}$, ${}^3R$ and ${}^3g_{ij}$ are the Ricci tensor, the curvature scalar and the metric of the 3D space, respectively, and a "semicolon" means covariant derivative with respect to the 3-metric. One can prove the following: the 3D space is conformally flat *iff* the Cotton-York tensor is zero at any point. Y_{ij} will be used to characterize a tensor ("gravitational wave") perturbation on a homogeneous and isotropic background.

2.5. THE ENERGY-MOMENTUM TENSOR

In GR, the matter content is represented by a 2nd rank, symmetric and conserved tensor (T^{ab})

$$T^{ab} = T^{ba}, \quad T^{ab}{}_{;b} = 0. \tag{12}$$

If the content can be represented by a *fluid* with average velocity $u^a(x^b)$, $u^2 \equiv g_{ab}u^au^b = -1$, then T^{ab} can be decomposed:

$$T^{ab} = \rho u^a u^b + p\,h^{ab} + q^a u^b + q^b u^a + \pi^{ab}, \quad h^{ab} \equiv g^{ab} + u^a u^b, \tag{13}$$

where ρ, p, q^a and π^{ab} represent the energy density, pressure, energy flux and anisotropic pressure, respectively. The motion equations of the fluid can be derived from the conservation Equation (12).

If $q^a = \pi^{ab} = 0$, then we have a *perfect* fluid. This is approximately the case for the energy content of the Universe: During the radiation and matter eras, the equation of state is $p = \frac{1}{3}\rho$ and $p = 0$, respectively. In general, if the statistical state of the constituent particles of the Universe is described by a distribution function, then the energy-momentum tensor must be calculated in terms of this distribution (see Equation 16).

2.6. THE FIELD EQUATIONS

The evolution equations for the metric tensor g_{ab} were obtained by A. Einstein (1915). They involve the Einstein tensor S_{ab} and the energy-momentum tensor:

$$S_{ab} = T_{ab}. \tag{14}$$

They constitute a 2nd order, non-linear, partial differential system on g_{ab}, called the Einstein field equations. Obviously, unless one specifies T^{ab}, the system is undetermined. In general, if the system under study cannot be represented by a perfect fluid with a given equation of state, but can be described by a distribution satisfying the Boltzmann equation, one needs to solve the coupled Einstein-Boltzmann equations.

2.7. THE BOLTZMANN EQUATION

A complete description of the history of our Universe involves the study of all the constituents: photons, baryons, massless neutrinos and maybe cold dark matter (CDM) and hot dark matter (HDM). The statistical state of each of these particles is usually described by 1-particle distribution functions, $f(t, x^i, p^j)$, that satisfy the Boltzmann equation

$$\frac{Df}{D\lambda} \equiv p^0 \frac{\partial f}{\partial t} + p^i \frac{\partial f}{\partial x^i} - \Gamma^i_{ab} p^a p^b \frac{\partial f}{\partial p^i} = C[f], \tag{15}$$

where p^0 is the positive root of the equation $p^2 \equiv g_{ab} p^a p^b = -m^2$, m being the mass of the particle. The third term represents the interaction with the gravitational field, and $C[f]$ the interaction with other species. For the photons, $C[f]$ contains absorption and emission processes as well as Compton scattering with the baryons. Einstein's field equations couple to Boltzmann's equation through Γ^i_{ab} and T^{ab}, this latter tensor given by the formula

$$T^{ab} \equiv \int d^3\vec{p} f p^a p^b [p^0 (-g)^{1/2}]^{-1}, \tag{16}$$

where $g \equiv det(g_{ab})$.

2.8. KINEMATICS

The gradient of the average velocity of the matter, $u^a(x^b)$, can be decomposed in the following way

$$u_{a;b} = \frac{1}{3}\theta + \sigma_{ab} + \omega_{ab} - \xi_a u_b, \tag{17}$$

where $\theta, \sigma_{ab}, \omega_{ab}$ and ξ_a are the expansion scalar, the shear, the vorticity and the acceleration of the fluid. These quantities generalize the classical ones. If $\theta \neq 0$ at a point, a sphere of fluid surrounding the point expands (or contracts) with the same orientation; $\sigma \neq 0$ implies that such a sphere is distorted, leaving the volume and principal axes of shear unchanged; $\omega \neq 0$ gives a rotation, leaving a direction unchanged (the rotation axis). Finally,

$\xi \equiv u^a{}_{,b}u^b \neq 0$ is the 4-acceleration of the fluid. The interest of these quantities in cosmology is twofold: they allow a classification of cosmological models and, as we shall see in the next section, the evolution equations of these kinematical quantities can be solved in some simple cases (in particular, linear perturbations of FRW models). So, they complement the geometrical content that can be obtained through the metric.

2.9. THE FLUID-FLOW APPROACH

The fluid-flow approach is a technique to analize and classify cosmological models and to study linear perturbations of FRW models. It is based on a set of evolution equations and constraints on the kinematical quantities previously defined (θ, ω, σ and ξ) and on two geometrical fields closely related to the Weyl tensor ("electric" and "magnetic" components, see below). Hawking (1966) studied the evolution of linear perturbations in the Universe following this approach, and Ellis (1971) wrote a review on general properties of cosmological models using such a framework. The study of linear perturbations of FRW models using this formalism (instead of the usual metric perturbation approach of Lifshitz (1946) or the more recent gauge-invariant metric perturbation technique of Bardeen (1980)) made it simple and physically transparent. *Vorticity* and *gravitational waves* are described in a covariant and gauge-invariant way. Recently, Ellis & Bruni (1989) have introduced a couple of quantities (see below) to also describe *density perturbations*.

On the one hand, the conservation of the energy-momentum tensor leads to the following evolution equations for ρ and u^a

$$\rho^{\cdot} = -(\rho+p)\theta + NPF, \quad \xi^a = -(\rho+p)^{-1}h^{ab}p_{,b} + NPF, \qquad (18)$$

where a "dot" means a covariant derivative along the worldlines of the fluid (i.e., $F^{\cdot} \equiv F_{;a}u^a$). Hereafter, NL and NPF mean "non-linear terms" involving $\xi_a, \omega_a, \sigma_{ab}, E_{ab}$ and H_{ab} and "non–perfect fluid contributions" involving q_a and π_{ab}, respectively. Their expressions can be found in the review by Ellis (1971). We remark that if there are no pressure gradients then the flow is geodesic (i.e., $\xi = 0$).

Taking into account the Ricci identities as applied to the velocity vector, $u_{a;bc} - u_{a;cb} = R_{dabc}u^a u^b u^c$, one obtains

$$\theta^{\cdot} + \frac{1}{3}\theta^2 + \frac{1}{2}(\rho + 3p) - \xi^a{}_{;a} = NL \qquad (19)$$

$$h^{ab}\omega^{\cdot}_b + \frac{2}{3}\theta\omega^a - \frac{1}{2}\eta^{abcd}u_b\xi_{c;d} = NL \qquad (20)$$

$$h_a^c h_b^d \dot{\sigma}_{cd} + \frac{2}{3}\theta\sigma_{ab} - \frac{1}{2}h_a^c h_b^d(\xi_{c;d} + \xi_{d;c}) + \frac{1}{3}h_{ab}\xi_{;c}^c + E_{ab} = NL + NPF. \quad (21)$$

Equation (19) is due to Raychaudhuri (1955), and $\omega^a \equiv \frac{1}{2}\eta^{abcd}u_b\omega_{cd}$, while η is the Levi-Civitta pseudotensor normalized to $\eta^{0123} \equiv (-g)^{-1/2}$ and E^{ab} is the "electric" component of the Weyl tensor. This component and its "magnetic" counterpart are defined by

$$E_{ab} \equiv C_{acbd}u^c u^d, \quad H_{ab} \equiv \frac{1}{2}\eta_{ac}^{de}C_{debf}u^c u^f. \quad (22)$$

Taking into account the Bianchi identities $(R_{abcd;e} + R_{abde;c} + R_{abec;d} \equiv 0)$ and the Einstein equations, we obtain evolution equations for these geometrical quantities (they are "Maxwell"-like equations):

$$h^{ac}h^{bd}\dot{E}_{cd} + \theta E^{ab} + h_c^{(a}\eta^{b)def}u_d H_{e;f}^c = -\frac{1}{2}(\rho + p)\sigma^{ab} + NL + NPF, \quad (23)$$

$$h^{ac}h^{bd}\dot{H}_{cd} + \theta H^{ab} - h_c^{(a}\eta^{b)def}u_d E_{e;f}^c = NL + NPF, \quad (24)$$

where a parenthesis including two indices denotes symmetrization.

Moreover, one obtains constraint equations on the kinematical and geometrical quantities (see Ellis 1971). Recently, Bruni & Ellis (1989) have also derived the evolution equations for the gradients associated with the relative energy density $(\Delta^a \equiv \rho^{-1}h^{ab}\rho_{,b})$ and expansion scalar $(Z^a \equiv h^{ab}\theta_{,b})$:

$$\dot{\Delta}_a + (\frac{2}{3} + \frac{p}{\rho})\Delta_a + (1 + \frac{p}{\rho})Z_a = NL + NPF, \quad (25)$$

$$\dot{Z}_a + \theta Z_a + \frac{1}{2}\rho\Delta_a - \frac{1}{2}{}^3R\,\xi_a - h_a^b\xi_{;cb}^c = NL + NPF. \quad (26)$$

The last set of equations can be used to obtain the time behavior of Δ_a, regarding matter perturbations. This is all we need to calculate anisotropies in the CMB in a gauge-invariant way.

One last comment concerning the different characterizations of FRW models: for a perfect fluid, the following conditions are equivalent: i) $E = H = 0, p = p(\rho)$ (geometrical, gauge-invariant), ii) $\sigma = \omega = \xi = 0$ (kinematical, gauge-invariant) or iii) there exist comoving and synchronous coordinates such that

$$u^a = \delta_o^a, \quad ds^2 = -dt^2 + a^2(t)[1 + \frac{k}{4}x^2]^{-2}\delta_{ij}dx^i dx^j, \quad (27)$$

where k is a constant that can be normalized to $-1, 0, +1$ and $a(t)$ is an arbitrary function of its argument. A proof of this theorem can be undertaken through Equations (19-24).

3. Basics of Cosmology

Modern cosmology, in its classical aspects, is theoretically based on classical GR. The beginning can be attributed to A. Einstein (1917), whose work implies that the Universe can be studied with ordinary physics without contradictions or ambiguity. The introduction of a non-static metric with spherical symmetry to describe a matter-filled universe is due to A. Friedmann (1922, 1924). The closed model was introduced in the first paper, whereas the open one appeared in the second. On the observational side, E. Hubble (1924) confirmed that the *spiral nebulae* are other *galaxies* similar to our own Galaxy, and a new era began that culminated in the empirical distance-radial velocity relation for such "extragalactic" nebulae (E. Hubble, 1929). G. Lemaître (1929) studied the closed model and made observational predictions. H. P. Robertson (1929) found the general form of the metric under the assumption of homogeneity and isotropy at any instant of time, whereas A. G. Walker (1936) proved that isotropy at any space point implies homogeneity. G. Gamow (1948) elaborated the synthesis of light elements in the hot Universe and obtained a residual electromagnetic radiation that could be observed at the present time. The discovery of such a radiation by A. Penzias and R. Wilson (1965) is one of the major contributions to the developments of modern cosmology. R. H. Dicke et al (1965) interpreted such a radiation in the framework of the Big Bang model. For a good historical review see Peebles' book (1993).

During the last 15 years or so, there have been new developments in the field that some people call the "new cosmology". Regarding observations: i) the discovery of the large-scale structure of the Universe, ii) the Planckian spectrum and the anisotropies detected by the FIRAS and DMR instruments on board the COBE satellite, iii) the new observations of objects at intermediate-z by HST and iv) the recent observations of objects at high-z with terrestrial telescopes (Keck,...). All of these relevant observations are a new challenge for theory. On the other hand, from the theoretical point of view: i) the idea of inflation is alive and ii) theories of galaxy formation based on dark matter such as the "cold dark matter" model (CDM) have been the paradigm during more than a decade.

3.1. THE REDSHIFT

This is one of the most relevant physical quantities in cosmology. A covariant definition of the *redshift* (z) can be given as follows

$$1 + z \equiv \frac{(u.k)_e}{(u.k)_o}, \qquad (28)$$

where a "dot" means scalar product (i.e., $(u.k) \equiv g_{ab}u^a k^b$), u^a and k^b denote the tangent vector to the fluid flow lines and tangent vector to the geodesic connecting the emitting source (e) and the observer (o), respectively. So, the vector k^a satisfies Equations (2-3). The physical interpretation of z is clear: it represents the relative ratio of the wavelengths of the photons observed in locally inertial systems at the observer and source $(1+z = \lambda_o/\lambda_e)$. It is usual to distinguish three different types of redshift according to their physical origin (although such a division is not clear in the spirit of GR, because one can always describe the local effects in terms of a metric): i) Doppler, ii) gravitational and iii) cosmological, due to pure kinematics (velocity), local mass and the expansion, respectively. In the first case, Equations (2-3) with $g = \eta$ give the relativistic Doppler effect

$$1 + z = \frac{1 + \vec{n}.\vec{v}_e}{(1 - v_e^2)^{1/2}}, \qquad (29)$$

where $\vec{v}_e$ denotes the 3-velocity of the source at the emitting point and $\vec{n}$ is the unit vector in the direction of observation. The previous equation has been obtained for the case when the observer is at rest.

For the case of a static gravitational field with spherical symmetry generated by a mass m, the Schwarzschild solution applies. So Equations (2-3) give

$$1 + z = \frac{1 - \frac{2m}{r_o}}{1 - \frac{2m}{r_e}}, \qquad (30)$$

where r_o, r_e denote radial distance to the observer and emitting point (from the mass), respectively.

If the Universe is expanding according to the RW metric (see Equation (27)), integration of null geodesics gives

$$1 + z = \frac{a(t_o)}{a(t_e)}, \qquad (31)$$

which is called the cosmological redshift.

3.2. DISTANCES

The "proper" distance between two points in the 3-space orthogonal to the observer u^a is defined by

$$d_p \equiv \int_1^2 (h_{ab} dx^a dx^b)^{1/2}, \quad h_{ab} \equiv g_{ab} + u_a u_b. \tag{32}$$

In comoving coordinates $(u^i = 0)$: $h_{0a} \equiv 0, h_{ij} = g_{ij} - g_{00}^{-1} g_{0i} g_{0j}$.

If the observer sees a source over a solid angle $d\Omega_o$ with intrinsic area dS_e, the area distance is defined by

$$d_A \equiv \left(\frac{dS_e}{d\Omega_o}\right)^{1/2}. \tag{33}$$

If the intrinsic area can be described by a circle of diameter dD_e, the previous definition coincides with the standard angular diameter distance

$$d_A \equiv \frac{dD_e}{d\theta_o}, \tag{34}$$

θ_o being the angle seen by the observer.

Moreover, one can define the following operational distance, called "luminosity" distance:

$$d_L \equiv \left(\frac{L}{4\pi l}\right)^{1/2}, \tag{35}$$

where L, l are the absolute and apparent luminosities of the source, respectively. Counting photons at the source and at the observer, and taking into account the definition of redshift, one gets

$$d_L = \left(\frac{dS_o}{d\Omega_e}\right)^{1/2}, \tag{36}$$

where $d\Omega_e$ is the solid angle at the source that sees a cross-sectional area dS_o.

One can prove the following "reciprocity" theorem (Ellis 1971):

$$dS_o d\Omega_o = dS_e d\Omega_e (1+z)^2, \tag{37}$$

which implies the relation $d_L = d_A (1+z)^2$.

In addition, the idea of an "horizon" associated with cosmological models with a "bang" singularity is quite relevant. The idea dates back to Rindler (1956), who defined the *particle horizon* in FRW models as the intersection of the 3-surface $t = t_o$ with the border of the set formed by the worldlines of particles visible by an observer up until t_o, i.e., the horizon marks the region of the Universe that has been observed up until the time t_o. This idea also applies to more general spacetimes where the notion of cosmic time can be introduced (causal stability is satisfied).

3.3. THE COSMIC MICROWAVE BACKGROUND

The specific intensity is defined as the power radiated by the source $I_{\nu o}$ (received by the observer, $I_{\nu e}$) per unit surface and solid angle in the frequency band $(\nu, \nu + d\nu)$. So, one has the law

$$I_{\nu o} = \frac{I_{\nu e}}{(1+z)^3}. \tag{38}$$

For black-body radiation at temperature T, the Planck law is

$$I_\nu \propto \frac{\nu^3}{e^{h\nu/k_B T} - 1}. \tag{39}$$

If the black-body spectrum is preserved during the evolution of the Universe, the two previous equations imply that

$$T_o = \frac{T_e}{1+z}, \tag{40}$$

where T_o, T_e denote the temperature at the observer and emitting point, respectively.

3.4. THE FRIEDMANN-ROBERTSON-WALKER MODEL

a) *The cosmological principle*:
Modern cosmology is based on a geometry that arises assuming the "cosmological principle": there exist fundamental observers u^a in free-fall (i.e., $u_a^{\cdot} = 0$) that define 3D surfaces of "homogeneity" and "isotropy" at any instant of time. Then, one can prove (Robertson 1929) that the most general form for the metric is the RW one (see Equation (27)). This standard assumption can be realized in some inflationary scenarios, or the argument of simplicity may be assumed. On the other hand, one can ask: to what extent can the global isotropy of the model be deduced from the CMB data? The Ehlers-Geren-Sachs theorem proves that if all observers see an isotropic CMB, then the Universe is described by the RW metric. Maartens et al. (1995a, b) developed a covariant and gauge-invariant formalism for analyzing the direct relationship between anisotropies in the CMB and deviations of the Universe from the FRW form. They obtained limits on anisotropy and inhomogeneity assuming bounds on the temperature multipoles, their spatial gradients and their time derivatives. Martínez-González and Sanz (1995) assumed an exactly homogeneous, but anisotropic, cosmological model of Bianchi type I (which includes the flat FRW model) and proved that the small quadrupole component detected by COBE is enough to imply that such a model must represent only a small departure from the flat case. So, taking into account all of these results, one can conjecture

that the isotropy of the metric is a direct observational consequence of the observed CMB isotropy.

b) *Geometry of the 3D surfaces* $t = constant$:

The geometry of the 3D surfaces at any instant of time is closely related to the parameter k. A calculation of the Gaussian curvature gives: $K = k/a^2(t)$, i.e., they are spaces of constant curvature. The standard topologies associated with these 3D surfaces are: $k = 0$ represents the usual 3D Euclidean space R^3, $k = -1$ is the 3D hyperboloid contained in 4D Minkowski space ($H_3 \subset M_4$) and $k = +1$ is the 3D sphere immersed in 4D Euclidean space ($S_3 \subset R^4$). With this picture in mind, the corresponding 4D spaces are called: flat, open and closed models, respectively.

A useful form of the RW metric is obtained by changing to polar coordinates: $(r(\chi), \theta, \phi)$ with $r(\chi) = \chi, 2\tanh(\frac{\chi}{2})$, and $2\tan(\frac{\chi}{2})$ for $k = 0, -1, +1$, respectively. In this case:

$$ds^2 = -dt^2 + a^2(t)[d\chi^2 + S_k^2(\chi)d\Omega^2], \tag{41}$$

$S_k(\chi) \equiv \chi, \sinh\chi, \sin\chi$ for $k = 0, -1, +1$ and $d\Omega^2 \equiv d\theta^2 + \sin^2\theta\, d\phi^2$. The domain for the "radial" coordinate is: $\chi \in [0, \infty)$ for $k = 0, -1$, and $\chi \in [0, \pi]$ for $k = +1$.

c) *Dynamics of Friedmann-Robertson-Walker models*:

The dynamics of FRW models can be obtained by introducing the RW metric into the Einstein equations and assuming a perfect fluid for the energy-momentum tensor; this results in the Friedmann equations for the density evolution and the expansion rate:

$$\frac{d\rho}{dt} = -3(\rho + p)\frac{1}{a}\frac{da}{dt}, \quad (\frac{1}{a}\frac{da}{dt})^2 = -\frac{k}{a^2} + \frac{\rho}{3}. \tag{42}$$

If there exists a cosmological Λ-term, the 2nd equation must be modified with the substitution: $\rho \rightarrow \rho + \Lambda$. If we introduce the Hubble constant H_o and the total density parameter Ω

$$H_o \equiv (\frac{1}{a}\frac{da}{dt})_o, \quad \Omega \equiv \frac{\rho_o}{3H_o^2}, \tag{43}$$

where a subscript "o" means the present time, the 2nd Friedmann equation gives: $H_o^2(1 - \Omega) = -k/a_o^2$, i.e., the characterization of FRW models can be given in terms of $\Omega = 1, < 1, > 1$, respectively.

One of the most remarkable characteristics of FRW models is the existence of physical singularities under general conditions. In fact, assuming a barotropic equation of state $p = p(\rho)$ satisfying the stability and causality conditions – $0 \leq \frac{dp}{d\rho} \leq 1$ – and the weak observational inequalities – $\theta_o, \rho_o, (\rho + p)_o, (\rho + 3p)_o > 0$ – then, from the Friedmann equations, one obtains (Collins & Ellis 1979) the following: there exists a point, that we shall identify with $t \equiv 0$, such that $a(0) = 0$. At this point (called the "Big Bang"), all physical and geometrical quantities (density, pressure, expansion, Gaussian curvature,...) go to ∞.

Taking into account the notion of the particle horizon introduced at the end of Section 3.2, the proper distance, d_{Ho}, from the observer to any point of the horizon is given by

$$d_H = a_o \tau_o = a_o \int_0^{a_o} da \, a^{-1} \left(-k + \frac{1}{3}a^2 \rho\right)^{-1/2}. \tag{44}$$

d) *Popular models*:

The behavior of the scale factor, density and horizon distance during the radiation era $(p = \frac{1}{3}\rho)$ for the flat case $(k = 0 = \Lambda)$ is as follows:

$$a = 2t^{1/2}, \quad \rho = \frac{12}{a^4} = \frac{3}{4t^2}, \quad d_H(t) = 2t, \tag{45}$$

where units have been chosen such that the scale factor and the Hubble distance at the present time are $a_o = 2H_o^{-1} \equiv 1$. On the other hand, during the matter era $(p = 0)$, for the most popular backgrounds, we have

$$k = \Lambda = 0 : a = 3t^{2/3}, \quad \rho = \frac{12}{a^3} = \frac{4}{3t^2}, \quad d_H(t) = 3t; \tag{46}$$

$$k = 0, \frac{\Lambda}{3H_o^2} = 1 - \Omega : a = (\Omega^{-1} - 1)^{1/3} \sinh^{2/3}[3(1 - \Omega t)], \quad \rho = \frac{12\Omega}{a^3}; \tag{47}$$

$$d_H(t) = [2(1 - \Omega)^{1/2}]^{-1} e^{2(1-\Omega)t} \quad for \quad t \gg 1; \tag{48}$$

whereas for the open model,

$$k < 0, \Lambda = 0 : a = \frac{\Omega}{2(1 - \Omega)} \cosh\tau, \quad t = \frac{\Omega}{4(1 - \Omega)^{3/2}} (\sinh\tau - \tau), \tag{49}$$

$$\rho = \frac{12\Omega}{a^3}, \quad d_H(t) = t \ln t \quad for \quad t \gg 1. \tag{50}$$

The temperature of the CMB is $T \propto a^{-1}$ in all models, and the redshift $1 + z \propto a^{-1}$. Radial propagation of photons is easily studied with the FRW metric given by Equation (41). Proper distance (normalized to the Hubble distance) can be calculated in terms of redshift in the form

$$d_P = 1 - (1 + z)^{-1/2}, \tag{51}$$

for a flat model. The corresponding formulae for open and closed models can be found in Weinberg's book (1972). The luminosity distance is found to be

$$d_L = \Omega^{-2}[\Omega z + (\Omega - 2)((1 + \Omega z)^{1/2} - 1)]. \tag{52}$$

For $\Omega = 1$ one gets $d_L = (1+z)d_{Po}$. The angular-diameter distance is given by $d_A = d_L(1+z)^{-2}$.

3.5. THE INFLATIONARY PARADIGM

The idea that during a small period of time the Universe underwent an inflationary expansion, $i.e.$, the scale factor expanded exponentially, has been the paradigm for more than 15 years. Such an idea dates back to Starobinsky (1979), Guth (1981), Linde (1982) and others. From the Friedmann equations (42) one obtains: $\frac{d^2a}{dt^2} = -a(\rho + p)$; so if $\rho, p > 0$, then $\frac{da}{dt}$ decreases with time and matter decelerates due to gravity. However, if $\rho + 3p < 0$, then $\frac{da}{dt}$ increases with time and matter accelerates due to "anti-gravity". As a toy model, let us assume that $p = -\rho = constant$ during a period of time $[t_i, t_f]$; then the Friedmann equations can be integrated:

$$a = A_g e^{(\frac{\rho}{3})^{1/2}t} + A_d e^{-(\frac{\rho}{3})^{1/2}t}, \quad k = \frac{4}{3}\rho A_g A_d, \tag{53}$$

and the Gaussian curvature of the 3D surfaces $t = constant$, $K = k/a^2(t)$, exponentially decreases with time: $^tK \propto e^{-2(\frac{\rho}{3})^{1/2}t}$. So, the Universe exponentially inflates and becomes flat! In the very early Universe (close to the Planck time) there is a chance to have such an unphysical equation of state, if the energy density is dominated by the potential energy density $V(\phi)$ of some scalar field $\phi(t, \vec{x})$. In this case, during the very early stages of its evolution, the Universe expanded quasi–exponentially (inflationary era). Later, this potential energy density of the scalar field was transformed into thermal energy and, still later, the Universe was correctly described by the standard radiation dominated universe predicting the existence of the CMB. This rapid *inflationary* expansion left its imprints: i) it made the Universe flat (*i.e.*, $k \simeq 0$ or $\Omega \simeq 1$), ii) it made the Universe homogeneous and isotropic (*i.e.*, the initial conditions are forgotten), iii) it solves the horizon problem (*i.e.*, it explains the isotropy of the CMB), and iv) it generated the seeds for structure formation. The last result is one of the most relevant – one has a natural explanation for the origin and characteristics of the initial density fluctuations that later formed "structure". To be more specific: quantum fluctuations of the scalar field generated density fluctuations, $\delta_i(\vec{x})$, at early times that constitute an homogeneous and isotropic

Gaussian random field. Thus, the correlation function, $\xi_i(r)$, or equivalently the power spectrum, $P_i(k)$ (k represents wavenumber), characterizes the density field:

$$\xi_i(r) = \frac{1}{2\pi^2} \int dk\, k^2 P_i(k) j_0(kr), \qquad (54)$$

where $j_0(x) = \frac{\sin x}{x}$. Moreover, for *curvature* perturbations, inflation predicts a scale–invariant power spectrum:

$$P_i(k) \propto k, \qquad (55)$$

i.e. it has the Harrison-Zeldovich form.

Inflation remains the only theory that explains the homogeneity of the observable Universe. However, on much larger scales, the Universe should be extremely inhomogeneous, with the energy density varying from the Planck density to almost zero. Instead of a single Big Bang producing a single bubble universe, it is usual to speak now about many inflationary bubbles producing new bubbles, as in the chaotic inflation scenario (Linde 1983).

One of the most robust predictions of inflation is that the Universe becomes extremely flat, which corresponds to $\Omega = 1$. The only way to avoid this conclusion is to assume that the Universe inflated only by e^{60} times (this number depends on the details of the theory). Very recently, the possibility of an $\Omega \neq 1$ has been a subject of active research. In particular, the $\Omega < 1$ case is complicated: it is possible to realize such a scenario with a single scalar field, but it requires a fine–tunning of conditions. It is much easier to obtain an open universe in models with two scalar fields (Linde & Mezhlumian 1995). However, there are still many questions to be addressed; in particular, the bubbles created by tunneling are not absolutely homogeneous, and if one such region is identified with the entire visible part of the Universe, maybe at the end one cannot explain the isotropy of the CMB to the level observed today ($\simeq 10^{-5}$). A final remark: in some of these scenarios with $\Omega < 1$, the initial power spectrum generated by the quantum fluctuations of the field(s) can differ from the Harrison-Zeldovich form, and this "opens a Pandora's "box". Thus, recent developments in inflationary theory have drastically modified our cosmological paradigm, but the last word has not yet been said.

3.6. LINEAR PERTURBATION THEORY: THE GAUGE PROBLEM

Linear perturbations of a FRW model were considered by Lifshitz (1946) and Lifshitz & Khalatnikov (1963). Their approach was to assume small variations of the metric tensor. The relevant applications of cosmological perturbation theory are galaxy formation via gravitational instability and the study of the propagation of the CMB in such theories (generating spectral distortions and anisotropies). In fact, this was the approach taken by Sachs & Wolfe (1967), concerning anisotropies of the CMB, and by many people afterwards. The method has the disadvantage that the metric tensor is not a physically significant quantity, but only its 2nd derivatives (the Riemann tensor) are. The relevant physical quantity called the "density perturbation" suffers the same problem.

In fact, if we perform infinitesimal coordinate transformations $x^{a'} = x^a + \epsilon^a(x^b)$ (a choice of ϵ^a is a choice of a gauge), the behavior of a scalar ρ is

$$\rho'(x^{a'}) = \rho(x^{a'}) - \epsilon^c \rho_{,c}. \tag{56}$$

So, only a constant scalar is "gauge-invariant". In particular, if ρ and ρ_b denote the perturbed and background densities, then the density perturbation $\delta\rho \equiv \rho - \rho_b$ vanishes for a suitable choice of ϵ^0. For a vector field $u^a(x^b)$, one obtains

$$u^{a'}(x^{b'}) = u^a(x^{b'}) - \epsilon^c u^a{}_{,c} + \epsilon^a{}_{,c} u^c. \tag{57}$$

In particular, if the old coordinates are adapted, i.e., $u^a = \delta_0^a$, the new ones are not : $u^{a'}(x^{b'}) = \delta_0^a + \epsilon^a{}_{,0}$. Finally, the tensor components are not "gauge-invariant" in general:

$$t_{a'b'}(x^{c'}) = t_{ab}(x^{c'}) - \epsilon^d t_{ab,d} - \epsilon^d{}_{,a} t_{db} - \epsilon^d{}_{,b} t_{ad}. \tag{58}$$

Therefore, the components of the perturbed RW metric are (see Equation (27); we shall use a conformal time parameter τ: $dt = a(\tau)d\tau$)

$$ds^2 = a^2(\tau)\{-(1+2A)d\tau^2 + 2B_i d\tau dx^i + [(1+2C)\gamma_{ij} + 2F_{ij}]dx^i dx^j\}, \tag{59}$$

and transform as

$$A' = A + \epsilon^{0\prime} + \frac{a'}{a}\epsilon^0, \quad B_{i'} = B_i - D_i\epsilon^0 + \epsilon'_j, \tag{60}$$

$$C' = C + \frac{1}{3}D_i\epsilon^i + \frac{a'}{a}\epsilon^0, \quad F_{i'j'} = F_{ij} + D_{(i}\epsilon_{j)} - \frac{1}{3}\gamma_{ij}D_k\epsilon^k. \tag{61}$$

A, B_i, C and F_{ij} depend on (τ, x^i) and represent the perturbation to the metric tensor. F_{ij} is traceless, $F_i^i \equiv \gamma^{ij}F_{ij} = 0$. A "prime", on a quantity in the 2nd member of the previous equations, means $\frac{d}{d\tau}$. D_i denotes the covariant derivative with respect to the metric $\gamma_{ij} = (1 + \frac{k}{4}x^2)^{-2}\delta_{ij}$.

Hereinafter, the symbol (ij) means symmetrization with respect to the two indices.

The components of the perturbed energy-momentum tensor (given by a perfect fluid, $u_b^a = \delta_0^a a^{-2}, \rho_b, p_b$) are not gauge-invariant

$$\rho = \rho_b + \delta\rho, \quad \delta \equiv \frac{\delta\rho}{\rho_b}, \quad p = p_b + \delta p, \tag{62}$$

$$\delta u^0 = -u_b^0 A, \quad \delta u^i = u_b^0 v^i. \tag{63}$$

Finally, we remark that the only linear perturbations that are gauge-invariant are: a vanishing tensor, a constant scalar or a linear combination of δ_b^a's with constant coefficients (Stewart 1990).

To avoid the unphysical degrees of freedom, two different approaches have been considered: the metric gauge-invariant method (Bardeen 1980) and the fluid-flow approach (Hawking 1966; Ellis & Bruni 1989).

3.7. THE BARDEEN FORMALISM

Bardeen's (1980) idea was to consider linear combinations of the metric perturbations A, B_i, C and F_{ij} which are gauge-invariant. Hereafter, we shall work in real space, following Stewart (1990), and not in Fourier space as Bardeen did in his pioneering paper. Terms such as $A, D^i B, D^{ij} F$ will be called "scalars", whereas terms such as $B^i, D^{(i} B^{j)}$ (where B^i is solenoidal, i.e., $D_i B^i = 0$) will be called "vectors". Finally, transverse and traceless tensors, i.e., $D_i T^{ij} = 0 = T^i{}_i$, will be called "tensors". It is remarkable that, in any vector or tensor equation, the scalar, vector and tensor parts of each side are separately equal, taking into account the uniqueness of the previous decompositions under appropriate hypotheses on the boundary conditions. So, we can decompose the metric coefficients, given by Equation (59), in the form

$$B_i = D_i B + B_i^S, \quad F_{ij} = D_{ij}F + D_{(i}F_{j)}^S + F_{ij}^{TT}, \tag{64}$$

$$v_i = D_i v + v_i^S, \quad D_{ij} \equiv D_i D_j - \frac{1}{3}\gamma_{ij}D_k D^k, \tag{65}$$

where the superscript S on a vector means that it is solenoidal ($D_i f^{Si} = 0$) and TT tensors are transverse and traceless ($D_j F^{TTij} = 0, F^{TT}{}_i^i = 0$).

i) *Scalar perturbations*:

Two gauge-invariant quantities can be associated with the metric ($B_i = D_i B, F_{ij} = D_{ij}F$)

$$\phi_A \equiv A + (B' + \frac{a'}{a}B) - (F'' + \frac{a'}{a}F'), \tag{66}$$

$$\phi_H \equiv C - \frac{1}{3}\Delta F + \frac{a'}{a}(B - F'), \tag{67}$$

where $' \equiv \frac{d}{d\tau}$, and $\Delta F \equiv D_i D^i F$; ϕ_A can be interpreted as the lapse function and ϕ_H gives the intrinsic scalar curvature of the zero-shear hypersurfaces (these are characterized by normal unit vector with zero shear). For matter perturbations, one can introduce the following set of gauge-invariant quantities (we will restrict ourselves to a perfect fluid)

$$\epsilon_m \equiv \delta - 3(1 + \frac{p}{\rho})\frac{a'}{a}(v + B), \quad V = v + F'. \tag{68}$$

It is remarkable that there is no preferred choice of gauge-invariant density perturbation, because many other combinations are possible, e.g.,

$$\epsilon_g = \epsilon_m + 3(1 + \frac{p}{\rho})\frac{a'}{a}V; \tag{69}$$

ϵ_m represents the natural choice from the point of view of the matter, whereas ϵ_g is natural from the point of view of hypersurfaces representing a Newtonian slicing.

The Einstein equations imply the relations

$$\phi_A = -\phi_H, \quad (\Delta + 3k)\phi_A = \frac{1}{2}a^2\rho_b\epsilon_m, \quad (\Delta + 3k)V = \frac{k(\rho_b a^3 \epsilon_m)'}{(\rho + p)a^3}, \tag{70}$$

and the density perturbation ϵ_m must satisfy ($\epsilon \equiv \rho_b a^3 \epsilon_m$) the 2nd order differential equation

$$\epsilon'' + (1 + 3c_s^2)\frac{a'}{a}\epsilon' - [c_s^2(\Delta + 3k) + \frac{1}{2}(\rho + p)a^2]\epsilon = 0, \tag{71}$$

where $c_s^2 \equiv \frac{dp}{d\rho}$. A direct consequence of these equations is that there is a scalar perturbation iff $\epsilon_m \neq 0$.

One last comment concerning the evolution of the perturbations in the radiation and matter dominated eras, with equations of state $p = \frac{1}{3}\rho$ ($a = 2\tau$) and $p = 0$ ($a = \tau^2$), respectively. If we adopt a flat model, $k = 0$. For scalar perturbations, one finds the modes (k is the wavenumber of the perturbation)

$$\epsilon_m \propto a\{j_1(\frac{ka}{2\sqrt{3}}), y_1(\frac{ka}{2\sqrt{3}})\} \quad (p = \frac{1}{3}\rho); \quad \propto \{a, a^{-3/2}\} \quad (p = 0). \tag{72}$$

Taking into account Equation (70), ϕ_A is given by

$$\phi_A \propto a^{-1}\{j_1(\frac{ka}{2\sqrt{3}}), y_1(\frac{ka}{2\sqrt{3}})\} \quad (p=\frac{1}{3}\rho); \quad \propto \{1, a^{-5/2}\} \quad (p=0), \quad (73)$$

i.e., for scales very much greater than the Jeans length, $\lambda_J \equiv \frac{a}{2\sqrt{3}}$, the gravitational potential is approximately constant for the growing mode.

When the background curvature is different from zero, the general solution of all the previous equations is not known and some sort of approximation must be done. For open models, during the matter era, one finds $(x \equiv (\Omega^{-1} - 1)a)$

$$\epsilon_m \propto \{1+\frac{3}{x}+3x^{-3/2}[1+x]^{1/2}ln[(1+x)^{1/2}-x^{1/2}], x^{-3/2}(1+x)^{1/2}\} \quad (p=0).$$
$$(74)$$

Taking into account the Poisson-like equation (66), one obtains, for the growing mode during the matter era, the following time dependence:

$$\phi_A \propto const \quad (k=0); \quad \propto \frac{D(a)}{a} \quad (k<0), \quad (75)$$

where $D(a)$ represents the growing mode of Equations (72) or (73).

ii) *Vector perturbations*:
In this case $F_{ij} = D_{(i}F^S_{j)}, D_i F^{Si} = 0$, and we can define the gauge-invariant metric perturbation

$$\psi_i \equiv -D_i B + F^{S\prime}_i, \quad (76)$$

and the gauge-invariant velocity variable

$$V_i \equiv v^S_i + F^{S\prime}_i, \quad (77)$$

or, equivalently, $V_i^* \equiv V_i - \psi_i$. The Einstein equations give the relationship

$$(\Delta + 2k)\psi_i = -2(\rho + p)a^2 V_i^*, \quad (78)$$

and the basic differential equation

$$V_i^{*\prime} + (1 - 3\frac{p}{\rho})\frac{a'}{a}V_i^* = 0. \quad (79)$$

It is easy to see that a vector perturbation exists *iff* $V^* \neq 0$.
For vector perturbations, one finds the solutions

$$V_i^* \propto constant \quad (p=\frac{1}{3}\rho); \quad \propto a^{-1} \quad (p=0). \quad (80)$$

iii) *Tensor perturbations:*

In this case the TT component of the metric F_{ij}^{TT} is gauge invariant and the TT component of the anisotropic pressure is zero (perfect fluid). The Einstein equations give the following

$$F_{ij}^{TT\prime\prime} + 2\frac{a'}{a}F_{ij}^{TT\prime} + (2k - \Delta)F_{ij}^{TT} = 0. \tag{81}$$

So, the condition for the existence of a tensor perturbation is $F^{TT} \neq 0$. The tensor mode perturbations are, for $\Omega = 1$,

$$F_{ij}^{TT} \propto \{j_0(\frac{a}{2}), y_0(\frac{a}{2})\} \quad (p = \frac{1}{3}\rho); \quad \propto a^{-1/2}\{j_1(a^{1/2}), y_1(a^{1/2})\} \quad (p = 0). \tag{82}$$

In the previous formulae j_0, y_0, j_1 and y_1 are the Bessel functions of fractional order.

Finally, let us note the main drawbacks of this gauge-invariant formulation: i) there is no unique definition of the density $(\frac{\delta\rho}{\rho})$ or expansion $(\frac{\delta\theta}{\theta})$ perturbation; ii) there is not any clear physical meaning of the gauge-invariant quantities, *i.e.*, they are pure mathematical definitions; and iii) the definitions only apply to linear perturbations.

3.8. THE COVARIANT APPROACH

The "covariant approach" is an alternative approach to Bardeen's gauge-invariant formalism, to study the evolution of linear perturbations of a FRW model. First introduced by Hawking (1966), it did not receive much attention (apart from some review by Ellis (1971)) until a few years ago (Ellis & Bruni 1989, Lyth & Stewart 1990, Hwang & Hyun 1994). The idea is to use not the metric components, but the kinematical quantities – the electric and magnetic components of the Weyl tensor and the gradients of the energy density and expansion (these trivially vanish on the FRW background). The covariant approach can be proved to be formally equivalent to Bardeen's method (Dunsby 1992), but the advantages of the covariant approach are obvious: i) in this formalism all variables are tensors with a clear physical and geometrical meaning; ii) it is more convenient for dealing with gravitational waves; iii) it can be used to study non-linear evolution; iv) the physical and geometrical characterization of different types of perturbations appear in a clear way; and v) the basic differential equation for the evolution of the density (and also the expansion) perturbation can be easily obtained.

i) *Density perturbations:*

Let us assume an irrotational ($\omega = 0$) perfect fluid with a linear equation of state, $p = n\rho$. The linearization of the fluid-flow equations around the FRW background – $\rho = \rho_b(1+\delta), \theta = \theta_b(1+\chi)$ – leads to a set of equations that can easily be worked out in a comoving and orthogonal coordinate system. Due to the fact that we have no vorticity and a potential acceleration ($\xi^a = -\frac{n}{1+n}h^{ab}(\ln \rho)_{,b}$), there always exists a local set of comoving and orthogonal coordinates, *i.e.*,

$$u^a = e^U\delta^a_0, \quad ds^2 = -e^{2U}dt^2 + h_{ij}dx^idx^j. \tag{83}$$

For linear perturbations, $U \equiv \frac{n}{1+n}\delta$ and the coordinates are approximately synchronous. Choosing such a coordinate system, one finds

$$\chi = \frac{1}{3(1+n)}[3n\delta - a\frac{\partial\delta}{\partial a}], \tag{84}$$

$$\frac{\partial^2\delta}{\partial t^2} + (2-3n)\frac{a^{\cdot}}{a}\frac{\partial\delta}{\partial t} - \frac{n}{a^2\rho_b}\Delta\delta - [\frac{1}{2}(1+2n-3n^2)\rho_b + 2n\Lambda]\delta = 0. \tag{85}$$

The last equation, for $\Lambda = 0$, was obtained by Lyth & Stewart (1990). It is interesting to compare it with the one obtained through Newtonian cosmology: $\frac{\partial^2\delta}{\partial t^2} + 2\frac{a^{\cdot}}{a}\frac{\partial\delta}{\partial t} - \frac{n}{a^2\rho_b}\Delta\delta - \frac{1}{2}\rho_b\delta = 0$. So, except when we are treating perturbations in the matter era, the Newtonian approximation fails to reproduce the relativistic solutions. On the other hand, as δ represents the gauge-invariant density perturbation ϵ_m for the particular coordinates we have chosen, its solutions for the flat model are given by Equations (72-73).

Finally, from the evolution equations for the kinematical and geometrical quantities, one obtains in the matter era, for the flat model and $\Lambda = 0$,

$$E_{ij} = -\frac{\partial\sigma_{ij}}{\partial t}, \quad H_{ij} = -a^{-1}\epsilon_{(i}{}^{kl}\sigma_{j)k,l}, \tag{86}$$

$$(\Box + \frac{1}{2}a^2\rho_b)\sigma_{ij} = 3aa^{\cdot}[\chi_{,ij} - \frac{1}{3}\delta_{ij}\Delta\chi], \tag{87}$$

where $\Box \equiv -\frac{d^2}{d\tau^2} + \Delta$. The solution of the last equation for scalar perturbations gives two modes

$$\sigma_{ij} \propto \{a^{3/2}, a^{-1}\}, \quad E_{ij} \propto \{1, a^{-5/2}\}, \quad H_{ij} \propto \{a^{1/2}, a^{-2}\}, \tag{88}$$

which allow one to calculate the relative quantities

$$\frac{\sigma}{\theta}, \frac{E}{\rho}, \frac{aH}{\rho} \propto \{a, a^{-3/2}\} \quad (p=0). \tag{89}$$

For the growing mode ($k = 0 = \Lambda$, $p = 0$), choosing a gauge such that $B_i = F_{ij} = 0$, and denoting $-\phi_H = \phi_A \equiv \phi$, one can obtain the relationship

$$\delta = a\delta_o(\vec{x}), \quad \sigma_{ij} = -\frac{1}{3}a^{3/2}(\phi_{,ij} - \frac{1}{3}\delta_{ij}\Delta\phi), \tag{90}$$

where $\phi(\vec{x})$ is the gravitational potential at the present time ($\Delta\phi = 6\delta_o(\vec{x})$).

ii) *Vorticity perturbations*:

Substituting the 2nd of Equations (18) into Equation (20) one gets

$$[(\rho + p)a^5\omega]^{\cdot} = 0, \tag{91}$$

i.e., angular momentum is conserved (Sanz 1982). In the radiation dominated era ($p = \frac{1}{3}\rho$, $\rho_b \propto a^{-4}$, $\theta_b \propto a^{-2}$) $\omega \propto a^{-1}$, whereas in the matter era ($p = 0$, $\rho_b \propto a^{-3}$, $\theta_b \propto a^{-3/2}$) $\omega \propto a^{-2}$. Regarding the relative vorticity, one finds the behavior: $\frac{\omega}{\theta} \propto a$, $\frac{\omega^2}{\rho} \propto a^2$ (radiation era) and $\frac{\omega}{\theta} \propto a^{-1/2}$, $\frac{\omega^2}{\rho} \propto a^{-1}$ (matter era). Thus vorticity dies away as the Universe expands.

iii) *Gravitational wave perturbations*:

Assuming $\omega = \delta\rho = 0$, Equations (21, 22, 24) lead to

$$E_{ij} = -a^{-1}\sigma'_{ij}, \quad H_{ij} = -a^{-1}\epsilon_{(i}^{kl}\sigma_{j)k|l}, \tag{92}$$

$$(\Box + \frac{1}{2}a^2\rho_b)\sigma_{ij} = D_{(j}D^l\sigma_{i)l}, \quad D^j\sigma_{ij} = 0. \tag{93}$$

The solution can be expressed in terms of F_{ij}^{TT} given by Equation (81)

$$\sigma_{ij} = aF_{ij}^{TT\prime}, \quad E_{ij} = -\frac{1}{2}F_{ij}^{TT\prime\prime}, \quad H_{ij} = -a^{-2}D^l F_{(i}^{TT\ kl}\epsilon_{j)kl}. \tag{94}$$

The evolution of shear and gravitational–wave perturbations larger than the Hubble length was studied in FRW models in connection with the isotropy of the Universe (Sanz 1983). For flat models ($\Lambda = 0$), the behavior of the relative quantities in this limit is:

$$\frac{\sigma}{\theta}, \frac{E}{\rho} \propto \{a^2, a^{-1}\}; \quad \frac{H}{\rho} \propto a \quad (p = \frac{1}{3}\rho); \tag{95}$$

$$\frac{\sigma}{\theta}, \frac{E}{\rho} \propto \{a, a^{-3/2}\}; \quad \frac{H}{\rho} \propto const \quad (p = 0); \tag{96}$$

whereas for open models with a dust content

$$\frac{\sigma}{\theta} \propto (ax)^{-1}\{P_2(x), Q_2(x)\}, \frac{E}{\rho} \propto a^{-1}\{\frac{d\,P_2(x)}{d\,x}, \frac{d\,Q_2(x)}{d\,x}\}, \frac{H}{\rho} \propto const, \tag{97}$$

where $x \equiv [1 + \frac{\Omega}{(1-\Omega)a}]^{1/2}$ and P_2, Q_2 are the Legendre functions of degree 2: $P_2(x) \equiv \frac{1}{2}(3x^2 - 1), Q_2(x) \equiv \frac{1}{4}[-6x + (3x^2 - 1)ln(\frac{x+1}{x-1})]$.

A last comment concerning the formal equivalence of the two approaches. It can be proved (Dunsby 1992) that the kinematical and geometrical quantities are related via the equations

$$\sigma_{ij} = a[D_{ij}V + D_{(i}V_{j)}^S + F_{ij}^{TT}], \quad \omega_{ij} = a\, D_{(i}V_{j)}^*, \tag{98}$$

$$-2\,E_{ij} = D_{ij}(\phi_H - \phi_A) + D_{(i}\psi_{j)}' + F_{ij}^{TT''} + (\Delta - 2\,k)F_{ij}, \tag{99}$$

$$-2\,H_{ij} = a^{-2}[D^l D_{(i}\psi^k + 2\,D^l H^k_{(i}]\epsilon_{j)kl}. \tag{100}$$

Moreover, the Cotton-York tensor is given by (Goode 1989)

$$Y_i{}^j = (-\Delta + 2\,k)\epsilon^{jkl}D_l F_{ik}^{TT}. \tag{101}$$

This enables us to give a physical-geometrical characterization of the different perturbation types: only the scalar perturbations contribute to ϵ_m (and for this reason are called "matter" perturbations), only vector perturbations contribute to ω (they are usually referred to as "rotational" or "vorticity" perturbations), and only tensor perturbations contribute to the Cotton-York tensor Y_{ij}, so the presence of gravitational waves is closely related to this tensor.

4. Basics of Cosmic Microwave Background

The CMB is without any doubt the most powerful tool of cosmology. On the one hand, the FIRAS instrument on board the COBE satellite has confirmed the *black-body* spectrum at a temperature of $T = 2.726 \pm 0.017\,K$ to a high degree of precision (see the review by G. Smoot in these Proceedings). This has important implications for the past history of our Universe; in particular, it rules out the existence of a hot intergalactic medium producing the X-ray background.

On the other hand, the DMR instrument has detected for the first time intrinsic anisotropies in the CMB on large angular scales $- > 10°$ (Smoot et al 1992). This is consistent with our ideas about gravitational instability generating structure in the Universe from initial seeds created at a very early epoch in the history of the Universe. Other experiments, using different techniques and frequencies, seem to detect some level of anisotropy at smaller angular scales (Netterfield et al. 1995; Hancock et al. 1994; for a review see Cayón 1996).

In this chapter, I will try to set out the basic equations for the propagation of the photons on a perturbed FRW background. I shall assume simple hypotheses: i) the black-body character is preserved along the evolution,

ii) there is an instantaneous recombination, iii) no polarization is present. The idea is to give some insight into the generation of the anisotropies, particularly on large angular scale, not to develop the whole theory involving Einstein-Boltzmann equations including all the particles (photons, baryons, massless neutrinos, cold dark matter,,...) and also the complicated process of recombination. Anyone interested to get more refined calculations can see the appropriate references (see the review by T. Bunn in these Proceedings).

4.1. ANISOTROPIES IN THE CMB

The sources of anisotropy are usually divided in the following groups: i) *Primary* anisotropy, generated at the last scattering surface (photon fluctuations at recombination, Doppler effect due to the motion of the last scatters and the Sachs-Wolfe effect, i.e., the gravitational potential at recombination). ii) *Secondary* anisotropy, generated by different physical processes after recombination (the integrated Sachs-Wolfe effect due to the time variation of the gravitational potential for linear perturbations, the Rees-Sciama effect due to non-linear evolution, gravitational lensing of the photons; reionization can induce new Doppler contributions and non-linear couplings can produce new anisotropies through the Vishniac effect; local reionization -e.g., in clusters- can induce new anisotropies through the Sunyaev-Zeldovich effect and kinematic S-Z effect; finally the existence of dust early enough can also generate anisotropy). iii) *Foregrounds* (extragalactic sources emitting in radio and IR; galactic contributions, i.e., synchrotron and free-free radiation and dust; local contributions due to the solar system, atmosphere, noise,...). iv) *Extrinsic* anisotropy, a dipole contribution interpreted as due to our motion with respect to the CMB.

a) *The radiation power spectrum*:
A basic quantity associated with the description of the CMB are the C_l's, i.e., the 2D radiation power spectrum. The standard decomposition of the anisotropies $\frac{\Delta T}{T}$ in terms of the spherical harmonics

$$\frac{\Delta T}{T}(\vec{n}) = \sum_{l,m} a_{lm} Y_{lm}(\theta, \phi), \qquad (102)$$

leads to the correlation function $C(\alpha)$

$$C(\alpha) \equiv \langle \frac{\Delta T}{T}(\vec{n}_1) \frac{\Delta T}{T}(\vec{n}_2) \rangle = \frac{1}{4\pi} \sum_l (2l+1) C_l P_l(\cos \alpha), \qquad (103)$$

where $C_l \equiv < a_{lm}^2 >$, $\cos \alpha \equiv n_1 n_2$ and P_l is the Legendre polynomial of degree l. So, the previous equation allows to interpret the C_l's as the 2D radiation power spectrum.

b) *Geodesics in FRW backgrounds*:

The integration of the null geodesic equation (2,3) for the RW metric (27), rewritten with conformal time τ $(dt = a\,d\tau)$, gives for the tangent vector k^a

$$k^0 \propto a^{-2}, \quad \frac{k^i}{k^0} = \frac{d\,x^i}{d\tau} = -(1 + \frac{k}{4}\lambda^2)\vec{n}, \tag{104}$$

where $\vec{n}$ is the direction of observation and λ is the distance to the photon from the observer. For the most popular models

$$k = 0 = \Lambda: \quad \lambda = 1 - a^{1/2}, \tag{105}$$

$$k = 0, \quad \frac{\Lambda}{3\,H_o^2} = 1 - \Omega: \quad \lambda = \Omega^{-1/2} \int_{a^{1/2}}^1 dy\,[1 + (\Omega^{-1} - 1)y^6]^{-1/2}, \tag{106}$$

$$k < 0: \lambda = \frac{1 - q^{1/2}}{1 - (1 - \Omega)q^{1/2}}, \quad q \equiv \frac{a}{\Omega + (1 - \Omega)a}. \tag{107}$$

On the other hand, the trajectory of the photon is given by

$$\vec{x} = \lambda\vec{n}. \tag{108}$$

Taking into account Equations (28, 40) and $u^a = a^{-1}\delta_0^a$, one easily gets $T_o = T_r a_r / a_o$. Hereinafter, the subscripts "r" and "o" mean recombination and present time, respectively.

4.2. CALCULATION OF ANISOTROPIES IN A GAUGE

The propagation of photons done in the appropriate way implies solving the coupled system Einstein-Boltzmann for all the particles constituting the energy-content of the Universe, in particular polarization of the CMB must be included in the approach. As we have commented at the beginning of this chapter, we will try to understand how the anisotropies are generated using elementary techniques, so simple assumptions will be considered. The inclusion of the polarization has been considered in synchronous coordinates (Bond 1995) and in the gauge-invariant approach (Kodama & Sasaki 1984; Kosowski 1996). On the other hand, the Boltzmann equation has been considered in the covariant approach (Bonano & Romano 1994) but a full treatment of anisotropies is missed with such a technique.

Moreover, we shall assume an instantaneous recombination and afterwards free propagation of the photons into the gravitational field generated by the perturbations. This is not a problem for low multipoles but gives the wrong answer for the high l's. A damping factor is usually introduced to account for the process of recombination.

The inclusion of the polarization and of non-instantaneous recombination generates differences in the C_l's of $\leq 20\%$ according to estimations for standard models like CDM. The assumptions that we are going to undertake are reasonable as far as a fine description is not required.

a) *Density perturbations*:

We can always choose a gauge with $B_i = 0 = F_{ij}$. Assuming a perfect fluid, then $A = -C \equiv \phi$, and the metric in the longitudinal gauge is written

$$ds^2 = a^2(\tau)[-(1+2\phi)d\tau^2 + (1-2\phi)\gamma_{ij}dx^i dx^j], \qquad (109)$$

where $|\phi| \ll 1$. Integrating the null geodesic equation -up to the first order-between recombination and the present time and taking into account $u^2 = -1$, one obtains

$$u^0 \simeq a^{-1}(1-\phi), \quad \frac{k_o^0}{k_r^0} \simeq \left(\frac{a_r}{a_o}\right)^2 [1+2(\phi_r - \phi_o) + 2\int_r^o d\tau \frac{\partial \phi}{\partial \tau}(\tau, \vec{x}(\tau))], \quad (110)$$

where the integral must be performed along the photon trajectory in the FRW background and $a_o \equiv a(\tau_o), a_r \equiv a(\tau_r)$. Taking into account Equation (28)

$$1+z = \frac{a_o}{a_r}[1 - \phi_r + \phi_o + (1 + \frac{k}{4}\lambda_r^2)\vec{n}.\vec{v}_r - \vec{n}.\vec{v}_o - 2\int_r^o d\tau \frac{\partial \phi}{\partial \tau}, \qquad (111)$$

where $\vec{n}.\vec{v} \equiv \gamma_{ij}n^i v^j$ and $\lambda_r = [1 - (1 + \Omega z_r)^{-1/2}][1 - (1-\Omega)(1 + \Omega z_r)^{-1/2}]$, $z_r \simeq 10^3$. So, taking into account Equation (40) and deleting local terms, the anisotropy generated is

$$\left(\frac{\Delta T}{T}\right)_o \simeq \left(\frac{\Delta T}{T}\right)_r - \vec{n}.\vec{v}_r^* + \frac{1}{3}\phi_r + 2\int_r^o d\tau \frac{\partial \phi}{\partial \tau}. \qquad (112)$$

This equation has been explicitly given by Martinez-González, Sanz & Silk (1990). The first term corresponds to the photon fluctuations at recombination, the 2nd one is a Doppler effect due to the motion of the last scatters ($\vec{n}.\vec{v}_r^* = -\frac{1}{6(1+z_r)}n^i \partial_i \phi_r$), the 3rd one is the Sachs-Wolfe effect due to the gravitational potential and the 4th one is an integrated effect along the trajectory of the photon from recombination to the present. To obtain the previous formula, we have changed from the longitudinal to the synchronous-comoving gauge through the transformation $x^{a'} = x^a + \epsilon^a$

$$\epsilon^0 = 2(a^3\rho_b)^{-1}(a\phi)^{\cdot}, \quad \vec{\epsilon} = 2(a^3\rho_b)^{-1}\left(1 + \frac{k}{4}x^2\right)^2 \vec{\nabla}(a\phi). \tag{113}$$

For the flat case the perturbed 3-metric has the Sachs-Wolfe form:

$$ds^2 = a^2\{-d\tau^2 + [(1 - \frac{10}{3}\phi)\delta_{ij} - 4a(a^3\rho_b)^{-1}\phi_{,ij}]d\,x^i d\,x^j\}. \tag{114}$$

A final remark, changing from Newtonian (τ) to synchronous time (t) introduces a factor $-\frac{2}{3}f_r\phi_r \simeq -\frac{2}{3}\phi_r$ ($f_r \equiv \left(\frac{dlnD}{dlna}\right)_r \simeq 1$ if $\Omega > 0.1$ and recombination happened at the matter dominated era) that partially cancels the contribution, calculated in the Newtonian gauge, coming from the gravitational potential ϕ_r.

The calculation of the Doppler, Sachs-Wolfe and integrated terms for the growing solution $D(a)$ associated with ϕ is

$$\phi(\tau, \vec{x}) = \frac{D(a)}{a}\phi_o(\vec{x}), \quad \Delta\phi_o = 6\Omega\delta_o(\vec{x}), \tag{115}$$

where $\delta_o(\vec{x})$ is the linear density perturbation at the present time that can be given in terms of the transfer function and the initial power spectrum of the density fluctuations.

a) *Vorticity perturbations*:
This type of perturbation dies away when the Universe expands, so it is considered not relevant from the point of view of dynamics and anisotropy.

a) *Gravitational wave perturbations*:
In this case, one can always choose a gauge such that $A = B_i = 0$ and the remaining 3-metric is generated by a TT-tensor. The integration of the geodesic equation gives the anisotropy

$$\left(\frac{\Delta T}{T}\right)_o \simeq \left(\frac{\Delta T}{T}\right)_r - \frac{1}{2}n^i n^j \int_r^o d\tau \frac{\partial F_{ij}^{TT}}{\partial\tau}(\tau, \vec{x}(\tau)). \tag{116}$$

The growing modes during the radiation and matter eras are given by Equation (81): $F_{ij}^{TT} \propto j_0(a/2)$ and $\propto a^{-1/2}j_1(a^{1/2})$, respectively. The anisotropy generated by linear gravitational waves, before and after recombination, can be estimated with the previous equation once the initial power spectrum of the waves is set out.

One last remark concerning calculation of anisotropies in the gauge-invariant approach. It is an easy task to rewrite all the previous formulae for $\frac{\Delta T}{T}$ in terms of the gauge-invariant quantities introduced by Bardeen (Abbott & Schaeffer 1986; Gouda & Sasaki 1989).

4.3. CALCULATION OF ANISOTROPIES IN THE COVARIANT APPROACH

Calculations of CMB anisotropies using the covariant approach have not received much attention but for a few papers. Goicoechea & Sanz (1985) calculated the expression for the anisotropy along two directions in terms of all the kinematical, geometrical quantities and their gradients for a general inhomogeneous and anisotropic model and applied it to isolated structures. In a recent paper (Martinez-González & Sanz 1995) have obtained a formula for the temperature as an integrated effect calculated through the shear and the density gradient. The idea was to rewrite the equation

$$T_o = T_r \frac{(u.k)_o}{(u.k)_r} = T_r exp\{\int_r^o dp \, \frac{1}{(u.k)} \frac{d\,(u.k)}{d\,p}\} = T_r exp\{\int_r^o dp \, \frac{u_{a;b}k^a k^b}{(u.k)}\},$$

(117)

and then using the kinematical decomposition (17) and taking into account that for a dust $\rho^{\cdot} = -\rho\theta$

$$T_o = T_r (\frac{\rho_o}{\rho_r})^{1/3} exp\{\int_r^o dp\,(u.k)[\sigma_{ab}n^a n^b - \frac{1}{3}\Delta_a n^a]\}$$

(118)

where $n^a \equiv u^a + (u.k)^{-1}k^a$ and $\Delta_a \equiv \rho^{-1}h^{ab}(\rho)_{,b}$ is the fractional gradient introduced by Ellis & Bruni (1989) (see the evolution equation (25)). The previous equation is an exact, covariant and gauge-invariant formula where only physical quantities appear. It has been applied to homogeneous but anisotropic (Bianchi I) models -that include the flat FRW model- to conclude that the level of the quadrupole component detected by COBE implies that the metric must be a perturbed FRW universe.

For linear perturbations, $\omega = 0$ and $p = 0$, if one chooses the synchronous-comoving gauge one gets the anisotropy

$$(\frac{\Delta T}{T})_o = (\frac{\delta s}{s})_r - n^i n^j \int_r^o dt\,\sigma_{ij} + \frac{1}{3}n^i \int_r^o dt\,\Delta_i,$$

(119)

where now n^i is approximately the unit vector in the direction of observation and $(\frac{\delta s}{s})_r \equiv \frac{1}{4}\delta_\gamma - \frac{1}{3}\delta_r$ the entropy fluctuation at recombination.

For density perturbations, it is a straightforward calculation to see that the previous equation leads to the one given in terms of the gravitational potential (see Equation (111)) if one considers the time behavior of σ_{ij} and Δ_i, as given by Equations (72, 87), and integrates Equation (118) by parts.

The main conclusion of this covariant approach is that the anisotropy in $\frac{\Delta T}{T}$ can be given as an integrated effect over the shear and density gradient if the fluctuations are approximately *adiabatic* at recombination. There is also an extra term if entropy fluctuations are present at that time.

4.4. SECONDARY ANISOTROPIES: 2ND ORDER GRAVITATIONAL EFFECT

We have already obtained (Martinez-González, Sanz & Silk 1990) an expression for the secondary anisotropies generated by the linear gravitational potential $\varphi(t, \vec{x})$ associated with non-linear density fluctuations $\Delta(t, \vec{x})$. For a flat or open universe with $p = 0$

$$\left(\frac{\Delta T}{T}\right)_o^{sec} = 2 \int_r^o dt \frac{\partial \varphi}{\partial t}, \quad \Delta \varphi = 6\Omega^{-1} \frac{\delta \rho}{\rho}(t, \vec{x}). \tag{120}$$

The previous expression is valid in the open case only for scales below the curvature scale. The line integral must be performed along the FRW geodesic (see Equations (105-107)).

On the other hand, perturbation theory up to 2nd order gives the following

$$\frac{\delta \rho}{\rho}(t, \vec{x}) \simeq D\delta + D^2[\frac{5}{7}\delta^2 + \vec{\nabla}\delta.\vec{\nabla}\xi + \frac{2}{7}\xi_{ij}\xi^{ij}], \quad \Delta \xi = \delta(\vec{x}), \tag{121}$$

where $\delta(\vec{x}) \equiv \delta_r(1 + z_r)$ and $D(a)$ is the growing mode of the density fluctuations normalized to 1 at the present time. The C_l's can be calculated in terms of the 2nd order perturbation power spectrum (Sanz et al. 1996)

$$C_l = \frac{1152}{\pi} \int dk \, k^{-2} P_2(k) R_l^2(k), \tag{122}$$

$$R_l(k) \equiv \int_0^1 d\lambda \frac{1 - (1 - \Omega)\lambda}{(1 - \lambda)^2} D^2 (2f - 1) j_l(k\lambda), \tag{123}$$

where j_l is the Bessel function of fractional order and $f \equiv \frac{dlnD}{dlna}$. $P_2(k)$ can be calculated in terms of the first order power spectrum (Goroff et al. 1986).

We have applied all of this to flat and open CDM models with either Harrison-Zeldovich or Ratra-Peebles primordial density fluctuation power spectrum. For low-Ω models, the contribution of the secondary multipoles to the radiation power spectrum is negligible both for standard recombination and for reionized scenarios, with the 2-year COBE-DMR normalization. For a flat universe this contribution is $\simeq 1 - 10\%$ depending on the reionization history of the Universe.

4.5. SECONDARY ANISOTROPIES: WEAK GRAVITATIONAL LENSING

Let us consider a perturbed FRW model with dust $(p = 0)$ matter content. For scalar perturbations, the metric in the conformal Newtonian (or longitudinal) gauge is given in terms of a single potential $\phi(\tau, \vec{x})$ that satisfies the Poisson equation: $\Delta \phi = \frac{1}{2}a^2\rho_b\delta$ where δ is the density perturbation.

After straightforward calculation, the i-component of the geodesic equation can be integrated

$$\vec{x} = \lambda\vec{n} + A\vec{n} + \vec{\alpha}_\perp, \quad \nabla^i_\perp \equiv (\delta^{ij} - n^i n^j)\partial_j, \tag{124}$$

$$\vec{\alpha}_\perp = -2\int_0^\lambda d\lambda' W(\lambda, \lambda')\vec{\nabla}_\perp\phi(\lambda', \vec{x} = \lambda'\vec{n}), \tag{125}$$

where $\vec{n}$ is the unit vector in the direction of observation, λ is the distance to the photon and $W(\lambda, \lambda')$ is a window function. The previous equation has been recently given by us (Martinez-González et al. 1996). The first term represents the photon propagation in the FRW universe, the second one is a perturbation in the direction of observation and the third one represents the perturbation of the trajectory in the plane orthogonal to $\vec{n}$. For photons that are propagated from recombination, λ_r, to the observer, $\lambda_o = 0$, the *lensing vector* is defined in the usual way

$$\vec{\beta} \equiv \vec{n} - \frac{\vec{x}_r - \vec{x}_o}{|\vec{x}_r - \vec{x}_o|}, \tag{126}$$

so an easy calculation leads to

$$\vec{\beta} = -\frac{1}{\lambda_r}\vec{\alpha}_\perp(\lambda_r). \tag{127}$$

This lensing vector has been given by Kaiser (1992) for a flat universe. For the open case, Seljak (1996) has used a window function that agrees with ours after a straightforward calculation. It has been proved in our paper that the lensing vector in the synchronous-comoving gauge (that is the appropriate one from the point of view of the observations) is approximately given by Equations (124-126).

On the other hand, assuming that anisotropies and lensing are uncorrelated, the difference between the correlation function including and not including lensing is given in terms of the bending correlation matrix

$$Q_{ij} \equiv \langle[\beta_i(\vec{n}) - \beta_i(\vec{n}')][\beta_j(\vec{n}) - \beta_j(\vec{n}')]\rangle, \tag{128}$$

that can be decomposed into the trace and an anisotropic component

$$Q^k_k \equiv 2\sigma^2(\theta), \quad \xi(\theta) \equiv Q_{ij}\frac{\beta_i\beta_j}{\beta^2} - \sigma^2, \tag{129}$$

$\sigma(\theta)$ being the bending dispersion and $\xi(\theta)$ the anisotropic correlation. These quantities can be calculated in terms of the linear power spectrum of the density fluctuations, the C_l^{gl}'s can be calculated once the C_l's without lensing are given.

The general result for flat as well as for open CDM models is that lensing slightly smoothes the radiation power spectrum. For a given multipole the effect of lensing increases with Ω but for the same acoustic peak it decreases with Ω. The maximum contribution of lensing to the radiation power spectrum ($l < 2000$) is $\simeq 5\%$ for Ω values in the range $0.1-1$. Therefore, this effect should be considered in future analyses of CMB anisotropy data provided by very sensitive experiments.

5. References

Abbott, L. F. & Schaeffer, R. K. (1986) *Ap. J.* , **308**, 546

Bardeen, J. M. (1980) *Phys. Rev. D*, **22**, 1882

Bonnano, A. & Romano, V. (1994) *Phys. Rev. D*, **49**, 6450

Bond, J. R. (1995) *Theory and observations of the CMB* in Cosmology & Large Scale Structure, Proc. Les Houches School, Session LX, ed. R. Schaeffer, Elsevier Sci. Publ. Netherlands

Cayòn, L. (1996) *Detections of CMB anisotropies at large and intermediate angular scales: data analysis and experimental results* in The universe at High-z, Large-Scale Structure and the CMB (1996), Proc. Laredo Summer School, eds. E. Martinez-González & J. L. Sanz, Springer-Verlag.

Collins, C. B. & Ellis, G. F. R. (1979) *Phys. Reports*, **56**, 65

Dicke, R. H., Peebles, P. J. E., Roll, P. G. & Wilkinson, D. (1965) *Ap. J. Lett.*, **142**, 17

Dunsby, P. K. S. (1992) *The covariant approach to density perturbations*, preprint

Einstein, A. (1917) *Sitzungsber. Preuss. Akad. Wiss.*, **1**, 142

Ellis, G. F. R. (1971) *Relativistic Cosmology*, XLVII Enrico Fermi Summer School Proc., ed. R. K. Sachs, Academic Press, N. Y.

Ellis, G. F. R. & Bruni, M. (1989) *Phys. Rev D*, **40**, 1804

Friedmann, A. (1922) *Zeits. f. Physik*, **10**, 377

Friedmann, A. (1924) *Zeits. f. Physik*, **21**, 326

Gamow, G. (1948) *Nature*, **162**, 680

Goicoechea, L. & Sanz, J. L. (1985) *Ap. J.* , **293**, 17

Goode, S. W. (1989) *Phys. Rev. D*, **39**, 2882

Goroff, H., Grinstein, B., Rey, S.-J. & Wise, M. B. (1986) *Ap. J.* , **311**, 6

Gouda, N., Sasaki, M. & Suto, Y. (1989) *Ap. J.*, **341**, 557

Guth, A. H. (1981) *Pys. Rev. D*, **23**, 347

Hancok, S. et al. (1994) *Nature*, **367**, 333

Hawking, S. W. (1966) *Ap. J.* , **145**, 544

Hubble, E. P. (1924) *Annual Reports of the Mount Wilson Observatory*

Hubble, E. P. (1929) *Proc. Nat. Acad. Sci.*, **15**, 168

Hwang, J. & Hyun, J. J. (1994) *Ap. J.*, **420**, 512

Kaiser, N. (1992) *Ap. J.* , **388**, 272

Kodama, H. & Sasaki, M. (1984) *Prog. Theor. Phys.*, **78**, 1

Kosowsky, A. (1996) *CMB polarization*, preprint

Lemaitre, G. (1927) *Ann. Soc. Sci. Bruxelles*, **47A**, 49

Lifshitz, E. M. (1946) *J. Phys. (Moscow)*, **10**, 116

Lifshitz, E. M. & Khalatnikov, L. M. (1963) *Sov. Phys. Usp.*, **6**, 495

Linde, A. D. (1982) *Phys. Lett.*, **108B**, 389

Linde, A. D. (1983) *Phys. Lett.*, **129B**, 177

Linde, A. D. & Mezhlumian, A. (1995) *Phys. Rev. D*, **52**, 6789

Lyth, D. H. & Stewart, E. D. (1990) *Ap. J.*, **361**, 343

Maarteens, R., Ellis, G. F. R. & Stoeger, W. R. (1995), *Phys. Rev. D*, **51**, 525

Maarteens, R., Ellis, G. F. R. & Stoeger, W. R. (1995), *Phys. Rev. D*, **51**, 5942

Martinez-González, E., Sanz, J. L. & Silk, J. (1990) *Ap. J.*, **355**, L5

Martinez-González, E., Sanz, J. L. & Cayón (1996) *Ap. J.*, preprint

Martinez-González, E. & Sanz, J. L. (1995) *A& A*, **300**, 348

Netterfield, C. B. et al. (1995) *Ap. J. Lett.*, **445**, L69

Peebles, P. J. E. (1993) *Physical Cosmology*, Princeton Univ. Press, Princeton, N. Y.

Penzias, A. A. & Wilson, R. W. (1965) *Ap. J.*, **142**, 419

Raychaudhuri, A. (1955) *Phys. Rev. D*, **98**, 1123

Rindler, W. (1956) *M.N.R.A.S.*, **116**, 662

Robertson, H. P. (1929) *Proc. Nat. Acad. Sci.*, **15**, 822

Sachs, R. K. & Wolfe, A. N. (1967) *Ap.J.*, **147**, 73

Sanz, J. L. (1982) *J. Math. Phys.*, **23**, 1732

Sanz, J. L. (1983) *A& A*, **120**, 109

Sanz, J. L., Martinez-González, E., Cayón, L., Silk, J. & Sugiyama, N. (1996) *Ap. J.*, **467**, 489

Smoot, G. F. (1992) *Ap.J.*, **396**, L1

Starobinsky, A. A. (1979) *JETP Lett*, **30**, 682

Seljak, U. (1996) *Ap. J.*, **463**, 1

Stewart, J. M. (1990) *Class. Quantum Grav.*, **7**, 1169

Walker, A. G. (1935) *Proc. Lond. Math. Soc. (2)*, **42**, 90

Weinberg, S. (1972)*Gravitation & Cosmology*, John Wiley & Sons, N. Y.

York, J. W. (1971) *Pys. Rev. Lett.*, **26**, 1656

Acknowledgments: I would like to thank E. Martinez-González & L. Cayón for useful comments and a careful reading of the manuscript.

STATISTICS AND RANDOM FUNCTIONS IN ASTROPHYSICS

BERNARD J.T. JONES

Imperial College, London

1. Introduction

There are many texts and papers on random functions and statistics - probably more than there are texts and papers in astronomy. However, relatively few of these focus on the use of statistics and random functions in astronomy. The task facing the astronomer is to extract from that plethora of information something that is relevant in the astronomical context. In this short lecture I will try to provide some background and a first order guide to some things that the astronomer should be aware of. In order to make life simple I shall focus on the a few noteworthy textbooks to which the reader can turn for more details. The book by Papoulis, *Probability, Random Variables and Stochastic Processes* is an excellent starting point, and so I will try to maintain some level of consistency with his notation and approach. Indeed, an alternative title for this lecture could be "*A Papoulis Primer*".

1.1. COSMOLOGY IS MODEL-FITTING

In astrophysics, as in many branches of physics, we generally handle data sets that are only experimentally determined samples taken from some greater distribution. We cannot observe every galaxy or quasar: the sample would be too big or too faint to measure with available instruments. Likewise, our numerical simulations are limited by the power of the computers we use: they generate data that is limited in spatial and temporal resolution. It is also often the case that the phenomena we are trying to measure are partially masked by other processes, which themselves are poorly understood.

C. H. Lineweaver et al. (eds.), The Cosmic Microwave Background, 67–110.

Our datasets and simulations are merely representations of the reality we seek to understand. In science we try to express that understanding in terms of mathematical models which are defined in terms of relatively few parameters and the goal is to determine these parameters by experiment. No model is perfect, but we may be able to say that some are better than others. The parameters of a good model themselves become the subject of our scientific discussion.

Thus in cosmology we use the Friedmann models for the cosmic expansion, and the models are parametrized by the cosmic density parameters, Ω. The determination of Ω, and identifying the constituents that give it that value, has been one of the main quests for cosmology over many past decades.

1.2. ENTER STATISTICS

The nature of our data, the fact that it is incomplete and often ill-determined, means that parameter determination is a statistical problem: we are interested in making reliable estimates for the values of our parameters. Here we encounter another aspect of data modelling: not only do we model the phenomenon we are observing, we also model the means by which it is measured and in particular the errors and possible biases that arise from the measurement technique.

The standard approach (one might say the "naive" approach) is to assume that our measurement errors follow the Gaussian distribution law and that if we make many repeated measurements we will get to the true values of the parameters. Getting to the true values generally consists of plotting two variables against one another (on logarithmic axes) and fitting a least squares straight line to the data plot. The axis intercept and slope of the line are the sought after parameters.

On a slightly more sophisticated level we may try to simulate the data on a computer and try to identify which of a series of computer models most looks like the observed data set. The parameters of the best look-alike are then attributed to the real world data. The statistics enters at the point where we try to identify the look-alike: this is frequently done through a set of fitting parameters which are determined (statistically) for both the model and the data.

1.3. ENTER RANDOM FUNCTIONS

Characterising distributions of points is of fundamental importance in modern cosmology. The notion is that there is some underlying density distribution which is being sampled in some unknown (but hopefully deterministic) way by galaxies. We have to use the given distribution of points (galaxies) to determine a set of parameters that define the underlying density distribution. The density field is uniform on the largest scales, but quite inhomogeneous on smaller scales where then points are evidently clustered.

Here we see two aspects of random functions. Firstly we assume that we can characterise the density field in terms of some model. Secondly we assume that the discrete distribution of our observed sample is representative of that field and that it too can be modelled.

1.4. THE APPROACH TAKEN IN THIS ARTICLE

The approach I will take will be, start with some basic concepts from probability and build quickly towards the concept of random functions. Some of the concepts will be defined accurately, but many of the details are left to Papoulis' book. Most of the formal discussion will be confined to random functions in one dimension, though some of the examples will be in three spatial dimensions (especially where the examples are relevant to cosmology).

A word about books. I have focused on the Papoulis text largely because it is simply an excellent textbook, largely self-contained with a wealth of detail. However, there are topics it does not cover and I have used the following texts as sources in these areas. The discussion of random fields in physics (and largely in condensed matter physics) is covered by Ziman's *Models of Disorder*. The mathematics of random fields in three spatial dimensions is covered in Adler's *Geometry of Random Fields*. This develops some important results in the theory of Gaussian random fields. The characterization of random spatial structure is a topic that I had neither time nor space to cover, but this is covered in Ripley's excellent monograph *Spatial Statistics*. The classical book on the subject is the compendium of articles edited by Wax: *Selected Papers on Noise and Stochastic Process"* - somewhat old-world in style, but worth a read. Finally, a classic in the making is Jaynes' *Probability Theory: The Logic of Science*. This book provides a superb insight into Bayes' approach to inference.

2. Probability Densities

Perhaps surprisingly, the notion of what "probability" is and what it measures is highly controversial. Jaynes in his remarkable text *Probability Theory: The Logic of Science* says that "People who believe in physical probabilities are like those who believe in astrology: they never ask what would constitute a controlled experiment capable of proving or disproving their belief". This may be a good point, but it is not very helpful in coming to terms with our naive intuition that probability measures something that describes the frequencies with which particular outcomes of some observation or experiment occur. This naive notion is the basis of the "frequentist" interpretation of probability. Mathematicians, and notably Kolmogorov, have tried to formalise the concept in an abstract way. But like so many formal mathematical systems it is often difficult to discern in this approach the elements of our naive (frequentist) perception of what a probability is.

So in the present context it is best to simply regard probability as a measure on a space of events and to interpret that measure formally in your favourite way. I will take the view that probability is an attribute of the set of states of a physical system that can be achieved under specified circumstances. It is important that this attribute can be modelled for these given circumstances. If experiments are performed on that system in a way that is consistent with the premises of the model, then this attribute can be measured in the experiments and compared with the model. If the model allows it (for example by providing a means of calculating the time the system will spend at each part of its phase space), we can interpret the attribute in a frequentist sense.

Note the phrases "... the set of states ..." and "... under specified circumstances". Note also that this approach does not say which specific attribute of the set of states we are discussing, nor indeed that any such attribute is unique. For a spatial random process we shall see that there is a whole hierarchy of such attributes defined by considering the random field at one, two, three or any number of points.

2.1. PROBABILITY - THE RULES

Denote events by the symbols $A, B, ...$ We can use the notation

$A \cap B$ both A and B occur

$A + B$ either A or B or both occur

These correspond to the intersection [1] and union of the events. Then we can denote probabilities for events as follows

$P(A)$ probability that A occurs

$P(B)$ probability that B occurs

$P(A|B)$ probability that A occurs, given B

$P(A \cap B)$ probability that A and B both occur

and so on. It is axiomatic (or obvious - depends how you look at it) that

$$P(A + B) \;=\; P(A) + P(B) - P(A \cap B) \qquad (1)$$
$$P(A \cap B) \;=\; P(B)P(A|B) = P(A)P(B|A) \qquad (2)$$

---

[1] The $\cap$ symbol is generally used in mathematical approaches to probability, however, we frequently see $P(A, B)$ in place of $P(A \cap B)$

The last of these equations is of considerable importance.

Finally, events A and B are said to be *independent* if

$$P(A \cap B) = P(A)P(B) \tag{3}$$

and they are said to be *mutually exclusive* if

$$P(A \cap B) = 0 \tag{4}$$
$$\text{ie:} \quad P(A + B) = P(A) + P(B) \tag{5}$$

2.2. SOME CONCEPTS

The convention is that capital letters denote random variables, while the corresponding lower case letters denote a particular value for that random variable. For a random variable X, its *(Cumulative) Distribution Function* $F(x)$ is defined by

$$P[X < x] = F(x), \quad F(-\infty) = 0, \quad F(+\infty) = 1 \tag{6}$$

$F(x)$ is nondecreasing: $F(x_1) \leq F(x_2)$ for $x_1 < x_2$. The derivative of this function

$$f(x) = \frac{dF(x)}{dx} \tag{7}$$

if it exists, is called the *Probability Density Function* (often abbreviated "pdf") or the Frequency Function. The set of x-values for which $f(x) > 0$ make up the domain of the random variable X. When we talk about Cauchy, Gaussian or other distributions we are in fact talking about probability density functions.

We can generalise to many random variables by writing

$$P[X < x \cap Y < y] = F(x, y), \quad F(\infty, \infty) = 1 \tag{8}$$

for the distribution function of two events occurring. The *marginal distributions* of X and Y are

$$F_x(x) = F(x, \infty) \tag{9}$$
$$F_y(y) = F(\infty, y) \tag{10}$$

These are not by themselves sufficient to determine $F(x, y)$.

The *Joint Density*, if it exists, is

$$f(x, y) = \frac{\partial^2 F(x, y)}{\partial x \partial y} \tag{11}$$

We will discuss the use of these functions in calculating properties of random functions below.

2.3. PROBABILITY DISTRIBUTIONS

Consider a field ϕ whose value is defined at all points of a three dimensional space $\mathbf{r}$, and such that at each point $\mathbf{r}$ the probability of getting the value ϕ is $P(\phi, \mathbf{r})$. The field is said to be *statistically homogeneous* if the function P is the same for all points: $P(\phi, \mathbf{r}) = P(\phi)$ and it is said to be *ergodic* if

$$\frac{1}{V} \int_V \phi(\mathbf{r}) d\mathbf{r} = \int \phi P(\phi) d\phi \tag{12}$$

If it is ergodic then we refer to this quantity as the *mean* of the field ϕ and denote its value by the symbol $\langle \phi \rangle$.

$P(\phi)$ is by itself inadequate to describe the field ϕ: we could cut up the volume into equal cells and shuffle the cells without changing $P(\phi)$. We need information about how the field ϕ is distributed at pairs of points, triples of points and so on:

$$\begin{aligned} P_2(\phi_1, \mathbf{r}_1; \phi_2, \mathbf{r}_2), \\ P_2(\phi_1, \mathbf{r}_1; \phi_2, \mathbf{r}_2; \phi_3, \mathbf{r}_3), \\ \dots \\ P[\phi(\mathbf{r})] \end{aligned} \tag{13}$$

The first of these describes the probability of getting the values ϕ_1 at the point $\mathbf{r}_1$ and ϕ_2 at the point $\mathbf{r}_2$. The last entry represents the limit as the number of points becomes infinite and is the *functional description* of the random process $\phi(\mathbf{r})$.

These N-point probability functions are not all independent. There is a hierarchy of relationships, the first of which is

$$P_1(\phi_1) = \int P_2(\phi_1, \phi_2) d\phi_2. \tag{14}$$

This relates the two-point and one-point distributions and can be regarded as a consequence of the definition of probability. In solid state physics there is an important way of looking at this equation: it can be viewed as an *integral equation* for P_2, the two-point function, given the 1-point function P_1. Of course, being able to solve it depends on the expression of the physics of the problem through suitable boundary conditions on the hierarchy of functions P_i.

The process $\phi(\mathbf{r})$ is said to be *isotropic* if

$$P_2(\phi_1, \mathbf{r}_1; \phi_2, \mathbf{r}_2) = P_2(\phi_1, \phi_2, r), \quad r = |\mathbf{r}_1 - \mathbf{r}_2|. \tag{15}$$

In other words, the two-point distribution function depends only on the distance between the points, not the direction of the vector joining them.

It is necessary, again as a consequence of the interpretation of the P_i as probabilities, that

$$P_2(\phi_1, \phi_2, r) \rightarrow \begin{cases} \delta(\phi_1 - \phi_2)P(\phi_1) & |\mathbf{r}| \rightarrow 0 \\ P(\phi_1)P(\phi_2) & |\mathbf{r}| \rightarrow \infty \end{cases} \tag{16}$$

This says that the two-point function tends to the one point function as the points get arbitrarily close, and that it tends to the product of the 1-point functions as the points get arbitrarily far apart.

The *correlation function* is then

$$C(r) = \int \int \phi_1^\star \phi_2 P_2(\phi_1, \phi_2, r)d\phi_1 d\phi_2 \tag{17}$$

where $\phi^\star$ represents the complex conjugate of ϕ in the case where ϕ is a complex valued function. We note that $C(r)$ is merely a simple descriptor of the function $P_2(\phi_1, \phi_2, r)$: it contains some (but not all) of the information contained in the two-point function.

2.4. A COUPLE OF EXAMPLES

It is worth looking at two examples describing the distribution of a field ϕ on a line (so $\phi = \phi(x)$). In both examples, the 1-point distribution of ϕ is a simple Gaussian: they differ in their two-point functions and yet they have the same correlation function! This shows clearly that the correlation function is only a partial descriptor of the two-point distribution.

2.4.1. *The Gaussian distribution*
The Gaussian 1-point distribution is

$$P(\phi) = \frac{1}{\sqrt{2\pi}\sigma} \exp\left(-\frac{\phi^2}{2\sigma^2}\right) \tag{18}$$

This has zero mean and variance σ. The standard two-point Gaussian distribution

$$P_2(\phi_1, \phi_2, r) = \frac{1}{2\pi\sigma^2\sqrt{1 - C^2(r)}} \exp\left\{-\frac{\phi_1^2 + \phi_2^2 - 2C(r)\phi_1\phi_2}{2\sigma^2[1 - C^2(r)]}\right\} \tag{19}$$

This describes a field with two-point correlation function $C(r)$.

2.4.2. *Non-Gaussian distribution*
There is an interesting example of a distribution in one spatial dimension whose 1-point distribution is a Gaussian, but whose two-point function is

74

not:

$$P(\phi) = \frac{1}{\sqrt{2\pi}\sigma} \exp\left(-\frac{\phi^2}{2\sigma^2}\right)$$ (20)

$$P_2(\phi_1, \phi_2, r) = P(\phi_1)\delta(\phi_1 - \phi_2)C(r) + P(\phi_1)P(\phi_2)[1 - C(r)]$$

This behaves as it should in the limits $r \to 0$ and $r \to \infty$, and it is easy to verify that $C(r)$ is indeed the correlation function.

The distribution describes a random function on a line (the Berry Step surface) whose value is constant over intervals of finite length. The constant values are Gaussian distributed and the size of the plateaux is random and depends on the correlation function. It should be noted that the two-point function P_2 is valid for any 1-point function, not only the Gaussian.

3. Characterising Random Variables

From the probability density function for a random variable, we can calculate quantities that characterise the variable in a simple way and we can introduce some techniques that can simplify the calculation of these quantities.

3.1. EXPECTATION, MEAN

The *Expectation Value* of a random variable X having probability density function $f(x)$ is

$$\langle x \rangle = \mathbf{E}(x) = \int_{-\infty}^{\infty} x f(x) dx$$ (21)

Both the $\langle . \rangle$ and $\mathbf{E}$ notations are used: I will favour the $\mathbf{E}$ notation since it looks like an operator. The expectation value of a random variable localises it as does the mass center of a body.

More generally the expectation of a function $g(x)$ of a random variable X is

$$\mathbf{E}[g(x)] = \int_{-\infty}^{\infty} g(x)f(x) dx$$ (22)

The operator $\mathbf{E}$ is additive:

$$\mathbf{E}[g_1(x) + ... + g_n(x)] = \mathbf{E}[g_1(x)] + ... + \mathbf{E}[g_1(x)]$$ (23)

and we can deduce from this that

$$\mathbf{E}[x|x \geq a] = \frac{\int_a^{\infty} x f(x) dx}{\int_a^{\infty} f(x) dx}.$$ (24)

The expectation value of the random variable X is also called the *mean* which is denoted variously as $\langle x \rangle, \bar{x}, \mu, ...$ and so on.

3.2. VARIANCE AND HIGHER MOMENTS

The expectation $\mathbf{E}(x)$ describes the location of X. The spread of values about the mean is described by the *Variance* or *Dispersion* of X, which if it exists is defined by the integral

$$\sigma^2 = \mathbf{E}[(x - \mu)^2] = \int_{-\infty}^{\infty} (x - \mu)^2 f(x) dx \tag{25}$$

where μ is the mean, i.e., $\langle x \rangle$. It is trivial to show that

$$\sigma^2 = \mathbf{E}[x^2] - \mathbf{E}[x]^2 \tag{26}$$

This generalises to the kth *moment* of the distribution

$$m_k = \mathbf{E}[x^k] = \int_{-\infty}^{\infty} x^k f(x) dx \tag{27}$$

and the kth *central moment*

$$\mu_k = \mathbf{E}[(x - \mu)^k] = \int_{-\infty}^{\infty} (x - \mu)^k f(x) dx \tag{28}$$

The following are obvious:

$$\mu_0 = 1, \mu_1 = 0, \mu_2 = \sigma^2 \tag{29}$$

Less obvious is the relationship between the moments and the central moments:

$$\mu_k = \sum_{r=0}^{k} (-1)^r \binom{k}{r} \mu^r m_{k-r} \tag{30}$$

$$m_k = \sum_{r=0}^{k} \binom{k}{r} \mu^r \mu_{k-r} \tag{31}$$

For example:

$$\mu_0 = m_0 = 1, \tag{32}$$
$$\mu_1 = m_1 - \mu = 0, \tag{33}$$
$$\mu_2 = m_2 - 2\mu m_1 + \mu^2 = m_2 - \mu^2 \tag{34}$$

For a Gaussian distribution:

$$\mathbf{E}[x^n] = \begin{cases} (n-1)\sigma^n, & n \text{ even} \\ 0, & n \text{ odd} \end{cases} \tag{35}$$

The odd moments are zero and all others are determined simply by the variance.

The low-order central moments have names: the 2nd central moment is called the *variance*, the 3rd the *skewness* and the 4th is the *kurotsis* or flatness. Note as a closing remark to this section that not all probability density functions have moments.

3.3. SOME IMPORTANT INEQUALITIES

There are some powerful results that are quite general and do not depend specifically on the distribution function, but are derived from quite general inequalities. Thus the following result is derived from the Tchebyshev inequality:

Useful result 1 *If $g(x) > 0$, then*

$$P[g(X) \geq t] \leq \frac{1}{t} \mathbf{E}[g(X)] \tag{36}$$

and, in particular, if $P[X < 0] = 0$ and $\mathbf{E}[X] = \mu$, then

$$P[X \geq t] \leq \frac{\mu}{t} \tag{37}$$

$$F(t) \geq 1 - \frac{\mu}{t} \tag{38}$$

Moreover, if $\mathbf{E}[(X - \mu)^2] = \sigma^2$

$$P[|X - \mu| \geq t\sigma] \leq \frac{1}{t^2} \tag{39}$$

$$F(\mu + t\sigma) - F(m - t\sigma) \geq 1 - \frac{1}{t^2} \tag{40}$$

If we impose further conditions on $F(x)$, such as $F'(\mu) > F'(x)$ for $x \neq x_0$, we can put even stricter bounds.

3.4. TRANSFORMS OF PROBABILITY DENSITY FUNCTIONS

[2] The *Characteristic Function* is simply the Fourier transform of the probability density function (if it exists):

$$\Phi(\omega) = \mathbf{E}[e^{i\omega x}] = \int_{-\infty}^{+\infty} e^{i\omega x} f(x) dx \tag{41}$$

---

[2]This section follows the development in Fry (1985)

The characteristic function always exists since $f(x)$ is absolutely integrable. The *Moment generating Function* is the transform

$$M(t) = \mathbf{E}[e^{tx}] = \int e^{tx} f(x) dx \tag{42}$$

In general, the argument t is an arbitrary complex variable. However, for some distributions, $M(t)$ is only defined for purely imaginary arguments $s = it$ (s real), in which case the restricted M is then just the characteristic function.

The *Cumulant Generating Function* is

$$K(t) = \ln M(t) \tag{43}$$

(There is an analogous function derived from the log of the characteristic function).

These functions have various uses in calculating properties of the distribution which would otherwise be tedious to calculate. In particular

$$\langle x^N \rangle = m_N = \frac{d^N}{dt^N} M(t) \Big|_{t=0} \tag{44}$$

giving the moments of the probability density function by repeated differentiation. This result follows directly from differentiating (42) and setting $t = 0$ in the integrand. Likewise, $K(t)$ can be used to get the central moments:

$$\begin{aligned}
\mu_1 &= m_1 = \langle x \rangle & (45) \\
\mu_2 &= m_2 - m_1^2 = \langle (x - \langle x \rangle)^2 \rangle & (46) \\
\mu_3 &= m_3 - 3m_1 m_2 + 2m_1^2 = \langle (x - \langle x \rangle)^3 \rangle & (47)
\end{aligned}$$

and so on.

Layzer, in his prescient article (Layzer 1965) on galaxy clustering, worked with the *probability generating function*, which is useful for positive-valued random variables. If X is a random variable taking only positive values with probability density function $f(x)$, then the probability generating function is

$$G_X(t) = \int_0^\infty f(x) t^x dx \tag{48}$$

(This is essentially the Mellin transform of the probability density function). With this the nth moment of the distribution is

$$\langle x^n \rangle = \left[\left(t\frac{d}{dt} \right)^n G_X(t) \right]_{t=1}, \tag{49}$$

as can be easily verified.

3.5. SOME IMPORTANT PROBABILITY DENSITY FUNCTIONS

3.5.1. *General Gaussian*

$$f(x) \;=\; \frac{1}{\sqrt{2\pi}\sigma} exp\left[-\frac{(x-\mu)^2}{2\sigma^2}\right],$$ (50)

$$K(t) \;=\; \mu t + \frac{1}{2}\sigma^2 t^2.$$ (51)

3.5.2. *Poisson*

$$f(x) \;=\; \sum_{n=0}^{\infty} \frac{1}{n!}\mu^n e^{-\mu}\delta(x-n)$$ (52)

$$K(t) \;=\; \mu(e^t - 1).$$ (53)

3.5.3. *Exponential*

$$f(x) \;=\; \lambda e^{-\lambda x}$$ (54)

$$K(t) \;=\; -\log(1 + t/\lambda)$$ (55)

3.6. DISCRETE POINT DISTRIBUTIONS

One of the important uses of these various functions is to generate new distributions for old. The observed galaxy distribution can be regarded as a discrete sample drawn from an underlying continuous density field. On this assumption we can determine the statistical properties of the discrete sample from the properties of the underlying distribution.

If we think of the underlying distribution as being a continuous (but spatially random) density field $n(\mathbf{r})$, the distribution of galaxies in each elemental volume δV_i can be thought of as a Poisson process whose mean is the (local) underlying density. In other words, if we denote the cell count in the elemental volume δV_i by N_i, we have

$$\langle N_i \rangle = n(\mathbf{r})\delta V_i$$ (56)

It is clear from the definition (42) of the moment generating function that the moment generating function for the distribution of points in the ensemble of elemental volumes δV_i is the product of the generating functions for the individual δV_i:

$$M(t) = \prod_i \exp\left[n(\mathbf{r})\delta V_i(e^t - 1)\right]$$ (57)

and so

$$K(t) = \sum_i \left[n(\mathbf{r})\delta V_i (e^t - 1) \right] \tag{58}$$

Here the field $n(\mathbf{r})$ is a given realization of a random field, and so $M(t)$ and $K(t)$ are stochastic functions. We can calculate the moments of the cell counts by differentiating $K(t)$, setting $t = 0$ and taking averages. This leads directly to a set of equations that is derived in a somewhat different way later on (cf. Equations (141) and (142)).

4. What is a Random Function?

To keep things simple in discussing random functions, let us focus on functions of one variable: the variable might be a timelike variable so that there is a sense of evolution, or a spacelike variable in which case one has a sense of a picture of a wiggly curve on some interval on a line. We have the naive notion that a random function is one whose value at each point is determined by a random process, and hence its value at any point is unpredictable. This brings with it the notion that the function oscillates randomly, so much so that we cannot interpolate to find the value even between two arbitrarily close points, and nor can we differentiate it. These naive notions are in fact only partially correct, and we shall see that the key factor in the discussion is the so-called *autocorrelation function* of the process.

We shall for the moment consider random functions that take on random values at different successive instants of time. We shall also consider the time variable to be continuous, there is no loss of generality in that. They are therefore specified by giving the distribution function of some time dependent random variable X:

$$F(x;t) = P[X(t) < x] \tag{59}$$

Thus we could think of $X(t)$ as being brightness, temperature, or position along a line. We speak of *the random process $X(t)$*.

We shall suppose further that the probability density function for the process $X(t)$ exists:

$$f(x;t) = \frac{\partial F(x;t)}{\partial x} \tag{60}$$

The values taken on by a random variable at different times can be considered as two random variables (though not necessarily independent). Thus we can talk about the joint distribution function of the two random variables $X(t_1)$ and $X(t_2)$:

$$F(x_1, x_2; t_1, t_2) = P[X(t_1) < x_1, X(t_2) < x_2] \tag{61}$$

The joint probability density function is then

$$f(x_1, x_2; t_1, t_2) = \frac{\partial^2 F(x_1, x_2; t_1, t_2)}{\partial x_1 \partial x_2} \tag{62}$$

if it exists.

4.1. STATIONARY RANDOM PROCESSES

An important class of random process are the *Stationary Random Processes*.

Definition 1 *A random process $X(t)$ is said to be stationary in the strict sense if its statistics are not affected by a shift in the time origin:*
$$x(t) \text{ and } x(t + \epsilon) \text{ have the same statistics for all } \epsilon$$

It follows from this that $f(x;t) = f(x;t + \epsilon)$ and hence

$$f(x;t) = f(x) \tag{63}$$

the probability density is independent of t. Note that this concerns only the 1-point probability density function, it makes no statement about the higher point functions.

Consequently, for a stationary random process

$$\mathbf{E}[x(t)] = \mu = \text{constant} \tag{64}$$

The autocorrelation function $R(t_1, t_2)$ (defined below, see Equation (68)) then depends only on $t_1 - t_2$ and we can write

$$R(\tau) = \mathbf{E}[x(t + \tau)x(t)] = R(-\tau) \tag{65}$$

From now on in this article, we shall assume that the random processes under discussion are stationary.

4.2. MEAN, AUTOCORRELATION AND AUTOCOVARIANCE

The *mean* of the random process $X(t)$ is

$$\langle x(t) \rangle = \mu(t) = \mathbf{E}[x(t)] = \int_{-\infty}^{+\infty} x f(x;t) dx \tag{66}$$

Note that this is little different from the previous definition of the mean, except that the time dependence of the random variable has been included specifically. At this point of our discussion, the mean could be time-dependent.

The *Autocorrelation Function* $R(t_1, t_2)$ of the random process $X(t)$ is just the joint moment of the random variables $X(t_1)$ and $X(t_2)$:

$$R(t_1, t_2) \;=\; \langle x(t_1)x(t_2) \rangle = \mathbf{E}[x(t_1)x(t_2)] \tag{67}$$

$$=\; \int_{-\infty}^{+\infty} x_1 x_2 f(x_1, x_2; t_1, t_2) dx_1 dx_2 \tag{68}$$

The *Autocovariance* of $x(t_1)$ and $x(t_2)$ is

$$C(t_1, t_2) \;=\; \mathbf{E}[(x(t_1) - \mu(t_1))(x(t_2) - \mu(t_2))] \tag{69}$$

$$=\; R(t_1, t_2) - \mu(t_1)\mu(t_2) \tag{70}$$

Note that the variance of $x(t)$ is then

$$\sigma^2_{x(t)} = C(t, t) = R(t, t) - \mu^2(t). \tag{71}$$

4.3. DERIVATIVES OF RANDOM FUNCTIONS

Random functions are clearly not differentiable in the same sense that a continuous function is differentiable, yet it is clearly important to have a notion of what the derivative is. We might, for example, wish to determine the properties of the acceleration of a particle by looking at the time derivative of its velocity. We have to introduce a probabilistic definition of the derivative. The obvious try is to say that the derivative exists if

$$\lim_{\epsilon \to 0} \mathbf{E}\left\{ \left[\frac{x(t+\epsilon) - x(t)}{\epsilon} - x'(t) \right]^2 \right\} = 0 \tag{72}$$

but this is not useful since we need to know what $x'(t)$ is in order to use it. The definition that avoids this circular argument is

$$\lim_{\epsilon_1, \epsilon_2 \to 0} \mathbf{E}\left\{ \left[\frac{x(t+\epsilon_1) - x(t)}{\epsilon_1} - \frac{x(t+\epsilon_2) - x(t)}{\epsilon_2} \right]^2 \right\} = 0 \tag{73}$$

If this limit exists for the random process $X(t)$, the process is said to be *differentiable in the r.m.s. sense*. From this definition we can prove the most important theorem that

Important Theorem 1 *A stationary process $X(t)$ is differentiable in the r.m.s. sense if its autocorrelation function $R(\tau)$ has derivatives of up to order two.*

If additionally $R(\tau)$ is analytic, then $X(t)$ can be expanded in a Taylor series.

The entire concept of non-differentiable functions was viewed with horror by many mathematicians of the early 19th century, but they were put on a firm footing by Lebesgue and Stjeltjes.

A famous example of a "reasonable" function that is nowhere differentiable is due to Weierstrass:

$$f(x) = \sum_{n=0}^{\infty} a^n \cos(m^n x), \qquad \begin{array}{l} 0 < a < 1 \\ m \text{ a positive odd integer} \end{array} \tag{74}$$

This is uniformly convergent for all real x and since m^n is an integer this has period 2π. Term by term differentiation yields a badly divergent series, though the proof that the derivative does not exist for any x is relatively tedious.

Another oft-quoted example is a saw-tooth fractal curve that is the limit of a set of right-angled isosceles triangles placed end to end with their hypotenuse on the x-axis. As the size of the triangles shrinks and the number of teeth increases the limit is a non-differentiable function. However, the derivative of the limit function (which is zero everywhere) certainly exists.

4.4. INTEGRALS OF RANDOM FUNCTIONS

Given a random process $x(t)$, we would like to give a meaning to

$$s = \int_a^b x(t)\,dt \tag{75}$$

This would be useful if, say, $x(t)$ were the velocity of a particle, for the integral would be its displacement in an interval of time. If $x(t)$ were an ordinary function, we would do this by dividing up the interval $[a, b]$ into subintervals and counting the lower and upper bounds of the contribution to the area under the curve from each subinterval: the standard Riemann integral. For a stochastic function this is not possible.

Formally, the way it is done is to divide the interval $[a, b]$ up into n subintervals $[t_i, t_i + \Delta t_i]$. If

$$\lim_{\Delta t_i \to 0} \mathbf{E}\left\{ \left[s - \sum_{i=0}^{n} x(t_i)\Delta t_i \right]^2 \right\} = 0 \tag{76}$$

then we say s is the mean square limit of the sum and is the value of the integral. It can be shown that this limit exists provided the autocorrelation function $R(\tau)$ satisfies mild integrability conditions.

If $\mathbf{E}[x(t)] = \eta(t)$, then it is trivial to show that

$$\mathbf{E}[s] = \int_a^b \mathbf{E}[x(t)]dt = \int_a^b \eta(t)dt \qquad (77)$$

In other words: *expectation and stochastic integration commute.*

4.5. VARIANCE OF STOCHASTIC INTEGRALS

The last result shows that the mean of a stochastic integral is the integral of the mean of the function. Calculating the variance is a little more tricky. Starting with the stochastic integral (75) we can write

$$s^2 = \int_a^b x(t_1)dt_1 \int_a^b x(t_2)dt_2 = \int_a^b \int_a^b x(t_1)x(t_2)dt_1 dt_2 \qquad (78)$$

Take the expectation value of this:

$$\mathbf{E}[s^2] = \int_a^b \int_a^b \mathbf{E}[x(t_1)x(t_2)]dt_1 dt_2 = \int_a^b \int_a^b R(t_1, t_2)dt_1 dt_2 \qquad (79)$$

The integral contains the autocorrelation function $R(t_1, t_2)$. Change to the autocovariance function using the relationship $C(t_1, t_2) = R(t_1, t_2) - \eta(t_1)\eta(t_2)$ and the fact that $\sigma^2 = \mathbf{E}[(x - \mu)^2]$:

$$\sigma_s^2 = \int_a^b \int_a^b C(t_1, t_2)dt_1 dt_2 \qquad (80)$$

The statistical properties of the integral are therefore described by the autocorrelation or autocovariance functions. Later we shall make use of the fact that for a stationary process $C(t_1, t_2) = C(t_1 - t_2)$, in which case

$$\sigma_s^2 = \int_a^b \int_a^b C(t_1 - t_2)dt_1 dt_2 \qquad (81)$$

4.6. BOX-CAR AVERAGING

A special instance in which the integral of a random function plays a crucial role is in averaging the process $x(t)$ over a finite interval $[-T, +T]$:

$$s = \frac{1}{2T} \int_{-T}^{+T} x(t)dt \qquad (82)$$

All the points along the interval have equal weight. The mean value of s is just the mean value of $x(t) = \mu$:

$$\langle s \rangle = \mathbf{E}[s] = \mu \qquad (83)$$

The variance is given by Equation (81)

$$\sigma_s^2 = \frac{1}{4T^2} \int_{-T}^{+T} \int_{-T}^{+T} C(t_1 - t_2) dt_1 dt_2 \tag{84}$$

This can be manipulated using the variables $\tau = t_1 - t_2$ and t_2 for doing the integrals to give

$$\sigma_s^2 = \frac{1}{2T} \int_{-2T}^{+2T} \left(1 - \frac{|\tau|}{2T}\right) C(\tau) d\tau. \tag{85}$$

The function multiplying $C(\tau)$ in the integrand has the shape of a triangle spanning $[-2T, +2T]$, it is the convolution of the two box-cars.

The usual example of this is a process having zero mean and autocorrelation function $R(\tau) = \exp(-2\lambda|\tau|)$ for which

$$\sigma_s^2 = \frac{1}{2\lambda T} - \frac{1 - e^{-4\lambda T}}{8\lambda^2 T^2} \tag{86}$$

This shows how the variance of the box-car sampled function varies with box-car size. For box-car widths $T \ll \lambda^{-1}$ the two terms almost cancel and the variance is constant, independent of T. When $T \gg \lambda^{-1}$ the first term dominates, and the variance gets smaller as the box-car width increases: the box-car averaged function gets smoother.

4.7. ERGODICITY

The notion of ergodicity is often used loosely to imply that different kinds of averaging of a random process yield consistent results. We frequently think in terms of the spatial average of a stochastic quantity being the same as the time average of the same quantity at some fixed point. We shall now formalise this in order to show precisely what is meant.

The *Time Average* of the random process $x(t)$ can be defined as

$$n = \lim_{T \to \infty} \frac{1}{2T} \int_{-T}^{+T} x(t) dt \tag{87}$$

This is the limit of bax-car smoothing as the size of the bax-car becomes infinite. It is important to realise that this time average n is a random variable - it depends on a given realization $x(t)$ of the process and its value will, in principle, differ from one realization to another.

The same is true of the box-car average of $x(t + \tau)x(t)$:

$$r(\tau) = \lim_{T \to \infty} \frac{1}{2T} \int_{-T}^{+T} x(t + \tau)x(t) dt \tag{88}$$

This is a random variable: it varies from one realization of the process to another.

The fact that the box-car averages of a random process and its correlation function are both random variables is very important in cosmology where the phenomenon has become endowed with the name: *cosmic variance*. We only have one Universe to observe, and our sample consists of a finite sample volume in that. Our estimates of the mean density and the correlation function for the distribution of galaxies are thus estimates of the sample values. Great care is obviously needed in comparing these values with theoretical models and it would obviously be better to use alternative measures that were independent of the particular realization.

It is not necessarily the case that we can use the random variable n and $r(\tau)$ to define the mean and autocorrelation of the process $x(t)$ from which they were derived. However, we can reassure ourselves that

Important Theorem 2 n *is a random variable with mean*

$$\mathbf{E}[n] = \mathbf{E}[x(t)] = \mu \tag{89}$$

if and only if the variance

$$\sigma_n^2 = \frac{1}{T} \int_0^{2T} \left(1 - \frac{\tau}{2T}\right) [r(\tau) - \mu^2] d\tau \tag{90}$$

has

$$\lim_{T \to \infty} \sigma_n^2 = 0 \tag{91}$$

This is the *Ergodic Theorem* for $\mathbf{E}[x(t)]$. Essentially this says what we wanted: $x(t)$ is ergodic if time averages equal expected values (i.e., ensemble averages).

This tells us the condition that must be satisfied for ergodicity, but it does not tell us under what circumstances that condition is likely to be satisfied. Consider the box-car average

$$R_T(\lambda) = \frac{1}{2T} \int_{-T}^{+T} x(t + \lambda) x(t) dt. \tag{92}$$

It is clear that

$$\mathbf{E}[R_T(\lambda)] = R(\lambda) \tag{93}$$

but it is not clear under what conditions

$$\lim_{T \to \infty} R_T(\lambda) = R(\lambda) \tag{94}$$

This is a tricky question and there are many theorems dealing with this, often involving averages of products of $x(t)$ at four points and thus displacing the problem from the correlation function to some higher order object.

5. The Power Spectrum - or Spectral Density

5.1. FOURIER REPRESENTATIONS

Definition 2 *The Power Spectrum $S(\omega)$ of the process $x(t)$ is the Fourier transform of its autocorrelation function:*

$$S(\omega) = \int_{-\infty}^{+\infty} e^{-i\omega\tau} R(\tau) d\tau \tag{95}$$

$S(\omega)$ is real since $R(\tau) = R^*(-\tau)$. Moreover, if the process $x(t)$ is real, $R(\tau)$ is real and even and so $S(\omega) = S(-\omega)$.

Inverting the definition (95)

$$R(\tau) = \frac{1}{2\pi} \int_{-\infty}^{+\infty} S(\omega) e^{i\omega t} d\omega. \tag{96}$$

Hence

$$R(0) = \frac{1}{2\pi} \int_{-\infty}^{+\infty} S(\omega) d\omega = \mathbf{E}[x(t)^2] \geq 0 \tag{97}$$

We call $R(0)$ the *total power* in the process. Equation (97) says that $R(0)$ is the second moment of the random variable $x(t)$ (and the variance of $x(t)$ if the process has zero mean), and it is positive.

It can be proved that

Important Theorem 3 *The power spectrum of an arbitrary process $x(t)$ (real or complex) is non-negative: $S(\omega) \geq 0$.*

Since not all functions have everywhere positive Fourier transforms, it is clear that not every function can be a correlation function.

Equation (97) tells us that the total variance of the process $x(t)$ can be considered as being made up of contributions from individual frequency components. Thus we can say, in one dimension, that $S(\omega)$ is the contribution to the total variance from modes of frequency ω in the Fourier representation of the process. (As we shall see, there is a similar interpretation in three spatial dimensions except that the power spectrum in three dimensions is to be interpreted as the contribution to the variance per unit logarithmic wavenumber.)

5.2. PERIODIC STOCHASTIC PROCESSES

A stochastic process is said to be *periodic in the mean square sense* if its autocorrelation function $R(\tau)$ is periodic:

$$R(\tau + T) = R(\tau) \qquad \text{for some } T \text{ and every } \tau \tag{98}$$

The smallest such T is called the *period* of the process.

If the autocorrelation function is periodic we can take its Fourier series:

$$R(\tau) = \sum_{n=-\infty}^{+\infty} \alpha_n e^{in\omega_0\tau}, \quad \omega_0 = \frac{2\pi}{T}. \tag{99}$$

with

$$\alpha_n = \frac{1}{T} \int_0^T R(\tau) e^{-i\omega_0 n\tau} d\tau. \tag{100}$$

We can take the Fourier transform of $R(\tau)$ to get the power spectrum in terms of the coefficients α_n:

$$S(\omega) = 2\pi \sum_{n=-\infty}^{+\infty} \alpha_n \delta(\omega - n\omega_0) \tag{101}$$

where $\delta(x)$ is the Dirac delta-function. We see that the power spectrum consists of a set of equidistant impulses of height α_n. The frequency separation of the impulses is just the fundamental frequency ω_0.

If $x(t)$ is periodic in the mean square sense, then it has a representation

$$x(t) = \sum_{n=-\infty}^{+\infty} a_n e^{in\omega_0 t}, \quad \omega_0 = \frac{2\pi}{T}. \tag{102}$$

where the a_n are given by

$$a_n = \frac{1}{T} \int_0^T x(s) e^{-i\omega_0 ns} ds. \tag{103}$$

Note that since $x(t)$ is a random process the a_n are random numbers that depend on the realization of $x(t)$. We can calculate their mean and variance from (103):

$$\mathbf{E}[a_n] = \begin{cases} \mathbf{E}[x(t)], & n = 0 \\ 0, & n \neq 0 \end{cases} \tag{104}$$

$$\mathbf{E}[a_m a_n^\star] = \begin{cases} \alpha_n, & m = n \\ 0, & m \neq n \end{cases} \tag{105}$$

$$\tag{106}$$

The coefficients a_n have zero expectation value for any $n \neq 0$, while for $n = 0$ the coefficient a_0 has the same mean value as the process itself (often referred to as the "DC value"). The coefficients are statistically orthogonal (they are said to be *uncorrelated*) and in particular $\mathbf{E}[a_n a_n^\star] = \alpha_n$, is the quantity that represents the contribution to $S(\omega)$ from modes of frequency $n\omega_0$.

5.3. NONPERIODIC PROCESSES

Whether or not the process $x(t)$ is periodic, we can still calculate the coefficients a_n as in (103), except that there is now no prior reason to choose any particular value of T. The difference is that for nonperiodic processes, the coefficients a_n are no longer statistically orthogonal. Equations (105) do not hold - the coefficients are correlated.

In order to achieve a representation of a nonperiodic process $x(t)$ in which the coefficients are uncorrelated it is necessary to use the Karhunen-Loève expansion of $x(t)$. An alternative to that is to use the notion of an *approximate Fourier expansion*.

Consider a nonperiodic stationary process $x(t)$ and calculate the coefficients

$$c_n = \int_{-\infty}^{+\infty} x(t) \frac{\sin(\frac{1}{2}\omega_0 t)}{\pi t} e^{-in\omega_0 t} dt \qquad (107)$$

for some unspecified but small ω_0. It can be shown that

$$\begin{align}
\mathbf{E}[c_n c_m^*] &= 0, \quad n \neq m \qquad &(108) \\
\mathbf{E}[c_0] &= \mathbf{E}[x(t)] \qquad &(109) \\
\mathbf{E}[c_n] &= 0, \quad n \neq 0 \qquad &(110)
\end{align}$$

In other words these coefficients are uncorrelated as required. Moreover

$$\mathbf{E}[|c_n^2|] = \frac{1}{2\pi} \int_{(n-\frac{1}{2})\omega_0}^{(n+\frac{1}{2})\omega_0} S(\omega) d\omega \qquad (111)$$

and so the mean square value of the coefficient c_n is the contribution to the power spectrum from the pass band around frequency $n\omega_0$.

Now reconstruct the random process using these coefficients by computing the quantity

$$\hat{x}(t) = \sum_{n=-\infty}^{+\infty} c_n e^{in\omega_0 t} \qquad (112)$$

The process $\hat{x}(t)$ has the properties

$$\mathbf{E}[\hat{x}(t)] = \mathbf{E}[c_0] = \mathbf{E}[x(t)] \qquad (113)$$

and

$$\mathbf{E}[|x(t) - \hat{x}(t)|^2] \to 0, \quad \omega_0 \to 0 \qquad (114)$$

In other words $\hat{x}(t)$ is a mean square approximation to $x(t)$ for small enough ω_0.

6. Fitting and Predicting

The fitting of models to data and the prediction of behaviour are two of the prime goals of random function theory. When we fit models to data, we wish to determine the likely values of the parameters of the theory and hopefully accept or reject the theory at a given level of confidence. In this section we shall touch on the question of paramter determination, leaving the general discussion of testing theories to a later section.

The wish to predict the future has been a part of human ambition since humans were able to think and wish. Today, one of the main driving forces seems to be the wish to predict activities on the World's Stock Markets and so there is a lot of research activity in this area. Mathematically, the goal of prediction theory can be summarized easily: given a random process $s(t)$ and a constant λ, we want an estimate of $s(t + \lambda)$ given $s(t)$.

6.1. LINEAR MEAN SQUARE ESTIMATION

Suppose we wish to construct an estimate for a random variable s in terms of a linear combination of some other variables $x_1, \ldots, x_N$. Specifically, the estimator will be of the form

$$\hat{s} = a_1 x_1 + \ldots + a_N x_N \qquad (115)$$

and we wish to choose the a_i so as to minimise some measure of the likely error between the estimator $\hat{s}$ and the true value s. The most common such measure is the mean square error:

$$e = \mathbf{E}[(s - \hat{s})^2]. \qquad (116)$$

Substituting (115) into (116), gives an equation for $e(a_1, \ldots, a_N)$ and by differentiating with respect to each of the a_i in turn we generate a set of N equations that can be solved for the N parameters a_i that minimise the expected error. These equations are

$$\mathbf{E}[(s - \hat{s})x_i] = 0, \quad i = 1, \ldots, N \qquad (117)$$

This equation expresses the important *orthogonality principle*: the parameter set that minimises the expected error makes the error statistically orthogonal to the data used in making the estimate.

The value of the minimum error is found by substituting (117) into (116):

$$e_{min} = \mathbf{E}[(s - \hat{s})s] \qquad (118)$$

It is important to note that this is a very general result: we have made no assumptions about the distribution of the variables $\{x_i\}$ and we have chosen a specific model, the linear model (115). If we know the mean values of the variables $\{x_i\}$ we can improve the estimate by simply adding a to-be-determined constant b to (115). This means that we get an additional equation by differentiating (116) with respect to b: $\mathbf{E}[s - \hat{s}] = 0$. Although prediction of random time sequences has manifestly been of more interest in simulations of the stock market than in cosmology, it is nonetheless an important part of the theory of random functions and contains some interesting insight into the subject as a whole.

6.2. PREDICTION BASED ON A SINGLE DATUM

Look for an estimator of the form

$$s(t + \lambda) = as(t) \tag{119}$$

and let us try to determine the best value of a. The principle we shall apply is that the error in our estimate should be uncorrelated with the given value:

$$\mathbf{E}[\underbrace{(s(t + \lambda) - as(t))}_{error}s(t)] = 0 \tag{120}$$

This is a logical approach to the problem, it yields an estimate that is statistically independent of the given data value. This makes sense since if we found an estimator whose error did correlate with the given value, we could in principle use that information to improve our estimate. We shall see later that this assumption is in fact equivalent to minimizing the expected mean square error in the estimator.

By expanding this last equation we get

$$\mathbf{E}[s(t + \lambda)s(t)] \quad - \quad a\mathbf{E}[s(t)s(t)] = 0 \tag{121}$$

$$R(\lambda) \quad - \quad aR(0) = 0 \tag{122}$$

which give the required value for the parameter a:

$$a = \frac{R(\lambda)}{R(0)} \tag{123}$$

Note that this tells us that $a < 1$, and so we always predict a smaller value than the given one. As an aside, this is one way of seeing that $R(\lambda)$ can be regarded as a shape parameter - the autocorrelation function provides the scaling factor as you move away from the given value.

The mean square error of this estimate is

$$e = R(0) \left[1 - \frac{R(\lambda)^2}{R(0)^2} \right] \tag{124}$$

and hence the prediction is poor over scales λ such that $|R(\lambda)| \ll R(0)$. If as an example we take a double exponential for the autocorrelation function: $R(\tau) = e^{-\alpha|\tau|}$ then over displacements λ such that $e^{-2\alpha\lambda} \ll 1$ knowledge of $s(t)$ is of no use in estimating $s(t + \lambda)$.

6.3. PREDICTION BASED ON PAST HISTORY

Suppose now that the entire past history of $s(t)$ is known and that again we wish to predict $s(t + \lambda)$. We can once again look for a solution in which information at each past time $s(\beta)$ is multiplied by some weighting factor $h(t - \beta)$ which depends on how long ago β was:

$$s(t + \lambda) = \int_{-\infty}^{t} s(\beta)h(t - \beta)d\beta \tag{125}$$

where

$$h(t) = 0, \quad t < 0 \tag{126}$$

This last condition ensures that the system is *causal*. Making a simple shift of the integration parameter this can be rewritten

$$s(t + \lambda) = \int_{0}^{\infty} s(t - \alpha)h(\alpha)d\alpha \tag{127}$$

As before we will determine the multiplier $h(t)$ by the condition that the error should be statistically orthogonal to the data at every past instant of time

$$\mathbf{E}\left[\left\{s(t + \lambda) - \int_{0}^{\infty} s(t - \alpha)h(\alpha)d\alpha\right\} s(\xi)\right] = 0, \quad \xi < t. \tag{128}$$

Some work reduces this to an integral equation for $h(\alpha)$ in terms of the autocorrelation function $R(\tau)$ of the process:

$$R(\tau + \lambda) = \int_{0}^{\infty} R(\tau - \alpha)h(\alpha)d\alpha, \quad \tau > 0 \tag{129}$$

This is a form of the *Weiner-Hopf Equation*. (It is the half-infinite range of integration that is the key issue here and that makes the equation less than trivial.) Solving the Weiner-Hopf equation is an industry in itself: the method of solution using rational function approximation to the power spectral density is discussed in Papoulis' book.

7. The Galaxy Distribution

The galaxy distribution appears to be a spatially homogeneous and isotropic random field and can be characterised statistically in many ways. The particular characterisation that is chosen is determined by the kind of information that one wishes to convey. The statistical properties such as the mean density and autocorrelation function are good starting places. There are several peculiar factors that influence discussions of the galaxy distribution: there is the fact that the data sample consists of discrete points (galaxies) and there is the fact that we have only one realization - our Universe.

The first point is dealt with by assuming that there is a continuous underlying density distribution that is being Poisson sampled. The underlying continuous distribution is a particular realization of a continuous random field. There is little we can do about the second point, though we can come to terms with it by careful analysis, or by performing numerical simulations of the Universe that give rise to a diversity of realizations[3].

7.1. THE GALAXY TWO-POINT CORRELATION FUNCTION

Suppose we have two elemental volumes δV_1 and δV_2 located at $\mathbf{r}_1$ and $\mathbf{r}_2$. The probability δP_1 of finding a galaxy in δV_1 is simply

$$\delta P_1 = n \delta V_1 \qquad (130)$$

where n is the expectation value of the galaxy density, i.e., the mean density.

If the distribution of galaxies were Poissonian, then the event of finding a galaxy in δV_1 would be independent of the event of finding a galaxy in δV_2 and the probability of finding a galaxy in both of δV_1 and δV_2 would be the product of the elemental probabilities (130): $n^2 \delta V_1 \delta V_2$. However, this would not be correct if the galaxy distribution were not uniform: if galaxies were clustered the events would not be independent. In the general case we can write

$$\delta^2 P_{12} = n^2 \delta V_1 \delta V_2 [1 + \xi(\mathbf{r}_1, \mathbf{r}_2)] \qquad (131)$$

where the function $\xi(\mathbf{r}_1, \mathbf{r}_2)$ measures the deviation from the Poissonian result due to clustering.

In cosmology, the function $\xi(\mathbf{r}_1, \mathbf{r}_2)$ is called *the galaxy two-point correlation function*. Since $\delta^2 P_{12}$ is a probability, we must have

$$\xi(\mathbf{r}_1, \mathbf{r}_2) > -1 \qquad (132)$$

and if we require that the volumes become statistically independent at very large separations then

$$\xi(\mathbf{r}_1, \mathbf{r}_2) \to 0, \qquad |\mathbf{r}_1 - \mathbf{r}_2| \to \infty \qquad (133)$$

---

[3]This is not as trivial as it sounds since the way most numerical simulations are done forces the mean density of the sample volume to be the global cosmic density.

If the galaxy distribution is statistically homogeneous and isotropic the galaxy correlation function will depend only on the magnitude of the vector joining the volumes:

$$\xi(\mathbf{r}_1, \mathbf{r}_2) = \xi(r), \qquad r = |\mathbf{r}_1 - \mathbf{r}_2| \tag{134}$$

This is all a matter of definitions, we have not shown formally that the galaxy two-point correlation defined in this way is indeed a correlation function in the sense of Equation (68).

7.2. THE POISSON MODEL

To make this link we suppose that the distribution of galaxies is a consequence of two processes. Firstly there is an underlying continuous stationary random field $\rho(\mathbf{r})$ whose value at each point is the local density. Our Universe contains one realization of this field. Secondly there is a Poisson process that assigns a particle to an elemental volume δV at each point with probability proportional to the local density $\rho(\mathbf{r})$.

With this model, the probability of finding a point in an elemental volume δV at the point $\mathbf{r}$ is

$$\delta p_1 \propto \rho(\mathbf{r})\delta V \tag{135}$$

and the joint probability of finding points in both of two elemental volumes δV_1 and δV_2 located at $\mathbf{r}_1$ and $\mathbf{r}_2$ is

$$\delta p_{12} \propto \rho(\mathbf{r}_1)\delta V_1 \rho(\mathbf{r}_2)\delta V_2 \tag{136}$$

Recall that the values of the density appearing here are from one realization of a random process: they are stochastic variables. We need to calculate the expectation value of δp_{12}. If the underlying field is statistically homogeneous and isotropic p_{12} will depend only on $|\mathbf{r}_1 - \mathbf{r}_2|$.

We average over all pairs of points separated by distance r in the realization:

$$\delta P_{12} = \langle \delta p_{12} \rangle \quad \propto \quad \langle \rho(\mathbf{r}_1)\rho(\mathbf{r}_2) \rangle \delta V_1 \delta V_2 \tag{137}$$

$$\propto \quad R(r)\delta V_1 \delta V_2, \quad r = |\mathbf{r}_1 - \mathbf{r}_2| \tag{138}$$

$$\propto \quad \bar{\rho}^2 \left[1 + \frac{1}{\bar{\rho}^2} C(r) \right] \delta V_1 \delta V_2 \tag{139}$$

Here, in terms of our previous notation (Equation (68)), we have denoted the autocorrelation function of the underlying random process by $R(r)$ and the autocovariance function by $C(r)$. The last equality comes from Equation (70).

The comparison of (139) and (131) shows that what we in cosmology call the galaxy two-point correlation function is in fact the normalized autocovariance function of the underlying continuous process from which the sample is constructed.

The fact that $\xi(r)$ is a renormalized autocovariance function leads to some semantic difficulties, especially when talking with scientists who do not work in cosmology. It is therefore perhaps fortunate that we call $\xi(r)$ the "galaxy two-point correlation function" and refer to its Fourier transform as the "Cosmic Power Spectrum'. That way we can keep the terms "autocovariance function" and "spectral density" for what they are generally taken to represent in the literature of stochastic processes.

7.3. COUNTS IN CELLS

The most direct way of characterising the galaxy distribution is via the counts in cells. A sample volume is divided up into identical cells of volume V. The occupancy of the cell at position $\mathbf{r}$ in the sample volume is denoted by N, this obviously depends on V. We can then characterise the statistics of the random process N. Later on we shall look at this as a function of V.

To make things easy we can choose an arbitrarily fine subdivision of the sample volume into micro-cells of volume δV such that each of these micro-cells contains at most one galaxy. If the count in the ith micro- cell is n_i, then the cell count is

$$N = \sum_V n_i \tag{140}$$

where the sum extends over all micro-cells contained in the volume V. Note that since $n_i = 0$ or 1, this sum is just the sum over occupied cells and equals the number of occupied cells. The expectation value of the cell count is

$$\langle N \rangle = nV \tag{141}$$

where n is the mean density[4].

The mean square count in the cells is

$$\langle N^2 \rangle = \left\langle \left(\sum_{i \in V} n_i \right)^2 \right\rangle \tag{142}$$

$$= \sum_{i \in V} \langle n_i^2 \rangle + \sum_{i \neq j \in V} \langle n_i n_j \rangle \tag{143}$$

Each of the terms in the sums is 0 or 1. In the first sum the number of terms is just the number of occupied cells, and so this sum is just the occupancy of the volume V, i.e., nV.

---

[4] n is the ensemble average mean density, not the sample density.

Now look at the quantity $\langle n_i n_j \rangle_{i \neq j}$ where the product contains only products of cells $i \neq j$, such that both cells are occupied. The probability that two cells separated by a distance r are occupied is given by the two-point correlation function (Equation (131)) and so we can write

$$\langle N^2 \rangle = nV + (nV)^2 + n^2 \int_V \xi_{12} dV_1 dV_2 \qquad (144)$$

Note that $\langle N \rangle = nV$, and so the second central moment, the variance, of the counts in cells is

$$\mu_2 = \langle (N - \langle N \rangle)^2 \rangle = nV + n^2 \int_V \xi_{12} dV_1 dV_2 \qquad (145)$$

In the case $\xi_{12} = 0$ this reduces to $\mu_2 = nV$, which is the mean for a Poisson distribution, as it should be.

If we define the volume averaged J-point correlation function to be

$$\bar{\xi}(V) = \frac{1}{V^J} \int_V \xi_J(\mathbf{r}_1, \ldots, \mathbf{r}_J) d\mathbf{r}_1 \ldots d\mathbf{r}_J \qquad (146)$$

and if we write $\bar{N} = nV$ for the expectation count in a cell of volume V, then the central moments of the cells counts are

$$
\begin{array}{rcl}
\langle n \rangle & = & \bar{N} \qquad\qquad\qquad\qquad\qquad\qquad\qquad (147)\\
\langle (\Delta n)^2 \rangle & = & \bar{N}^2 \bar{\xi}_2 + \bar{N} \qquad\qquad\qquad\qquad\qquad (148)\\
\langle (\Delta n)^3 \rangle & = & \bar{N}^3 \bar{\xi}_3 + 3\bar{N}^2 \bar{\xi}_2 + \bar{N} \qquad\qquad\qquad (149)\\
\langle (\Delta n)^4 \rangle & = & \bar{N}^4 \bar{\xi}_4 + 6\bar{N}^3 \bar{\xi}_3 + 7\bar{N}^2 \bar{\xi}_2 + \bar{N} + 3\langle (\Delta n)^2 \rangle^2 \quad (150)
\end{array}
$$

$$\cdots$$

In principle such relationships enable us to test the scaling of correlation functions of various orders.

7.4. SCALING THE MOMENTS - INTERMITTENCY

Spatial random process generated by different underlying probability density functions can look very different. One might say that they have different "textures": some are "spiky", some are "sporadic" or "episodic" and these can even be quasi-periodic. Finding descriptors of this texture is rather difficult: this is in part a pattern classification problem.

Local decriptors like the moments of the underlying one-point probability density function can hardly be adequate, so one approach is to look how these moments vary as a function of scale. If we write the probability that a random variable X takes on a particular value x when measured in a volume of size L as $P(x; L)$ then the pth moment on scale L is

$$m_q(L) = \sum_{cells} P(x; L)x^q \qquad (151)$$

If these moments scale as a power law over some range of scales L, then we can make progress in visualizing the process.

Suppose that there exists a monotonic function $D(q)$ such that

$$\sum_{cells} P(x; L)x^q \propto L^{(q-1)D(q)} \qquad (152)$$

The distribution of X is then said to have scaling properties characterised by dimension $D(q)$. The exponent is written in this way since the case $q = 1$ simply counts the total number of particles, which is obviously independent of cell size. The case $q = 0$ simply counts the number of occupied cells in the volume and $D(0)$ is loosely referred to as the Hausdorff Dimension. The case $q = 2$ is related to the variance and to the integral of the two-point correlation function.

Equation (152) does not describe arbitrary processes, but it does describe a large and important set of distributions that have the property known as *multifractal scaling*. If $D(q)$ is a constant we refer to the distribution as being a *simple fractal* since it is characterised by a single dimension.

A quantity that is often used to characterise a random process is the *Intermittency*:

$$\nu_n = \frac{\langle x^n \rangle}{\langle x \rangle^n} \propto L^{-(q-1)D(q)} \qquad (153)$$

where the scaling behaviour occurs if the underlying moments scale as in (152). This essentially tells how successive moments of the distribution are related on different physical scales. Processes like the log-normal are highly intermittent.

7.5. ESTIMATING THE GALAXY CORRELATION FUNCTION

The galaxy clustering two-point correlation function has been one of the most important tools in the study of large scale structure. In the twenty or so years since it was first used on galaxy catalogues, many papers have been devoted to how best to estimate it. In the present context it is only necessary to bring a flavour of that discussion for the purpose of completeness.

The very definition of the two-point galaxy clustering correlation function $\xi(r)$ gives us a procedure: $\xi(r)dr$ is the excess number of pairs of points (galaxies) in a catalogue having separation between r and $r+dr$, relative to a Poisson distributed catalogue of the same density. So select a point in the catalogue and draw around it a shell of radius r having thickness dr. Count the number of points in the shell, noting that if the shell intersects the boundary of the sample the count will be smaller than otherwise. Repeat for all points in the sample volume and let the total number of pairs thus found be $N_{GG}(r)$. Repeat this for a range of r values. Repeat the entire process for a Poisson catalogue having the same number of points within the boundary and let the corresponding pair count be $N_{RR}(r)$. Then our estimator is

$$1 + \xi_{est}(r) = \frac{N_{GG}(r)}{N_{RR}(r)} \tag{154}$$

The fact that some shells in both the galaxy sample and the Poisson sample cross the sample boundary accounts, to some extent, for the boundary effects.

There are some simple and obvious refinements of this. There is no reason that the random sample should have the same size as the galaxy sample. A larger random sample gives a better estimate of $N_{RR}(r)$, though instead of counting the number of pairs of random galaxies, we count the number of galaxies around each random galaxy $N_{GR}(r)$. A normalisation factor has to be included to account for the different populations:

$$1 + \xi_{est}(r) = \frac{\bar{n}_R}{\bar{n}_G} \frac{N_{GG}(r)}{N_{GR}(r)} \tag{155}$$

Here $\bar{n}_R$ and $\bar{n}_G$ are the densities of the random and galaxy samples.

There are other details to take care of in practise. Often, the galaxy sample used is not uniform and so the counts in shells need an additional weighting factor depending on the distance of the galaxy from us. This brings in the selection function of the catalogue.

Recently, Hamilton has produced yet another estimator for $\xi(r)$ which is claimed to be more robust:

$$1 + \xi_{est}(r) = \frac{N_{GG}(r)N_{RR}(r)}{N_{GR}(r)^2} \tag{156}$$

This is certainly less sensitive to errors in estimating the mean density of the sample.

8. Data Reconstruction - Bayes

8.1. BAYES "THEOREM"

Go back to the fundamental statement that if A and B are events with probabilities $P(A)$ and $P(B)$, then

$$P(A \cap B) = P(A|B)P(B) = P(B|A)P(A) \tag{157}$$

From this we have

$$P(A|B) = P(B|A)\frac{P(A)}{P(B)} \tag{158}$$

This result is one expression of what is referred to as *Bayes Theorem*. Bayes wrote about this in 1763, and it became an important tool throughout the nineteenth century until it came under savage criticism by Fisher in the 1920's.

The key issue is that Bayes Theorem provides a relationship between $P(A|B)$ and $P(B|A)$. A simple example shows the power of this. Suppose that galaxies are either field galaxies (F) or Cluster galaxies (C), and that the probability of being a field galaxy $P(F) = 0.9$ (and so $P(C) = 0.1$). Suppose in addition that galaxies can be classified either as Spirals (S) or Ellipticals (E), and that 5% of the Field galaxies are Ellipticals, while 45% of the cluster galaxies are Ellipticals. This comes out as $P(E|F) = 0.05$ and $P(E|C) = 0.45$. If I observe an elliptical galaxy, what do I conclude?

Using Bayes theorem we have

$$P(F|E) = P(E|F)P(F)/P(E) \tag{159}$$
$$P(C|E) = P(E|C)P(C)/P(E) \tag{160}$$

We know everything on the right hand side, though we have not said anything about $P(E)$. In any case, with the given numbers

$$\frac{P(F|E)}{P(C|E)} = \frac{0.05 \times 0.9}{0.45 \times 0.1} = 1.0 \tag{161}$$

So, given that you are observing an elliptical, it is as likely that this is a cluster elliptical as a field elliptical. (Of course, we didn't really need Bayes to sort that out - it is a simple counting exercise!)

We had some prior knowledge of $P(F)$ and $P(C)$ and we were given the likelihood that a galaxy in a cluster was an elliptical or a galaxy in the field was an elliptical. In fact, in Bayes theorem as written in Equation (158), $P(A)$ is the *prior distribution* of the parameter we are trying to estimate and $P(A|B)$ is its *posterior distribution*.

Knowledge of the prior distribution is an important aspect of the Bayesian approach. That knowledge corresponds to the fact that you can obviously make better estimates of a parameter if you have some prior knowledge about it. The question is where that knowledge comes from: usually it comes from previous examination of relevant data, so in the example of the field versus cluster elliptical galaxies, we derive the prior distribution by noting in our data what fractions of the field and cluster populations are in fact ellipticals.

8.2. EVALUATING THEORIES USING BAYES THEOREM

Now let us consider some theoretical models $\mathcal{M}$ that predict data values D which we can measure experimentally. We propose to use the data to assess the relative merits of these models. We might interpret the symbols

$P(D|\mathcal{M})$ probability that this particular data set D would arise on the basis of the model

$P(\mathcal{M}|D)$ probability that the model $\mathcal{M}$ is correct given that our experiment gave the data set D

and so write the *likelihood* of our theory, given the data, as

$$P(\mathcal{M}|D) = P(D|\mathcal{M})\frac{P(\mathcal{M})}{P(D)} \tag{162}$$

Hence if we have a set of models or theories $\mathcal{M}_i$, given a dataset D we can in principle give relative weights for the acceptability of the theories given this data. Using this equation we can calculate the ratio of $P(\mathcal{M}_i|D)$ and $P(\mathcal{M}_j|D)$ and the ratio does not depend on $P(D)$ since we are comparing on the basis of the same data. We appear to have a mechanism for deciding which is the best theory.

The argument is however not quite as straightforward as has been stated. How are we to understand the quantity $P(\mathcal{M})$? How can we assign prior probabilities to models if not via an intuition gained of looking at data? After all, it is observation of patterns in data that motivates the creation of a model in the first place. Clearly, $\mathcal{M}$ and D do not have the same status vis a vis A and B in (158).

There has been an enormous debate about this (see *Jaynes'* book), a substantial part of it misleading. There appear to be three situations concerning the prior distribution:

a: it is a frequency distribution
b: it is a representation of what it is rational to believe about the parameter (usually in a state of ignorance)
c: it is a subjective measure of the scientist's own belief and credulity

The first case (*the frequentist approach*) was what happened in our example of cluster and field elliptical galaxies: we knew the key numbers as a consequence of some prior observational data. Often, however, it will not be that easy to disentangle the required prior distribution from effects of error or experimental bias. This is an uncontroversial use of Bayes Theorem, and indeed it is an excellent approach when the prior information can be established. Because of that, the approach is rather limited.

The second case (*rational degree of belief*) arises when we know nothing about a theory and we wish to perform an experiment to acquire data

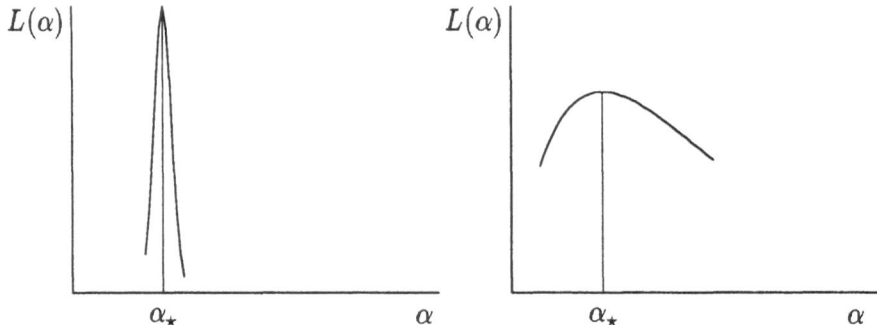

$L(\alpha)$ $L(\alpha)$

$\alpha_\star$ α $\alpha_\star$ α

Figure 1. Maximum Likelihood estimation: on the left the parameter α is depicted as being sharply defined, whereas on the right the uncertainty is larger

that will add to our knowledge (it will lessen our ignorance). The posterior distribution then measures the evidence provided by the data. In complete ignorance we should logically assign all prior possibilities equal probability, to do otherwise would pretend some knowledge. The difficulty then comes when it later turns out that one of the prior possibilities should be divided into two. This changes all the prior probabilities.

The third case (*subjective degree of belief*) is the most interesting because it opens up a door to drawing inferences under very diverse circumstances. It is also the most controversial. In this approach, you are, in essence, invited to back your conviction by making a serious bet. Much has been written as to how to formulate this and what strategies are available. One argument for such an approach is that the evaluation of an idea does not depend solely on acquiring one data set, but is part of a longer scientific history of investigation and indeed a large scientific context. You may suspect, for reasons that have nothing to do with the current set of experiments, that one particular option is going to fail. The approach to inference based on subjective degrees of belief is an attempt to quantify prior knowledge in terms of experience.

8.3. MAXIMUM LIKELIHOOD PARAMETER ESTIMATION

Notwithstanding such arguments, Equation (162) provides one of the most important tools we have for estimating parameters in models - the method of *maximum likelihood*. If we have one model with several free parameters that are to be determined from data, we can choose the parameter values that maximize $L = P(\mathcal{M}|D)$. There an implicit assumption that all parameter sets are equally likely until we have some data that allows us to express a preference for certain values.

Suppose we are trying to determine a single parameter α in a theory, we can make a plot of the likelihood function $L(\alpha)$ that the data turned out the way it did. Such a function will generally be bell-shaped, hopefully with a well defined maximum at some value $\alpha_\star$ of the parameter. $\alpha_\star$ is the *maximum likelihood estimator* of α.

If the likelihood function is sharp, we can argue that we have a good estimator of α. The width of the function $L(\alpha)$ can be calculated by

$$\Delta\alpha = \left[\frac{\int(\alpha - \alpha_\star)^2 L(\alpha)d\alpha}{\int L(\alpha)d\alpha}\right]^{\frac{1}{2}} \tag{163}$$

and this can be regarded as the standard error of the estimate. Note that the normalisation of $L(\alpha)$ does not come into this.

Maximum likelihood estimation is not without its difficulties and there are many examples in the literature of relatively simple (though often contrived) instances where it fails. Perhaps the most difficult situation arises where the maximum is not unique.

8.4. MAXIMUM LIKELIHOOD ESTIMATORS

Suppose that we have a theory involving some to-be-determined parameters a_j $(j = 1, \ldots, n)$. The theory predicts values of measurable quantities ξ_i $(i = 1, \ldots, N)$ in terms of the a_j: $\xi_i = \xi_i(a_1, \ldots, a_n)$. Now perform an experiment the outcome of which is a measurement of the N quantities ξ_i, yielding values x_i with Gaussian distributed errors having standard deviation σ_i $(i = 1, \ldots, N)$. How do we estimate the parameters a_j of the theory?

The likelihood function is proportional to the probability of getting this data set, and given the Gaussian nature of the error distribution, this is

$$L(a_1, \ldots, a_n) = \frac{1}{(2\pi)^{N/2}} \exp\left[-\sum_{i=1}^{N} \frac{(x_i - \xi_i)^2}{2\sigma_i^2}\right] \tag{164}$$

We wish to select the parameters a_j so as to maximize this expression. This is obviously equivalent to minimizing the exponent

$$\frac{1}{2}\chi^2(a_j) = \sum_{i=1}^{N} \frac{(x_i - \xi_i)^2}{2\sigma_i^2} \tag{165}$$

Because we are minimizing a sum of squares, this is referred to as a *Least Squares Fit*. Differentiating with respect to each of the a_j yields a set of equations

$$\sum_{i=1}^{N} \frac{(x_i - \xi_i)}{\sigma_i^2} \frac{\partial\xi_i}{\partial a_j} = 0, \quad j = 1\ldots, n. \tag{166}$$

For a general function $\xi(a_1, \ldots, a_n)$ this is a nonlinear set of equations, and the fitting procedure is referred to more precisely as a *nonlinear least squares fit*.

We can make easy progress if the function $\xi(a_1, \ldots, a_n)$ is linear in the a_j. Write this linear relationship in matrix form for convenience:

$$\xi = \mathbf{Ca} \qquad (167)$$

(where the matrix $\mathbf{C}$ is not necessarily square). Define a new matrix $\mathbf{D}$ whose elements are

$$D_{ij} = \frac{C_{ji}}{\sigma_j^2} \qquad (168)$$

and with this calculate the *data vector* $\mathbf{X}$:

$$\mathbf{X} = \mathbf{Dx} \qquad (169)$$

This involves known quantities on the right hand side. It is then easy to show that

$$\mathbf{X} = \mathbf{Ma}, \qquad \mathbf{M} = \mathbf{DC} \qquad (170)$$

Even though $\mathbf{C}$ is not square, the *measurement matrix* $\mathbf{M}$ is square and symmetric and depends only on the errors.

The maximum likelihood estimator of the parameters a_j is then

$$\mathbf{a} = \mathbf{M}^{-1}\mathbf{X} \qquad (171)$$

It can be shown that the standard error for the parameter a_j is given by the jth diagonal element of the inverse of $\mathbf{M}$:

$$\Delta a_j = \sqrt{(\mathbf{M}^{-1})_{jj}} \qquad (172)$$

Hence $\mathbf{M}^{-1}$ is called the *error matrix*. Note that $\mathbf{M}^{-1}$ will in general have off-diagonal terms: this reflects the fact that the parameters a_j are not statistically independent and that their errors are correlated.

8.5. A SIMPLE ASTRONOMICAL EXAMPLE

It is instructive to give a simple example of how maximum likelihood is used in practise. We will show how to estimate the velocity $\mathbf{U}$ of the Galaxy relative to a sample of galaxies distributed around the sky. It will be supposed that we have redshift independent distance estimates d_l for the galaxies in the sample. We can therefore estimate the radial component of the peculiar velocity of each galaxy, u_l over and above the Hubble flow: $u_l = cz - Hd_l$ (cz is the observed recession velocity).

Suppose that in a sample of galaxies, galaxy l is observed in direction $\hat{\mathbf{r}}_l$ and that it is assigned a radial component of peculiar velocity u_l. Suppose further that the probable error in measuring u_l is σ_l (the error is a consequence of the uncertainty in the distance estimate).

The component of our velocity $\mathbf{U}$ relative to the sample in the direction of galaxy l is $\hat{\mathbf{r}} \cdot \mathbf{U}$. The radial velocity of galaxy l relative to the sample is therefore $u_l - \hat{\mathbf{r}} \cdot \mathbf{U}$. Hence on the assumption that the errors are Gaussian, the likelihood of the entire data set is

$$L(U_1, U_2, U_3) = \prod_l \frac{1}{\sqrt{2\pi}\sigma_l} \exp\left[-\frac{(u_l - \hat{\mathbf{r}}_l \cdot \mathbf{U})^2}{2\sigma_l^2} \right] \qquad (173)$$

We wish to chose the components of $\mathbf{U}$ that maximize this. To this end we take the logarithm of this expression, thus turning the product into a sum, and then differentiate with respect to the components U_i of $\mathbf{U}$. This gives

$$\mathbf{U} = \underline{\mathbf{A}}^{-1} \cdot \sum_l \frac{u_l \hat{\mathbf{r}}}{\sigma_l^2}, \qquad A_{ij} = \sum_l \frac{\hat{r}_i \hat{r}_j}{\sigma_l^2} \qquad (174)$$

The matrix $\underline{\mathbf{A}}$ contains only information about the directions in which the galaxies are observed and the errors in measuring a radial velocity.

It is the inverse of $\underline{\mathbf{A}}$ which comes into the solution for $\mathbf{U}$. For a sample of 1000 galaxies, this means inverting a 1000- square matrix. However, if the distribution of the sample on the sky is roughly spherical, it is an almost diagonal matrix and there are special techniques to deal with that efficiently.

Since it is harder to measure the distances of the furthest galaxies in the sample, the error σ_l increases with distance. The most distant galaxies in the sample therefore have the least weight. The error analysis is, however, very complicated in part because the vectors $\hat{r}_i$ are not in fact randomly distributed on the sky: there is a zone of avoidance to contend with, and we know the vectors are correlated since galaxies lie in clusters and the clusters themselves are correlated.

This is an important problem in cosmology because we would like to know what our motion is relative to the most distant systems of galaxies. We can obtain an alternative measure of this by observing the dipole anisotropy of the cosmic microwave background radiation. The two estimates should agree in magnitude and direction.

9. Data Reconstruction

In this section we discuss attempts to reconstruct data that has been somehow degraded, and to which noise of some sort has been added. This is a vast subject in its own right with special techniques having been developed to cope with rather specific models.

We shall restrict attention to a fairly generic model in which the real world data $s(\mathbf{x})$ is multiplicatively degraded by a function $h(\mathbf{x}, \mathbf{y})$ and noise $\nu(\mathbf{x})$ added to give an observed dataset $d(\mathbf{x})$:

$$d(\mathbf{x}) = \int h(\mathbf{x}, \mathbf{x}')s(\mathbf{x}')d\mathbf{x}' + \nu(\mathbf{x}). \tag{175}$$

In imaging, s would represent the scene, h would represent the point spread function of the detector, and ν the detector noise. We shall assume that the degradation function h is translation invariant:

$$h(\mathbf{x}, \mathbf{y}) = h(\mathbf{x} - \mathbf{y}) \tag{176}$$

The spatial coordinates $\mathbf{x}, \mathbf{y}, \dots$ could be 1-, 2- or 3- dimensional.

9.1. NAIVE RECONSTRUCTION

In the absence of noise and with the degradation function of the form (176), the simplest thing to do is take the Fourier transform of (175):

$$D(\mathbf{u}) = H(\mathbf{u})S(\mathbf{u}) \tag{177}$$

The notation used here is that capital letters represent the Fourier transform of the function having the corresponding lower case letter, and $\mathbf{u}$ represents the Fourier space coordinate variable:

$$F(\mathbf{u}) = \int e^{-i\mathbf{u}\mathbf{x}} f(\mathbf{x})d\mathbf{x} \tag{178}$$

From Equation (177) we can write, in the absence of noise, the Fourier transform of the reconstructed scene as

$$S(\mathbf{u}) = \frac{D(\mathbf{u})}{H(\mathbf{u})} \tag{179}$$

Taking the inverse Fourier transform reconstructs the original scene s from the data d provided we know H. Even if $H(\mathbf{u})$ has no more than a countably infinite number of zeros, this works since $D(\mathbf{u})$ will also have zeros at the same points.

The situation is however unrealistic and one must consider the presence of the noise, in which case the same argument yields

$$\frac{D(\mathbf{u})}{H(\mathbf{u})} = S(\mathbf{u}) + \frac{N(\mathbf{u})}{H(\mathbf{u})} \tag{180}$$

N here is the Fourier transform of the noise field.

We write it in this way because we measure $d(\mathbf{x})$. Presumably we know $h(\mathbf{x} - \mathbf{y})$ and so we can calculate the left hand side and take its Fourier transform: this is no different from the noiseless case. However, D contains, as this last equation shows, a contribution from a particular realization of the noise, which we cannot know. Moreover, H can be small or even zero where N is not, and then H/N may dominate the contribution from S at such points in the Fourier space. The left side of (180) could be dominated by noise spikes and we need techniques to deal with these. It is possible to take a direct approach and to clip these noise spikes "by hand"; however, this would be an *ad hoc* procedure and we would like something less subjective.

9.2. SPECTRAL REPRESENTATION OF THE SOURCE AND DATA

The solution to our reconstruction problem will lie not in dealing with a realization of the noise that we cannot know about, but in understanding the statistical properties of the source, data and noise as expressed through their correlation functions and spectral densities. We start with an important relationship between these as expressed in the following theorem:

Important Theorem 4 *Let homogeneous random fields $s(\mathbf{x})$ and $d(\mathbf{x})$ be related by*

$$d(\mathbf{x}) = \int h(\mathbf{x}, \mathbf{x}')s(\mathbf{x}')d\mathbf{x}' + \nu(\mathbf{x}). \tag{181}$$

where the field $\nu(\mathbf{x})$ represents additive noise. Suppose also that the fields $s(\mathbf{x})$ and $\nu(\mathbf{x})$ have zero mean values and are uncorrelated. Then the spectral densities S_{dd}, S_{ss} and $S_{\nu\nu}$ of the fields d, s and ν are related by

$$S_{dd}(\mathbf{u}) = S_{ss}(\mathbf{u})|H(\mathbf{u})|^2 + S_{\nu\nu}(\mathbf{u}) \tag{182}$$

where $H(\mathbf{u})$ is the Fourier transform of $h(\mathbf{x})$.

The result follows directly from (181) by multiplying by $d(\mathbf{y})$ and taking expectation values.

9.3. LEAST SQUARES FILTERING AND WEINER FILTERS

We cannot know what in our data d is source s and what is noise ν, but we can attempt to find a statistical reconstruction s' of the source that is likely to differ from the original by as little as possible. We can measure the difference in a mean square sense:

$$e^2 = \mathbf{E}[(s(\mathbf{x}) - s'(\mathbf{x}))^2] \tag{183}$$

The constraint on the minimization is the data $d(\mathbf{x})$.

We can simplify this task considerably if we insist that the reconstructed source $s'(\mathbf{x})$ be a linear function of the data $d(\mathbf{x})$. This is entirely analogous to what we did earlier (see Equation (115)). This model is referred to as the linear least squares reconstruction. (It is possible to avoid this assumption and do a nonlinear least squares reconstruction).

We can represent this linear dependence as

$$s'(\mathbf{x}) = \int m(\mathbf{x} - \mathbf{x}')d(\mathbf{x}')d\mathbf{x}', \tag{184}$$

where the function m gives a relative weighting to data values. Expressing m as a function of $(\mathbf{x} - \mathbf{y})$ is motivated by an assumption that the noise is homogeneous (as assumed for the degradation function h).

With this we can substitute the model (184) into the error expression (183) and differentiate to find the m that minimizes the error. The weighting function m in fact satisfies

$$\mathbf{E}\left\{\left[s(\mathbf{x}) - \int m(\mathbf{x} - \mathbf{x}')d(\mathbf{x}')d\mathbf{x}'\right] d(\mathbf{y})\right\} = 0 \tag{185}$$

for all pairs of points $(\mathbf{x}, \mathbf{y})$. This equation is entirely analogous to Equation (117) and expresses the fact that the data is statistically orthogonal to the errors.

This last equation can be rewritten as

$$\int m(\mathbf{x} - \mathbf{x}')\mathbf{E}[d(\mathbf{x}')d(\mathbf{y})d\mathbf{x}'] = \mathbf{E}[s(\mathbf{x})d(\mathbf{y})] \tag{186}$$

and hence

$$\int m(\mathbf{x} - \mathbf{x}')R_{dd}(\mathbf{x}' - \mathbf{y})d\mathbf{x}' = R_{sd}(\mathbf{x} - \mathbf{y}) \tag{187}$$

for all point pairs $(\mathbf{x}, \mathbf{y})$. The function R_{dd} is the autocorrelation function of the data with itself, and the function R_{sd} is the cross-correlation function of the data with the source. Their arguments have been written in a way that assumes that the fields are statistically homogeneous.

Finally, doing some coordinate shifts, we obtain

$$\int m(\mathbf{x} - \mathbf{y})R_{dd}(\mathbf{y})d\mathbf{y} = R_{sd}(\mathbf{x}) \tag{188}$$

and we can take the Fourier transform of this to give the simple expression

$$M(\mathbf{u}) = \frac{S_{sd}(\mathbf{u})}{S_{dd}(\mathbf{u})} \tag{189}$$

where S_{dd} is the spectral density of the data (the degraded source) and S_{sd} is the cross spectral density of the data and source.

This is all well and good, but we do not know the cross spectral density of the data and source since we do not yet know the source! To progress beyond this point we need further assumptions: that the source $s(\mathbf{x})$ and noise $\nu(\mathbf{x})$ are uncorrelated and that the mean value of either the noise or the source is zero:

$$\mathbf{E}[s(\mathbf{x})\nu(\mathbf{x})] = E[s(\mathbf{x})]E[\nu(\mathbf{x})] = 0 \qquad (190)$$

Given this quite reasonable assumption, we deduce from (175) that

$$R_{sd}(\mathbf{x}, \mathbf{y}) = \mathbf{E}[s(\mathbf{x})d(\mathbf{x})] = \int h(\mathbf{y} - \mathbf{x}')\mathbf{E}[s(\mathbf{x})s(\mathbf{x}')] \qquad (191)$$

which reduces to

$$\int h(\mathbf{x} - \mathbf{y})R_{ss}(\mathbf{y})dy = R_{sd}(\mathbf{x}) \qquad (192)$$

Equation (191) should be compared with (188): they show how the kernels h and m transform the autocorrelation functions of the source and data into the cross-correlation of source and data. Our last equation (192) Fourier transforms to

$$S_{sd}(\mathbf{u}) = H^{\star}(\mathbf{u})S_{ss}(\mathbf{u}) \qquad (193)$$

We are still not done since we need to bring in the system noise explicitly, this is done via our key result (182). Substituting this and (192) into (188) we get the result for the Fourier transform data weighting function $m(\mathbf{x})$:

$$M(\mathbf{u}) = \frac{H^{\star}(\mathbf{u})S_{ss}(\mathbf{u})}{S_{ss}(\mathbf{u})|H(\mathbf{u})|^2 + S_{\nu\nu}(\mathbf{u})} \qquad (194)$$

$$= \frac{1}{H(\mathbf{u})}\frac{|H(\mathbf{u})|^2}{|H(\mathbf{u})|^2 + [S_{\nu\nu}(\mathbf{u})/S_{ss}(\mathbf{u})]} \qquad (195)$$

This is generally referred to as the *Weiner restoration filter*, or *Weiner filter* for short. As this discussion shows, it is predicated on a linear least squares fit to the data. In the absence of noise ν this turns back into the simple case given in Equation (179), as it should. The contribution in square brackets has the effect of smoothing $1/H(\mathbf{u})$ in order to provide optimal restoration (in the mean square sense) in the presence of noise.

Often, we know nothing about the random processes involved and it is customary to approximate (195) by

$$M(\mathbf{u}) = \frac{1}{H(\mathbf{u})}\frac{|H(\mathbf{u})|^2}{|H(\mathbf{u})|^2 + \Gamma} \qquad (196)$$

where Γ is a constant approximating the signal to noise power density ratio. Obviously the choice of the value of this constant reflects some knowledge

108

about the relative magnitudes of the signal and noise based on other criteria. One point that is not always appreciated is that the Weiner Filter cannot be used iteratively: we cannot assume the noise is zero ($\Gamma = 0$), do a reconstruction and estimate the noise from the reconstructed image - the process does not converge to the source.

The Weiner Filter was one of the first attempts at image reconstruction and although it works well, there are now many better approaches to data reconstruction. We shall discuss some of these briefly below. For practical implementation of the Weiner Filter and some fine insight into the way it works, see Section 13.3 et seq. of *Numerical Recipes* (Press et al., 1992).

9.4. BAYESIAN IMAGE RECONSTRUCTION

The goal of image reconstruction is to reconstruct from data about a scene (which may be 2-dimensional or 3-dimensional) an image of the "true" scene. The data may fail to correctly represent the scene for a variety of reasons: instrumental noise, abberations in the optical system and so on. In this case we will consider the scene S to be transformed to measurable data D by an acquisition system that we can model by a model M.

In the Bayesian spirit we can write the joint probability $P(D, S, M)$ in a number of ways:

$$
\begin{aligned}
p(D, S, M) &= P(D|S, M)P(S, M) & (197)\\
&= P(D|S, M)P(S|M)P(M) & (198)\\
&= P(S, M|D)P(D) & (199)\\
&= P(S|D, M)P(D|M)P(M) & (200)
\end{aligned}
$$

From variants (200) and (198) we can write

$$
P(S|D, M) = \frac{P(D|S, M)P(S, M)}{P(D, M)} \tag{201}
$$

$P(S, M)$ makes no reference to the data, it represents our *prior* knowledge and, in principle, we know what it is even before we take any data. If we have no bias concerning the scene or the instrumentation we would take this as a constant. However, this is not the only possibility. It is possible to take this to be related to the *image entropy, S*:

$$
P(S, M) = e^{\alpha S} \tag{202}
$$

The constant α in this case has to be chosen carefully.

The quantity $P(D|S, M)P(S, M)$ measures the likelihood of the data given a particular image and acquisition model. A simple choice for this is the *chi-square* goodness of fit statistic

$$P(D|\mathcal{S}, M)P(\mathcal{S}, M) = e^{-\frac{1}{2}\chi^2} \tag{203}$$

Minimising this ensures fidelity, but in practise results in the generation of artifacts.

Maximum Entropy Reconstruction optimizes

$$P(\mathcal{S}, M) = \exp\left[-\frac{1}{2}\chi^2 + \alpha S\right] \tag{204}$$

for a suitable α. If we put $\alpha = 0$ we have simple least squares reconstruction: the image is the one that minimizes the sum of the squares of the deviations between the data and the scene.

There is an important alternative to Equation (201) derived from (199) and (198):

$$P(\mathcal{S}, M|D) = \frac{P(D|\mathcal{S}, M)P(\mathcal{S}|M)P(M)}{P(D)} \tag{205}$$

This is an important alternative to (201) since it allows us to model the instrument at the same time as the image. It gives what is called the *maximum a posteriori image/model pair*, and is called MAP for short.

10. Closing Comments

Statistical methods play a key role in modern astrophysics, and in cosmology in particular. Unfortunately, there appears to be relatively little formal teaching of the concepts in either graduate physics or graduate astrophysics courses and the situation becomes one of "learn-as- you-go". Consequently, well known ideas are often rediscovered, sometimes incorrectly and often in a more limited form than is known already. Part of the blame of course lies in the fact that many of these well-known ideas are expressed more abstractly than appears at first useful in Journals that are not readily available to the astrophysicist (*Biometrika, Journal of Applied Probability, ...*).

This lecture has been an attempt to scratch the surface of this vast subject. It has been made easier by having a fine text, that of Papoulis, to lean on. However, many issues are left untouched. I have not, for example, discussed nonparametric techniques and nor have I discussed the use and measurement of power spectra in cosmology. That has to await another summer school.

Acknowledgements I thank many of my friends who through the years have helped me come to grips with various aspects of this subject. The article was written while I enjoyed being a long term guest at Imperial College. I thank Michael Rowan Robinson for that. I also thank the organisers of the Strasbourg NATO school for making my participation possible.

110

References

1. Adler, R.J., 1981, *The Geometry of Random Fields*, Wiley.
2. Jaynes, E.T., 1995, *Probability Theory: The Logic of Science*
3. Layzer, D., 1965, in *Galaxies in the Universe: Star and Stellar Systems Vol. IX* ed. A. and M. Sandage. Univeristy of Chicago Press.
4. Papoulis, A., 1965, *Probability, Random Variables and Stochastic Processes*, McGraw Hill.
5. Press, W.H., Teukolsky, S.A., Vetterling, W.T. and Flannery, B.P., 1992, *Numerical Recipes in C: the art of scientific computing* Cambridge University Press (2nd ed).
6. Ripley, B.D., 1981, *Spatial Statistics* Wiley.
7. Wax, N., 1954, *Selected Papers on Noise and Stochastic Processes*, Dover
8. Ziman, J.M., 1979, *Models of Disorder*, Cambridge University Press.

STRUCTURE FORMATION

JOSEPH SILK
Departments of Astronomy and Physics, and
Center for Particle Astrophysics
University of California, Berkeley, CA 94720

1. Introduction

The theory of gravitational instability of small density fluctuations in the expanding Universe has met with considerable success. Several important properties of large-scale structure and of galaxies can be explained, and the primordial fluctuations have been measured, at least on large scales. However, there are also some outstanding features that challenge the current theory, especially on the scales over which galaxies form.

In these lectures, I present an overview of the formation of structure and gravitational instability in both linear and nonlinear regimes. I will describe the interaction between theory and observations, and present some ideas on how the luminous component of galaxies formed.

2. Gravitational Instabilities: Linear Theory

In a cold, static cloud, density fluctuations grow exponentially rapidly: $\delta\rho/\rho \propto \exp(t\sqrt{G\rho})$. However, in the expanding Universe the growth rate becomes a power-law since the background density decreases with time (as t^{-2}). The Jeans length, $L_J \sim v_s(G\rho)^{-1/2}$ still demarcates the critical transition between stable and unstable modes.

Consider first the evolution of small density perturbations , superimposed on a Friedmann background. I present a qualitative discussion, followed by a formal Newtonian treatment.

2.1. QUALITATIVE ANALYSIS.

Denote mass fluctuations averaged over spheres containing mass M by $\frac{\delta M}{M} \propto M^{-\frac{n+3}{6}}$. While $n = 0$ yields a Poisson distribution, $n = 1$ is the

C. H. Lineweaver et al. (eds.), The Cosmic Microwave Background, 111–133.

requisite scale-invariant index. This follows from the definition of metric, or equivalently over small scales, gravitational potential fluctuations associated with density fluctuations δM :

$$|h_{ij}| \equiv |\delta\phi| = |\frac{G\delta M}{rc^2}|$$

over a scale containing mass M. One has $\delta\phi = constant$, if $\delta M/M \propto M^{-\frac{2}{3}}$, that is, if $n = 1$.

After setting $M = \frac{4}{3}\pi\rho r^3$, it is instructive to rewrite $|\delta\phi|$ as

$$|\delta\phi| \approx \frac{\delta M}{M}\left(\frac{r}{a}\right)^2\left(\frac{a}{ct}\right)^2,$$

where I have set $\rho \sim 1/Gt^2$. Here r is the physical scale of a fluctuation, and $\lambda \equiv r/a(t)$ is the comoving scale.

One can now write, if $|\delta\phi|$ is constant,

$$\frac{\delta M}{M} \propto \lambda^{-2}t \propto M^{-\frac{2}{3}}t$$

in the radiation-dominated regime ($a \propto t^{1/2}$), and

$$\frac{\delta M}{M} \propto \lambda^{-2}t^{\frac{2}{3}} \propto M^{-\frac{2}{3}}t^{\frac{2}{3}}$$

in the ensuing matter-dominated epoch ($a \propto t^{\frac{2}{3}}$).

More generally,

$$|\delta\phi| \propto M_h^{\frac{1-n}{6}}$$

where M_h is the mass contained within a horizon-scale density fluctuation. The scale-invariant spectrum is now seen to explicitly specify the same amplitude for density fluctuations at horizon crossing.

Hence constant metric fluctuations, $|h_{ij}| = $ constant, are equivalent to gravitational potential fluctuations, $\delta\phi$, evaluated at horizon crossing. I deduce that for a comoving sphere of physical diameter $\lambda(1+z)^{-1}$ containing the invariant mass

$$M(\lambda) = 1.54 \times 10^{11}\Omega h^2 (\lambda/1\text{Mpc})^3\,\text{M}_\odot,$$

one has

$$|\delta\phi| = \left|\frac{G\delta M}{\lambda c^2}\right|_h = \text{constant}.$$

The resulting fluctuations δM over the sphere M are given by

$$\frac{\delta M}{M} \propto \left(\frac{ct}{\lambda}\right)^2 \propto M^{-2/3}\left(\frac{t}{1+z}\right)^2.$$

I conclude that a scale invariant spectrum has mass fluctuations $\propto M^{-2/3}$, and moreover that the fluctuations grow as either $t^{2/3}$ (matter-dominated) or t (radiation-dominated). These remarks apply provided pressure effects are neglected (a concern on sub-horizon scales). Their validity on super-horizon scales is evident once one realizes that arbitrary linear perturbations of a de Sitter model ($k = 0$) can be decomposed into a superposition of plane wave curvature perturbations, which can in turn be regarded as a comparison of two Friedmann models with slightly different curvature.

2.2. PARAMETRIC ANALYSIS ($P = 0, K = 0$).

The $k = 0$ background with superimposed fluctuations may be examined by expanding the parametric solutions for the Friedmann model with arbitrary curvature:

$$a(t) = A(1 - cos\theta) \approx \frac{A}{2}\theta^2 \left(1 - \frac{\theta^2}{12}\right),$$

$$t = B(\theta - sin\theta) \approx B\frac{\theta^3}{6}.$$

This immediately yields the growing mode

$$\frac{\delta\rho}{\rho} = -3\frac{\delta a}{a} \propto t^{2/3},$$

in the matter–dominated regime. In general, there is also a decaying mode; one can write more generally

$$t - \tau = B(\theta - sin\theta) \approx B\frac{\theta^3}{6},$$

whence

$$\frac{\delta\rho}{\rho} \propto t^{2/3}(1 - \frac{2}{3}\frac{\tau}{t}).$$

The decaying mode corresponds physically to non-simultaneous (on some initial hypersurface at $t \to 0$) initiation of fluctuations, where as the growing mode describes curvature or gravitational potential fluctuations.

2.3. NEWTONIAN ANALYSIS ($K = 0$).

Inclusion of pressure gradients is expected to stabilize gravitational instability, and is most simply described in the Newtonian approximation, valid over small scales ($\ell \ll ct$) and velocities ($v \ll c$). The following general equations are linearized about the Friedmann equation, including expansion in lowest order: the equation for mass conservation,

$$\frac{\partial\rho}{\partial t} + \nabla \cdot (\rho v) = 0;$$

the momentum equation,

$$\frac{dv}{dt} = \nabla\phi - \nabla p/\rho;$$

and the generalized Poisson equation,

$$\nabla^2\phi = 4\pi G(\rho + 3p/c^2),$$

together with an appropriate equation of state, $p = p(\rho)$.

After linearizing these equations about the Friedmann background, one can make the following statements about the first order perturbed quantities of interest. Decomposition of the perturbed velocity field into rotational v_{rot} and irrotational parts leads to a decaying mode,

$$\nabla \cdot \mathbf{v}_{rot} = 0,$$

and a compressible mode,

$$(\nabla \cdot \mathbf{v}_{irrot})/a = -\dot{\delta},$$

where a comoving coordinate $x = r/a$ has been introduced. Rotational velocities therefore decay as $d(a\omega)/dt = 0$ or $\omega \propto a^{-1}$, where $a^{-1}\nabla_x \times \mathbf{v}_{rot} = \omega$. The irrotational component of velocity drives compressions and rarefactions of the density fluctuations, which are governed by

$$\frac{\partial^2}{\partial t^2}\delta + \frac{2\dot{a}}{a}\frac{\partial\delta}{\partial t} = 4\pi G\rho\delta + \frac{1}{a^2}\frac{dp}{d\rho}\nabla^2\delta.$$

This equation explicitly reveals the damping effect of the expansion, the destabilizing role of self-gravity, and the stabilizing effect of pressure gradients. Definition of the physical wavelength $\lambda = 2\pi a/k$, with arbitrary perturbations described by

$$\delta(\mathbf{x}, t) = \sum_k \delta_k e^{i\mathbf{k}\cdot\mathbf{x}},$$

leads to the generalized Jeans criterion for instability:

$$\lambda > \lambda_J \equiv \left(\frac{\pi v_s^2}{G\rho}\right)^{1/2}.$$

The "Jeans mass" is $M_J = \frac{\pi}{6}\lambda_J^3\rho$. If pressure is negligible ($v_s^2 = dp/d\rho = 0$), we recover the zero-pressure solutions: $\delta \propto t^{2/3}, t^{-1}$. Mass scales above M_J are gravitationally unstable.

In the radiation era, $v_s = c/\sqrt{3}$, and $M_J \sim M_{horizon} \propto t$. Note that at phase transitions during the early Universe, the sound speed can drop substantially for a brief period. This is of greatest significance during the quark-hadron transition. After matter-radiation equality at $z_{eq} = 4 \times 10^4 \Omega h^2$, the Jeans mass is approximately constant, since

$$v_s = \left(\frac{d(p_m + p_r)}{d(\rho_m + \rho_r)} \right)^{1/2} \approx \frac{c}{3^{1/2}} \left(1 + \frac{3}{4} \frac{\rho_m}{\rho_r} \right)^{-1/2} \propto (1 + z).$$

The maximum value of the Jeans mass is

$$M_J^{max} = 10^{16} (\Omega h^2)^{-2} \, M_\odot,$$

and drops abruptly after decoupling to $\sim 10^6 (\Omega h^2)^{-1} \, M_\odot$ (assuming $T = T_r$).

3. Fluctuation Modes

At late times, after the last scattering epoch, the growing mode of density fluctuations corresponds to a scalar metric perturbation. However, at much earlier times, other modes may also be present. The mode mix depends on the initial conditions at the epoch of fluctuation generation, associated with an early phase transition or with inflation. Vector perturbations are decaying modes and unimportant at late times. Tensor modes are gravitational waves , and do not involve any density compression. Hence, they are unimportant for structure formation, but can contribute to microwave background fluctuations.

There are two types of scalar mode. The adiabatic mode corresponds to metric perturbations, and conserves entropy. The second mode involves no metric (or curvature) perturbations but consists of entropy fluctuations. In the adiabatic mode, one has

$$\frac{\delta \rho_r}{\rho_r} = \frac{4}{3} \frac{\delta \rho_m}{\rho_m},$$

whereas in the isocurvature mode,

$$\delta \rho_r + \delta \rho_m = 0 \quad ; \qquad \frac{\delta s}{s} = -\frac{\delta \rho_m}{\rho_m} \left(1 + \frac{3}{4} \frac{\rho_m}{\rho_r} \right).$$

Isocurvature perturbations are initially produced as baryon number fluctuations. Once the Universe is matter-dominated, the isocurvature mode generates a pressure gradient at horizon crossing. This couples the radiation and matter to drive curvature fluctuations.

4. The Role of Dark Matter

The sound speed controls the growth of baryon fluctuations. Prior to the epoch of last scattering, the sound speed of the coupled baryon-radiation fluid is high and baryon fluctuations behave like acoustic oscillations. They are described by

$$\frac{\delta\rho}{\rho} \propto \cos\left(\frac{kv_s t}{a}\right) \quad ; \quad k = 2\pi/\lambda,$$

where a is the cosmological scale factor, λ is the comoving wavelength and v_s $\left(\approx c/\sqrt{3}\right)$ is the sound speed. Only the cosine mode is involved, since as $k \to 0$ at the last scattering epoch, this gives the required initial condition component on superhorizon scales of primordial curvature, or energy density, fluctuations.

Cold dark matter in the form of weakly interacting particles dominates the mass density. Nucleosysthesis of H, ^{2}H, and ^{7}Li constrains $\Omega_B \approx 0.02(\pm 0.01)\ h^{-2}$, whereas $\Omega_{mat} \geq 0.2$. Adiabatic energy density perturbations involve ρ as well as ρ_b, and prior to decoupling, fluctuations in the cold dark matter, unlike the baryon component, are Jeans unstable. Throughout matter domination, from matter-radiation equality at $z_{eq} \equiv 4 \times 10^4 \Omega_{mat}\ h^2$, until the epoch of last scattering, at $z_{LS} \sim 1000$, the cold dark matter fluctuations grow. Only after last scattering, when the matter and radiation decouple, does the baryon component of the fluctuations grow, by falling into the cold dark matter potential wells that are growing by gravitational instability.

The cold dark matter prevents two disasters from occurring that vitiate a purely baryonic adiabatic model. Radiative diffusion smooths out baryon fluctuations, up to a scale of order $10^{15}\,M_\odot$. In the absence of cold dark matter, this would result in a top-down model of structure formation, in contradiction with the observation that galaxy clusters are young, currently forming, objects. Moreover, such a model produces excessive microwave background temperature fluctuations. Cold dark matter allows fluctuations to develop at all scales, so that galaxies can form first, and the growth between z_{eq} and z_{LS}, as well as logarithmic growth in the radiation era, reduces the amplitude of temperature fluctuations by an order of magnitude. The minimum fluctuation scale is determined by the ability of baryons to cool and condense in the cold dark matter potential wells. This sets a limit of about $10^4\,M_\odot$. A bottom–up formation sequence for structure development results if the primordial fluctuation spectrum is approximately scale-invariant. This is the actual prediction of inflationary models, as well as models that are seeded by topological defects such as cosmic strings .

It is instructive to compare three alternate representations of the density fluctuation spectrum. Define first the Fourier decomposition of the density fluctuation

$$\delta\rho(r)/\rho = \int \delta_k\, e^{i\mathbf{k}\cdot\mathbf{r}} d^3k.$$

The power spectrum $|\delta_k|^2$ specifies the variance in $\delta\rho/\rho$:

$$\langle(\delta\rho/\rho)^2\rangle = \int |\delta_k|^2\, d^3k,$$

so that

$$\delta\rho/\rho \equiv \langle(\delta\rho/\rho)^2\rangle^{1/2} \approx \left(|\delta_k|^2\, k^3\right)^{1/2}.$$

The gravitational potential fluctuations $\delta\phi$ may be defined by

$$\nabla^2\delta\phi = 4\pi\mathrm{G}\, a^2\, \delta\rho,$$

so that

$$\delta\phi = \left(\int d^3k|\,\delta_k|^2\, k^{-4}\right)^{1/2}.$$

The power spectrum itself may be expressed as a power-law in k:

$$|\delta_k|^2 \propto k^n,$$

where $n = 1$ for a primordial scale-invariant spectrum. Note that $\delta\phi \propto k^{(n-1)/2}$, so that scale-invariance is equivalent to constant potential fluctuations; indeed, this properly defines scale-invariance. Also, since wave-number $k = 2\pi/\lambda$, and is comoving, one can write

$$\delta\rho/\rho \propto k^{\frac{n+3}{2}} \propto M^{-\frac{n+3}{6}}$$

as a mass spectrum, where M represents the mass in a sphere of comoving diameter λ.

The primordial scale-invariant fluctuation spectrum develops a feature after the epoch of matter-radiation equality that corresponds to a peak in the power spectrum at the corresponding horizon scale, $\lambda_{eq} = 13\,(\Omega h^2)^{-1}$ Mpc. Subhorizon growth is suppressed for smaller scale fluctuations in the radiation-dominated era: hence, only larger scale fluctuations retain the primordial shape. The suppression flattens the fluctuation spectrum on smaller scales by the fourth power of k in power. In terms of $\delta\rho/\rho$, this means that on large scales, a scale-invariant spectrum satisfies $n \approx 1$ and

$$\delta\rho/\rho \propto M^{-\frac{n+3}{6}} \propto M^{-2/3},$$

whereas on scales below λ_{eq}, $n \approx -3$ and $\delta\rho/\rho \propto$ constant. In fact, $\delta\rho/\rho$ rises logarithmically towards smaller scales because $(\delta\rho/\rho)^2 \propto \int k^{n+3}\, dk/k$. This suffices to result in a bottom-up sequence of structure formation. In

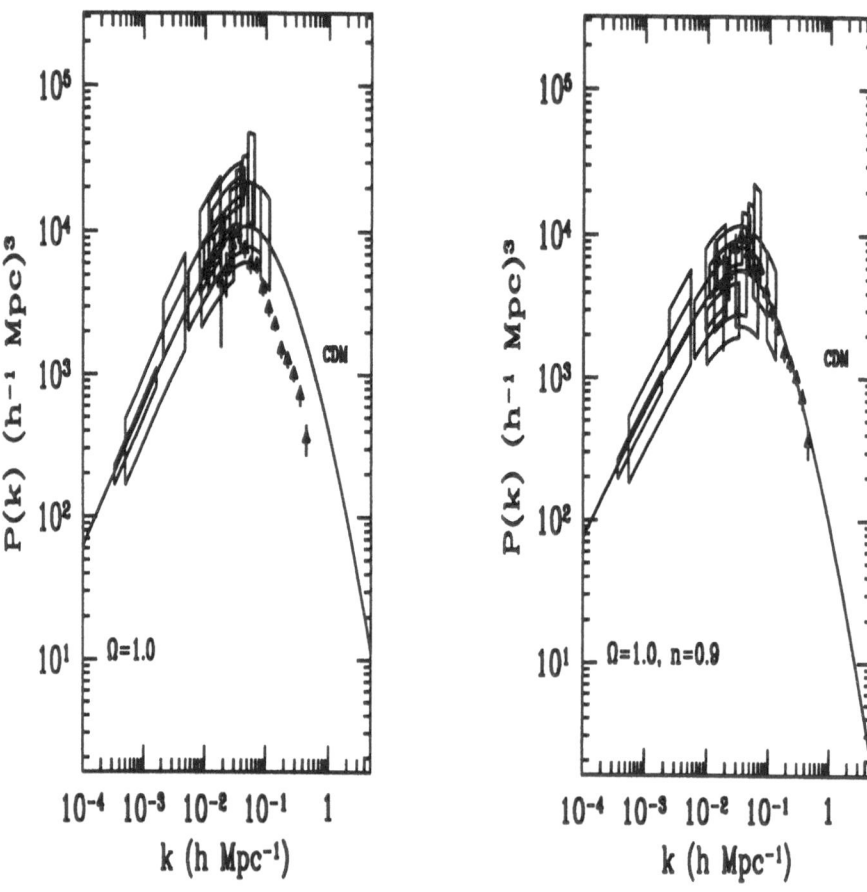

Figure 1. The matter power spectrum, $P(k)$, as reconstructed from large–scale structure and cosmic microwave background data (updated from [1]). Boxes are $\pm 1\sigma$ values of $P(k)$ inferred from CMB measurements, assuming CDM ($\Omega_0 = 1$, $h = 0.5$, $\Omega_B = 0.03$) with $n = 1$ and $n = 0.9$. The overall amplitude of the LSS data (triangles), taken from a compilation by Peacock and Dodds [2], is uncertain to $\sim 20\%$.

fact, the effective value of n only approaches -3 on dwarf galaxy mass scales, so that larger scales are well separated in the clustering hierarchy as gravitational instability operates.

Cosmic microwave background fluctuations directly probe the linear regime of $\delta\rho$, at the epoch of last scattering. Attempts to reconstruct the primordial power spectrum have met with mixed success because of uncertain systematics due, in particular, to foregrounds. However, there is unambiguous confirmation of a feature in the power spectrum that corresponds to λ_{eq} when measures of the local density fluctuations inferred from redshift surveys are compared with the density fluctuations that are reconstructed at last scattering, by inversion of the Boltzmann equation that couples matter and radiation (Figure 1). Important assumptions that enter into this comparison concern the cosmological model parameters, since one is comparing power at $z \sim 1000$ with power at $z \sim 0$, and the bias factor, since the galaxy surveys generally probe only the luminous component of matter. Nevertheless, one can see that while the standard COBE-normalized CDM power spectrum results in excessive small-scale power, a modest tinkering of parameters, such as introducing a slight spectral tilt, yields satisfactory agreement.

5. Non-Linear Evolution and the Galaxy Mass Function

The linear theory of gravitational instability in the expanding Universe provides an adequate description of fluctuation growth until self-gravity becomes significant. One can describe the non-linear evolution by numerical simulations, but a simple analytic description captures the salient features.

Consider a simple spherical top-hat model for the non-linear growth of a cold dark matter fluctuation. It is described by the Friedmann equation for a bound spherical shell taken to be embedded in, for simplicity, a $k = 0$ background. The shell is described parametrically by

$$a_s = a_{max} (1 - \cos \theta);$$

$$t/t_{max} = (\theta - \sin \theta)/\pi,$$

while the background satisfies

$$a = a_{max} (\theta - \sin \theta)^{2/3} \, 3^{2/3} \, 2^{-4/3}.$$

The density contrast within the shell relative to the background is

$$\rho_s(t)/\rho(t) = (a/a_s)^3,$$

which may be evaluated at the time of maximum shell radius t_{max} to be $9\pi^2/16$. A dissipationless shell of matter collapses by a factor of 2 in radius, when it rapidly reaches equilibrium. The collapse time from maximum radius is t_{max}. We can identify epoch $2\,t_{max}$ with the virialization epoch for the shell. At this epoch, the overdensity is $18\pi^2$. Linear theory would give an overdensity at this epoch that is given by

$$\delta(t) = -3\frac{\delta a}{a} = \frac{3}{20} \theta^2 = \frac{3}{20} \left(6\pi \frac{t}{t_{max}} \right)^{2/3}$$

evaluated at $2\,t_{max}$, or $\delta_c = \frac{3}{20}\,(12\pi)^{2/3} = 1.686$.

One now has the machinery to be able to calculate the collapsed mass fraction. Take the matter density to be described by a random Gaussian field. For cold dark matter, with power on arbitrarily small scales, one needs to smooth the density field in order to be able to compute the mean density and its variance. Smoothing is effected by introducing a filter function, the simplest form of which is a spherical top-hat filter of comoving radius R. Implementing the filter, we can compute the rms density fluctuations on scale R:

$$\sigma^2(R,\,t) = \left\langle \left(\frac{\delta\rho}{\rho}(x) \right)^2 \right\rangle \equiv D^2(t)\,\sigma_0^2(R),$$

where, for $\Omega = 1$, the linear theory growth factor $D(t) \propto t^{2/3}$ and $\sigma_0(R) = (R_{n\ell}/R)^{(n+3)/2}$ for a power-law spectrum. The fraction of mass in spheres of radius R with overdensity $\delta > \delta_c$, the linear overdensity at virialization, is

$$F(R,\,t) = \int_{\delta_c}^{\infty} \frac{d\delta}{\sqrt{2\pi}\,\sigma}\, e^{-\delta^2/2\sigma^2}.$$

The normalization of $\sigma_0(R)$ is such that if mass traces light, $R_{n\ell} = 8h^{-1}$ Mpc, the scale over which galaxy count fluctuations have unit variance. In terms of mass,

$$\sigma(M) = (M_{n\ell}/M)^{\frac{n+3}{6}},$$

where $M_{n\ell} = 5.9 \times 10^{14}(1+z)^{-6/(3+n)}\Omega h^{-1}$ $\mathrm{M_\odot}$.

To proceed further, one has to confront the following issue: counting spheres of radius aR is not necessarily equivalent to counting lumps of mass $M = 4\pi\bar\rho a^3 R^3/3$, because mass lumps can contain substructure in the form of smaller spheres that are no longer distinct entities. This uncertainty has to be addressed with numerical simulations. Assuming that one can identify the fraction of mass in virialized spheres with the fraction of mass in lumps of the equivalent scale, one can then infer the mass function of newly virialized lumps,

$$\frac{dN}{dM}\,(M,t) = -2\frac{\bar\rho}{M}\frac{\partial F}{\partial R}\frac{dR}{dM} \tag{1}$$

$$= -\sqrt{\frac{2}{\pi}}\,\frac{\bar\rho}{M^2}\,\frac{\delta_c}{M}\,\frac{d\ln\sigma(M)}{d\ln M}\, e^{-\delta_c^2/2\sigma^2(M,t)} \tag{2}$$

A factor of 2 has been added to account for the fact that there is an equal amount of matter in underdense as in overdense regions, relative to the background, and the underdense matter is presumably accreted by the mass lumps. Remarkably, despite these assumptions, the expression for the mass function of newly formed objects is found to agree with numerical simulations of structure formation in the expanding Universe.

One can also deduce the spherically-averaged properties of the galaxies and clusters, or more precisely, of their dark halos, that form by hierarchical clustering. From linear theory

$$\delta\rho/\rho \propto M^{-\frac{n+3}{6}}\, t^{2/3},$$

and "formation" occurs at $\delta\rho/\rho = 1.67$. Hence the formation time t_f scales as $t_f \propto M^{(n+3)/4}$. The redshift at which an object of present mass M has on average acquired half its mass is [3]

$$z_f = \left(2^{\frac{n+3}{3}} - 1\right)^{1/2}(M/M_{n\ell})^{-\frac{n+3}{6}}$$

Applying the condition of virial equilibrium, one then infers, since the mean density is approximately $180\bar{\rho}$, where $\bar{\rho} = 1/6\pi\, Gt_f^2$, that velocity dispersion $V^2 \propto M^{(1-n)/6}$, surface density $\epsilon \propto M^{-(n+2)/3}$ and density $\rho \propto V^{-3(n+3)/(1-n)}$. The dispersion in mean properties is large, of order unity, and the scaling relations are valid provided that the effective spectral index lies in the range $-3 < n < 1$. This is always satisfied for primordial spectra that are already scale invariant. For example, the effective index on galaxy scales for a scale-invariant initial spectrum is approximately -2.

6. Comparison with Observations

To the extent that luminosity tracks mass, luminous galaxies should be associated with massive dark halos, and dwarfs with smaller halos. The luminosity function of galaxies is well described by the Schechter function

$$\frac{dN}{dL} = \frac{\phi_*}{L_*}\left(\frac{L_*}{L}\right)^{\alpha} e^{-L/L_*},$$

where $\phi_* = 0.01\, h^3$ Mpc^{-3}, $L_* = 10^{10}\, h^{-2}\, L_\odot$, and $\alpha \approx 1.1$.

Comparison with the Press-Schechter mass function immediately raises two questions. The general form is similar. However the current epoch scale at which luminous mass structures are becoming non-linear, as inferred from the rms fluctuations in galaxy counts, is $\sim 5\, h^{-1}$ Mpc, equivalent to a present epoch mass scale $M_{n\ell} \sim 4 \times 10^{13}\, \Omega h^{-1}\, M_\odot$. Moreover, the slope of the predicted mass function at the low mass end is steeper by about one power in mass than the equivalent slope of the galaxy luminosity function for luminosities below L_*. Reconciliation of characteristic mass with luminosity and mass function slope with luminosity function slope requires additional physics that incorporates the effects of dissipative matter (baryons) and of star formation. Another distinctive feature of galaxies is rotation: disks are rotationally supported, but spheroids are not.

7. The Characteristic Luminosity L_*

Since L_* refers to the stellar mass, one can immediately ask what baryonic mass associated with $M_{n\ell}$ can have cooled to have formed stars. This presumably is a necessary condition to form the luminous mass of a galaxy. Moreover the baryonic matter dissipation must have occurred at an epoch corresponding to the redshift of galaxy formation. Observations of damped Lyman alpha absorption line systems towards quasars , considered to be disk precursors, and of high redshift galaxies suggest that the bulk of galaxy formation occurred at $z \sim 2 - 3$. Consider a $\nu\sigma$ fluctuation, where σ is the rms density fluctuation so that a galaxy precursor satisfies $\delta\rho/\rho = \nu\sigma_0(1+z)^{-1}$ if $\Omega = 1$. Since $M_{n\ell} \propto [\nu/(1+z)]^{6/(n+3)}$, I infer that typical 2σ galaxy scale fluctuations undergoing collapse at $z \approx 2.5$, say, have mass $M = (10^{12} - 10^{13})\,\Omega h^{-1} M_\odot$, since $-n_{eff} = 1.5 - 2$. It is encouraging that this mass scale, that of dark halos, lies in the expected range. The associated baryonic mass is

$$M_b = \left(2 \times 10^{10} - 2 \times 10^{11}\right) \left(\Omega_B h^2/0.02\right) h^{-3} f_b\, M_\odot,$$

where f_b allows for a possible baryon enhancement on galactic scales over the primordial value that I have scaled as $\Omega_B h^2 = 0.02$: this has an uncertainty of about a factor of 2.

The baryonic mass represents an upper limit on the luminous mass since there is no guarantee that all of the baryons have condensed into galactic stars. One may expect cooling to be a necessary precursor, and in particular, one could argue that cooling with a dynamical time-scale is required in order for star formation to occur efficiently within a pregalactic structure. However, the mass of cooled gas that condenses within a dark halo is found to increase without limit as the potential well depth increases. Cooling does not therefore account for L_*, there being an effective upper limit to the potential well depth of luminous galaxies that corresponds to a central velocity dispersion for an L_* elliptical of about 270 km s^{-1}.

To limit the mass in cooled gas to $M_* \sim 10^{11} h^{-1} M_\odot$ as expected for the stellar mass in L_* galaxies, one has to appeal to feedback from star formation and death. Supernovae provide an attractive means of feedback from luminous protogalaxies, because of the inference from intracluster gas abundance studies that significant ejection of iron and other heavy elements occurred early in the history of the early-type galaxies that dominate rich clusters . For a nominal supernova rate of one per $250\,M_{250}$ solar masses per year that forms stars, corresponding to a solar neighborhood initial mass function for which $M_{250} \approx 1$, one finds that the protogalactic gas can radiate away the injected supernova remnant kinetic energy provided that the protogalactic potential well satisfies

$$\sigma \; < \; 270 \left(\epsilon_{0.2} E_{51} M_{250}^{-1} \right)^{1/2} \, \mathrm{km\,s^{-1}},$$

where $E_{SN} \equiv 10^{51} E_{51}$ ergs is the initial injected energy of a supernova and $\epsilon \equiv 0.2\,\epsilon_{0.2}$ denotes the fraction of gas turned into stars per protogalactic dynamical time. This demonstrates that despite the uncertain efficiency of star formation and uncertainty in the early IMF, supernova feedback more than suffices to constrain M_* within the observed bound, given that the protogalactic scale length (or central surface density) complies with the empirical (M, σ) scaling that is found in the Faber-Jackson or Tully-Fisher relations.

8. Surface Brightness

Central surface density is presumably determined by rotational support for disks and by dynamical relaxation for spheroids. Central surface brightness peaks for L_* galaxies and declines both towards high and low luminosities. We lack an understanding of the central surface brightness of galaxies. The problem is primarily that of understanding disks, since spheroid formation can be satisfactorily simulated by mergers of disks.

Analytic collapse calculations appear to explain the scale of disks, via accounting for the origin of disk angular momentum. Tidal torques between neighboring protogalaxies generate an initial amount of angular momentum, expressible in terms of a dimensionless parameter $\lambda \equiv |E|^{1/2} J / G M^{1/2}$, where E is the potential energy and J is the angular momentum of a halo of mass M, that spans the range $0.01 \le \lambda \le 1$ but has a median initial value

$$\lambda \approx 0.05 \, \Omega^{0.1},$$

at turn-around.

In rotationally supported disks with rotational velocity V_{rot} and halo velocity dispersion σ, one can express λ in the form

$$\lambda \approx 0.4 \, V_{rot} / \sigma \approx 0.4 \,.$$

Self-gravitating non-dissipative collapse fails to bridge the gap between initial and current values of λ in disks, since in this case $\lambda \propto R^{-1/2}$. Disks typically have $R_{disk} \approx J / M V_{rot} \approx 5$ kpc, requiring collapse from 500 kpc, an absurdly large critical disk radius. Dissipative baryonic collapse within a dark non-dissipative halo effectively transfers angular momentum via tidal torquing against the halo dark matter. Since specific angular momentum is conserved ($V_{rot} R \approx$ constant) within an isothermal halo ($\sigma \approx$ constant), one now obtains $\lambda \propto R^{-1}$, which implies an initial radius of 50 – 100 kpc at maximum extent of the protogalaxy.

An initial gas extent of order 50 kpc is consistent with the interpretation of damped Lyman alpha absorption clouds, seen in absorption towards high redshift quasars , as being protodisks. Gas collapse within a dark halo can also explain the shapes of galaxy rotation curves, observed to be approximately flat outside a disk scale-length, provided that disk self-gravity plays a role in helping account for the flattening of the inner rotation curve. The dominant dark matter distribution produces a nearly flat rotation curve at large radii. Such a conspiracy between baryonic and dark matter components is a natural outcome of simulations of disk galaxy formation.

However the simulations have revealed a serious problem. In hierarchical clustering, the initial halos are clumpy. The substructure results in efficient dynamical friction of infalling lumps, and the resulting baryonic disk is found to be far too small. The analytic prediction $\sigma_{disk} \approx \lambda \, \sigma_{halo}$ for uniform spherical collapse overpredicts the disk size by approximately a factor of 5. For spheroids, such efficient collapse is exactly what is required to account for the observed centrally concentrated light profiles, provided that the final structure is not rotationally supported. This is more or less a natural outcome of disk mergers . Feedback from star formation may heat the gas sufficiently during disk formation to avert this catastrophe, but there is as yet no detailed modelling of collapse with energy feedback.

9. Successes and Failures of the Hierarchical Collapse Model

Bottom-up structure formation has been extensively simulated, usually in the context of a cold dark matter-dominated universe at critical density. The theory is well-formulated for dark matter, and modelled via N-body simulations, and has been extended to include the baryonic component, with inclusion of smoothed particle hydrodynamics. Gas cooling has been included on galaxy formation scales, but the theory lacks any fundamental prescription for star formation, and is consequently even more seriously deficient in the ability to include effects of feedback from star formation.

The successes of hierarchical structure formation are numerous. One can account for galaxy clustering: simulations of large-scale structure are indistinguishable from actual surveys. On the largest scales where effective comparison is made, $10 \, h^{-1}$ – $100 \, h^{-1}$ Mpc, and structure is in the linear regime, one can measure the shape of the power spectrum of luminous matter. This is expressible as $P(k) \propto k^{-1.4}$, and corresponds to a CDM model with $\Omega h \approx 0.2 - 0.3$, as naturally occurs either in a flat, vacuum-dominated or open cosmological model if the primordial index is scale-invariant ($\lambda = 1$) at least over these scales. Bulk flows can, in principle, measure the dark matter power spectrum over similar scales, although the results do not discriminate between rival cosmological models. The amplitude of the bulk flows is determined by the parameter $\sigma_8 \Omega^{0.6}$, where σ_8 is the value of $\sigma_0(M)$ at $8h^{-1}$ Mpc, the scale at which the galaxy number counts in spheres have unit variance. The observed value of this parameter [4] (~ 0.8, but with considerable uncertainty) provides the strongest evidence that supports a high value of Ω.

The abundance of galaxy clusters effectively probes the power spectrum shape at a scale of $\sim 10h^{-1}$ Mpc, the comoving scale from which rich clusters condensed. A best fit value $\sigma_8 \approx 0.6$ (± 0.1) gives a reasonably robust measure of the degree of biasing, or the ratio of dark to luminous matter, on the largest non-linear scales that have been usefully probed to date. The present epoch abundance of luminous galaxies is a further natural outcome of a model with primordial index $n \approx 1$, although at high redshift ($z = 3 - 4$), recent observations are beginning to discriminate between cosmological models. A significant abundance of luminous star-forming galaxies has been found in this redshift range, as have damped Lyman alpha clouds that appear to have rotational velocities of 200 km s^{-1} or more. Models with diminished power on subgalactic scales, due to a component of hot or warm dark matter, are most strongly constrained by such observations. In general, however, galaxy formation at redshifts $z \approx 1 - 5$ is a natural outcome of hierarchical models.

Mergers are also a natural outcome of hierarchical models. Major mergers , between nearly equal mass systems are rare today, but more common at high redshift. Major mergers lead to formation of spherical systems that are not rotationally supported, assuming high star formation effeciency as is needed to account for the old stellar populations of spheroids. Minor mergers incorporate substructures in which the gas fraction is easily disrupted by stellar feedback, and so are expected to form stars inefficiently. Such systems are the logical precursors of disk galaxies, for which the predicted rotation curves appropriate to self-gravitating massive gas disks embedded within dark halos, constitute another success of the bottom-up theory.

Evolutionary and morphological studies of high redshift galaxies fit well into a hierarchical formation scheme. Deep redshift surveys reveal that blue, star-forming galaxies have evolved by about 0.5 magnitude to $z \sim 1$. HST observations reveal many of these to be disks. However, red galaxies, identified as E's and S0's, show little evidence of any evolution in luminosity. Field studies show a significantly increasing population of blue, irregular galaxies towards fainter magnitudes, these galaxies dominating the very faint galaxy counts. All of this is consistent with early formation via mergers and strong tidal interactions, and may be considered to be a qualitative success of the hierarchical galaxy formation theory.

The quantitative failures of the hierarchical formation theory are a consequence of its failure to provide a unique recipe for star formation and associated feedback. The excess of faint galaxies at low luminosities, the upper limit on galaxy luminosity characterized by L_*, and the upper limit on disk galaxy surface brightness specified by the Freeman law for spiral galaxies, are all issues whose explanation seemingly demands incorporation of feedback effects.

Bottom-up formation predicts that massive galaxies form more recently. This produces the following dilemma: luminous ellipticals have deeper potential wells, and hence more massive halos, on the average, than do luminous spirals . Yet ellipticals have predominantly old stellar populations. Resolution of this difficulty has been achieved in a somewhat *ad hoc* manner, by assuming that the major mergers , elliptical precursors, form stars efficiently over a dynamical time-scale, leaving little gas behind for late star formation, whereas the minor mergers and slow infall accumulation of gas into disks, in low density environments, result in a continued gas-rich star formation at low efficiency for a Hubble time. Such a scheme may be said to "work," in the sense that ellipticals are red and spirals are blue, but clearly leaves something to be desired in the sense of extracting more from a model than one inputs into the model.

10. How to Really Form Galaxies, or Minding Your P's and Q's

Numerical simulations of structure formation adopt what may be called the "forwards" approach to galaxy formation. One assumes initial conditions that have a plausible cosmological ancestry, adopts an *ad hoc* prescription for star formation, and runs a numerical simulation. Now, star formation in our local patch of the Universe, let alone at remote locations and epochs, depends on many parameters, including the gas density, molecular abundances, dust opacity, ionization, magnetic field strength, turbulence, protostellar outflows, and feedback from dying stars. One can more readily aspire to make long-term predictions of the weather from first principles than develop a predictive theory of star formation. As with meteorology, only a highly phenomenological approach that incorporates as much local data as is available is likely to have even a remote chance of success. For the study of galaxy formation, incorporation of star formation knowledge acquired locally results in a "backwards" approach.

One may readily outline the ingredients of the backwards approach to galaxy formation. Commence with a semi-phenomenological theory for the instability of cold, self-gravitating disks, and apply that to the Milky Way galaxy. The phenomenological ingredients with which one begins are that the star formation rate has been approximately constant over the past 10

Gyr, to within a factor of 2, and that the star formation rate surface density (per unit disk area) is proportional to the total gas surface density above a threshold value. The threshold is determined by either of the following arguments. Within the disk corotation radius, clouds passing through the spiral density wave acquire a non-circular component of velocity proportional to $\Omega(R) - \Omega_p$, where Ω_p is the density wave pattern angular velocity and $\Omega(R)$ is the disk rotation rate at radius R.

Alternatively, one can argue that the disk is unstable to non-local gravitational instabilities as well as being locally Jeans unstable if the Toomre parameter $Q \leq 1$, where Q is defined to be given by

$$Q \equiv f \kappa \sigma_g / \pi G \mu_{gas} \equiv \mu_{cr} / \mu_{gas}$$

for a gas disk, and σ_g is the gas velocity dispersion, μ_{gas} is the gas surface density (κ is the epicycle frequency ($\approx \sqrt{2}\Omega(R)$) for a flat rotation curve), and f is a correction factor of order unity, that allows for the contribution of the stellar component to the self-gravity of the gas. One can show that the linear instability growth rate is approximately equal to $\kappa(1 - Q^2)^{1/2}/Q$ if $Q < 1$, and equal to zero if $Q > 1$.

One can now write the star formation in the following physically motivated forms [5, 6]:

$$\text{SFR} = \epsilon \Sigma_{gas} (\Omega(R) - \Omega_p)$$

or

$$\text{SFR} = \epsilon \Sigma_{gas} \kappa (1 - Q^2)^{1/2}/Q .$$

In either case, one has, for a flat rotation curve, in the star-forming region of the disk that

$$\text{SFR} \propto \Sigma_{gas}/R .$$

Hence the models predict that disks form inside out. This is a generic feature of disk models [7]. Figure 2 shows the predicted surface brightness profiles for spiral galaxies as viewed at $z = 0$ and $z = 3$, both as a function of physical radius and as viewed by the HST.

One can adjust the model parameters, in essence ϵ and the initial gas surface density, with the rotation curve being directly determined by observation and assumed not to vary over disk history, to fit the following characteristics of the Milky Way: the radial distributions of star formation rate, stellar surface density and gas surface density, the metallicity gradients of gas and of stars, and the metallicity distribution of stars near the Sun. One finds that the star formation decreases exponentially with an e-folding time of $\epsilon \Omega^{-1} \sim 3\,\text{Gyr}$, so that $\epsilon \sim 0.01$ at the solar neighborhood.

An additional ingredient is infall of metal-poor gas. This may be required in order to account for the lack of metal-poor disk stars, since most

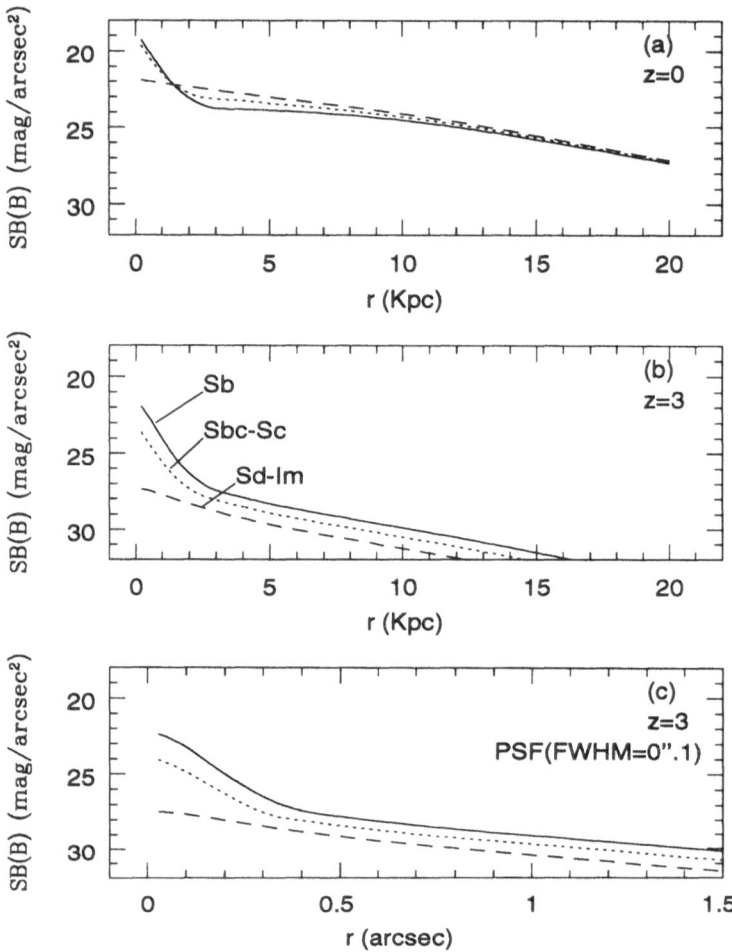

Figure 2. Disks form inside-out: apparent surface brightness profiles as a function of physical radius for Sb (solid line), Sbc-Sc (dotted line) and Sd-Im (dashed line) model galaxies with infall at (a) $z = 0$ and (b) $z = 3$, for $\Omega = 1$, $H_0 = 50$ km s^{-1}Mpc^{-1}, and a formation redshift $z_f = 10$. (c) same as (b) but as a function of angular radius and after convolving the model profile with a PSF of $FWHM = 0.1$ arcs ec. From [8].

of the disk is then formed after the early enrichment has occurred. However pre-enrichment of the disk to a level of 0.1 of the solar metallicity with infall from the halo may provide an alternative prescription. In general, infall from the halo, both before and during disk formation, is a natural ingredient of hierarchical galaxy formation.

It is interesting to note that disks are only marginally unstable: Q is approximately, but slightly less than unity, in the inner disk. In the solar neighborhood, for example, $\mu_{cr} \approx 7\ \mathrm{M_\odot\ pc^{-2}}$ whereas $\mu_{gas} \approx 15\ \mathrm{M_\odot\ pc^{-2}}$. An explanation of the inefficiency of star formation, or why ϵ is of order a percent, so that only ~ 1 percent of the disk gas forms stars per dynamical time, is as follows. If supernovae remnants are responsible for cloud acceleration, the cloud velocity dispersion σ, is generated by the specific momentum available from remnants, namely $V_{SN} \equiv E_{SN}/V_c M_{SN}$, where E_{SN} ($\approx 10^{51}$ ergs) is the initial kinetic energy per supernova, V_c is the velocity at which an initially adiabatic expanding remnant enters the approximately momentum-conserving regime, and M_{SN} is the mass of gas undergoing star formation in order to produce a single Type II supernova (or star of mass $> 8\ \mathrm{M_\odot}$). It follows that

$$\epsilon \approx \frac{\sigma_g}{V_{SN}} = 0.02 \left(\frac{\sigma_g}{10\ \mathrm{km\ s^{-1}}} \right) \left(\frac{10^{51}\mathrm{ergs}}{E_{SN}} \right) \left(\frac{M_{SN}}{250\mathrm{M_\odot}} \right) \left(\frac{V_c}{400\ \mathrm{km\ s^{-1}}} \right) .$$

Why are disks marginally unstable throughout the star forming region? Evidently there is a conspiracy that keeps Q near unity. The answer must lie in self-regulation [9]. As Q decreases, the star formation rate increases, feedback increases and enhances σ_g, and consequently Q now increases. The interstellar medium plays a crucial role in coupling disk instability and the rate of star formation. If the feedback indeed is controlled by supernovae remnants, long believed to be the primary source of interstellar cloud turbulence and heating, then a relevant parameter is the porosity P of the interstellar medium to the hot, $\sim 10^6$ K gas associated with the interiors of supernova remnants. The porosity is defined to be the product of supernova rate and supernova 4-volume (age multiplied by volume at maximum expansion). The fraction of volume of the interstellar medium filled by the hot phase (gas at $\sim 10^6$ K) is $1 - e^{-P}$. Effective feedback requires $P \sim 1$, and $P \gg 1$ results in a supernova-driven galactic wind. In fact, $P \sim 0.1 - 1$ is observed for the local interstellar medium. Even if $P \sim 1$, one expects occasional hot bubbles to break out of the cold interstellar gas in the disk, since V_{SN} exceeds the disk escape velocity. This phenomenon manifests itself in the form of the chimneys that are observed in the local interstellar medium.

The self-regulation is expected to operate as follows. As Q decreases, stars form and die at an enhanced rate, leading to more supernovae, and P increases. At fixed P, the cold gas fraction is inversely proportional to Q ($\equiv \mu_{cr}/\mu_{gas}$) but at fixed Q, the cold gas fraction decreases as P increases. Hence one can show that the cold gas fraction remains constant while the cold, unstable gas layer is compressed, occupying a volume proportional to e^{-P}, as P increases and Q decreases, as long as $P \lesssim 1$. The star formation rate, which depends primarily on the surface density of the cold gas, should be unchanged. Perhaps the competition between P and Q accounts for the universality and low dispersion in the Tully-Fisher relation.

11. And Now for Elliptical Galaxies

A semiphenomenological theory fares well for forming disk galaxies. One has the basic ingredients of the gravitational instability of disks to nonaxisymmetric perturbations well under control, and it is possible to embed this framework into a scheme that incorporates the rich data available in the Milky Way and other nearby disk galaxies.

Elliptical and, more generally, spheroid formation presents rather more of a challenge. There is no theory of star formation in dynamically hot systems. A purely phenomenological aproach is essential in the absence of any robust rules. Formation of spheroids is approached as follows, using the philosophy of "backwards" evolution. One can apply population synthesis techniques, adopting a universal form for the initial stellar mass function, to model the spectral energy distribution of a spheroid. One finds that most of the stars formed within the first $1 - 2$ Gyr of the birth of a stellar population that is now at least as old as the oldest globular star clusters . Such a starburst is a purely empirical model for the current epoch spectrum of a typical elliptical galaxy. There is evidence for some intermediate age stars in nearby ellipticals, and this can be modelled in terms of a starburst as recently as $4 - 5$ Gyr ago.

Starbursts are naturally explained in a hierarchical model of galaxy formation. In a galaxy merger, any gas is rapidly concentrated via inelastic cloud encounters into the center of the resulting potential well. The massive concentration of gas should provide a fertile environment for a starburst. In a sufficiently deep potential well, with escape velocity $\gtrsim 100$ km s^{-1}, the debris from supernova explosions should be trapped within the cloud, so that star formation should be capable of efficiently consuming the available gas. The contrast between protodisk and protoelliptical star formation is likely to be [10] that the star-forming units in a disk galaxy have masses comparable to giant molecular clouds ($\sim 10^5 - 10^6$ M$_\odot$) with escape velocities of a few km s^{-1}, whereas stars form efficiently in the massive substructures that characterize protoellipticals. In the smaller, shallow potential well, substructures, star formation is inevitably a highly inefficient process.

The dynamics of major mergers are consistent with the hypothesis of spheroidal formation. Relaxation is rapid, with ongoing mergers already having developed, when azimuthally averaged, a de Vaucouleurs light profile. The high central surface density of a spheroid is attained via baryonic dissipation and settling, and the associated star formation and enrichment is capable of generating the observed gradients in metallicity. In the nearby Universe, major starbursts , which are extremely rare at the present epoch, are shrouded by dust, and most of the star formation luminosity is emitted at far infrared wavelengths. This is consistent with the merger hypothesis, which predicts a strongly decreasing merger rate with cosmic epoch.

12. Galaxies at High Redshift

Armed with a theoretical description of nearby galaxies, one can attempt to project the models back in time. For disks, the model predictions are dramatic. Only modest evolution, by a magnitude or two, is found to red-shift unity, as is seen in the deep redshift surveys. However disk angular sizes are greatly reduced, because of the inside-out evolution. One implication is that galaxy angular size cannot be used as a cosmological probe of geometry because of the dominant role of evolution.

Protodisks constitute a more speculative area for model predictions. Damped Lyman alpha clouds, seen in absorption towards quasars especially at high redshift, have been long conjectured to be protodisks on the basis of column density, HI dominance, and spatial extent. The number density evolves strongly with redshift, and translates directly into the HI gas fraction (Ω_{HI}) in such systems, which is found to peak at a redshift of about 3. The observed ratio of Ω_{HI} at this redshift corresponds to the total mass in stellar disks at present, and has been inferred to therefore be a representative measure of protodisks in the early Universe. The substantial decrease of Ω_{HI} between $z = 3$ and the present epoch can only be reconciled with the chemical evolution of disk galaxies, inferred to be relatively modest over the past 10 Gyr of disk evolution, if [11] considerable amounts of dust are present that result in undercounting of quasars and therefore underestimates of Ω_{HI} at redshifts between 1 and 3.

The kinematics of the damped Lyman alpha clouds, as probed by high resolution optical spectrometry, substantiate the hypothesis that these clouds are massive protospiral galaxies, rotational velocities of 100 – 300 km s^{-1} being measured [12]. However, their identification as protodisks remains elusive, for several reasons. The kinematic rotation signature of line distribution asymmetry can only be produced by rotating systems with a substantial scale-length, of at least several kpc. Moreover, the abundances are low, although with a large dispersion especially at low redshift [13]. At high redshift, the abundances are generally around a percent of the solar value, about an order of magnitude below the oldest disk star abundances. The abundance ratios are found to reveal the pattern of nucleosynthetic yields associated with Type II supernovae. The abundances of damped Lyman alpha clouds resemble those of the galactic halo rather than the old disk stars. A halo origin may be more appropriate for these clouds, although it is also possible that in the outer disk, where stellar abundance ratio determinations are unavailable, the scale height is large, and the inner disk metallicity gradient could continue, may serve as an evolutionary endpoint for the damped Lyman alpha systems.

The lack of luminosity and spectral evolution for ellipticals found in deep galaxy redshift surveys favors an origin for ellipticals at high redshift. Theory predicts that the elliptical formation phase should be luminous. One has to assemble $\sim 10^{11}$ $M_\odot$ of stars on a dynamical time-scale (~ 1 Gyr) in order to reproduce the morphological and kinematic characteristics of an elliptical, and this assembly must involve a substantial amount of star formation as well as of dynamical merging. Population synthesis confirms the time-scale for the star formation, but cannot distinguish between a series of small bursts and a single luminous starburst. The well-known difficulty has been that optical and near infrared surveys for luminous protogalaxies at high redshift have been unsuccessful.

There are two possibilities for elliptical formation. Assembly by many mergers, with an associated sequence of many ministarbursts, or a coherent collapse in which the luminous starburst is shrouded by dust and hence only visible in the far infrared. The former hypothesis is consistent with some interpretations of the faint blue galaxy counts, and the latter with detections of ultra-luminous starbursts detected by the IRAS survey that may represent relatively low redshift examples of a large population of such systems. Either hypotheses is consistent with the discovery of vast amounts of intracluster iron and other heavy elements. The intracluster iron mass has an abundance of about 1/3 of the solar value, and requires a nucleosynthetic yield that is about 5 times that in the solar neighborhood. The abundance ratios are consistent with those of Type II supernova ejecta. Explanations of the enrichment of the intracluster gas appeal to galactic winds from a population of dwarfs, now mostly disrupted, or to early protogalactic winds from luminous E and S0 galaxies. Some support for the dwarf hypothesis may come from evidence for a large population of low surface brightness dwarfs in clusters. However, these dwarfs are metal-poor, and a more logical origin for the enrichment is from metal-rich galaxies. Several observations lend credence to this interpretation: the intracluster iron abundance as measured in groups and clusters is proportional to the luminosity in early-types, as opposed to all galaxies, one sees a similar abundance pattern for Mg/Fe ratios in the central regions of ellipticals, and the stellar Mg abundance is found to be proportional to the local escape velocity, suggestive of regulation by an early wind [14].

A resolution of the puzzle of elliptical formation may come with sub-millimeter measurements. There already are indications via the tentative discovery of a diffuse background at submillimeter wavelengths [15] that the integrated emission from dust-shrouded starbursts at high redshifts may have been detected, and protoellipticals provide an attractive interpretation of the diffuse flux. Confirmation will come with mapping of blank fields at high latitudes by ISO (out to 200 μ) and by submillimeter arrays (at 400, 800 μ). If elliptical starbursts are responsible for the diffuse background, deep imaging should reveal their presence. Alternatively, if ellipticals are assembled by mergers of many smaller star-forming units, studies of deep fields in the optical and near infrared with HST should reveal the conclusive signature that is inevitable in a merging model.

References

1. Scott, D., Silk, J. and White, M., *Science*, **268**, 829 (1995)
2. Peacock, J. A. & Dodds, D. D., *Mon. Not. R. astr. Soc.* **267**, 1020 (1994)
3. Lacey, C. and Cole, S., *MNRAS* **262**, 627 (1993)
4. Kolatt, T., & Dekel, A. , *Astrophys. J.*, 479, 592 (1997)
5. Wyse, R. F. G. and Silk, J., *Astrophys. J.* **339**, 700 (1989)
6. Wang, B. and Silk, J., *Astrophys. J.* **427**, 759 (1993)
7. Prantzos, N. and Aubert, O., *Astr. Astrophys* **302**, 69 (1995)
8. Cayon, L., Silk, J. and Charlot, S. *Astrophys. J.*, **467**, L53 (1996)
9. Silk, J., *Astrophys. J.*, 481, 703 (1997)
10. Silk, J. and Wyse, R. F. G., in preparation (1996).
11. Fall, S. M. and Pei, Y. C., *Astrophys. J.* **402**, 479 (1993)
12. Wolfe, A. M., in *QSO Absorption Lines*, ed. G. Meylan (Heidelberg: Springer) (1996)
13. Lu, L., Sargent, W. L. W., Barlow, T. A., Churchill, C. W. and Vogt, S., *Astrophys. J. Suppl.*, 107, 475 (1996)
14. Zepf, S. and Silk, J. *Astrophys. J.* **466** 114–121 (1996)
15. Puget, J.-L. et al., *Astr. Astrophys*, **308**, L5 (1996)

CALCULATION OF COSMIC BACKGROUND RADIATION ANISOTROPIES AND IMPLICATIONS

EMORY F. BUNN
Astronomy Department
University of California, Berkeley

Abstract. We review the physical processes that are thought to produce anisotropy in the cosmic microwave background, focusing primarily (but not exclusively) on the effects of acoustic waves in the early Universe. We attempt throughout to supply an intuitive, physical picture of the key ideas and to elucidate the ways in which the predicted anisotropy depends on cosmological parameters such as Ω_0 and h. The second half of these lectures is devoted to a discussion of microwave background data analysis techniques, with an emphasis on the analysis of the COBE DMR data. In particular, the Karhunen-Loève method of data compression is described in detail.

1. Introduction

Since the discovery four years ago of cosmic microwave background (CMB) fluctuations (Smoot *et al.* 1992), the data from anisotropy experiments have improved in both quality and quantity at a very rapid pace. CMB data already provide stringent constraints on cosmological models, and with a plethora of balloon-borne and ground-based experiments underway and two planned satellite missions, we can expect further dramatic improvement over the next decade. In fact, there is a very real possibility that we will accurately measure many of the most important cosmological parameters via the CMB anisotropy spectrum (Jungman *et al.* 1996, Kosowsky *et al.* 1996).

In order to realize this promise, we must take great care in developing tools for comparing observational data with theoretical predictions. Even with existing data, this process is far from trivial, and with the much larger data sets of the near future the task will become trickier. There are at least two independent problems to be faced: we must be able to make accurate

C. H. Lineweaver et al. (eds.), The Cosmic Microwave Background, 135–183.

predictions of the anisotropy spectrum for any particular theory, and we must develop adequate statistical techniques to facilitate the comparison of these predictions with observations.[1]

These lectures are concerned with these two subjects. We will first review the primary physical mechanisms that are thought to be responsible for generating CMB anisotropies. The emphasis in this half of the lectures will be on building an intuitive picture of the relevant physical effects. We will therefore give ourselves free rein to make physically motivated approximations, rather than trying to treat the rather involved subject of anisotropy formation with complete precision. This section of the lectures will draw heavily on the work of Wayne Hu and Naoshi Sugiyama (Hu & Sugiyama 1994, 1995a, 1995b, 1996; Hu 1995), as well as on a review article by Hu, Sugiyama, & Silk (1996) and two previous summer-school proceedings on the subject (Hu 1996, Tegmark 1996c).

The second half of these lectures is devoted to issues of statistics and data analysis. We will study various ways in which theoretical predictions of CMB anisotropy may be compared with data sets. Our primary focus will be on methods for analyzing the COBE DMR data, since this is the largest and most powerful CMB data set in existence; however, many of the issues that arise in analyzing the COBE data are directly relevant to analyses of other experiments, both present and future. For example, we will pay special attention to the issue of *data compression*; this subject was fairly important in analyzing the COBE data, and its importance will only increase as CMB data sets get larger and larger. In particular, the planned MAP and PLANCK missions will both return data sets several orders of magnitude larger than COBE, and their analysis will therefore require extensive data compression.

These lectures are organized as follows. Section 2 provides an overview of the key physical processes that produce CMB anisotropy. Section 3 discusses the primary anisotropy, including the Sachs-Wolfe effect (Sachs & Wolfe 1967) and anisotropies produced by acoustic oscillations of the photon-baryon fluid (Peebles & Yu 1970; Doroshkevich, Zel'dovich, & Sunyaev 1978; Bond & Efstathiou 1984), as well as the diffusive damping of fluctuations (Silk 1968). In Section 4 we discuss anisotropies produced after last scattering, such as the integrated Sachs-Wolfe effect (Sachs & Wolfe 1967, Rees & Sciama 1968), the effect of gravitational lensing (Blandford & Narayan 1992, Seljak 1996b), and reionization (Sunyaev 1977, Silk 1982). Section 5 attempts to synthesize the main ideas of the previous sections and concludes the first half of these lectures.

---

[1] Not to mention the far more difficult task of actually gathering the data!

The second half, which concerns issues of statistics and data analysis, begins with Section 6, in which we establish some basic results and notation having to do with Gaussian random processes on the sphere. Section 7 presents a series of idealized thought experiments designed to introduce some of the key issues of CMB data analysis. This section also contains a digression on Bayesian and frequentist statistical techniques. In Section 8, we apply what we have learned to an analysis of the four-year COBE DMR data, and Section 9 contains some brief concluding remarks.

2. An Overview of Anisotropy Formation

CMB anisotropies encode large amounts of information about the Universe. Physical processes around the redshift of last scattering (typically $z \simeq$ 1100) produce the *primary anisotropy*, which can be significantly altered by *secondary* processes between the last-scattering surface and the present. In addition, the angular scale subtended by a particular source of anisotropy depends on the spatial geometry as well as the distance to the last-scattering surface.

With the exception of some effects at very low redshift, and ignoring topological defect models, calculations of CMB anisotropy are done in linear perturbation theory. All of the relevant quantities are small perturbations about a homogeneous Friedmann-Robertson-Walker solution. Nonetheless, making accurate numerical predictions of the CMB anisotropy in a particular theory is a daunting numerical task. In a typical cold dark matter (CDM) model, the variables one must keep track of include

- $\delta_B \equiv \delta\rho_B/\rho_B$, the baryon density perturbation;
- $\delta_{CDM} \equiv \delta\rho_{CDM}/\rho_{CDM}$, the perturbation in the CDM density;
- $\mathbf{v}_B$, the baryon peculiar velocity field;
- $\mathbf{v}_{CDM}$, the CDM peculiar velocity field;
- Ψ, essentially the Newtonian gravitational potential;
- Φ, the perturbation to the spatial curvature;[2]
- f_γ, the photon phase-space distribution function;
- f_ν, the neutrino phase-space distribution function.

All of these quantities depend on position $\mathbf{x}$ and time t, and f_γ and f_ν are also momentum-dependent. Their evolution is governed by a nasty set of coupled partial differential equations. For the nonrelativistic species, we must keep track of the usual equations of perturbation theory, namely the

---

[2]We will work throughout in Newtonian gauge. For our purposes Ψ and Φ are the only important perturbations to the metric. Ψ is related to the perturbation to the time-time component g_{00} of the metric, and Φ has to do with the perturbation to the spatial part g_{ij}. For more information on gauges, see the contribution of J.-L. Sanz to this volume, and also Hu (1995, 1996) and references therein.

continuity equation, the Euler equation, and the Poisson equation. For the CDM, these equations look like

$$\dot{\delta}_{\text{CDM}} + \nabla \cdot \mathbf{v}_{\text{CDM}} = 0, \tag{1}$$

$$\dot{\mathbf{v}}_{\text{CDM}} + 2\frac{\dot{a}}{a}\mathbf{v}_{\text{CDM}} = -\frac{1}{a^2}\nabla\Psi, \tag{2}$$

$$\nabla^2\Psi = 4\pi G\bar{\rho}\delta. \tag{3}$$

Here a is the scale factor, $\bar{\rho}$ is the average density, and a dot denotes a time derivative. All spatial derivatives are taken with respect to comoving coordinates. In the last equation, δ represents the total density perturbation, although we will generally consider models that are gravitationally dominated by CDM, so that we can replace δ with δ_{CDM}. There are also continuity and Euler equations for the baryons, the latter containing a pressure term.

The relativistic species (photons and neutrinos) are not characterized by a simple velocity field, but by a distribution function whose evolution is governed by the Boltzmann equation,

$$\frac{Df}{Dt} \equiv \frac{\partial f}{\partial t} + \frac{\partial f}{\partial x^i}\frac{dx^i}{dt} + \frac{\partial f}{\partial p}\frac{dp}{dt} + \frac{\partial f}{\partial \gamma^i}\frac{d\gamma^i}{dt} = C[f]. \tag{4}$$

Here p is the magnitude of the momentum, γ^i is a direction cosine of the momentum, and C is a collision term having to do with scattering. This equation applies to both f_γ and f_ν, although at the epochs we are interested in the neutrino collision term is zero.

In order to make accurate predictions of the CMB anisotropy in a particular model, it is necessary to solve this system of equations numerically. If we work in Fourier space, we find that different fluctuation modes are uncoupled and the solution is therefore greatly simplified. We write

$$\delta(\mathbf{x}, t) = \sum_{\mathbf{k}} \delta_{\mathbf{k}}(t)\exp(i\mathbf{k} \cdot \mathbf{x}), \tag{5}$$

and similarly for the other quantities. [For the distribution functions, it is convenient to make a second expansion in Legendre polynomials $P_l(\hat{\mathbf{k}} \cdot \hat{\mathbf{p}})$.] The fact that different k-modes decouple makes the problem computationally tractable. Furthermore, as we shall see, the fact that we can work with one mode at a time makes it easier to get a conceptual understanding of anisotropy formation.

In recent years excellent codes have been developed for integrating these equations. [See Hu *et al.* (1995) and Bond (1996) for fairly recent discussions of the state of the art, and Seljak & Zaldarriaga (1996) for an important subsequent development.] We will not discuss the details of such precise calculations here; rather, we will follow a less precise but more intuitive picture of the formation of anisotropies, based on a series of physically motivated approximations. This approach makes it easier to see what the important physical processes are and also gives us an understanding of how various features in the anisotropy spectrum depend on key cosmological parameters.

We will begin by discussing the sources of primary anisotropy: the Sachs-Wolfe effect (Sachs & Wolfe 1967), which describes gravitational red- and blueshifts due to potential differences on the surface of last scattering; the Doppler effect due to bulk motions of the last-scattering surface (Sunyaev & Zel'dovich 1970); and intrinsic temperature variations from point to point (Silk 1967). We will then discuss some sources of secondary anisotropy, the most important of which is the integrated Sachs-Wolfe (ISW) effect, which describes energy changes in photons as they pass through time-varying potentials. [This effect was also treated by Sachs & Wolfe (1967), as well as by Rees & Sciama (1968) at nearly the same time.] Other secondary sources of anisotropy include scattering by reionized matter and gravitational lensing.

At first, we will consider the evolution of only one Fourier mode at a time; however, we will eventually need to synthesize all of the different Fourier modes together to see what the total CMB anisotropy on the sky looks like. To do that, we will need to know the *power spectrum* of the density perturbation. This is simply the mean-square amplitude of the various Fourier modes:

$$P(k) = \langle |\delta_{\mathbf{k}}|^2 \rangle. \tag{6}$$

(As long as space is isotropic, P depends only on the magnitude of $\mathbf{k}$.) The angle brackets here denote an ensemble average, although it is frequently acceptable to assume δ is ergodic, in which case the angle brackets can equally well be regarded as a spatial average.[3] We often assume that the initial power spectrum is a power law in k: $P(k) \propto k^n$. As we will see below, the analogous quantity for describing the observed CMB anisotropy is the *angular power spectrum*:

$$C_l = \langle |a_{lm}|^2 \rangle. \tag{7}$$

Here a_{lm} is a coefficient of an expansion of a spherical harmonic expansion of the temperature anisotropy (spherical harmonic expansions being the natural analogue of Fourier expansions for data sets that live on the sphere). A mode with spherical harmonic index l probes an angular scale on the sky of $\theta \sim l^{-1}$. In any particular cosmological model, the angular power spectrum C_l is related linearly to the matter power spectrum $P(k)$. The

---

[3]Beware: When we describe $\Delta T/T$ as a random field on the sphere, we may *not* assume ergodicity: $\Delta T/T$ is never ergodic.

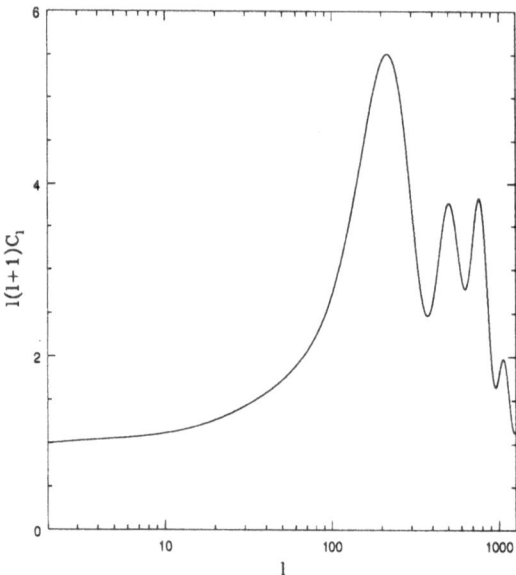

Figure 1. The angular power spectrum $l(l+1)C_l$ for a standard cold dark matter model. The parameters of this model are as follows: $n = 1$, $h = 0.5$, $\Omega_0 = 1$, $\Omega_B h^2 = 0.013$. This power spectrum was computed by N. Sugiyama.

angular power spectrum for a CDM model is shown in Figure 1.[4] The primary goal of Section 3 will be to explain the multiple peaks in this spectrum.

3. Primary Anisotropies

3.1. THE GRAVITATIONAL POTENTIAL

We will begin by assuming that, after the end of the radiation epoch, most of the mass in the Universe is in the form of cold dark matter:

$$\Omega_B \ll \Omega_{CDM}. \tag{8}$$

Then the gravitational potential is completely determined by the CDM, and three equations (1–3) can be solved for Ψ and δ without worrying about what the other species are doing. Then, once we know the gravitational potential Ψ, we can solve for the evolution of the photons and baryons.

---

[4]The prefactor $l(l + 1)$ in Figure 1 (and all of the other power spectrum plots we will see) is traditional. In a flat cosmological model with an $n = 1$ power spectrum, the Sachs-Wolfe contribution to the power spectrum is proportional to $1/l(l + 1)$. The Sachs-Wolfe effect dominates on large scales, explaining the flatness of Figure 1 at low l. The quantity $l(l + 1)C_l$ is also approximately proportional to the total power per logarithmic interval in l. (To make this proportionality exact, one would use $l(l + \frac{1}{2})C_l$ instead.)

Equations (1–3) can be combined into a single second-order equation for δ,

$$\ddot{\delta} + 2\frac{\dot{a}}{a}\dot{\delta} - 4\pi G\bar{\rho}\delta = 0. \tag{9}$$

At early times, when the Universe is radiation dominated, the last term in this equation is negligible, and the two linearly independent solutions are $\delta = \text{const.}$ and $\delta \propto \ln t$. There is therefore little growth during the radiation era.

If the Universe is matter dominated (meaning that both radiation and curvature are negligible in the Friedmann equation), then we have $a \propto t^{2/3}$, and the solutions are $\delta \propto t^{2/3} \propto a$ and $\delta \propto t^{-1}$. At late times, of course, the growing mode is the one that matters. If we plug the matter-dominated growing-mode solution into the Poisson equation (3), we find that Ψ *is independent of time*. This is a key fact, to which we will return repeatedly.

3.2. THE PHOTON-BARYON FLUID

Now that we know what the gravitational potential is doing, we are ready to study the evolution of the photons and baryons. We do this by making another approximation: we assume *tight coupling* between photons and baryons. Specifically, we assume that the mean free time τ between photon collisions is small compared to the other important time scales:

$$\tau \ll H^{-1}, (ck)^{-1}, (c_s k)^{-1}. \tag{10}$$

Here H^{-1} is the expansion time scale, $(ck)^{-1}$ is the light-travel time across a Fourier mode, and $(c_s k)^{-1}$ is the sound-travel time across a mode (c_s being the sound speed). This is an excellent approximation right up until around the time of last scattering.

In the tight-coupling approximation, frequent scattering isotropizes the photon distribution function f_γ: at any particular point, f_γ is isotropic in the rest frame of the baryons at that point. In fact, f_γ is completely characterized by the temperature distribution. Furthermore, the photon and baryon densities are coupled adiabatically: $n_\gamma \propto n_B \propto T^3$. The behavior of the photon-baryon fluid is therefore characterized by a single variable: if we know, say, $\delta_B(\mathbf{x}, t)$, we can determine $\mathbf{v}_B$, T, and f_γ. We will find it convenient to take as our variable the fractional temperature fluctuation, which is simply one third of the baryon density fluctuation:

$$\Theta(\mathbf{x}, t) \equiv \frac{\Delta T}{T}(\mathbf{x}, t) = \frac{1}{3}\delta(\mathbf{x}, t). \tag{11}$$

With these approximations, the dynamics of the photon-baryon fluid is described by the single equation

$$\frac{d}{d\eta}\left[(1+R)\dot\Theta\right] + \frac{k^2}{3}\Theta = F(\eta). \tag{12}$$

This equation comes from the Euler and continuity equations for the fluid. We are working in units in which $c = 1$. For a derivation of this equation, see Hu (1995). In this equation, η is the conformal time,

$$\eta = \int^t \frac{dt}{a(t)}, \tag{13}$$

and $R \equiv 3\rho_B/4\rho_\gamma$ is essentially the baryon-to-photon energy ratio. The overdot denotes a derivative with respect to conformal time. This equation is in Fourier space, so $\Theta = \Theta_k$ represents a single Fourier mode with wavenumber $\mathbf{k}$.[5] The right-hand side $F(\eta)$ is a gravitational driving term,

$$F(\eta) = -\frac{k^2}{3}(1+R)\Psi - \frac{d}{d\eta}\left[(1+R)\dot\Phi\right]. \tag{14}$$

The rest of this section will be devoted almost entirely to a discussion of the solution of equation (12). We begin by making some useful observations. First,

$$R = \left(\frac{450}{1+z}\right)\left(\frac{\Omega_B h^2}{0.015}\right), \tag{15}$$

where h is the Hubble parameter in units of $100\,\mathrm{km\,s^{-1}\,Mpc^{-1}}$, z is the redshift, and Ω_B is the baryonic contribution to the density parameter. So for standard recombination at $z \simeq 1000$ and baryon densities around the nucleosynthesis range, $R \simeq \frac{1}{2}$ at the time of last scattering.

With the approximations that we're making, there are no anisotropic stresses, so the two gravitational potentials are simply related to each other:

$$\Phi = -\Psi. \tag{16}$$

Furthermore, we have seen that during the matter-dominated epoch, if linear theory is valid, Ψ is independent of time. The gravitational driving term therefore simplifies to

$$F(\eta) = -\frac{k^2}{3}(1+R)\Psi. \tag{17}$$

---

[5] It has become standard practice in cosmology to denote functions and their Fourier transforms by the same symbol, relying on context to tell the difference. [For the only recent exception I know about, see Tegmark (1996c).] Odious as this practice is, I have bowed to convention in these lectures.

3.3. ACOUSTIC OSCILLATIONS

To develop an intuitive feel for the solutions to equation (12), we will start by making some excessive and unwarranted approximations. We will then gradually relax those approximations to get a more accurate picture. First, let's assume that R and Ψ are independent of time. Then

$$(1+R)\ddot{\Theta} + \frac{k^2}{3}\Theta = -\frac{k^2}{3}(1+R)\Psi. \tag{18}$$

This is the equation for a simple harmonic oscillator, with solution

$$\Theta(\eta) = -(1+R)\Psi + K_1\cos(kc_s\eta) + K_2\sin(kc_s\eta). \tag{19}$$

Here K_1 and K_2 are constants to be fixed by the initial conditions and $c_s = (3(1+R))^{-1/2}$ is the sound speed. In this approximation, then, each Fourier mode represents an acoustic plane wave propagating at speed c_s.

There is a simple physical picture underlying this result. The baryon-photon fluid wants to fall into the potential wells, but it is supported by radiation pressure. The balance between pressure and gravity sets up acoustic oscillations. The three terms in equation (12) come from the inertia of the fluid, the radiation pressure, and the gravitational field.

In fact, let's make things even simpler and set $R = 0$. Then

$$\Theta(\eta) = -\Psi + K_1\cos(kc_s\eta) + K_2\sin(kc_s\eta). \tag{20}$$

In many theories, the initial perturbation is *adiabatic*, meaning that the matter and radiation fluctuations are the same at any particular point. With these initial conditions, $\dot{\Theta} = 0$ at very early times, and $\Theta(0) = -2\Psi/3$, so

$$\Theta(\eta) = -\Psi + \tfrac{1}{3}\Psi\cos kc_s\eta. \tag{21}$$

Continuing to focus our attention on a single Fourier mode, let us determine what kind of anisotropy we would expect to see on the sky. As we have mentioned, the three sources of primary anisotropy are gravity, the Doppler effect, and intrinsic temperature variations,

$$\frac{\Delta T}{T} = [\Psi - \hat{\mathbf{r}}\cdot\mathbf{v} + \Theta]_{\eta=\eta_{\mathrm{LS}}}, \tag{22}$$

where η_{LS} is the time of last scattering and $\hat{\mathbf{r}}$ is a unit vector in the direction of observation.

Ignoring the Doppler term for the moment, note that the other two terms give a pure cosine oscillation,

$$\Psi + \Theta = \tfrac{1}{3}\Psi\cos kc_s\eta, \tag{23}$$

so the r.m.s. $\Delta T/T$ is large when $kc_s\eta_{\rm LS}$ is an integer multiple of π. Therefore, if the initial conditions have a smooth power spectrum, $\Delta T/T$ will have a harmonic series of peaks in k-space, leading to a harmonic series in the angular power spectrum of anisotropy on the sky. This is the origin of the so-called "Doppler peaks" in Figure 1. Ironically, the peaks have nothing to do with the Doppler effect. In fact, the peaks are caused by modes that have reached maxima of compression and rarefaction at the time of last scattering; the Doppler contribution to the anisotropy in these modes is zero!

The first peak is caused by modes that have had time to oscillate through exactly one half of a period before last scattering; the modes that cause the second peak have oscillated through a full period, and so on. The physical scale of the first peak is therefore $\lambda \sim k^{-1} = c_s\eta_{\rm LS}/\pi \sim 30\,{\rm Mpc}$. The distance to the last-scattering surface is $D \equiv \eta_0 - \eta_{\rm LS} \sim 6000\,{\rm Mpc}$, so the angular scale of the first peak is $\lambda/D \sim 0\overset{\circ}{.}25$. We will be more precise about the correspondence between physical scales and angular scales later.

Earlier, we threw out the Doppler term in equation (22) for no particular reason. We had better put it back. Using the continuity equation (1) and the relation $\delta = 3\Theta$, we find that

$$\mathbf{v} = \frac{3i}{k}\dot{\Theta}\hat{\mathbf{k}}. \tag{24}$$

Here $\hat{\mathbf{k}}$ is a unit vector in the direction of $\mathbf{k}$ and $\mathbf{v}$ and δ are still in Fourier space. Differentiating equation (21) and using the fact that $c_s = 1/\sqrt{3}$ for $R = 0$, we find that $\dot{\Theta} = -\frac{1}{3\sqrt{3}}\Psi \sin kc_s\eta$. Since the r.m.s. value of $\hat{\mathbf{r}} \cdot \hat{\mathbf{k}}$ is $1/\sqrt{3}$, the r.m.s. Doppler contribution to equation (22) is

$$\left[\frac{\Delta T}{T}\right]_{\rm Doppler} = \frac{i}{3}\Psi \sin kc_s\eta. \tag{25}$$

This has the same amplitude as the $(\Theta + \Psi)$ contribution, but is 90° out of phase in both time (it goes like a sine instead of a cosine) and space (it has an extra factor of i). This has the rather disastrous consequence of completely erasing the Doppler peaks: the total $\Delta T/T$ is the quadrature sum of (23) and (25):

$$\left(\frac{\Delta T}{T}\right)^2 \propto \sin^2 kc_s\eta + \cos^2 kc_s\eta = 1. \tag{26}$$

The k dependence, which led to the peaks, is gone.

The problem, of course, is that we have taken our approximations too far. Specifically, the culprit is the limit $R \to 0$. Physically, taking the limit

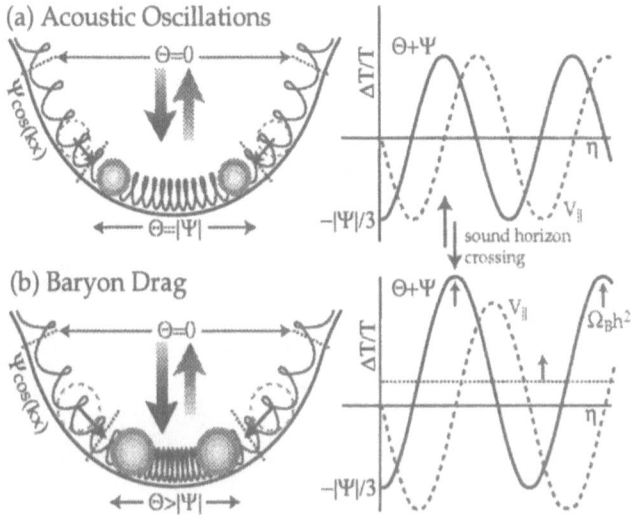

Figure 2. A simple mechanical model for a single mode of acoustic oscillation of the photon-baryon fluid. The behavior of the fluid inside of a potential well is shown; the behavior atop a potential hill would be the reverse. The springs represent the restoring force of the photon pressure and the balls represent the effective mass of the system. The top panel shows the case where the baryon contribution to the effective mass can be neglected, and the lower panel shows the effect of including baryons. Baryons increase the mass of the fluid, causing a displacement of the zero point of the oscillations. In addition, the sound speed is lowered. This has two effects, both of which may be seen in the plots on the right: baryons make the oscillations proceed more slowly and also reduce the Doppler contribution to $\Delta T/T$ relative to the intrinsic and Sachs-Wolfe contributions. Reprinted from Hu (1996).

$R \to 0$ means ignoring the dynamical effects of the baryons. Let us remove that assumption, but keep the approximation that R is time-independent. Then the solution for $\Theta(\eta)$ changes in two ways. The sound speed gets smaller by a factor $1/\sqrt{1+R}$, and the driving term $F(\eta)$ gets bigger by a factor $1+R$. The adiabatic solution to equation (12) is now

$$\Theta(\eta) = \tfrac{1}{3}(1 + 3R)\Psi \cos kc_s\eta - (1 + R)\Psi. \qquad (27)$$

By allowing R to be nonzero, we have increased the amplitude of the cosine oscillations by a factor $(1+3R)$. Furthermore, there is now an offset in the combined Sachs-Wolfe and adiabatic contributions to $\Delta T/T$: in the limit $R \to 0$, we found that $\Theta + \Psi$ oscillated symmetrically about zero; now it oscillates about $-R\Psi$. Most important, a nonzero R reduces the

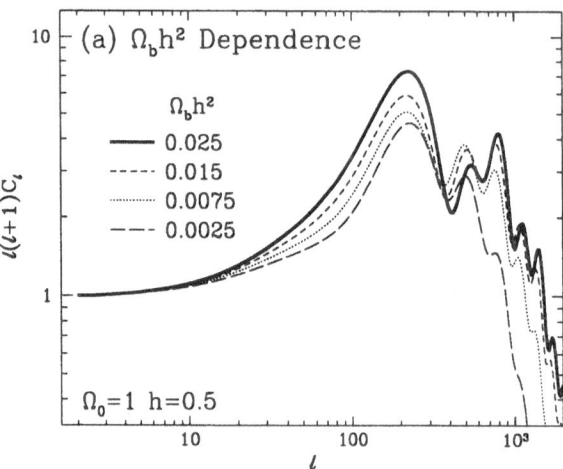

Figure 3. Angular power spectra for CDM models with varying values of the baryon density $\Omega_B h^2$. Reprinted from Hu (1996).

amplitude of the Doppler contribution to the anisotropy, relative to the Sachs-Wolfe contribution, since v is proportional to $c_s \Theta$ and c_s has gotten smaller. Since the cosine oscillations are now larger in amplitude than the sine oscillations, we do indeed expect to see a series of peaks at $k c_s \eta_{LS} = m\pi$.

Why does including the dynamical effect of the baryons effect these changes in the solution? The essential reason is that baryons contribute to the *effective mass* of the photon-baryon fluid, but not to the *pressure*. (This is clear from looking at equation (12): the first term, representing the effective mass, depends on R, but the second term, representing pressure support, does not.) The effect of the baryons, therefore, is to slow down the oscillations, and also to make the fluid fall deeper into the potential wells. This explains all three of the key effects we have just mentioned: the increased oscillation amplitude, the offset in the center of the oscillations, and the reduction in importance of the velocity term relative to the other terms. These effects are represented pictorially in Figure 2.

Based on this analysis, we can predict that the height of the peaks in the CMB anisotropy spectrum should depend on the baryon density: the larger the baryon density, the larger R, and the greater the amplitude of the oscillations. Furthermore, because of the offset in the oscillations, we expect the odd-numbered peaks to be enhanced relative to the even-numbered ones. (In the language of Figure 2, the compressions produce larger anisotropies than the rarefactions. Of course, if we had chosen to draw a potential peak instead of a potential well in Figure 2, we would

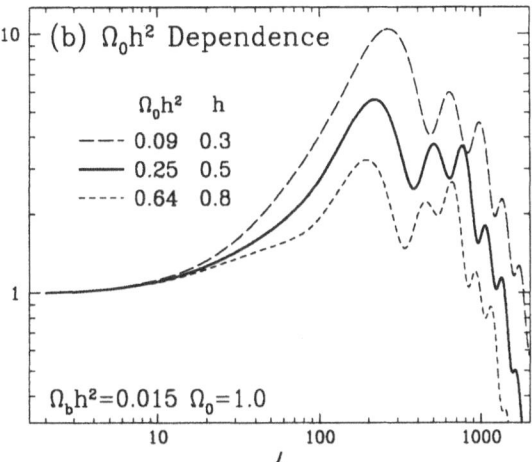

Figure 4. Angular power spectra for CDM models with varying values of h. All of these models have $\Omega_0 = 1$. For lower values of $\Omega_0 h^2$, matter domination occurs later. The driving effect of the decay in the gravitational potential is therefore more significant, increasing the peak height. Reprinted from Hu (1996).

make precisely the opposite statement.)

Both of these effects are found in detailed calculations and can be seen in Figure 3.

We can make further refinements to these approximations without too much difficulty. For instance, we can allow R to vary with time. The time scale on which R varies is of order a Hubble time and is much longer than the period of the acoustic waves. We can therefore treat the variation of R (and the concomitant variation in c_s) in the WKB approximation. There are two main results. First, the phase of the oscillation changes from $k c_s \eta$ to $k \int c_s \, d\eta$. Second, the amplitude of the oscillations grows with time in proportion to $c_s^{1/2}$, or $(1 + R)^{-1/4}$.[6]

3.4. DRIVING

We can also relax the approximation that $F(\eta)$ is constant in time. This has interesting consequences. A constant term on the right-hand side of an oscillator equation merely offsets the center of the oscillations; in contrast, a time-varying term genuinely drives oscillations. In particular, if the driving

---

[6]The easiest way to see this is to note that $m \omega A^2$ is an adiabatic invariant for a harmonic oscillator. Here m is the mass, A is the amplitude, and ω is the frequency. Of course, the result can also be derived directly from the WKB approximation.

148

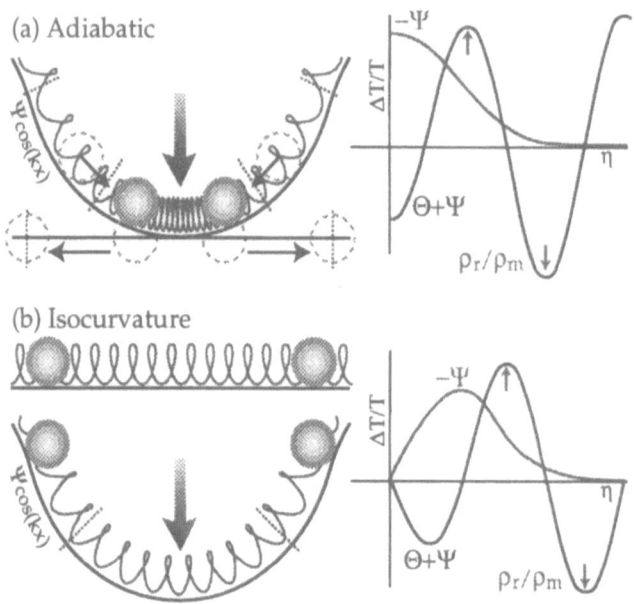

Figure 5. Driving effects on the acoustic oscillations. (a) In the adiabatic case, the gradual decay in the potential causes a relatively small increase in the amplitude of the oscillations. (b) For isocurvature initial conditions, the initial perturbation $\Theta + \Psi$ is zero, and the growth (and subsequent decay) of Ψ is entirely responsible for driving the oscillations. Reprinted from Hu, Sugiyama, & Silk (1996).

term varies significantly on a time scale comparable to the period of the oscillations, resonant driving can occur.

We have seen that Ψ (and hence F) is constant during matter domination, but it decays during the radiation epoch. For modes that enter the horizon before matter domination, Ψ decays while that mode is undergoing its oscillations. The decay in Ψ therefore boosts the amplitude of those short-wavelength modes. The modes that receive the largest boost are those that entered the horizon before matter-radiation equality at a redshift

$$z_{\text{eq}} = 24000\Omega_0 h^2. \tag{28}$$

These modes are characterized by wavenumbers

$$k \gtrsim k_{\text{eq}} = (14\,\text{Mpc})^{-1}\Omega_0 h^2. \tag{29}$$

The effect of the driving term becomes evident if we look at power spectra for critical-density models with different values of the Hubble parameter: for low h, matter domination occurs later and the boosting effect is greater. This effect is shown in Figure 4.

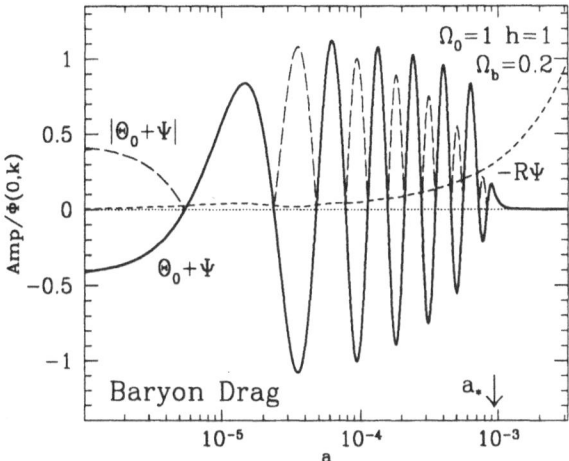

Figure 6. The time evolution of a single Fourier mode. a is the scale factor, normalized to unity today. a_* is the scale factor at recombination. At early times potential decay increases the amplitude of the oscillations. The heights of the positive and negative peaks are offset by $-R\Psi$ with respect to each other. The decline in amplitude at late times is due to diffusion damping. Reprinted from Hu (1996).

We have been focusing on models with adiabatic initial conditions. If we instead consider *isocurvature* models, the effect of the driving term becomes even more evident. In isocurvature models, the total density perturbation vanishes at early times:

$$\delta\rho_{\text{total}} = \delta\rho_B + \delta\rho_\gamma + \delta\rho_{\text{CDM}} + \ldots = 0. \qquad (30)$$

Clearly $\Theta(0) = 0$ in these models. As time passes, $\delta\rho_\gamma$ redshifts away, leaving genuine density perturbations and hence nonzero potentials Φ and Ψ. Oscillations are therefore driven in Θ. In contrast to the adiabatic case, these isocurvature oscillations are proportional to $\sin kc_s\eta$ rather than $\cos kc_s\eta$. The peaks in an isocurvature spectrum are therefore different in phase from adiabatic peaks. The peak locations in the CMB anisotropy spectrum can distinguish quite robustly between adiabatic and isocurvature models. Figure 5 illustrates the origin of the peaks in isocurvature models.

3.5. DAMPING

We have been assuming so far that the tight-coupling approximation holds perfectly right up until the moment η_{LS}, and that the photons are instantaneously released at that moment. In fact, the failure of the tight-coupling approximation, especially around the time of last scattering, causes significant damping of fluctuations as photons diffuse out of hot, overdense regions. Furthermore, the last-scattering "surface" is really a shell of some thickness. Oscillations on scales smaller than this thickness do not show up as observable anisotropies on the sky, since any particular line of sight will look at multiple peaks and troughs of that mode.

To get a rough estimate of the importance of diffusion damping (also known as Silk damping), consider a photon undergoing a random walk through the photon-baryon fluid. If the mean free path is λ, then at a time η, a typical photon has scattered about $N \sim \eta/\lambda$ times and has diffused through a distance $\lambda_{\rm D} \sim \sqrt{N}\lambda \sim \sqrt{\eta\lambda}$. If a particular Fourier mode has a wavelength less than this diffusion length, then the photons will have diffused from overdense to underdense regions, and the mode will be damped away. Diffusion damping thus occurs for modes with $k^{-1} \gtrsim \lambda_{\rm D}$. Most of the damping occurs around the time of last scattering, since that is when the mean free path λ becomes large.

In Figure 6 we show the time evolution of a particular mode, including the damping at the end, and in Figure 10 below we show the net effect of diffusion damping on a CMB power spectrum.

3.6. PROJECTION

In order to complete the story of primary anisotropies, we need to specify precisely how a particular plane wave is projected onto a specific angular scale on the sky. It is clear that a mode with wavelength λ will show up on an angular scale $\theta \sim \lambda/R$, where R is the distance to the last-scattering surface, or in other words, a mode with wavenumber k shows up at multipoles $l \lesssim k$. Consequently, tilting the spectral index n of the primordial matter power spectrum essentially just tilts the angular power spectrum. Let us now make this rough observation mathematically precise.

If we are looking in a direction $\hat{\mathbf{r}}$ in the sky, then (ignoring the thickness of the last-scattering surface), the anisotropy we see is simply $\Delta T/T(\hat{\mathbf{r}}) = \Theta^{(\rm tot)}(R\hat{\mathbf{r}})$, where $\Theta^{(\rm tot)}$ includes all three terms in equation (22). For a single Fourier mode, this is simply

$$\frac{\Delta T}{T}(\hat{\mathbf{r}}) = \Theta_{\mathbf{k}}^{(\rm tot)} \exp(i\mathbf{k} \cdot \hat{\mathbf{r}}R). \qquad (31)$$

To quantify the amount of power this produces on different angular scales, we expand in spherical harmonics Y_{lm}. The relevant identity is (Jackson 1975)

$$\exp(ikR\hat{\mathbf{k}} \cdot \hat{\mathbf{r}}) = 4\pi \sum_{l,m} i^l j_l(kR) Y_{lm}^*(\hat{\mathbf{k}}) Y_{lm}(\hat{\mathbf{r}}). \qquad (32)$$

Combining equations (31) and (32), we find that

$$\frac{\Delta T}{T}(\hat{\mathbf{r}}) = \sum_{l,m} a_{lm} Y_{lm}(\hat{\mathbf{r}}), \tag{33}$$

where

$$a_{lm} = 4\pi \Theta_{\mathbf{k}}^{(\text{tot})} i^l j_l(kR) Y_{lm}^*(\hat{\mathbf{k}}). \tag{34}$$

The total power produced by this mode in the multipole l is

$$a_l^2 \equiv \sum_{m=-l}^{l} |a_{lm}|^2 = 4\pi(2l+1) \left|\Theta_{\mathbf{k}}^{(\text{tot})}\right|^2 j_l^2(kR). \tag{35}$$

The spherical Bessel function $j_l(x)$ peaks at $x \sim l$, so a single Fourier mode $\mathbf{k}$ does indeed contribute most of its power around multipole $l_k = kR$, as expected. However, as Figure 7 shows, j_l does have significant power beyond the first peak; that is, $j_l(kR)$ can be non-negligible even when kR is significantly greater than l. Turning this statement around, we can say that the power contributed by a Fourier mode of wavenumber k "bleeds" to l-values lower than l_k. This is due to the fact that a mode appears to have a longer wavelength when looked at along a line of sight nearly perpendicular to the wavevector.

These formulae assume that the Universe is spatially flat. If there is curvature, then the correspondence between physical scales at last scattering and angular scales on the sky changes. In an open Universe, for example, geodesics focus in such a way that a particular angular scale corresponds to a much larger physical scale on the last-scattering surface. A particular Fourier mode in an open Universe projects to multipoles $l \sim kR_A$, where R_A is the *angular-diameter distance* to the last-scattering surface, given by

$$R_A = \frac{1}{\sqrt{|K|}} \sinh\left(\sqrt{|K|}R\right). \tag{36}$$

Here K is the curvature. When $|K|$ is small, $R_A \to R$, but for large $|K|$ (low Ω_0), R_A grows exponentially with metric distance.

This projection effect is easy to see in predictions of the CMB anisotropy. In an open Universe, features such as the acoustic peaks and the damping scale are shifted towards smaller angular scales, *i.e.*, towards higher l. (See Figure 8.)

Note that the approximate linear relation between l and k holds only for primary anisotropies. The secondary anisotropies, which we discuss below, tend to occur at a wide range of distances (in contrast to the relatively thin last-scattering surface). Thus for secondary anisotropies, each k-mode can contribute to a wide range of l's.

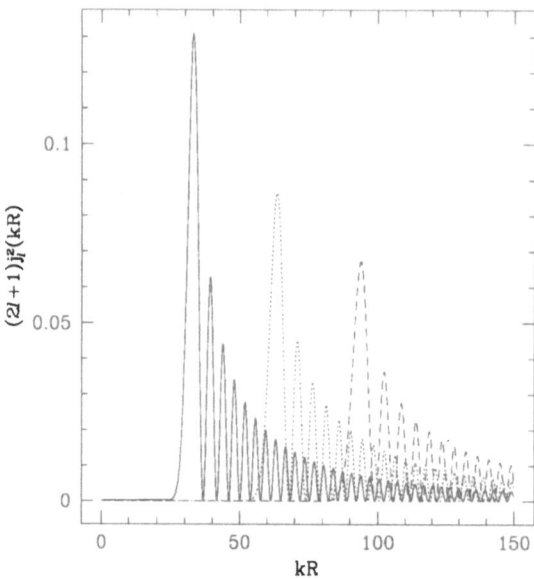

Figure 7. The quantity $(2l + 1)j_l^2(kR)$ is plotted for $l = 30$, $l = 60$, and $l = 90$. This quantity determines how much power a Fourier mode with wavenumber k contributes to multipole l. Note that, while most of the power is deposited at $l \simeq kR$, there is significant "bleeding" to lower l.

4. Secondary Anisotropies

After last scattering, the photons and baryons are no longer tightly coupled. In fact, if the effects of reionization are negligible, there is no coupling at all. In this case, the photons simply propagate freely along spacetime geodesics from last scattering to the observer. The causes of secondary anisotropy are then entirely gravitational, the dominant effect being the ISW effect. Weak gravitational lensing can also distort the anisotropy spectrum, although this effect is generally small.

If the intergalactic medium reionized at a sufficiently early redshift, then some fraction of the photons will interact again after the time of "last" scattering. The main result is that primary fluctuations are erased, and in addition new fluctuations can be generated from the new last-scattering surface. However, the last-scattering surface in a reionized model is extremely thick (since the photon-baryon coupling is weak), so the nature of the regenerated anisotropy is quite different from the primary anisotropy.

4.1. INTEGRATED SACHS-WOLFE EFFECT

As Sachs & Wolfe (1967) showed, fluctuations in the spacetime curvature produce CMB anisotropy in two distinct ways. The "ordinary" Sachs-Wolfe

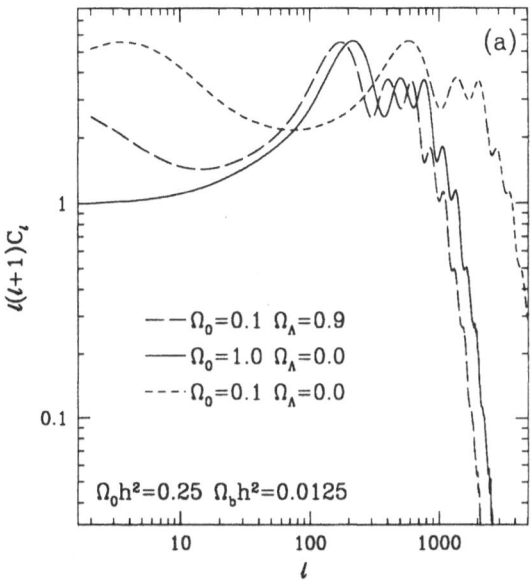

Figure 8. Ω_0-dependence of the angular power spectrum. In open models, the angular-diameter distance to the last-scattering surface is large, so the features in the power spectrum are shifted to small angular scales. In a flat model with a cosmological constant, the distance to the last-scattering surface is larger than in an $\Omega_0 = 1$ model, but the size of the sound horizon also increases, producing little net effect on the location of the peaks. The structure at low l in the low-density models is due to the integrated Sachs-Wolfe effect. Reprinted from Hu & White (1996).

effect is simply the gravitational red- or blueshift due to the potential difference between the points of emission and reception of a photon. In addition, if the gravitational potential changes with time, there is an "integrated" Sachs-Wolfe effect.

Imagine a photon falling into a potential well, and then climbing out the other side. If the potential does not vary with time, the photon suffers no net change in energy. However, if the potential well decays while the photon is passing through it, then the redshift upon climbing out of the well is smaller than the blueshift upon falling in. The photon therefore gains energy. The magnitude of the ISW effect is given by an integral along the photon's path:

$$\left(\frac{\Delta T}{T}\right)_{ISW} = \int \left(\dot{\Psi}(\mathbf{x}, \eta) - \dot{\Phi}(\mathbf{x}, \eta)\right) d\eta. \tag{37}$$

We observed earlier that the gravitational potential is time-independent if certain conditions are satisfied:

• The Universe is matter-dominated ($\rho_{\text{matter}} \gg \rho_{\text{rad}}$).

- Spatial curvature is negligible ($\Omega_0 = 1$).
- Linear perturbation theory is valid ($\delta \ll 1$).

If all of these conditions are satisfied, there is no ISW effect. However, in any realistic cosmological model some or all of these conditions are violated at some point.

4.2. EARLY ISW EFFECT

In a typical model, the epoch of matter-radiation equality occurs before the time of last scattering, but not long before. The matter-dominated limit is therefore not quite correct around the time of last scattering and shortly thereafter. The decay in the potential shortly after last scattering gives rise to the early ISW effect. This effect is largest when the matter density $\Omega_0 h^2$ is low.

The early ISW effect is most important on large scales. Specifically, the scales that are most affected are those with k^{-1} comparable to the time scale on which the potential decays. Modes with wavelengths much shorter than this oscillate many times while the potential is decaying, causing both positive and negative ISW contributions, which tend to cancel each other out. The time scale for potential decay is of order the horizon size at last scattering, so the early ISW effect shows up on large angular scales $l \lesssim 200$.

4.3. LATE ISW EFFECT

In models with $\Omega_0 \neq 1$, the potential decays at late times, typically at redshifts $z \lesssim \Omega_0^{-1}$. This potential decay, which occurs whether or not there is a cosmological constant, gives rise to an ISW effect at late times. As with the early ISW effect, modes with wavelengths comparable to the time scale for the potential to decay are most affected. The relevant time scale is the horizon size at the time of potential decay, so the late ISW effect also leaves its imprint on large angular scales.

4.4. OTHER ISW EFFECTS

At very late times, nonlinear structure forms, causing the potential to grow with time. The ISW effect due to nonlinear structure is often called the *Rees-Sciama effect* (Rees & Sciama 1968). In standard models, the Rees-Sciama effect is typically much weaker than the other effects we have discussed (Seljak 1996a).

A background of primordial gravity waves, if there is one, produces its own ISW effect. Gravity waves redshift once they enter the horizon, so modes that enter the horizon well before last scattering leave no imprint on the CMB. The gravity-wave contribution to the CMB anisotropy therefore occurs on large angular scales $l \lesssim 100$. Because of the quadrupolar nature of the spacetime distortion caused by a gravity wave, the gravity-wave contribution to the CMB quadrupole is enhanced relative to other modes.

There may be other sources of spacetime distortion besides linear density fluctuations and gravity waves. In particular, topological defects cause spacetime curvature and hence an ISW effect. We will not discuss topological defects further; for more information, see Brandenberger (1996) and the references therein.

4.5. GRAVITATIONAL LENSING

The ISW effect may be thought of as gravity imparting a "kick" to a photon forward or backward along the direction of motion. Gravity can also kick the photons in the transverse directions, changing their directions of motion but not their energies. The result of this weak gravitational lensing is that our image of the last-scattering surface is slightly distorted, as if we were looking at it through an irregular refracting medium. This distortion of the last-scattering surface results in a slight smearing of the angular power spectrum, with power from the peaks being moved into the valleys. The effect is typically weak, resulting in changes at the few-percent level in the power spectrum (Seljak 1996b).

4.6. REIONIZATION

We will not undertake a detailed discussion of reionized models here. Instead, we refer the interested reader to Hu $et\ al.$ (1994), Dodelson & Jubas (1995), and references therein. We will, however, make some general comments.

The Gunn-Peterson test (Gunn & Peterson 1965) tells us that the intergalactic medium is ionized out to redshifts of a few. In CDM-like models of structure formation, reionization is generally thought to occur at such moderate redshifts, with the formation of the earliest nonlinear structures. If this is correct, then reionization does not dramatically alter the CMB anisotropy predictions. If, on the other hand, reionization somehow happened earlier, say at $z \gtrsim 100$, then a significant fraction of the CMB photons have been scattered by the reionized matter after the so-called epoch of last scattering.

The main effect of such early reionization is to erase anisotropy on degree scales. The reason is quite simple: if we have early reionization, then a photon that comes toward us from a particular direction need not have originated from that direction. Rather, as Figure 9 illustrates, each direction on the sky contains photons that originate from a variety of different locations at the time of "last" scattering. In severely reionized models, the

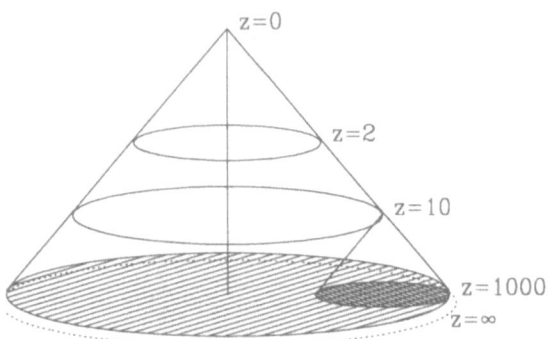

Figure 9. Our backward light cone. The vertical axis represents conformal time, and the horizontal axes are two of the three spatial directions. In the absence of reionization, each line of sight corresponds to a particular point on the last-scattering surface at $z \simeq 1000$. In a reionized model in which a typical photon last scattered at $z = 10$, a photon arriving from a particular direction may have originated from any point in the shaded circle. Reprinted from Tegmark (1996c).

peaks are completely washed away. Such models may already be ruled out by degree-scale CMB experiments (Scott, Silk, & White 1995).

Inhomogeneities and bulk motions of the reionized matter induce new CMB anisotropies, which must generally be treated to second order in perturbation theory (Ostriker & Vishniac 1986; Hu, Scott, & Silk 1994; Dodelson & Jubas 1995), but we will not discuss these regenerated anisotropies here. We also neglect to discuss the effect of nonuniform or patchy reionization, including the Sunyaev-Zel'dovich effect (Sunyaev & Zel'dovich 1970).

5. Summary of Anisotropy Formation

We have now concluded our tour of the mechanisms of anisotropy formation. Figure 10 illustrates some of the key points. The dominant features in a typical CDM power spectrum are the peaks due to acoustic oscillations of the photon-baryon fluid. The peaks correspond to modes that are undergoing maximum compression and rarefaction at the time of last scattering. Modes that are out of phase with these modes produce anisotropy via the Doppler effect, partially filling in the valleys between the acoustic peaks. The effect of damping on small scales is evident, and the rise at $l \sim 500$ in the undamped spectrum shows the driving effect of the decaying gravitational potential at early times.

We can use what we have learned to determine how the predicted anisotropy spectrum should depend on the key cosmological parameters:

- In models with *spatial curvature* ($\Omega_0 \neq 1$), the position of the acoustic

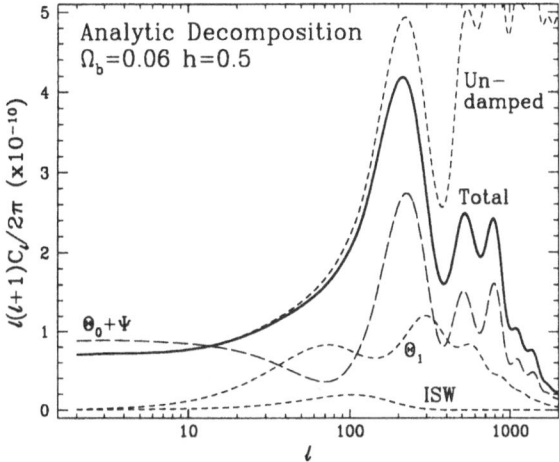

Figure 10. Analytic decomposition of anisotropies. The solid line shows the angular power spectrum of a critical-density CDM model. The upper dashed curve shows the spectrum that would be seen in the absence of diffusion damping. Note that the undamped peak heights increase at scales small enough to have crossed the horizon before matter-radiation equality. (See Section 3.4.) The other curves show the relative importance of the Sachs-Wolfe, integrated Sachs-Wolfe, and Doppler (Θ_1) contributions. Reprinted from Hu (1996).

peaks shifts due to geodesic deviation. In addition, the late ISW effect boosts the large-scale power.

- If there is a *cosmological constant*, then the position of the peaks shifts slightly due to the increased distance to the last-scattering surface, and again, the late ISW effect boosts the large-scale power.
- Lowering the *Hubble parameter* (for fixed Ω_0) reduces the matter density. The gravitational driving of oscillations is enhanced, and the peaks increase in height.
- The higher the *baryon density* $\Omega_B h^2$, the greater the peak amplitude. Odd-numbered peaks in particular are enhanced.
- The *spectral index n* of the primordial power spectrum essentially just tilts the angular power spectrum.
- If we add *gravity waves* to a model, we increase the quadrupole, and in addition the whole "plateau" at low l rises relative to the acoustic peaks.

Although we have made many approximations in deriving these conclusions, all of them are borne out by detailed Boltzmann calculations. Because the CMB anisotropy predictions depend sensitively on the various parameters, an experiment that could map out the acoustic peaks would be able to measure these cosmological parameters accurately. The spatial curvature in particular should be relatively easy to pick out, thanks to the shift in position of the first peak. The relative positions of successive peaks also provide a robust way of determining whether the initial conditions are isocurvature or adiabatic.[7] The dependence of the power spectrum on other parameters such as h and Ω_B is somewhat more subtle, but if we manage to detect and measure the heights of two or three peaks, we should be able to do quite well (Jungman *et al.* 1996), assuming, of course, that the general paradigm sketched above is correct and the multiple peaks are really there.

6. Statistical Properties of $\Delta T/T$

Before we discuss methods for comparing theories with data, we need to discuss briefly the statistical properties of the CMB anisotropy as it appears on the sky. As we have mentioned, it is convenient to expand the observed anisotropy in spherical harmonics:

$$\frac{\Delta T}{T}(\hat{\mathbf{r}}) = \sum_{l,m} a_{lm} Y_{lm}(\hat{\mathbf{r}}). \tag{38}$$

We have focused on the anisotropy produced by an individual plane wave; the observed anisotropy is of course a superposition of contributions from all of these plane waves:

$$a_{lm} = \sum_{\mathbf{k}} a_{lm}^{(\mathbf{k})}. \tag{39}$$

Since all of the relevant physics is described by linear perturbation theory — as we know, everything in nature is linear[8] — each $a_{lm}^{(\mathbf{k})}$ is proportional to the initial density perturbation $\delta_{\mathbf{k}}^{(\text{init})}$.

One often assumes that the initial conditions have "random phases," meaning that different Fourier modes are uncorrelated,

$$\langle \delta_{\mathbf{k}} \delta_{\mathbf{k}'} \rangle = 0 \qquad \text{when} \qquad \mathbf{k} \neq \mathbf{k}'. \tag{40}$$

In this case, the $a_{lm}^{(\mathbf{k})}$ are also uncorrelated. The mean-square power in a particular multipole is then simply the sum of the contributions from the various Fourier modes:

$$\langle |a_{lm}|^2 \rangle = \sum_{\mathbf{k}} \langle |a_{lm}^{(\mathbf{k})}|^2 \rangle. \tag{41}$$

---

[7]For isocurvature initial conditions, the second and third peaks occur at three and five times the location of the first; in adiabatic models, they occur at two and three times.

[8]to first order.

And, of course, the left-hand side of this equation is simply the angular power spectrum C_l. (This quantity is independent of the azimuthal index m as long as space is isotropic.)

We often go beyond the assumption of random phases and assume *Gaussian initial conditions*. This is a prediction of inflationary scenarios, but one often assumes Gaussian initial conditions even in non-inflationary phenomenological models such as isocurvature baryon models (Peebles 1987). When we talk about a Gaussian theory, we simply mean that at some initial time t_i the density perturbation δ was a realization of a Gaussian random field. Bernard Jones has provided a detailed discussion of Gaussian random processes elsewhere in this volume; for our purposes, all we need to know is that the assumption of Gaussian initial conditions, together with homogeneity and isotropy, implies that each Fourier coefficient $\delta_{\mathbf{k}}$ is an independent Gaussian random variable of zero mean.[9] In other words, the real-space density perturbation $\delta(\mathbf{x})$ is a stochastic superposition of plane waves of all different wavelengths. Since a Gaussian random variable is completely determined by its mean and variance, and since $\langle \delta_{\mathbf{k}} \rangle = 0$, the statistical properties of our Gaussian random field are completely determined by the power spectrum $P(k) \equiv \langle |\delta_{\mathbf{k}}|^2 \rangle$.

If we assume Gaussian initial conditions, then each coefficient a_{lm} is a Gaussian random variable, since it is a linear combination of the Gaussian variables $\delta_{\mathbf{k}}$. The statistical properties of $\Delta T/T$ are therefore completely specified by the means,

$$\langle a_{lm} \rangle = 0, \tag{42}$$

and the covariances,

$$\langle a_{lm} a^*_{l'm'} \rangle = C_l \delta_{ll'} \delta_{mm'}, \tag{43}$$

of the coefficients a_{lm}. In other words, for Gaussian initial conditions, the angular power spectrum C_l tells us everything we need to know.

Even when the initial conditions are not Gaussian, it often suffices to treat the CMB anisotropy as Gaussian, at least on sufficiently large angular scales. The CMB fluctuation on large angular scales is typically due to a superposition of many incoherent fluctuations. Even if the individual fluctuations fail to be Gaussian, the central limit theorem guarantees that the superposition will be approximately Gaussian. When comparing the COBE data with the predictions of a cosmic string model, for example, it is perfectly adequate to treat $\Delta T/T$ as Gaussian, even though the underlying perturbations are highly non-Gaussian.

---

[9]That is, Gaussian initial conditions (together with homogeneity and isotropy) imply random phases, but not conversely.

7. An Introduction to CMB Data Analysis

7.1. AN IDEALIZED EXPERIMENT

We will explore the key issues in CMB data analysis by first considering an absurdly idealized experiment (the sort of thing only a theorist could dream up). We will gradually introduce real-world complications to see what the main issues are.

Imagine, then, an experiment that measured $\Delta T/T$ at many pixels that cover the entire sky completely and uniformly. Furthermore, imagine that each data point is a perfect, noise-free measurement. With this data set, we could determine each coefficient a_{lm} with essentially perfect accuracy by inverting equation (38):

$$a_{lm} = \int \frac{\Delta T}{T}(\hat{r}) Y_{lm}^*(\hat{r})\, d\Omega \approx \frac{4\pi}{N_{\text{pix}}} \sum_{p=1}^{N_{\text{pix}}} \frac{\Delta T}{T}(\hat{r}_p) Y_{lm}^*(\hat{r}_p). \qquad (44)$$

Here $d\Omega$ is an element of solid angle in the direction of $\hat{r}$, N_{pix} is the total number of pixels, and $\hat{r}_p$ is a unit vector in the direction of the pth pixel.

Even in this hopelessly idealized experiment, we still can't measure the angular power spectrum C_l perfectly. The reason is that C_l is an ensemble-average quantity: it is the variance of the distribution from which a_{lm} is drawn. We have only a finite number, $2l+1$, samples of this distribution at each l. This fact, generally called *cosmic variance*, sets a fundamental limit on how well we can ever hope to measure the angular power spectrum.[10]

If we assume Gaussian statistics, then the best estimator of C_l is simply the average of $|a_{lm}|^2$ over m:

$$\hat{C}_l \equiv \frac{1}{2l+1} \sum_{m=-l}^{l} |a_{lm}|^2. \qquad (45)$$

This quantity is chi-squared distributed with $2l+1$ degrees of freedom, and so it has a fractional uncertainty of

$$\frac{\text{Var}^{1/2}(\hat{C}_l)}{C_l} = \sqrt{\frac{2}{2l+1}}. \qquad (46)$$

The unfortunate fact, therefore, is that even in a perfect experiment we will never know C_l with a fractional uncertainty better than $(l+\frac{1}{2})^{-1/2}$. We are stuck with a 63% uncertainty in the quadrupole power C_2 and a 30%

[10]Cosmic variance is closely related to the failure of ergodicity. If $\Delta T/T$ were ergodic, then the average value of $|a_{lm}|^2$, measured in different orientations over the sphere, would be the ensemble-average quantity C_l. But $\Delta T/T$ isn't ergodic, so this doesn't work.

uncertainty in C_{10}, although we can in principle hope to determine C_{1000} to 0.3%.

7.2. NOISE

Let's mess up our nice, clean experiment by adding noise. Each pixel is no longer a perfect measurement of $\Delta T/T$: the ith data point d_i consists of a sum of signal and noise,

$$d_i = \frac{\Delta T}{T}(\hat{\mathbf{r}}_i) + n_i. \tag{47}$$

Let us assume that the noise n_i in each pixel is independent and Gaussian distributed, with some standard deviation σ. For the moment we will assume *homoscedasticity*, that is, that σ is the same in all pixels.

We can still try to estimate a_{lm} using equation (44),

$$\hat{a}_{lm} = \frac{4\pi}{N_{\text{pix}}} \sum_{p=1}^{N_{\text{pix}}} d_p Y_{lm}^*(\hat{\mathbf{r}}_p), \tag{48}$$

and average over m to get an estimate of C_l,

$$\hat{C}_l = \frac{1}{2l+1} \sum_{m=-l}^{l} |\hat{a}_{lm}|^2, \tag{49}$$

but this quantity will no longer be a good estimate of the true C_l; it will be biased upward. Using equations (48) and (47), together with the fact that $\langle n_p n_{p'} \rangle = \sigma^2 \delta_{pp'}$, it is straightforward to check that

$$\langle |\hat{a}_{lm}|^2 \rangle = C_l + \frac{4\pi}{N_{\text{pix}}} \sigma^2. \tag{50}$$

The estimator $\hat{C}_l$ is the average of these quantities, so it too is biased upward by $4\pi\sigma^2/N_{\text{pix}}$.

We can of course get a better estimate of C_l by subtracting off the noise bias,

$$\hat{C}_l' \equiv \hat{C}_l - \frac{4\pi}{N_{\text{pix}}} \sigma^2. \tag{51}$$

We now have an unbiased estimator, but unfortunately the uncertainty of $\hat{C}_l'$ has increased:

$$\text{Var}^{1/2}(\hat{C}_l') = \sqrt{\frac{2}{2l+1}} \left(C_l + \frac{4\pi}{N_{\text{pix}}} \sigma^2 \right). \tag{52}$$

7.3. A DIGRESSION ON STATISTICAL METHODS IN GENERAL

The problem we just considered was a classic example of statistical parameter estimation. We had some *data*, $\{d\}$, from which we wanted to estimate a *parameter*, C_l. We did it by choosing an *estimator*, $\hat{C}_l'$, which we could compute from the data, and which we hoped would be close to the true value of the parameter.

In the problem above, there was a fairly natural choice of an estimator, but in general, for a more complicated problem, there may be no obvious choice. There is no universal, "correct" way to choose an estimator, but in many situations the *maximum-likelihood* estimator is a good choice. We will illustrate maximum-likelihood estimators with a simple example.

Suppose that we have M data points x_i, each of which is the sum of a signal s_i and some noise n_i. We will take both s_i and n_i to be Gaussian random variables with zero mean. The variances of the signal and noise are

$$\langle s_i^2 \rangle = S, \qquad\qquad \langle n_i^2 \rangle = N, \qquad (53)$$

And everything is uncorrelated:

$$\langle s_i s_j \rangle = \langle n_i n_j \rangle = \langle s_i n_j \rangle = 0, \qquad (54)$$

where the first two expressions assume $i \neq j$. Let us suppose we know the noise variance N, and we want to estimate the unknown quantity S, using a maximum-likelihood estimator.[11]

The first step is to compute the probability density of the data for fixed S. We want to know $p(\{x\} \mid S)$, where $p(\{x\} \mid S)\, d^M x$ is the probability of getting a set of data that lie within an infinitesimal volume $d^M x$ at the location of the actual data $\{x\}$. In this case, each x_i is an independent Gaussian with variance $S + N$,

$$p(x_i \mid S) = \frac{1}{\sqrt{2\pi(S + N)}} \exp\left(-x_i^2/2(S + N)\right), \qquad (55)$$

and the joint probability density is the product

$$p(\{x\} \mid S) = \prod_{i=1}^{M} p(x_i \mid S) \qquad (56)$$

$$= (2\pi(S + N))^{-M/2} \exp\left(\frac{-\sum_{i=1}^{M} x_i^2}{2(S + N)}\right). \qquad (57)$$

---

[11] The astute reader will have noticed that this is precisely the same problem we considered at the end of the last subsection. We have simply changed all of the notation for no good reason. To be specific, the correspondence with the previous problem goes like this: $M \to 2l + 1$, $S \to C_l$, $s_i \to a_{lm}$, $N \to 4\pi\sigma^2/N_{\text{pix}}$.

The probability density we have computed is a function of the data $\{x\}$ for fixed S. But the data are known, and S is what we want to know. We therefore choose to regard this probability density as a function of S and call it the *likelihood*.

$$L(S) = p(\{x\} \mid S). \tag{58}$$

When working with Gaussian probability distributions, it is often convenient to work with the quantity $\mathcal{L} \equiv -2\ln L$ instead. The maximum-likelihood estimator, as its name suggests, is the value $\hat{S}$ of S for which L is maximized (or $\mathcal{L}$ is minimized). In other words, it is the value of the parameter for which it would have been most likely for us to get the data we actually did.

In the problem at hand, the maximum-likelihood estimator is found by differentiating

$$\mathcal{L} = M\ln 2\pi + M\ln(S+N) + \frac{\sum x_i^2}{S+N} \tag{59}$$

with respect to S, setting the result equal to zero, and solving for S. The result is

$$\hat{S} = \frac{1}{M}\sum_{i=1}^{M} x_i^2 - N. \tag{60}$$

That is, we compute the mean-square value of the data points and subtract off the noise bias. This is precisely what we did when we computed $\hat{C}_l'$ in equation (51). Although we didn't know it at the time, we were using a maximum-likelihood estimator.

In this case, the maximum-likelihood estimator turned out to be unbiased: its ensemble average $\langle \hat{S} \rangle$ is equal to the correct value S. In general, there is no guarantee that this will happen. To take a simple example, suppose that we had chosen to estimate the quantity S^{289} instead of S. The maximum-likelihood estimator would be $\hat{S}^{289}$, and it is easy to see that this quantity is highly positively biased.

Now we know how to estimate parameters. But in most cases an estimator isn't much good without a way of quantifying the uncertainty in it. Methods for doing this generally fall into two categories: the classical or *frequentist* approach (*e.g.*, Rice 1995) and the *Bayesian* approach (*e.g.*, Berger 1985, Gull & Daniell 1978, Press 1996). We will discuss each in turn.

In the frequentist picture, we look at one value of the parameter S at a time, and try to determine if that value is so far from our estimator $\hat{S}$ that it is ruled out. Specifically, for each S, we compute the probability distribution of the estimator $\hat{S}$. We use this probability distribution to determine how likely it is that we would have gotten a value of $\hat{S}$ as far off as we did, or worse. If the actual value of $\hat{S}$ is far off in the tail of the probability distribution, then this probability will be low. If the probability lies below some *significance level* (say 5%), we say that that value of S is ruled out with 5% significance.[12] We repeat this process for a range of values of S, and we say that the set of values that are not ruled out form a 95% *confidence interval* for the parameter.

For a frequentist, a value of the parameter is ruled out if there is a low probability of getting data that fits as badly as the actual data. The Bayesian approach is quite different in spirit: A Bayesian attempts to determine the subjective probability distribution that characterizes her knowledge of the parameter given the data. Armed with that probability distribution, she can calculate how likely the parameter is to lie in any particular range.

In order to implement the Bayesian strategy, we want to turn the likelihood function $L(S) = p(\{x\} \mid S)$, which represents the probability of the data given a value of the parameter, into $p(S \mid \{x\})$, the probability of the parameter given the data. The way to do this is to invoke Bayes's theorem:

$$p(S \mid \{x\}) \propto p(\{x\} \mid S)p(S), \tag{61}$$

with the constant of proportionality chosen to make the integral of the left-hand side equal one. The left-hand side of this equation is the *posterior probability distribution*, and it is precisely what we are looking for: it tells us the probability of a particular parameter value, given the data. On the right-hand side we have the product of the likelihood function and the *prior distribution* of the parameter S. The latter represents our state of knowledge of S before we looked at the data.

A Bayesian characterizes the uncertainty in a parameter estimate by drawing a *credible region* around the estimate. A 95% credible region, for example, is an interval $S_{min} < S < S_{max}$ such that there is a 95% posterior probability that S lies in that interval,

$$\int_{S_{min}}^{S_{max}} p(S \mid \{x\}) \, dS = 0.95. \tag{62}$$

The boundaries S_{min} and S_{max} of the credible region are typically chosen to have equal values of the posterior probability density.

Although the frequentist approach is the one most people think of when they think of statistics, and although most scientists profess to prefer it, many if not most error bars in cosmology are determined using Bayesian techniques.

---

[12] Astrophysicists often phrase that same statement differently, saying that the value is ruled out at *95% confidence*.

The main objection people raise to the Bayesian is that the final results depend on the prior distribution $p(S)$. For a true, orthodox Bayesian, this is not really a problem: the Bayesian view is that all probabilities represent our subjective knowledge, and that prior distributions are therefore secretly built into all statistical reasoning. It is better, the argument goes, to have the prior out in the open for all to see.

Whether or not you like this argument, there is no denying that in practice choosing a prior can be tricky. If one has essentially no prior knowledge about the parameter, then the prior distribution should be broad and flat. [For a flat prior, we can see from equation (61) that the posterior probability distribution is simply the likelihood function.] But even in this situation, it is not generally obvious which "flat" prior to choose. For example, if we are trying to estimate an element of the power spectrum C_l, should we choose a prior that is flat in C_l or one that is flat in $\sqrt{C_l}$? (C_l is after all a mean-square amplitude; maybe the r.m.s. amplitude is a more "natural" choice.) Perhaps we should even choose a prior that is flat in $\ln C_l$, since such a prior avoids choosing a preferred scale. It would be hard to say that any of these choices is "wrong," but in some situations the result of a calculation may depend on which choice is made. For an example, see Bunn et al. (1994).

The situation is not as bad as it appears, however. If the data set in question contains a good, strong detection of the parameter of interest, then the likelihood function is sharply peaked, and the shape of the posterior probability (61) is determined mostly by the likelihood rather than the prior. Prior dependence is thus typically weak in the case of strong detections. The situations where prior dependence is a serious problem are typically those in which someone is trying to coax a value out of a data set that is capable of only a weak constraint anyway.

7.4. INCOMPLETE SKY COVERAGE

We now return to our hypothetical CMB experiment. The next complication we need to consider has to do with the fact that no actual experiment ever achieves complete sky coverage. In the case of COBE, pixels close to the Galactic plane are contaminated, leaving only about two thirds of the sky usable. All other experiments to date have covered even smaller patches of sky.

This fact requires us to completely change our approach. As much as we would like to estimate each a_{lm} and hence each C_l individually, in the absence of complete sky coverage it is impossible to do so. There is in fact no estimator of a particular C_l that is "uncontaminated," i.e., that is independent of all of the other $C_{l'}$.

We may decide that it is important to estimate each C_l individually, with the minimum possible contamination from other multipoles. Tegmark (1996a) has devised power-spectrum estimators with this property in mind and has applied them to both galaxy surveys (Tegmark 1995) and the four-year COBE data (1996b). For instance, suppose we have our hearts set on knowing the value of C_{17} as well as possible. Since the power spectrum is quadratic in $\Delta T/T$, it is natural to choose a quadratic estimator,

$$\hat{C}_{17} = \sum_{i,j} A_{ij} d_i d_j - B. \tag{63}$$

Here d_i is a data point and we want to choose the matrix elements A_{ij} and the bias correction B in order to get as good an estimator as possible. Tegmark (1996a) proposes that we choose these quantities to make our estimator unbiased and to minimize the dependence of $\hat{C}_{17}$ on all of the other C_l's. He shows that it is impossible to completely remove contamination from other multipoles and that in general the "spectral resolution" Δl of an experiment is approximately the reciprocal of the angular scale $\Delta\theta$ covered by the sky map. In particular, for an experiment like COBE, $\Delta\theta \sim 1$ radian, and it turns out that it is possible to estimate a particular C_l with significant contamination only from modes with $\Delta l \approx 2$ (Tegmark 1996a, 1996c).

7.5. MAXIMUM-LIKELIHOOD PARAMETER ESTIMATION

We may, however, decide that it isn't so important to estimate each C_l individually. Often, a more fruitful approach is to parameterize the power spectrum C_l with a small number k of parameters,

$$C_l = C_l(q_1, q_2, \ldots, q_k), \tag{64}$$

and use maximum-likelihood methods to estimate those parameters. This is in fact the usual approach in CMB data analysis. Specific choices of the parameters $\{q\}$ include the following:

- We may assume a *shape* for the power spectrum and estimate the *normalization*. In this case, there is only one free parameter, which is conventionally taken to be the quadrupole amplitude $\langle Q \rangle \equiv \sqrt{5C_2/4\pi}$.[13] Most degree-scale experiments are only powerful enough to determine a single number, the total power. One therefore frequently assumes a

---

[13] A bewildering variety of notations exist in the literature. We choose to call this quantity $\langle Q \rangle$ to emphasize the fact that it is a theoretical ensemble-average quantity. In particular, it is not the same as the local quadrupole $Q_{\text{rms}} \equiv \sum_{m=-l}^{l} |a_{2m}|^2/4\pi$. The COBE group generally denotes its estimators of $\langle Q \rangle$ by $Q_{\text{rms}-\text{PS}}$.

"flat" power spectrum $l(l+1)C_l = $ const. and estimates the normalization, which in this context is often called Q_{flat}.

- Both the normalization $\langle Q \rangle$ and the spectral index n may be chosen as free parameters. For a large-angle experiment like COBE, the predicted power spectrum depends only weakly on many of the other parameters.

- White & Bunn (1995) have suggested a phenomenological parameterization of the power spectrum. At large angular scales, many popular theoretical models are well approximated by power spectra that are quadratics in $\log l$. To be specific, we may set

$$l(l+1)C_l = D_1(1 + D'(\log_{10} l - 1) + \tfrac{1}{2}D''(\log_{10} l - 1)^2) \qquad (65)$$

and work with a three-parameter family (D_1, D', D'') of power spectra.

- We may choose to divide the power spectrum over the range probed by a particular experiment into a small number of "bands." We then estimate the power in each band, assuming that $l(l+1)C_l$ is constant in each band. This has been done for COBE (Hinshaw et al. 1996) and Saskatoon (Netterfield et al. 1996), although the latter uses a completely different method.

No matter what parameterization we adopt, we need a way to compute the likelihood L for a given power spectrum. As long as we assume Gaussian statistics, it is relatively easy to write down a formula for the likelihood, although as we shall see it can be cumbersome to compute it in practice.

We begin by introducing some notation. Each data point d_i is as usual the sum of the signal $\Delta T/T(\hat{\mathbf{r}}_i)$ and noise n_i. Expanding $\Delta T/T$ in spherical harmonics, we have

$$d_i = \sum_{l,m} a_{lm} Y_{lm}(\hat{\mathbf{r}}_i) + n_i. \qquad (66)$$

Let us denote a pair of indices (lm) by a single Greek index μ. The correspondence is $\mu = l(l+1) + m$, so that μ ranges from 1 to ∞ as (lm) take on all of their allowed values. Then we can write equation (66) more compactly as

$$\vec{d} = \mathbf{Y}\vec{a} + \vec{n}, \qquad (67)$$

where $\vec{d} = (d_1, d_2, \ldots, d_{N_{\text{pix}}})$ is the data vector,[14] $\vec{n} = (n_1, \ldots, n_{N_{\text{pix}}})$ is the noise vector, and the infinite-dimensional vector $\vec{a} = (a_1, a_2, \ldots, a_\mu, \ldots)$ contains the spherical harmonic coefficients. The $N_{\text{pix}} \times \infty$-dimensional spherical harmonic matrix $\mathbf{Y}$ has elements

$$Y_{i\mu} = Y_\mu(\hat{\mathbf{r}}_i). \qquad (68)$$

---

[14]We denote vectors that live in abstract spaces such as "pixel space" by arrows, and vectors in real three-dimensional space are written in boldface.

The statistical properties of $\vec{d}$ are determined by the properties of $\vec{a}$ and $\vec{n}$. Assuming Gaussian statistics, both are Gaussian random vectors with zero mean and covariances given by

$$\langle a_\mu a_\nu^* \rangle = C_\mu \delta_{\mu\nu} \equiv C_{\mu\nu}, \tag{69}$$

$$\langle n_i n_j \rangle = \sigma_j^2 \delta_{ij} \equiv N_{ij}, \tag{70}$$

$$\langle a_\mu n_i \rangle = 0. \tag{71}$$

($C_\mu \equiv C_l$ where l is the index corresponding to μ, and $\mathbf{C}$ and $\mathbf{N}$ are diagonal matrices.) Since $\vec{d}$ is a linear combination of $\vec{a}$ and $\vec{n}$, it too is a multivariate Gaussian, and the likelihood function therefore has the form

$$L(C_l) \equiv p(\vec{d} \mid C_l) = \frac{1}{(2\pi)^{N_{\mathrm{pix}}/2} \det^{1/2} \mathbf{M}} \exp\left(-\tfrac{1}{2} \vec{d}^{\mathrm{T}} \mathbf{M}^{-1} \vec{d}\right). \tag{72}$$

The T denotes a transpose, and the covariance matrix $\mathbf{M}$ is given by

$$\mathbf{M} \equiv \langle \vec{d}\vec{d}^{\mathrm{T}} \rangle = \langle (\mathbf{Y}\vec{a} + \vec{n})(\mathbf{Y}\vec{a} + \vec{n})^{\mathrm{T}} \rangle = \mathbf{Y}\mathbf{C}\mathbf{Y}^{\mathrm{T}} + \mathbf{N}. \tag{73}$$

In principle, we are now ready to estimate parameters. Equation (72) tells us how to compute the likelihood for any particular power spectrum C_l, so all we need to do is hunt through our parameter space for the parameters that maximize the likelihood.

In fact, for a typical degree-scale experiment with tens or at most hundreds of pixels, this is essentially what is done. For a large data set such as COBE, though, there are too many pixels for this to be convenient: each time we wish to compute a likelihood, we must invert the $N_{\mathrm{pix}} \times N_{\mathrm{pix}}$ matrix $\mathbf{M}$. For COBE, therefore, we must implement some form of "data compression" to make the analysis tractable. (Data compression will be even more essential for a future satellite experiment with orders of magnitude more pixels than COBE.)

7.6. BEAM-SMOOTHING AND CHOPPING

Before we discuss data compression, though, we need to discuss one more issue. The hypothetical experiment we have been discussing is still overly idealized in one important way. We have assumed that the signal measured by the experiment is the temperature anisotropy $\Delta T/T$ at a point. In reality, no experiment has perfect resolution, so the observed signal is actually the convolution of $\Delta T/T$ with some beam pattern or point-spread function. Furthermore, many experiments chop their beams between two (or more) points on the sky, with the measured signal being a difference between these points.

The effect of the beam pattern on our analysis is fairly simple. Let $B(\alpha)$ represent the response of the instrument to a point an angular distance α from the line of sight. (We assume that the beam pattern is azimuthally symmetric.) Then what the experiment actually measures is the convolution of the anisotropy with the beam pattern,

$$\left(\frac{\Delta T}{T} \star B\right) = \sum_{l,m} \bar{a}_{lm} Y_{lm}. \tag{74}$$

The coefficients $\bar{a}_{lm}$ are related to the true anisotropy coefficients a_{lm} like this:[15]

$$\bar{a}_{lm} = B_l a_{lm}, \tag{75}$$

where B_l is the expansion in Legendre polynomials of B,

$$B_l = \int_{-1}^{1} d\cos\alpha\, B(\alpha) P_l(\cos\alpha). \tag{76}$$

If the beam pattern happens to be a Gaussian,

$$B(\alpha) \propto \exp(-\alpha^2/2\sigma^2), \tag{77}$$

then the Legendre coefficients are

$$B_l = \exp(-\tfrac{1}{2}\sigma^2 l(l+1)). \tag{78}$$

Note that as expected B_l is very small for $l \gg \sigma^{-1}$, i.e., for angular scales $\theta \ll \sigma$.

We can adapt all of the previous results of this section to take beam-smoothing into account by simply saying that our experiment is measuring the beam-smoothed power spectrum,

$$\bar{C}_l \equiv C_l B_l^2, \tag{79}$$

instead of C_l.

We can account for the effect of beam-switching in a similar way. Consider an experiment that chops between two points with spherical coordinates $(\theta, \phi + \tfrac{1}{2}\alpha)$ and $(\theta, \phi - \tfrac{1}{2}\alpha)$. Ignoring beam-smoothing, the observed signal d is the difference in the anisotropy between these two points:

$$\begin{aligned} d &= \frac{\Delta T}{T}(\theta, \phi + \tfrac{1}{2}\alpha) - \frac{\Delta T}{T}(\theta, \phi - \tfrac{1}{2}\alpha) && (80) \\ &= \sum_{l,m} a_{lm}\left(Y_{lm}(\theta, \phi + \tfrac{1}{2}\alpha) - Y_{lm}(\theta, \phi - \tfrac{1}{2}\alpha)\right). && (81) \end{aligned}$$

[15]This result is simply the spherical version of the convolution theorem for Fourier transforms, $\widetilde{f \star g} = \tilde{f}\tilde{g}$.

The azimuthal dependence of Y_{lm} is $\exp(im\phi)$, so

$$d = \sum_{l,m} a_{lm} Y_{lm}(\theta, \phi) \left(\exp(\tfrac{1}{2} im\alpha) - \exp(-\tfrac{1}{2} im\alpha) \right) \qquad (82)$$

$$= \sum_{l,m} a_{lm} Y_{lm}(\theta, \phi) 2i \sin \tfrac{1}{2} m\alpha. \qquad (83)$$

The net result is that a_{lm} is replaced by $2i a_{lm} \sin \tfrac{1}{2} m\alpha$, so modes with low $|m|$ are suppressed. Since m ranges from $-l$ to l, this suppression affects primarily modes with low l.[16]

This suppression is conventionally quantified by computing a "window function" that represents the sensitivity of the experiment to different multipoles. To do this, we compute the mean-square signal,

$$\langle d^2 \rangle = \sum_{l,m} C_l |Y_{lm}(\theta, \phi)|^2 \left(2 \sin \tfrac{1}{2} m\alpha \right)^2 \equiv \sum_l \left(\frac{2l+1}{4\pi} \right) C_l W_l. \qquad (84)$$

The window function W_l is small for low l, indicating that chopping has rendered this experiment insensitive to the largest angular scales.

Note that we have not included beam-smoothing in equation (84). The correct window function, including beam-smoothing, is obtained by multiplying this result by B_l^2.

Equation (84) gives the window function for the particularly simple case of a single-difference experiment. There are more complicated switching strategies, including sinusoidal chops and triple-beam experiments. For a more detailed discussion of window functions, see White & Srednicki (1995).

8. Likelihood Analysis of the COBE Data

In the previous section, we discussed various issues of CMB data analysis from a general point of view. We will now apply what we have learned to a specific example, namely the COBE DMR data. We will not describe the COBE instrument in detail; the interested reader is referred to George Smoot's contribution to this volume, as well as to the papers reporting the four-year DMR data (Bennett *et al.* 1996, Górski *et al.* 1996, Hinshaw *et al.* 1996, Banday *et al.* 1996) and references therein. We will content ourselves with mentioning a few of the most relevant facts.

The COBE DMR produced all-sky maps of the microwave radiation at three frequencies, 31 GHz, 53 GHz, and 90 GHz, with a beam size of 7°

---

[16]The fact that modes with low $|m|$ are suppressed depends on the fact that we have oriented our coordinate system with the chop in the azimuthal direction. In contrast, the statement that, on average, modes with low l are suppressed is independent of the orientation of the coordinate system.

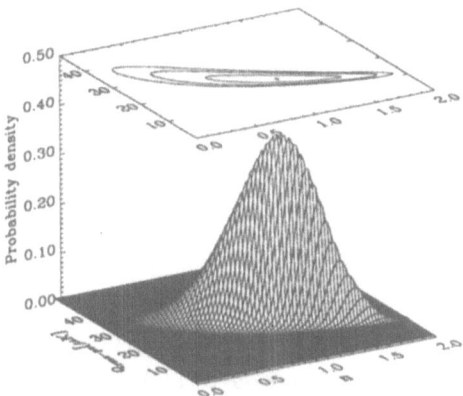

Figure 11. The likelihood function for the two-year COBE DMR data, based on a brute-force analysis involving the entire pixel covariance matrix. Only the Sachs-Wolfe contribution to the anisotropy is included. See Tegmark & Bunn (1995) for further details.

(FWHM). The maps consist of 6144 pixels, although only about 4000 of them are at high enough Galactic latitude to be used for studying the CMB. Although the DMR is a differencing instrument, the data have been used to produce sky maps of $\Delta T/T$, so we do not need to worry about beam-switching in our analysis. We do, however, have to worry about the fact that the maps are insensitive to the monopole and dipole of the anisotropy.[17]

The noise in the COBE maps appears to be Gaussian, and different pixels have noise that is approximately uncorrelated (Lineweaver *et al.* 1994). Therefore, as long as the CMB anisotropy obeys Gaussian statistics, equation (72) applies:

$$\mathcal{L} \equiv -2\ln L = \ln\left((2\pi)^{N_{\mathrm{pix}}} \det \mathbf{M}\right) + \vec{d}^{\mathrm{T}} \mathbf{M}^{-1} \vec{d}, \qquad (85)$$

where $\mathbf{M} = \mathbf{Y}\bar{\mathbf{C}}\mathbf{Y}^{\mathrm{T}} + \mathbf{N}$ with $N_{ij} = \sigma_i^2 \delta_{ij}$. The matrix $\mathbf{M}$ is $\sim 4000 \times 4000$, which is a size that can be inverted, with sufficient patience, on a workstation. Tegmark & Bunn (1995) have performed such a brute-force analysis on the two-year COBE DMR data for a two-parameter family of power spectra, with results shown in Figure 11. However, if we wish to explore a larger parameter space, we must find a more efficient way to compute likelihoods.

---

[17] Actually, COBE is in principle perfectly sensitive to the dipole; however, the intrinsic CMB dipole is impossible to distinguish from the much larger dipole due to our own motion with respect to the CMB center-of-momentum frame.

8.1. DATA COMPRESSION

All likelihood analyses of the COBE data, with the exception of the brute-force analysis mentioned above, have involved some form of data compression. That is, the pixel data vector $\vec{d}$ has been mapped to some smaller-dimensional data vector, which has been used for computing likelihoods. We will focus on *linear* methods of data compression, in which the compressed data vector $\vec{x}$ is linear in $\vec{d}$,

$$\vec{x} = \mathbf{A}\vec{d}, \tag{86}$$

for some $K \times N_{\text{pix}}$ matrix $\mathbf{A}$ with $K < N_{\text{pix}}$. $\vec{x}$ is a Gaussian random vector, so we can use equation (72) to compute the likelihood of $\vec{x}$ in terms of the covariance matrix

$$\mathcal{M} \equiv \langle \vec{x}\vec{x}^{\mathrm{T}} \rangle = \mathbf{A}\mathbf{M}\mathbf{A}^{\mathrm{T}}. \tag{87}$$

Of course, the likelihood computed in this way will not be the same as the true likelihood computed from $\vec{d}$, but we can hope that, if we perform our data compression wisely, we will get a reasonable approximation to the true likelihood.

In effect, linear data compression is equivalent to expanding the sky map in a set of normal modes, namely the rows of $\mathbf{A}$. Each element of the compressed data vector $\vec{x}$ is approximately the integral of the sky map, multiplied by some function: heuristically, we can write

$$x_i = \sum_{j=1}^{N_{\text{pix}}} A_{ij}d_j \approx \int A_i(\hat{\mathbf{r}})d(\hat{\mathbf{r}}) \, d\Omega. \tag{88}$$

If our pixels uniformly covered the whole sky, we would choose these mode functions to be the spherical harmonics by setting $A_{\mu j} = Y_\mu(\hat{\mathbf{r}}_j)$. Then x_μ would be an estimate of a_μ (up to an overall normalization). In fact, we would be performing precisely the analysis described in Section 7.2.

Even though we do not actually have complete sky coverage, there is still nothing stopping us from choosing the rows of $\mathbf{A}$ to be the spherical harmonics. This is in fact the technique described by Górski (1994), which has been applied to the DMR data by Górski *et al.* (1994, 1996).[18] By cutting off the spherical harmonic expansion at $l = 30$, Górski *et al.* compress the data from ~ 4000 to ~ 1000 numbers, with little loss of cosmological information. This is possible because the cosmic signal in the data drops off rapidly with increasing l (due to both the beam cutoff and the shape of

[18]Górski's method involves the additional step of orthogonalizing the spherical harmonics with an algorithm like Gram-Schmidt. Orthogonalizing with respect to the monopole and dipole is an excellent way to render the data insensitive to these modes, but orthogonalizing the modes with $l \geq 2$ with respect to each other has no effect on the likelihoods.

the anisotropy power spectrum), while the noise has approximately equal power in all modes.

8.2. THE KARHUNEN-LOÈVE TRANSFORM

The Karhunen-Loève transform (Karhunen 1947), which is also known as optimal subspace filtering or expansion in signal-to-noise eigenmodes, is another prescription for linear data compression. It was first introduced to CMB data analysis by Bond (1994, 1995, 1996), and has been used extensively on the COBE data (Bunn, Scott, & White 1995; Bunn 1995; White & Bunn 1995; Bunn & Sugiyama 1996; Bunn, Liddle, & White 1996; Bunn & White 1996) as well as in analyzing galaxy catalogues (Vogeley & Szalay 1996).

Let us consider a one-parameter family of power spectra $C_l(q)$, where the true value of q is q_0. We wish to choose our method of data compression (*i.e.*, the matrix $\mathbf{A}$) to enable us to estimate q as well as possible. Specifically, we choose $\mathbf{A}$ to maximize our ability to reject incorrect values of q.

On average, the likelihood function $L(q)$ has a peak at the true value $q = q_0$, so $\langle L'(q_0) \rangle = 0$. The average rejection power is determined by the rate at which the likelihood declines when we move away from this peak. The figure of merit for describing rejection power is therefore

$$\gamma \equiv \left\langle \left. \frac{d^2 \mathcal{L}}{dq^2} \right|_{q=q_0} \right\rangle. \tag{89}$$

The Karhunen-Loève transform consists of choosing the compression matrix $\mathbf{A}$ to maximize γ (for a fixed value of K, the dimension of the compressed data vector).

To solve this optimization problem, we write down the likelihood in terms of the reduced data vector $\vec{x}$,

$$\mathcal{L} = K \ln 2\pi + \mathrm{Tr} \left(\ln(\mathbf{A}\mathbf{M}\mathbf{A}^{\mathrm{T}}) + (\mathbf{A}\mathbf{M}\mathbf{A}^{\mathrm{T}})^{-1} \vec{x}\vec{x}^{\mathrm{T}} \right). \tag{90}$$

Then we compute γ, vary a matrix element A_{ij}, and set $\delta\gamma = 0$. After some algebra, we find that each row $\vec{\alpha}_a$ of $\mathbf{A}$ must satisfy an eigenvalue equation,

$$\mathbf{M}_0' \vec{\alpha}_a = \lambda_a \mathbf{M}_0 \vec{\alpha}_a. \tag{91}$$

Here $\mathbf{M}_0$ is the covariance matrix $\mathbf{M}$ corresponding to the correct parameter value $q = q_0$, and

$$\mathbf{M}_0' = \left. \frac{d\mathbf{M}}{dq} \right|_{q=q_0} \tag{92}$$

The rejection power γ is simply the sum of the squares of the eigenvalues λ_a.

This completes our prescription for choosing the matrix $\mathbf{A}$. We should choose the rows of $\mathbf{A}$ to be the solutions of equation (91) with the largest values of $|\lambda_a|$. Furthermore, we know when it is safe to stop adding new rows: once all of the remaining eigenvalues λ_a are small, we will no longer significantly increase γ by adding more rows to $\mathbf{A}$.

To get an intuitive understanding of the Karhunen-Loève transform, consider the case where the parameter q is the normalization of the power spectrum, so $C_l(q) = qC_l^{(0)}$. Then we can rewrite the eigenvalue equation (91) as

$$\mathbf{M}_{\text{signal}}\vec{\alpha}_a = \hat{\lambda}_a \mathbf{M}_{\text{noise}}\vec{\alpha}_a, \qquad (93)$$

where $\hat{\lambda}_a = \lambda_a/(1 - \lambda_a)$, and $\mathbf{M}_{\text{signal}} = \mathbf{Y}\bar{\mathbf{C}}\mathbf{Y}^{\mathrm{T}}$ and $\mathbf{M}_{\text{noise}} = \mathbf{N}$ are the signal and noise contributions to $\mathbf{M}$. We can see from equation (93) that $\vec{\alpha}_a$ is an eigenvector of $\mathbf{M}_{\text{noise}}^{-1}\mathbf{M}_{\text{signal}}$. This is why Bond (1994, 1995) calls it an "eigenmode of the signal-to-noise ratio." In effect, the Karhunen-Loève transform tells us which directions in the N_{pix}-dimensional pixel space are most sensitive to the cosmic signal, and which are dominated by noise.

The reader, being an extraordinarily perceptive soul, is no doubt wondering at this point whether this whole procedure is worth the trouble. After all, our original goal was to avoid having to invert an $N_{\text{pix}} \times N_{\text{pix}}$ matrix. Now we find ourselves having to solve an N_{pix}-dimensional eigenvalue problem, which is much harder than simply inverting a matrix. Recall, however, that our objection to a brute-force likelihood analysis was that we didn't want to invert the large matrix $\mathbf{M}$ *repeatedly* as we varied the power spectrum. The Karhunen-Loève eigenvalue problem needs to be solved only once, with all future operations being performed on the K-dimensional compressed data vector. Furthermore, it turns out that we can save ourselves a lot of work by solving equation (91) in spherical harmonic space rather than real space (Bunn 1995). Once we choose some cutoff l_{max}, the dimension of the eigenvalue problem is reduced from N_{pix} to $\sim l_{\text{max}}^2$. It turns out that none of the high signal-to-noise eigenmodes have significant power beyond $l = 30$ or so, so we can safely choose l_{max} to be 40 or 50, resulting in a substantial saving in computational effort.

The Karhunen-Loève transform depends on a choice of power spectrum. Ideally, we would like to use the true power spectrum, but of course we don't know the true power spectrum.[19] We must therefore choose a *fiducial power spectrum* more or less arbitrarily. In principle, this could lead to trouble: we might find that the choice of fiducial power spectrum had a significant effect on our final results. There are two ways to address this question: we can repeat the analysis with different fiducial power spectra, and we

---

[19]If we did, there would be no need to perform the analysis!

can perform Monte Carlo simulations to check that the likelihood analysis returns unbiased estimates of the parameters of interest.[20]

In the case of the COBE data, extensive tests have revealed that sensitivity to the fiducial power spectrum is not a problem (Bunn 1995, Bunn & White 1996). For example, the maximum-likelihood normalization of an $n = 1$ Sachs-Wolfe spectrum is $\langle Q \rangle = 18.73 \pm 1.25 \, \mu K$ using an $n = 1$ fiducial power spectrum and $\langle Q \rangle = 18.74 \pm 1.25 \, \mu K$ using an $n = 1.5$ power spectrum. The maximum-likelihood value of n also does not change when we change the fiducial power spectrum. Furthermore, Monte Carlo simulations show that our estimates of $\langle Q \rangle$ and n are unbiased to an accuracy much better than the statistical uncertainty ($\lesssim 0.03 \, \mu K$ and $\lesssim 0.05$ respectively). See Bunn & White (1996) for further details.

8.3. MONOPOLE AND DIPOLE REMOVAL

Since the COBE data do not contain useful monopole and dipole information, it is customary to remove a best-fit monopole and dipole from the data before performing any further analysis. Unfortunately, since incomplete sky coverage destroys the orthogonality of the spherical harmonics, this procedure covertly removes part of the contribution of the higher multipoles. There are two ways to compensate for this.

The first option is to treat the monopole and dipole coefficients (a_{00} and a_{1m}) as "nuisance parameters," *i.e.*, quantities whose true values we neither know nor care about.[21] In the context of Bayesian analysis, the natural thing to do with nuisance parameters it to marginalize over them. Marginalizing over a nuisance parameter ζ means replacing the likelihood L with the marginal likelihood

$$L_{\mathrm{marg}} = \int d\zeta L(\zeta) p(\zeta). \tag{94}$$

Here $p(\zeta)$ is a prior probability density for ζ, which is usually taken to be constant. By marginalizing over the data, we are using a standard identity of probability theory,

$$p(x) = \int p(x \mid \zeta) p(\zeta) \, d\zeta, \tag{95}$$

---

[20]Even in methods that do not involve a choice of fiducial power spectrum, it is wise to perform simulations to test for bias. Even a brute-force likelihood analysis using the full pixel data is not guaranteed to return unbiased parameter estimates.

[21]A parameter may be a nuisance parameter at one time and an interesting parameter at another. For instance, if we want to estimate the spectral index n, we should probably compute $L(\langle Q \rangle, n)$ and treat $\langle Q \rangle$ as a nuisance parameter. At some other time, though, we may think $\langle Q \rangle$ is an interesting thing to know.

to remove all ζ-dependence from the likelihood.

From a frequentist point of view, the natural way to get rid of a nuisance parameter is to maximize with respect to it. That is, we replace L with $\max_\zeta L(\zeta)$. That way, a particular model is ruled out only if it is ruled out for all possible values of ζ.

If we are performing some sort of data compression, then we have a second option for dealing with the monopole and dipole. We can simply impose a constraint on our compression matrix $\mathbf{A}$, requiring it to be insensitive to the unwanted multipoles. This is in effect the approach of Górski (1994): by orthogonalizing the spherical harmonics, he makes his compression matrix insensitive to the monopole and dipole. This approach turns out to be mathematically equivalent to marginalizing over the unwanted modes.

People frequently remove the quadrupole information from the COBE data in the same way as the monopole and dipole, on the grounds that the quadrupole is particularly susceptible to Galactic contamination. It has also been known since the earliest days of COBE analysis that the quadrupole is anomalously low (compared to the prediction of a flat power spectrum normalized to the other multipoles). From a statistical point of view, this is a delicate situation: it is perfectly acceptable, and even wise, to throw away data if there is a reasonable fear of contamination, but throwing away data that is known *a priori* to be discordant with favored theories is a major statistical *faux pas*. On balance, it is probably better to leave the quadrupole information in in the interest of avoiding even the possibility of biased editing of the data.

There is another argument in favor of retaining the quadrupole. Even if the quadrupole is contaminated, it still contains useful information, and so it may be unwise to throw it away entirely. Since the quadrupole is a root-mean-square quantity, any contaminant would tend to bias the quadrupole up. In fact, if a particular theory is ruled out because it predicts too large a quadrupole, hypothesizing an additional quadrupolar contaminant cannot save that theory: as long as the contaminant is statistically independent of the cosmic signal, the net result of hypothesizing a contaminant is necessarily to *lower* the likelihood of that theory.

8.4. RESULTS

The main purpose of this section is to discuss data analysis techniques, not results; however, we will briefly present some results based on a Karhunen-Loève analysis of the four-year COBE DMR data. The reader is referred to Bunn & White (1996) for a more detailed discussion.

The data set used for this analysis consists of a weighted average of the

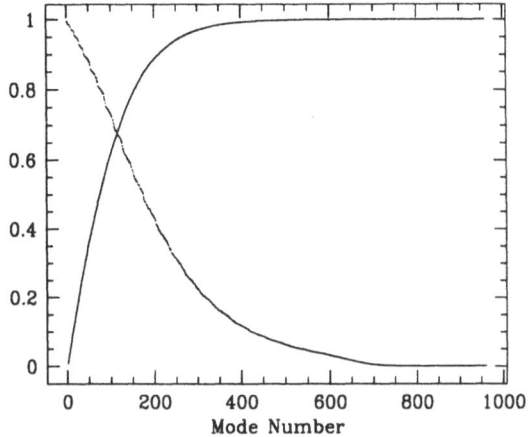

Figure 12. The points show the eigenvalues λ_a of the four-year COBE data, sorted in decreasing order. The solid curve is the running sum of λ_a^2, normalized to 1.

53 and 90 GHz maps from the four-year DMR data. The maps are averaged with weights inversely proportional to the noise variance, in order to minimize noise in the average map. (This is equivalent to performing a joint likelihood analysis of the individual maps.) We performed the Karhunen-Loève analysis using a flat fiducial power spectrum $l(l+1)C_l = \text{const.}$, and we retained the 500 most significant modes.

Figure 12 shows the eigenvalues λ_a, together with a running sum of the squares of the eigenvalues. (Recall that this sum is proportional to the rejection power γ.) This plot indicates that modes beyond the first 500 do not significantly increase our ability to discriminate among models.

Figure 13 shows the likelihood function for low-density CDM models, both with and without a cosmological constant. Figure 14 shows the maximum-likelihood power spectrum, found by allowing each C_l with $2 \leq l \leq 19$ to vary independently. The error bars shown in this figure are standard errors determined by approximating the likelihood near the peak as a Gaussian. The standard errors are then the square roots of the diagonal elements of the covariance matrix of this Gaussian. Error bars determined in this way should be viewed with extreme caution. First, the likelihood is not very well approximated by a Gaussian: on the contrary, it is strongly skew-positive at low l. Second, these standard errors contain no information about correlations between the errors. These correlations are largest for pairs of modes whose l-values differ by 2. (Coupling between modes with $\Delta l = 1$ is weak because the data have approximate reflection symmetry.) The deceptively small error bar on the estimate of C_2 is largely due to the failure of the Gaussian approximation for the likelihood, although the 15%

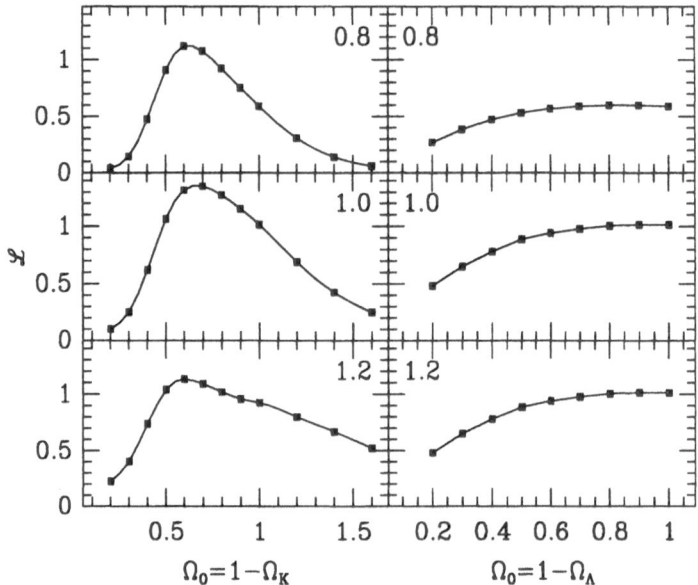

Figure 13.   Likelihood as a function of Ω_0 for CDM models with zero cosmological constant (left) and zero spatial curvature (right). The spectral index n increases from 0.8 to 1.2 from top to bottom. The likelihoods are normalized so that a flat spectrum has $L = 1$. See Bunn & White (1996) for further details.

anticorrelation between C_2 and C_4 also plays a role.

Finally, Table 1 shows values of the small-scale fluctuation amplitude σ_8 for various theoretical models. The observational constraint is approximately $0.5 \lesssim \sigma_8 \lesssim 0.8$ (*e.g.*, Viana & Liddle 1996).

8.5. WIENER FILTERING

Until now, we have focused on attempts to estimate the angular power spectrum C_l. While this is the most useful thing to do with a CMB data set, other complementary approaches can be interesting in certain contexts. For instance, we can assume that we know the angular power spectrum and try to determine the underlying cosmic signal from a noisy sky map. That is, we can attempt to *filter* a sky map, cleaning up the noise and leaving the signal. The Wiener filter (Wiener 1949) is an optimal linear filter for this purpose, in the sense of least squares. The recent use of Wiener filtering in astrophysics is largely due to Rybicki & Press (1992), and the filter has been applied to the COBE data by Bunn, Hoffman, & Silk (1996).

Suppose we have a data vector $\vec{d}$ containing signal and noise. We want to apply a linear filter $\mathbf{F}$ so that $\vec{y} \equiv \mathbf{F}\vec{d}$ approximates the true cosmic

	Ω_0	Ω_Λ	Ω_{HDM}	n	h	$\Omega_B h^2$	σ_8
standard CDM	1.0	0.0	0.0	1.0	0.50	0.0125	1.22
tilted CDM	1.0	0.0	0.0	0.8	0.50	0.0250	0.72
MDM	1.0	0.0	0.2	1.0	0.50	0.0150	0.79
ΛCDM	0.4	0.6	0.0	1.0	0.65	0.0150	1.07
Open CDM	0.4	0.0	0.0	1.0	0.65	0.0150	0.64
Low h CDM	1.0	0.0	0.0	1.0	0.35	0.0150	0.74

TABLE 1. The predicted fluctuation amplitude on scales of $8\,h^{-1}$ Mpc for various CDM-like models. MDM is a "mixed dark matter" model. All normalizations are from the four-year COBE DMR data. See Bunn & White (1996) for further details.

signal $\Delta T/T$ in such a way that the mean-square deviation,

$$\left\langle \left(y_i - \frac{\Delta T}{T}(\hat{\mathbf{r}}_i) \right)^2 \right\rangle, \qquad (96)$$

is as small as possible. The solution to this optimization problem is the Wiener filter,

$$\mathbf{F} = \mathbf{M}_{\mathrm{signal}}\mathbf{M}^{-1}, \qquad (97)$$

where $\mathbf{M}$ is as usual the data covariance matrix and $\mathbf{M}_{\mathrm{signal}}$ is the signal contribution to $\mathbf{M}$.

Under the assumption of Gaussian statistics, the Wiener-filtered data is also the maximum-likelihood estimator of $\Delta T/T$ at each point. Note that in regions of very high noise, where we have little information, the Wiener filter returns values near zero, because this is the most likely *a priori* value of a zero-mean Gaussian.

Figure 15 shows a Wiener-filtered COBE sky map. Although the signal-to-noise ratio in the raw pixel maps is typically less than one per pixel, the largest-amplitude features in the filtered map are significant at the five sigma level per pixel.

One of the main uses of the filtered maps is in making predictions for other experiments. Assuming Gaussian statistics, the full error covariance matrix of the Wiener-filtered map is known, and so we can produce maps with known uncertainties of a region of the sky. For predictions of the CMB sky as it should be seen by the Tenerife experiment, see Bunn, Hoffman, & Silk (1996).

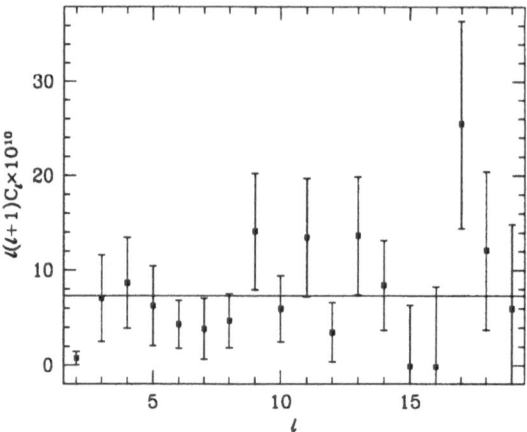

Figure 14. The points represent the maximum-likelihood power spectrum, obtained by letting all C_l's between 2 and 19 vary freely. A flat $\langle Q \rangle = 19\,\mu K$ power spectrum is plotted for comparison. The error bars are standard errors determined by approximating the likelihood by a Gaussian near the peak. Because the Gaussian approximation is poor, and because there are significant correlations between the errors, these error bars can be deceptive. The small formal error on C_2 is particularly misleading. See Bunn & White (1996) for further discussion.

9. Summary

The main lesson to be learned from this entire institute is that this is an exciting time in CMB research. The existing data are already telling us vast amounts about cosmology, and in the next few years the data should continue to improve dramatically. The high quality of present and future anisotropy observations presents us with some challenges. We must understand our theoretical models well enough to make accurate predictions, and we must develop statistical tools that enable us to determine which predictions are consistent with the data. Both of these challenges are currently being met with ever-increasing success.

The tools for making accurate predictions, at least in linear models like CDM, are by now quite well developed. Furthermore, in recent years analytic and semianalytic approximations have dramatically improved our understanding of the basic physical principles involved in anisotropy formation.

The problem of data analysis is also much better understood today than it was five years ago (before there were any actual detections to analyze). However, it is important to remember that analysis of future data sets will present challenges that make the COBE analysis look easy. When sky maps contain a million pixels instead of a few thousand, data compression will be absolutely essential. It is already time to start thinking about this difficult

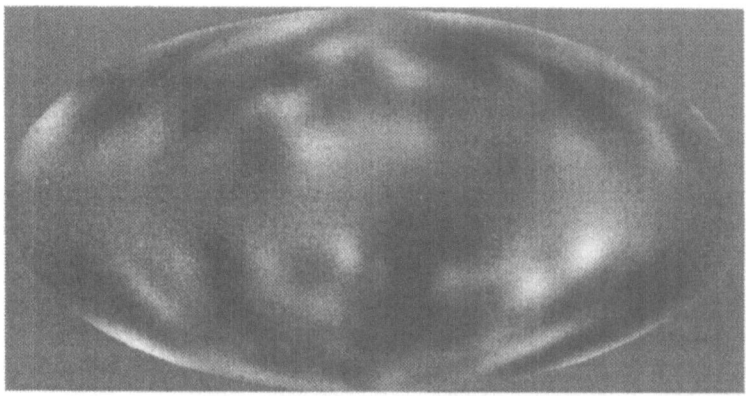

Figure 15. A sky map, in Aitoff projection, of the Wiener-filtered four-year DMR data. The relative lack of structure near the Galactic plane is due to the fact that no data from that region were used.

problem.

In addition, future experiments with higher resolution than COBE will be more susceptible to foreground contamination. In the case of COBE, it is believed that simply excising points too close to the Galactic plane is sufficient to remove most of the foreground contamination; for future high-sensitivity degree-scale experiments, more sophisticated methods will be necessary.

We have seen that CDM-like theoretical models predict that vast amounts of information are encoded in the CMB anisotropy power spectrum. There is a very real hope that the CMB will give us accurate values for all sorts of cosmological parameters. But even if the information is there, we will have to do a lot of work to wrest it from the data.

10. Acknowledgments

The first half of these lectures is based largely on the work of Wayne Hu and Naoshi Sugiyama. Much of the later material is based on work I performed in collaboration with Douglas Scott, Joseph Silk, Max Tegmark, and Martin White. I would like to thank all of these people for many helpful discussions. In addition, Wayne, Max, and Martin made some of the figures. Finally, I would like to thank the organizers of the meeting for their hard work and hospitality.

References

1. Banday, A. *et al.* 1997, Ap. J., 475, 393.
2. Bennett, C.L. *et al.* 1992, Ap. J. Lett., 396, 6.

3. Bennett, C.L. *et al.* 1996, Ap. J. Lett., 464, 1.
4. Berger, J.O. 1985, *Statistical Decision Theory and Bayesian Analysis* (Springer-Verlag).
5. Blandford, R.D. & Narayan, R. 1992, Ann. Rev. Astron. Astrophys., 30, 311.
6. Bond, J.R. 1994, in *Proceedings of the IUCAA Dedication Ceremonies*, (ed. T. Padmanabhan; John Wiley & Sons).
7. Bond, J.R. 1995, Phys. Rev. Lett., 74, 4369.
8. Bond, J.R. 1996, in *Cosmology and Large-Scale Structure* (1994 Les Houches summer school; ed., R. Schaefer; Elsevier, in press.)
9. Bond, J.R. & Efstathiou, G. 1984, Ap. J. Lett., 285, 45.
10. Bond, J.R. & Efstathiou, G. 1987, M.N.R.A.S, 226, 655.
11. Brandenberger, R. 1996, astro-ph/9604033, in *Pacific Conference on Gravitation and Cosmology* (World Scientific, in press).
12. Bunn, E.F. 1995, *Statistical Analysis of Cosmic Microwave Background Anisotropy*, Ph.D. thesis, Physics Department, U.C. Berkeley (ftp://pac2.berkeley.edu/pub/bunn/thesis).
13. Bunn, E.F., Hoffman, Y., & Silk, J. 1996, Ap. J., 464, 1.
14. Bunn, E.F., Liddle, A., & White, M. 1996, Phys. Rev. D, 54, R5917.
15. Bunn, E.F., Scott, D., & White, M. 1995, Ap. J. Lett., 441, 9.
16. Bunn, E.F. & Sugiyama, N. 1995, Ap. J., 446, 49.
17. Bunn, E.F. & White, M. 1997, Ap. J., 480, 6.
18. Bunn, E.F., White, M., Srednicki, M., & Scott, D. 1994, Ap. J., 429, 1.
19. Dodelson, S. & Jubas, J. 1995, Ap.J., 439, 503.
20. Doroshkevich, A.G., Zel'dovich, Ya.B., & Sunyaev, R. 1978, Sov. Astron., 22, 523.
21. Górski, K. 1994, Ap. J. Lett., 430, 85.
22. Górski, K. *et al.* 1994, Ap. J. Lett., 430, 89.
23. Górski, K. *et al.* 1996, Ap. J. Lett., 464, 11.
24. Gull, S.F. & Daniell, G.J. 1978, Nature, 272, 686.
25. Gunn, J.M. & Peterson, B.A. 1965, Ap. J., 142, 1663.
26. Hinshaw, C. *et al.* 1996, Ap. J. Lett., 464, 17.
27. Hu. W. 1995, *Wandering in the Background: A Cosmic Background Explorer*, Ph.D. Thesis, Physics Department, U.C. Berkeley (ftp://pac2.berkeley.edu/pub/hu/thesis).
28. Hu, W. 1996, in *The Universe at High-z, Large-Scale Structure and the Cosmic Microwave Background*, eds. E. Martinez-Gonzalez & J.-L. Sanz (Springer-Verlag) (astro-ph/9511130).
29. Hu, W., Scott, D., & Silk, J. 1994, Phys. Rev. D, 49, 648.
30. Hu, W., Scott, D., Sugiyama, N. & White, M. 1995, Phys. Rev. D., 52, 5498.
31. Hu, W. & Sugiyama, N. 1994, in *CWR CMB Workshop: Two Years After COBE*, eds. L. Krauss, P. Kernan (World Scientific, Singapore, p. 188).
32. Hu. W. & Sugiyama, N. 1995a, Phys. Rev. D, 51, 2599.
33. Hu, W. & Sugiyama, N. 1995b, Ap. J., 444, 489.
34. Hu. W. & Sugiyama, N. 1996, Ap. J., 471, 542.
35. Hu, W., Sugiyama, N., & Silk, J. 1996, Nature, 386, 37 (see also http://www.sns.ias.edu/~whu/physics/physics.html).
36. Hu, W. & White, M. 1996, in *Proceedings of the XXXIst Moriond Meeting, Microwave Background Anisotropies*, in press.
37. Jackson, J.D. 1975, *Classical Electrodynamics* (Wiley).
38. Jungman, G., Kamionkowski, M., Kosowsky, A., & Spergel, D. 1996, Phys. Rev. D, 54, 1332.
39. Karhunen, K. 1947, *Über Lineare Methoden in der Wahrscheinlichkeitsrechnung* (Kirjapaino oy. Sana, Helsinki).
40. Kosowsky, A., Kamionkowski, M., Jungman, G., & Spergel, D.N. 1996 preprint (astro-ph/9605147).

41. Lineweaver, C.H. *et al.* 1994, Ap. J., 436, 452.
42. Netterfield, C.B., Devlin, M.J., Jarosik, N., Page, L., & Wollack, E.J. 1997, Ap. J., 474, 47.
43. Ostriker, J. & Vishniac, E. 1986, Ap. J., 306, 51.
44. Peebles, P.J.E. 1987, Nature, 327, 210.
45. Peebles, P.J.E. & Yu, J.T. 1970, Ap. J., 162, 815.
46. Press, W.H. 1996, in *Unsolved Problems in Astrophysics* (ed. J.P. Ostriker, Princeton University Press, in press, astro-ph/9604126).
47. Rees, M. & Sciama, D. 1968, Nature, 519, 611.
48. Rice, J.A. 1995, *Mathematical Statistics and Data Analysis* (Duxbury).
49. Rybicki, G. & Press, W. 1992, Ap. J. Lett., 432, 75.
50. Sachs, R.K. & Wolfe, A.M. 1967, Ap. J., 147, 73.
51. Scaramella, R. & Vittorio, N. 1993, M.N.R.A.S, 263, 17.
52. Scott, D., Silk, J. & White, M. 1995, Science, 268, 829.
53. Seljak, U. 1994, Ap. J. Lett., 435, 87.
54. Seljak, U. 1996a, Ap. J., 460, 549.
55. Seljak, U. 1996b, Ap. J., 463, 1..
56. Seljak, U. & Bertschinger, E. 1994, Ap. J. Lett., 417, 9.
57. Seljak, U. & Zaldarriaga, M. 1996, Ap. J., 469, 437.
58. Silk, J. 1967, Nature, 215, 1155.
59. Silk, J. 1968, Ap. J. Lett., 151, 459.
60. Silk, J. 1982, Acta Cosmologica, 11, 75.
61. Smoot, G. *et al.* 1992, Ap. J. Lett., 396, 1.
62. Sunyaev, R.A. 1977, Sov. Astron. Lett., 3, 491.
63. Sunyaev, R.A. & Zel'dovich, Ya.B. 1970, Astrophys. Space Sci., 7, 3.
64. Sunyaev, R.A. & Zel'dovich, Ya.B. 1972, Comm. Astrophys. Space Phys., 4, 73.
65. Tegmark, M. 1995, Ap. J., 455, 429.
66. Tegmark, M. 1996a, M.N.R.A.S, 280, 299.
67. Tegmark, M. 1996b, Ap. J. Lett., 464, 35.
68. Tegmark, M. 1996c, to appear in Proc. Enrico Fermi, Course CXXXII, Varenna (astro-ph/9511148).
69. Tegmark, M. & Bunn, E.F. 1995, Ap. J., 451, 1.
70. Tegmark, M. & Efstathiou, G. 1996, M.N.R.A.S, 281, 1297.
71. Tegmark, M., Taylor, A., & Heavens, A. 1997, Ap. J., 480, 22.
72. Viana, P.T.P. & Liddle, A. 1996, M.N.R.A.S, 281, 323.
73. Vittorio, N. & Silk, J. 1984, Ap. J. Lett., 285, 39.
74. Vogeley, M.S. & Szalay, A.S. 1996, Ap. J., 465, 34.
75. White, M. & Bunn, E.F. 1995, Ap. J., 450, 477.
76. White, M., Scott, D., & Silk, J. 1994, Ann. Rev. Astron. Astrophys., 32, 319.
77. White, M. & Srednicki, M. 1995, Ap. J., 443, 6.
78. Wiener, N. 1949, *Extrapolation and Smoothing of Stationary Time Series* (Wiley).
79. Wilson, M. 1983, Ap. J., 273, 2.
80. Wilson, M. & Silk, J. 1981, Ap. J., 243, 14.
81. Wright, E.L. *et al.* 1992, Ap. J. Lett., 396, 11.
82. Wright, E.L. *et al.* 1994a, Ap. J., 420, 1.
83. Wright, E.L. *et al.* 1994b, Ap. J., 436, 443.
84. Zel'dovich, Ya.B. & Sunyaev, R., Astrophys. Space Sci., 4, 301 (1969).

THE CMB ANISOTROPY EXPERIMENTS

Cosmic Microwave Background

GEORGE F. SMOOT

Lawrence Berkeley National Lab & Physics Department
University of California
Berkeley CA 94720

1. Abstract

Anisotropies in the cosmic microwave background (CMB) encode information about the evolution and development of the Universe. It is understood that quality observations of the CMB anisotropies can provide a very strong test of cosmological models and provide high precision measurements of major cosmological parameters. This paper provides a review of the COBE DMR results, the current status of the measurements of the CMB anisotropy power spectrum and then focuses on the programs that are likely to provide additional results including both suborbital observations and the two selected satellite missions: the NASA MidEX mission MAP and the ESA M3 mission Max Planck Surveyor (formerly COBRAS/SAMBA). This review includes both a description of the experimental programs and the expected quality level of results.

2. Introduction

The observed cosmic microwave background (CMB) radiation provides strong evidence for the big bang model of cosmology and is the best probe we have for determining conditions in the early Universe as well as determining many important cosmological parameters. The angular power spectrum of the CMB contains information on virtually all cosmological parameters of interest, including the geometry of the Universe (Ω), the baryon density (Ω_b), the Hubble expansion rate (h), the cosmological constant (Λ), the number of light neutrinos (n_ν), the ionization history of the Universe, and the amplitudes and spectral indices of the primordial and tensor perturbation spectra. Precise CMB observations, data analysis, and interpretation can distinguish between cosmological models. They can be

185

C. H. Lineweaver et al. (eds.), The Cosmic Microwave Background, 185–240.

used to verify that the range of models under consideration is plausible and to distinguish between models with primordial perturbations (e.g. the inflationary big bang) or those with active perturbations (e.g. topological defects which must result from spontaneous symmetry breaking of unified forces). Once a model is thus singled out, its parameters can, in principle, be determined to accuracies of the order of a per cent [40].

Since the initial detection of CMB temperature anisotropies by the *COBE* DMR [74], over a dozen other balloon-borne and ground-based experiments have reported anisotropy detections on smaller angular scales. With the existence of anisotropies now firmly established, observational goals have shifted towards an accurate determination of the CMB anisotropy power spectrum over a wide range of angular scales. The reasons for this are two-fold. (1) If the processes producing the initial fluctuations are stochastic and random phase, then the power spectrum contains all the information of the underlying physical model. (2) It is observationally easier to obtain a power spectrum than a fully reliable map. Several technical advances, including improved receivers, interferometry, and long-duration balloon flights, hold great promise for high-precision maps in the next few years. Ultimately, the two approved satellites: NASA MidEX mission MAP and ESA M3 mission Planck are expected to provide high-angular-resolution high-sensitivity maps of the entire sky in multifrequency bands. Thus we can anticipate increasingly complex data sets requiring sophisticated analysis: COBE DMR 4-year maps (6144 pixels), the CfPA balloon experiments MAXIMA/BOOMERANG (26,000 to 130,000 pixels), MAP ($\approx$400,000 pixels), and Planck ($> 10^6$ pixels). These maps then hold the promise of revolutionizing cosmology in terms of making it significantly more precise in quantitative terms.

It is the goal of these lectures to provide the background necessary to understand the existing data, soon to be achieved data from experiments in progress, and finally the forthcoming data from the more advanced experiments and space missions. We proceed with some historical context, a review of the COBE observations both for the discovery of anisotropy and as a prototype for the next generation of space missions, a review of the current and proposed generations of balloon-borne experiments and interferometers, and finally a discussion of the new space-based experiments.

3. CMB Background

Primordial nucleosynthesis calculations require a cosmic background radiation (CBR) with a temperature $kT \sim 1$ MeV at a redshift of $z \sim 10^9$. Gamow, Alpher, & Herman [3] realized that this CBR was required and predicted its evolution to a faint residual relic radiation with a current temperature of a few degrees. Our more modern view of the hot big bang models gives the cosmic background radiation a very central role in the development of the Universe.

The CMB was serendipitously discovered by Penzias & Wilson [64] in 1964 (published in 1965) and they noted that it was isotropic to the sensitivity of their measurement ($<10\%$). The observed CMB spectrum is well characterized by a 2.73 ± 0.01 K blackbody spectrum. The hot big bang model predicts that the CBR should have a thermal spectrum and this is verified precisely. Combined with the observed spectrum of the dipole anisotropy, this precision thermal spectrum also provides us with the knowledge and ability to separate CMB anisotropies from the various foregrounds. Anisotropies will in general have a spectrum set by the derivative of the CMB spectrum. See my previous lectures for a more complete discussion of the expected anisotropy spectrum.

4. Theoretical Anisotropies

There are three primary threads of science that are pursued by CMB anisotropy measurements:

(1) Initial Conditions for Large Scale Structure Formation: The formation of galaxies, clusters of galaxies, and large scale structures is a key issue in cosmology. Theory indicates that whatever the seeds of structure formation, they will leave their imprint as anisotropies in the CMB. Different scenarios for structure formation will leave different anisotropies.

(2) Physics of the Early Universe: CMB anisotropy measurements are a probe of the ultra-high energy physics and processes that occur in the very early Universe. These observations are a probe of inflation or quantum gravity and a test of potential topological defects (monopoles, strings, domain walls, and textures) that must result from spontaneous symmetry breaking.

(3) Geometry and Dynamics of the Universe: Observations of the CMB anisotropies provide information on the metric and topology of the space-time, the isotropy of expansion of space-time, the curvature of space, and the possible rotation and shear of the Universe.

There is now a fairly extensive literature on the first two areas and the third is what we consider classical cosmology. In the early history of the field all measurements led only to upper limits on CMB anisotropy and these in turn led to limits on but not a measurement of these processes.

A major finding of the initial *COBE* DMR discovery [74] was that the CMB was anisotropic on all observed angular scales. A key question is what these anisotropies represent. Immediately, the interpretation focused upon the seeds of large scale structure formation. In the early 1970's the observed large scale structure and scaling arguments led to the prediction [62],[33],[88] that the primordial gravitational potential perturbations must have an equal *rms* amplitude on all scales. This corresponds to a matter density perturbation power-law spectrum, $P(k) \propto k^n$, where k is the comoving wavenumber, with $n = 1$. At that time there were no known mechanisms for producing such a scale-invariant power spectrum of fluctuations. In 1982 it was found that inflationary models predicted nearly scale-invariant perturbations as a result of quantum mechanical fluctuations at very early times. Even with the proliferation of inflationary models, it is found that essentially all reasonable inflationary models predict $n \approx 1$. Presumably, a more reasonable class of inflationary models will result in requiring a tie to particle physics. It is now known that topological defects naturally produce scale-invariant fluctuations. Thus there are at least two known mechanisms for producing a nearly scale invariant primordial perturbation spectrum.

The translation from a scale-invariant spectrum of perturbations to the CMB temperature anisotropies depends upon angular scale and the contents of the Universe. On large angular scales the results of most models are fairly similar. Including the effects of a standard cold dark matter model, a Harrison-Zeldovich $n = 1$ universe is consistent with the power spectrum measured by the *COBE* DMR data. The observed power spectrum of fluctuation amplitudes is also consistent with models of large scale structure formation based upon primordial seeds produced by quantum fluctuations or topological defects in the early Universe.

The physics of anisotropy caused by primordial density perturbations is usually divided into four generic areas although they are all treated properly in the full Sachs-Wolfe effect [69]. These effects are: the gravitational redshift which dominates at large angular scales, the (Rees-Sciama) effect on light propagating through a changing potential, the Doppler effect caused by the motion of the observer or the source, and temperature or entropy variations.

Models of the formation of structure in the Universe fall into two broad classes: inflationary and defect models. Each model predicts an angular power spectrum of CMB anisotropy described in terms of the amplitude of the spherical harmonic of multipole order ℓ.

The most detailed theoretical work has been carried out for inflationary models. Hu, Spergel, and White [39] have argued that all inflationary models produce an angular power spectrum with a unique set of "doppler" or "acoustic" peaks between $\ell = 100$ and $\ell = 1000$ ($11' < \theta < 1.8°$). The relative position and height and the detailed shape of the peaks provides more independent constraints than there are parameters in the inflationary scenario and allow their determination.

Defect models, including cosmic strings and textures, provide an alternative to inflation. Such models predict a non-gaussian distribution of temperature fluctuations and a power spectrum different from that of inflation.

Most cosmological models do not predict the exact CMB temperature pattern that would be observed in our sky, but rather predict a statistical distribution of anisotropies. In the context of such models, the CMB temperature observed in our sky is only a single realization drawn from the cosmic statistical distribution. Theoretical models most often predict a power spectrum in spherical harmonic amplitudes; as the physics of the models leads to primordial fluctuations that are Gaussian random fields, the power spectrum is sufficient to characterize the results. Observations of the sky can be expressed as a spherical harmonic temperature expansion $T(\theta, \phi) = \sum_{\ell m} a_{\ell m} Y_{\ell m}(\theta, \phi)$. If the original perturbations are Gaussian random fields, the $a_{\ell m}$ are Gaussianly distributed, and the power at each ℓ is $(2\ell + 1)C_\ell/(4\pi)$, where $C_\ell \equiv \langle |a_{\ell m}|^2 \rangle$, is sufficient to characterize the results. For an idealized full-sky observation, the variance of each measured C_ℓ is $[2/(2\ell + 1)]C_\ell^2$. This sampling variance (known as cosmic variance) comes about because each C_ℓ is chi-squared distributed with $(2\ell + 1)$ degrees of freedom for our observable volume of the Universe [83]. Thus, in addition to experimental uncertainties, we account for the *cosmic sample variance* uncertainties due to our observation of a single realization in our analyses of the DMR maps. Cosmic variance exists independently of the quality of the experiment. The power spectrum from the 4-year DMR map is cosmic variance limited for $\ell \lesssim 20$.

5. The Legacy of COBE

The Cosmic Background Explorer (*COBE*) was NASA's first satellite dedicated to cosmology and is a milestone for that and for the legacy of information that it has provided on the early Universe. Much of its results are from observations of the cosmic microwave background (CMB). The CMB is a pillar of the Big Bang model and encodes information about energy release in the early Universe, primordial perturbations, and the geometry of the Universe. As mentioned previously the CMB contains information on critical cosmological parameters such as Ω_0, Ω_b, Ω_Λ, and H_0. *COBE's* legacy of the precise measurement of the CMB spectrum and the discovery and early mapping of the CMB anisotropy low-ℓ power spectrum provides a position from which to carry out a program testing our cosmological theories and understanding the early Universe precisely.

The Differential Microwave Radiometers (DMR) experiment ([73]) discovered CMB anisotropies from analysis of its first year of data [74], [9], [85], [43]. The CMB temperature fluctuations were measured at an angular resolution of 7° at frequencies of 31.5, 53, and 90 GHz. These results were supported by a detailed examination of the DMR calibration and its uncertainties ([10]) and a detailed treatment of the upper limits on residual systematic errors ([43]). The *COBE* results were confirmed by the positive cross-correlation between the *COBE* data and data from balloon-borne observations at a shorter wavelength [31] and later by comparison of the *COBE* data and data from the ground-based Tenerife experiment [51] at longer wavelengths. The positive correlation at both longer and shorter wavelengths provides confidence in the results. The results from analysis of two years of DMR data [7] reconfirmed the results from the first year data.

This section summarizes the key and most recent results from COBE. Details can be found in the original references and in the most recent FIRAS paper [30] and in a set of DMR 4-year analysis papers [4], [5], [6], [32], [35], [36], [52], [45], [46], [47], [86].

5.1. THE *COBE* DMR INSTRUMENTS & DATA ANALYSIS

The DMR consists of 6 differential microwave radiometers: 2 nearly independent channels, labeled A and B, at frequencies 31.5, 53, and 90 GHz (wavelength 9.5, 5.7, and 3.3 mm respectively). Each radiometer measures the difference in power between two 7° fields of view separated by 60°, 30° to either side of the spacecraft spin axis [73]. Figure 1 shows a schematic signal path for the DMRs. *COBE* was launched from Vandenberg Air Force Base on 18 November 1989 into a 900 km, 99° inclination circular orbit, which precesses to follow the terminator (light dark line on the Earth) as the Earth orbits the Sun. Attitude control keeps the spacecraft pointed away from the Earth and nearly perpendicular to the Sun with a slight backward tilt so that solar radiation never directly illuminates the aperture plane. The combined motions of the spacecraft spin (75 s period), orbit (103 m period), and orbital precession ($\sim$ 1° per day) allow each sky position to be compared to all others through a highly redundant set of temperature difference measurements spaced 60° apart. The on-board processor box-car integrates the differential signal from each channel for 0.5 s, and records the digitized differences for daily playback to a ground station.

Ground data processing consists of calibration, extensive systematic error analyses, and conversion of time-ordered-data to sky maps [45]. Checks on the correlated noise in the maps [50] due to the map-making process indicate they are well below the 1% level. The DMR time-ordered-data include systematic effects such as emission from the Earth and Moon, the instrument's response to thermal changes, and the instrument's response to the Earth's magnetic field. The largest detected effects do not contribute significantly to the DMR maps; they are either on time scales long compared to

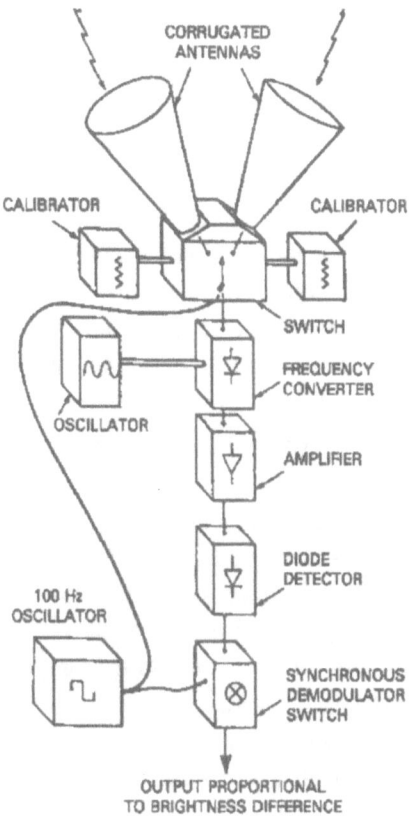

CORRUGATED
ANTENNAS

CALIBRATOR

CALIBRATOR

SWITCH

OSCILLATOR

FREQUENCY
CONVERTER

AMPLIFIER

DIODE
DETECTOR

100 Hz
OSCILLATOR

SYNCHRONOUS
DEMODULATOR
SWITCH

OUTPUT PROPORTIONAL
TO BRIGHTNESS DIFFERENCE

Figure 1. DMR signal flow schematic

the spacecraft spin sampling (e.g. thermal gain drifts) or have time dependence inconsistent with emission fixed on the celestial sphere (e.g. magnetic
effects). Detected and potential systematic effects were quantitatively analyzed in detail [45]. Data with the worst systematic contamination (lunar
emission, terrestrial emission, and thermal gain changes) were not used in
the map-making process and constitute less than 10% of the data in the 53
and 90 GHz channels. The remaining data were corrected using models of
each effect. The data editing and correction parameters were conservatively
chosen so that systematic artifacts, after correction, are less than 6 μK rms
(95% confidence upper limit) in the final DMR map in the worst channel.
This is significantly less than the levels of the noise and celestial signals.

A dipole $T_d = 3.356$ mK anisotropy signal (thermodynamic temperature
in Galactic coordinates Cartesian components $[X,Y,Z] = [-0.2173, -2.2451, +2.4853]$ mK) is subtracted from the time-ordered differential data prior
to forming the 4-year sky maps to reduce spatial gradients within a single
pixel. A small residual dipole remains in the maps from a combination of
CMB and Galactic emission. The mean signal-to-noise ratios in the 10°

smoothed maps are approximately 0.5, 1.5, and 1.0 for 31, 53, and 90 GHz, respectively. For a multi-frequency co-added map the signal-to-noise ratio is $\sim$ 2. This signal-to-noise level is adequate to portray an accurate overall visual impression of the anisotropy. Visual comparison of the full sky maps at each frequency, after averaging the A and B channels, removing the CMB dipole, and smoothing to 10° effective resolution show coincident features. Well off the Galactic plane these are clearly true CMB anisotropy features. Simulated data in combination with the noise appropriate to 1-, 2-, and 4-years of DMR 53 GHz observations show the convergence of the DMR maps with the input simulated data. Increasing years of data result in the emergence of the input large scale features. We can be confident that the large scale features in the 4-year DMR maps are real features rather than confusing noise.

Given the sensitivity of the 4-year DMR maps we have extended the cut made in our previous analyses to exclude additional Galactic emission. We use the *COBE* DIRBE 140 μm map as a guide to cut additional Galactic emission features. The full sky DMR maps contain 6144 pixels. An optimum Galactic cut maximizes the number of remaining pixels while minimizing the Galactic contamination. This cut leaves 3881 pixels (in Galactic pixelization) while eliminating the strongest $|b| > 20°$ Galactic emission. Moderate changes to this cut will cause derived CMB parameters to change somewhat, but this is consistent with the data sampling differences of real CMB anisotropy features and not necessarily Galactic contamination. Likewise, derived CMB parameters also vary by the expected amount when the maps are made in ecliptic rather than Galactic coordinates since about 1/2 of the noise is re-binned.

Kogut et al. [46] examine the Galactic contamination of the high Galactic latitude regions of the DMR maps which remain after the Galactic emission cut (described above). No significant cross-correlation is found between the DMR maps and either the 408 MHz synchrotron map or the synchrotron map derived from a magnetic field model [9]. This places an upper limit $T_{synch} < 11$ μK (95% confidence) on synchrotron emission at 31 GHz.

A significant correlation is found between the DMR maps and the dust-dominated DIRBE 140 μm map, with frequency dependence consistent with a superposition of dust and free-free emission. The correlation is really with a component of Galactic emission with a spectral index of about -2.1 which could be very flat spectrum synchrotron as expected where cosmic ray electron acceleration is actually occurring. We use the term free-free to stand for this component which corresponds to a 7° rms free-free emission component of 7.1 ± 1.7 μK at 53 GHz and a dust component of 2.7 ± 1.3 μK at 53 GHz. Since this emission is uncorrelated with CMB anisotropies it constitutes < 10% of the CMB power. The amplitude of the correlated free-free component at 53 GHz agrees with a noisier estimate of free-free emission derived from a linear combination of DMR data which includes *all* emission with free-free spectral dependence. The combined dust and free-free emission contribute $10 \pm 4 \ \mu K \ rms$ at both 53 and 90 GHz, well below the 30 μK cosmic signal. These Galactic signal analyses are consistent with the fact that the fitted cosmological parameters are nearly unaffected by removal of modeled Galactic signals [32], [35] with the notable exception of the quadrupole, which has significant Galactic contamination [46]. A search by Banday et al. [4] finds no evidence for significant extragalactic contamination of the DMR maps.

5.2. FOUR-YEAR DMR RESULTS

Monopole $\ell = 0$: Despite the fact that the DMR is a differential instrument, the known motion of the *COBE* spacecraft about the Earth and the motion of the Earth about the Solar System barycenter provide a means to determine the CMB monopole temperature from the DMR data. The CMB at millimeter wavelengths is well described by a blackbody spectrum [56],[30]. The Doppler effect from the combined spacecraft and Earth orbital motions creates a dipole signal $T(\theta) = T_0[1 + \beta \cos(\theta) + O(\beta^2)]$, where $\beta = v/c$ and θ is the angle relative to the time-dependent velocity vector. The satellite and Earth orbital motions are well known and change in a regular fashion, allowing their Doppler signal to be separated from fixed celestial signals. We fit the time-ordered data to the Doppler dipole and recover a value for the CMB monopole temperature, $T_0 = 2.725 \pm 0.020$ K [45].

Dipole $\ell = 1$: The CMB anisotropy is dominated by a dipole term attributed to the motion of the Solar System with respect to the CMB rest frame. A precise determination of the dipole must account for Galactic emission and the aliasing of power from higher multipole orders once pixels near the Galactic plane are discarded. One can account for Galactic emission by using a linear combination of the DMR maps or by cross-correlating the DMR maps with template sky maps dominated by Galactic emission [46]. The high-latitude portion of the sky is fitted for a dipole with a CMB frequency spectrum using a pixel-based likelihood analysis [35]. Accounting for the smoothing by the DMR beam and map pixelization, the CMB dipole has amplitude 3.353 ± 0.024 mK toward Galactic coordinates $(l, b) = (264°.26 \pm 0°.33, 48°.22 \pm 0°.13)$, or equatorial coordinates $(\alpha, \delta) = (11^h 12^m.2 \pm 0^m.8, -7°.06 \pm 0°.16)$ epoch J2000 [6].

A second analysis approach utilizes a phenomenological estimate of the Galactic foreground by examining the dipole fitted parameters as a function of cuts in Galactic latitude. It was found that the largest source of error in the dipole direction was reduced by using lower Galactic latitude cuts. Using the four year data set from all six channels of the COBE Differential Microwave Radiometers (DMR), the best-fit dipole amplitude $3.358 \pm 0.001 \pm 0.023$ mK amplitude in the direction $(\ell, b) = (264°.31 \pm 0°.04 \pm 0°.16, +48°.05 \pm 0°.02 \pm 0°.10)$ where the first uncertainties are statistical and the second are estimates of the combined systematics [52]. In celestial coordinates the dipole is RA $11^h 11^m 57^s \pm 23^s$ and Dec $-7.22° \pm 0.08°$ (J2000).

These dipole measurements are consistent with previous DMR and FIRAS results.

Quadrupole $\ell = 2$: On the largest angular scales (e.g., quadrupole), Galactic emission is comparable in amplitude to the anisotropy in the CMB. The quadrupole amplitude is found by a likelihood analysis which simultaneously fits the high-latitude portion of the DMR maps for Galactic emission traced by synchrotron- and dust-dominated surveys and a quadrupole anisotropy with a thermodynamic frequency spectrum [46],[35]. After correcting for the positive bias from instrument noise and aliasing, the CMB quadrupole amplitude observed at high latitude is $Q_{rms} = 10.7 \pm 3.6 \pm 7.1$ μK, where the quoted errors reflect the 68% confidence uncertainties from random statistical errors and Galactic modeling errors, respectively. The observed quadrupole amplitude, Q_{rms}, has a lower value than the quadrupole expected from a fit to the entire power spectrum, Q_{rms-PS}, but whether this is a chance result of cosmic variance or reflects the cosmology of the Universe cannot be determined from *COBE* data. The 68% confidence interval for the quadrupole amplitude, 6 μK $\leq Q_{rms} \leq 17$ μK, is consistent with the quadrupole normalization of the full power spectrum power-law fit (discussed below): $Q_{rms-PS} = 15.3^{+3.8}_{-2.8}$ μK.

Power spectrum $\ell \geq 2$: The simplest probe of the angular power spectrum of the anisotropy is its Legendre transform, the 2-point correlation function. The 2-point correlation function of the 4-year maps is analyzed by Hinshaw et al. [36], where it is shown that the 2-point data are consistent from channel to channel and frequency to frequency. The data are robust with respect to the angular power spectrum. A Monte Carlo-based Gaussian likelihood analysis determines the most-likely quadrupole normalization for a scale-invariant ($n = 1$) power-law spectrum. The results are summarized in Table 1 which also includes the results of 3 additional, independent power spectrum analyses, discussed below. The normalization inferred from the 2-point function is now in better agreement with other determinations than was the case with the 2-year data. The change is due to data selection: with the 2-year data, we only analyzed the 53 × 90 GHz cross-correlation function; with the 4-year data we have analyzed many more data combinations, including the auto-correlation of a co-added, multi-frequency map.

This latter combination is more comparable to the data analyzed by other methods, and the 2-point analysis yields consistent results in that case. The combined 31, 53 and 90 GHz CMB rms is 29 ± 1 μK in the 10° smoothed map [5], consistent with the level determined by the 2-point results.

It is logical to analyze the power spectrum directly in terms of spherical harmonics. However, this involves considerable subtlety since the removal of the Galactic plane renders the harmonics non-orthonormal, producing strong correlations among the fitted amplitudes. Wright et al. [86] found an angular power spectrum by modifying and applying the technique described by Peebles [61] and Hauser & Peebles [34] for data on the cut sphere. They compute a Gaussian likelihood on these data and calibrate their results with Monte Carlo simulations. Górski et al. [32] explicitly construct orthonormal functions on the cut sphere and decompose the anisotropy data with respect to these modes. They form and evaluate an exact Gaussian likelihood directly in terms of this mode decomposition. The results of these analyses are summarized in Table 1. Further details, including results from other data combinations are given in the respective papers.

Hinshaw et al. [35] evaluate a Gaussian likelihood directly in terms of a full pixel-pixel covariance matrix, a technique applied to the 2-year data by Tegmark & Bunn [79]. The results of the power-law spectrum fits are summarized in Table 1. Hinshaw et al. [35] also analyze the quadrupole anisotropy separately from the higher-order modes, to complement the analysis of Kogut et al. [46]. They compute a likelihood for the observed quadrupole Q_{rms}, nearly independent of higher-order power, and show that it peaks between 6 and 10 μK, depending on Galactic model, but that its distribution is so wide that it is easily consistent with the $Q_{rms-Ps} = 15.3^{+3.8}_{-2.8}$ μK, the value derived using the full power spectrum.

An important lesson from fitting with different cuts and configurations and from our Monte Carlo simulations is that the best fitted parameters depend both upon the random statistics of CMB fluctuations and on the choice of cuts and fitting parameters. Table 1 in Gorski et al.[32] is indicative of the range of results obtainable using a robust and stable approach.

Tests for Gaussian Statistics: It is important to determine whether the primordial fluctuations are Gaussian. The probability distribution of temperature residuals should be close to Gaussian if the sky variance is Gaussian and the receiver noise is Gaussian. The receiver noise varies somewhat from pixel to pixel because the observation times are not all the same, but when this is taken into account the data appear Gaussian [75]. There is no evidence for an excess of large deviations, as would be expected if there were an unknown population of point sources. A search for point sources in the 2-year maps found none [44]. Given the large beam of the instrument and the variance of both cosmic signals and receiver noise, it is still possible for interesting signals to be hidden in the data.

Kogut et al. [47] compare the 4-year DMR maps to Monte Carlo simulations of Gaussian power-law CMB anisotropy. The 3-point correlation function, the 2-point correlation of temperature extrema, and the topological genus are all in excellent agreement with the hypothesis that the CMB anisotropy on angular scales of 7° or larger represents a random-phase Gaussian field. A likelihood comparison of the DMR maps against non-Gaussian χ^2_N toy models tests the alternate hypothesis that the CMB is a random realization of a field whose spherical harmonic coefficients $a_{\ell m}$ are drawn from a χ^2 distribution with N degrees of freedom. Not only do Gaussian power-law models provide an adequate description of the large-scale CMB anisotropy, but non-Gaussian models with $1 < N < 60$ are five times less likely to describe the true statistical distribution than the exact Gaussian model.

5.3. SUMMARY OF 4-YEAR *COBE* DMR CMB MEASUREMENTS

(1) The full 4-year set of *COBE* DMR observations is analyzed and full sky maps have been produced [6]. The typical signal-to-noise ratio in a 10° smoothed frequency-averaged map is ~ 2, enough to provide a visual impression of the anisotropy.

(2) The DMR (despite its being a differential instrument) finds a CMB monopole temperature of $T_0 = 2.725 \pm 0.020$ K [45]. This is in excellent agreement with the *COBE* FIRAS precision measurement of the spectrum of the CMB, $T_0 = 2.728 \pm 0.002$ K [30].

(3) The CMB dipole from DMR has amplitude 3.358 ± 0.024 mK toward Galactic coordinates $(l, b) = (264°.31 \pm 0°.17, 48°.05 \pm 0°.10)$, or equatorial coordinates $(\alpha, \delta) = (11^h 11^m 57^s \pm 23^s, -7°.22 \pm 0°.08)$ epoch J2000. This is consistent with the dipole amplitude and direction derived by *COBE* FIRAS [30].

(4) The 95% confidence interval for the observed $\ell = 2$ quadrupole amplitude is $4\ \mu K \leq Q_{rms} \leq 28\ \mu K$. This is consistent with the value predicted by a power-law fit to the power spectrum yields a quadrupole normalization of: $Q_{rms-PS} = 15.3^{+3.8}_{-2.8}\ \mu K$ [46]; [35].

(5) The power spectrum of large angular scale CMB measurements is consistent with an $n = 1$ power-law [32], [35], [86]. If the effects of a standard cold dark matter model are included, *COBE* DMR should find $n_{eff} \approx 1.1$ for a $n = 1$ universe. With full use of the multi-frequency 4-year DMR data, including our estimate of the effects of Galactic emission, we find a power-law spectral index of $n = 1.2 \pm 0.3$ and a quadrupole normalization $Q_{rms-PS} = 15.3^{+3.8}_{-2.8}\ \mu K$. For $n = 1$ the best-fit normalization is $Q_{rms-PS}|_{n=1} = 18 \pm 1.6\ \mu K$. Differences in the derived values of Q and n

Technique	n [a]	Q_{rms-PS} [b] (μK)	$Q_{rms-PS\mid n=1}$ [c] (μK)
No Galaxy Correction [d]			
2-point correlation function [36]	—	—	$17.5^{+1.4}_{-1.4}$
Orthogonal functions [32]	$1.21^{+0.24}_{-0.28}$	$15.2^{+3.7}_{-2.6}$	$17.7^{+1.3}_{-1.2}$
Pixel temperatures [35]	$1.23^{+0.26}_{-0.27}$	$15.2^{+3.6}_{-2.8}$	$17.8^{+1.3}_{-1.3}$
Hauser-Peebles cut sky [86]	—	—	—
Internal Combination Galaxy Correction [e]			
2-point correlation function [36]	—	—	$16.7^{+2.0}_{-2.0}$
Orthogonal functions [32]	$1.11^{+0.38}_{-0.42}$	$16.3^{+5.2}_{-3.7}$	$17.4^{+1.8}_{-1.7}$
Pixel temperatures [35]	$1.00^{+0.40}_{-0.43}$	$17.2^{+5.6}_{-4.0}$	$17.2^{+1.9}_{-1.7}$
Hauser-Peebles cut sky [86]	$1.62^{+0.44}_{-0.50}$	—	$19.6^{+2.5}_{-2.5}$

[a] Mode & 68% confidence range of the projection of the 2-d likelihood, $L(Q, n)$, on n
[b] Mode & 68% confidence range of the projection of the 2-d likelihood, $L(Q, n)$, on Q
[c] Mode & 68% confidence range of the slice of the 2-d likelihood, $L(Q, n)$, at $n = 1$
[d] Formed from the weighted average of all 6 channels
[e] from a linear combination of 6 channel maps canceling free-free emission [46]

TABLE 1. Summary of DMR 4-Year Power Spectrum Fitting Results.

between various analyses of DMR data are much more dependent on the detailed data selection effects than on the analysis technique.

(6) The DMR anisotropy data are consistent with Gaussian statistics. Statistical tests prefer Gaussian over other toy statistical models by a factor of ~ 5 [47].

5.4. COBE CONCLUSIONS

The COBE-discovered [74] higher-order ($\ell \geq 2$) anisotropy is interpreted as being the result of perturbations in the energy density of the early Universe, manifesting themselves at the epoch of the CMB's last scattering. These pertubations are the seeds of large scale structure formation and are relics from processes occurring in the very early Universe at extremely high energies. In the standard scenario the last scattering of cosmic background photons takes place at a redshift of approximately 1100, at which epoch the large number of photons was no longer able to keep the hydrogen sufficiently ionized. The optical thickness of the cosmic photosphere is roughly $\Delta z \sim 100$ or about 10 arcminutes, so that features smaller than this size are damped. Observations of the CMB anisotropy power spectrum can reveal to us much of the interesting history of the early Universe and so a great deal of effort has gone into its observation.

198

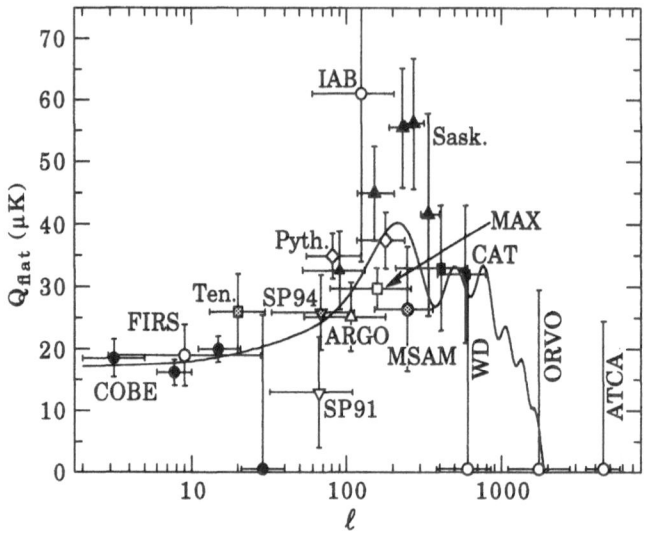

Figure 2. Current status of CMB anisotropy observations: Plotted are the quadrupole amplitudes for a flat (unprocessed scale-invariant spectrum of primordial perturbations, i.e., a horizontal line) anisotropy spectrum that would give the observed results for the experiment. The vertical error bars represent estimates of 68% CL, while the upper limits are at 95% CL. Horizontal bars indicate the range of ℓ values sampled. The curve indicates the expected spectrum for a standard CDM model ($\Omega_0 = 1$, $\Omega_b = 0.05$, $h = 0.5$), although true comparison with models should involve convolution of this curve with each experimental filter function.

6. Current Anisotropy Power Spectrum

On the order of ten experiments have now observed CMB anisotropies. Anisotropies are observed on angular scales larger than the minimum 10' damping scale (see Figure 3) and are consistent with those expected from an initially scale-invariant power spectrum of potential and thus metric fluctuations. It is believed that the large scale structure in the Universe developed through the process of gravitational instability where small primordial perturbations in energy density were amplified by gravity over the course of time. The initial spectrum of density perturbations can evolve significantly in the epoch $z > 1100$ for causally connected regions (angles $\lesssim 1° \; \Omega_{tot}^{1/2}$). The primary mode of evolution is through adiabatic (acoustic) oscillations, leading to a series of peaks that encode information about the perturbations and geometry of the Universe, as well as information on Ω_0, Ω_b, Ω_Λ (cosmological constant), and H_0 [70]. The location of the first acoustic peak is predicted to be at $\ell \sim 220 \; \Omega_{tot}^{-1/2}$ or $\theta \sim 1° \; \Omega_{tot}^{1/2}$ and its amplitude increases with increasing Ω_b.

Figure 4 shows the theoretically predicted power spectrum for a sample of models, plotted as $\ell(\ell+1)C_\ell$ versus ℓ which is the power per logarithmic

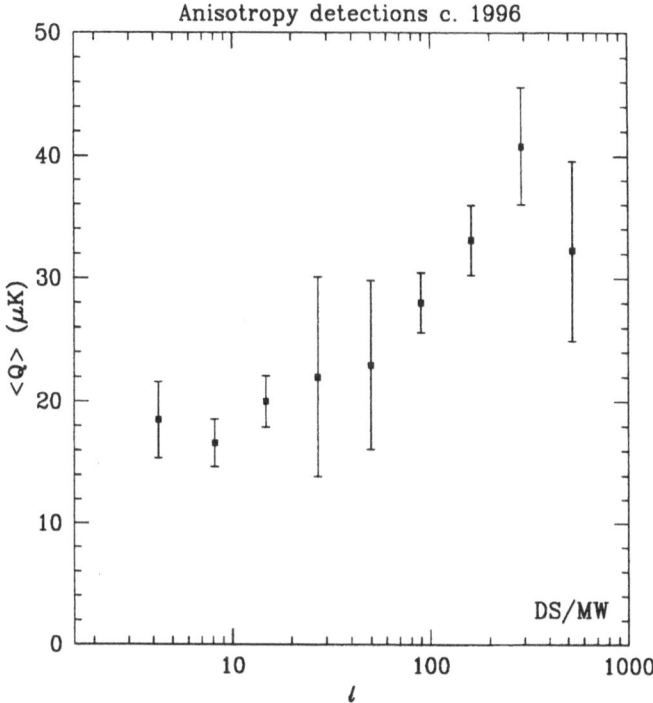

Figure 3. Current status of CMB anisotropy observations: Due to the overlapping sets of data Douglas Scott and Martin White have formed band-averages over experiments. Plotted are the quadrupole amplitudes for a scale-invariant spectrum that would give the experiment band averages. The vertical error bars represent estimates of 68% CL.

interval in ℓ or, equivalently, the two-dimensional power spectrum. If the initial power spectrum of perturbations is the result of quantum mechanical fluctuations produced and amplified during inflation, then the anisotropy spectrum and the fractional contribution from density (scalar) and gravity wave (tensor) perturbations are coupled. If the energy scale of inflation at the appropriate epoch is at the level of $\simeq 10^{16}$GeV, then detection of gravitons is possible, as well as partial reconstruction of the inflaton potential. If the energy scale is $\lesssim 10^{14}$GeV, then density fluctuations dominate and less constraint is possible. (See CMB theory lectures for more background.)

Fits to data over smaller angular scales are often quoted as the expected value of the quadrupole $\langle Q \rangle$ for some specific theory, e.g. a model with power-law initial conditions (primordial density perturbation power spectrum $P(k) \propto k^n$). The full 4-year COBE DMR data give $\langle Q \rangle = 15.3^{+3.7}_{-2.8}\,\mu$K, after projecting out the slope dependence, while the best-fit slope is $n = 1.2 \pm 0.3$, and for an $n = 1$ (scale-invariant potential perturbation) spectrum $\langle Q \rangle\,(n = 1) = 18 \pm 1.6\,\mu$K [6], [32]. The conventional notation is such that $Q^2_{\rm rms}/T^2_\gamma = 5C_2/4\pi$.

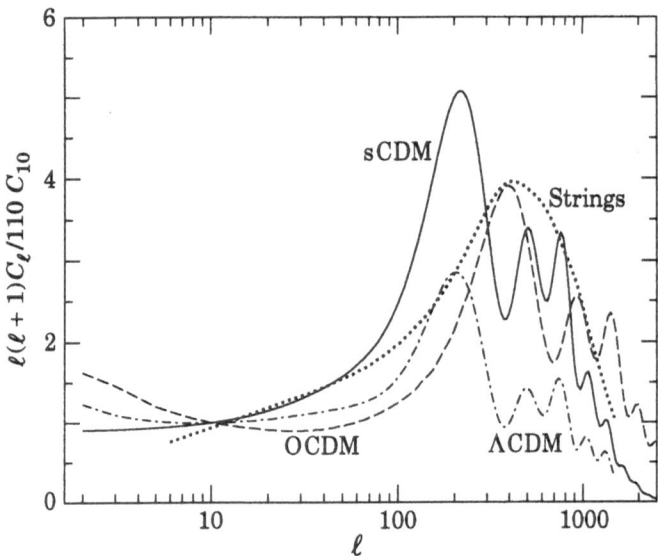

Figure 4. Examples of theoretically predicted $\ell(\ell + 1)C_\ell$ or CMB anisotropy power spectra. **sCDM** is the standard cold dark matter model with $h = 0.5$ and $\Omega_b = 0.05$. **ΛCDM** is a model with $\Omega_{tot} = \Omega_\Lambda + \Omega_0 = 1$ where $\Omega_\Lambda = 0.3$ and $h = 0.8$. **OCDM** is an open model with $\Omega_0 = 0.3$ and $h = 0.75$. (See White (1996) [84] for models). **Strings** is a model where cosmic strings are the primary source of large scale structure [1]. The plot indicates that precise measurements of the CMB anisotropy power spectrum could distinguish between current models.

Only somewhat weak conclusions can be drawn based on the current smaller angular scale data (see Figure 3). However, new data are being acquired at an increasing rate. With future experiments and the prospect of next generation satellite missions: MAP and Planck, a precise measurement of the CMB anisotropy power spectrum is possible and likely, allowing us to decode the information that it contains [42],[40].

7. Current and Near-Term Experiments

Many groups have been and are working to measure the anisotropy. Though some are focussed on large angular scales at frequencies not observed with DMR, most concentrate on smaller angular scales. Table 2 contains a list for recent, current and planned experiments modified from that compiled by L. Page [60].

Rather than review all the experiments, we focus here on a representative sample.

TABLE 2. Recently Completed, Current and Planned Anisotropy Experiments

Experiment	Resolution	Freq [GHz]	Detectors	Type	Groups
ACE(c)[89]	0.2°	25-100	HEMT	C/B	UCSB
APACHE(c)[90]	0.33°	90-400	Bol	C/G	Bologna, Bartol Rome III
ARGO(f)[91]	0.9°	140-3000	Bol	C/B	Rome I
ATCA[92]	0.03°	8.7	HEMT	I/G	CSIRO
BAM(c)[93]	0.75°	90-300	Bol	C/B	UBC, CfA
Bartol(c)[99]	2.4°	90-270	Bol	C/G	Bartol
BEAST(p)[89]	0.2°	25-100	HEMT	C/B	UCSB
BOOMERanG(p)[100]	0.2°	90-400	Bol	C/G	Rome I, Caltech, UCB, UCSB
CAT(c)[101]	0.17°	15	HEMT	I/G	Cambridge
CBI(p)[102]	0.0833°	26-36	HEMT	I/G	Caltech, Penn.
FIRS(f)[104]	3.8°	170-680	Bol	C/B	Chicago, MIT, Princeton, NASA/GSFC
HACME/SP(f)[105]	0.6°	30	HEMT	C/G	UCSB
IAB(f)[106]	0.83°	150	Bol	C/G	Bartol
MAT(p)[107]	0.2°	30-150	HEMT/SIS	C/G	Penn, Princeton
MAX(f)[108]	0.5°	90-420	Bol	C/B	UCB, UCSB
MAXIMA(p)[109]	0.2°	90-420	Bol	C/B	UCB, Rome I Caltech, UCSB
MSAM(c)[110]	0.4°	40-680	Bol	C/B	Chicago, Bartol, Brown, Princeton, NASA/GSFC
OVRO 40/5(c)[111]	0.033°, 0.12°	15-35	HEMT	C/G	Caltech, Penn
PYTHON(c)[112]	0.75°	35-90	Bol/HEMT	C/G	Carnegie Mellon Chicago, UCSB
QMAP(f)[113]	0.2°	20-150	HEMT/SIS	C/B	Princeton, Penn
SASK(f)[114]	0.5°	20-45	HEMT	C/G	Princeton
SuZIE(c)[115]	0.017°	150-300	Bol	C/G	Caltech
TOPHAT(p)[116]	0.33°	150-700	Bol	C/B	Bartol, Brown, DSRI,Chicago, NASA/GSFC
Tenerife(c)[117]	6.0°	10-33	HEMT	C/G	NRAL, Cambridge
VCA(p)[118]	0.33°	30	HEMT	I/G	Chicago
VLA(c)[119]	0.0028°	8.4	HEMT	I/G	Haverford, NRAO
VSA(p)[120]	–	30	HEMT	I/G	Cambridge
White Dish(f)[121]	0.2°	90	Bol	C/G	Carnegie Mellon

1. For "Type" the first letter distinguishes between configuration or interferometer, the second between ground or balloon.

2. An "f" after the experiment's name means it's finished; a "c" denotes current; a "p" denotes planned, building may be in progress but there is no data yet.

7.1. MAX/MAXIMA/BOOMERANG

The MAX/MAXIMA/BOOMERANG payloads are representative of current and currently planned balloon-borne missions.

7.1.1. *MAX*

The Millimeter-wave Anisotropy eXperiment (MAX) is a balloon-borne bolometric instrument which observes at multiple frequencies with high sensitivity on the 0.5° angular scale. MAX has completed five flights detecting significant CMB fluctuations [29], [2], [57], [24], [18], [77], [49].

The MAX instrument consists of an off-axis Gregorian telescope and a bolometric photometer mounted on an attitude-controlled balloon-borne platform which makes measurements at an altitude of 36 km. The Gregorian telescope consists of a 1-meter primary and a nutating elliptical secondary. The underfilled optics provide a 0.55° FWHM beam when focused and aligned. The 5.7 Hz nutation of the secondary modulates the beam on the sky sinusoidally though ±0.68° and the attitude control sweeps the beam over a 6° or 8° path and back in about 108 seconds, producing about 15 to 20 independent temperature differences on the sky. Depending upon the time of observation and location of the region under observation sky rotation can cause the observed region to be in the shape of a bow-tie.

On flights 4 & 5 the single-pixel four-band bolometric receiver featured negligible sensitivity to radio frequency interference and an adiabatic demagnetization refrigerator to cool the photometer to 85 mK. The dichroic photometer used for MAX has $(\delta\nu/\nu)$ of 0.57, 0.45, 0.35, and 0.25 filter bands at 3.5, 6, 9, and 15 cm^{-1} respectively. MAX covers the high frequency side of the window formed by Galactic dust emission rising at higher frequencies and Galactic synchrotron and free-free emission increasing at lower frequencies. The 15 cm^{-1} channel acts as a guard against Galactic dust and atmospheric emission. The multiple frequencies have sufficient redundancy to provide confidence that the signal is CMB and not a foreground or systematic effect.

MAX is calibrated both by an on-board commandable membrane and by observations of planets, usually Jupiter. The two techniques agree at roughly the 10% level. The calibration is such that the quoted temperature difference is the real temperature difference on the sky.

MAX makes deep CMB observations (typically one hour) on regions generally selected to be low in dust contrast and total emission and free from known radio sources. MAX has made observations on five flights. The data from most of the scans are in good agreement but the scan of the μ-Pegasi region is significantly lower than the rest. A combination of all the data seems to be coming out at an intermediate value between GUM and μ-Pegasi regions and may all be consistent with coming from a single parent population [78].

The center of the scan is the same for the three observations of GUM (the star Gamma Ursae Minoris) but the relative geometry is such that the three scans made bow-tie patterns which cross at the star. White and Bunn [82] have made use of this fact to construct a two dimensional map of the region which is roughly $10° \times 5°$. The title of their paper is "A First Map of the CMB at 0.5° Resolution". Since then a map covering 180 square degrees was generated by Tegmark et al. [81] using the Saskatoon data.

Making maps is clearly the appropriate approach for the current generation of new experiments. MAX is evolving to new systems MAXIMA and BOOMERANG, which are designed and constructed for the goal of getting the power spectrum around the first "Doppler" peak and further and making maps covering a significant portion of the sky.

7.1.2. *MAXIMA*

MAXIMA stands for MAX imaging system. The current one-dimensional scans are very useful data for the discovery phase of CMB anisotropy research. Soon progress will depend upon the availability of two-dimensional maps of low Galactic foreground regions (low dust in this case) with several hundred pixels so that sampling variance is less important (see Section 4). In addition one can look for properties of the sky which are not predicted by theories and could be overlooked in statistical analyses. It also makes it possible to catalog features for comparison to or motivation of other experiments.

Under the auspices of the NSF Center for Particle Astrophysics a collaboration consisting of groups from the University of California at Berkeley, Caltech, the University of Rome, and the IROE-CNR Florence have begun work and made good progress on the new systems for MAXIMA and BOOMERANG. The plan is to have a combination of northern hemisphere flights of MAXIMA and BOOMERANG and a Long Duration Balloon (LDB) flight of BOOMERANG from Antarctica.

To make an imager a new optical system was necessary. The primary mirror for MAXIMA is a 1.3-meter, off-axis, light-weight primary mirror. The primary will be modulated which allows a much larger beam chop angle on the sky with less spill over and thus more pixels in the focal plane. Cold secondary and tertiary mirrors provide a cold Lyot stop and the field-of-view required for the array of 12 arcminute pixels. The geometrical aberrations in the center of the field-of-view are less than 10 arcminutes.

A larger primary mirror requires a larger gondola which is now constructed (Figure 5). The chop angle can both be increased and varied al-

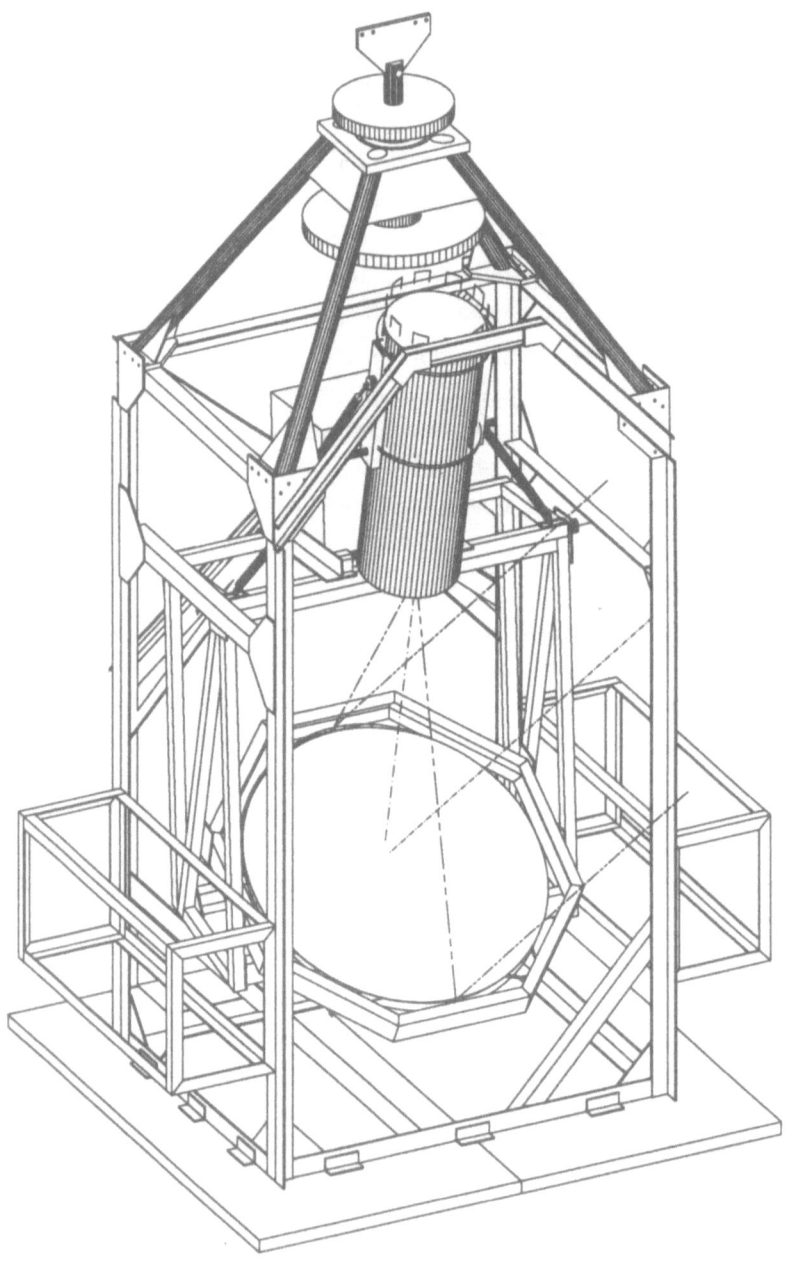

Figure 5. A schematic drawing of the MAXIMA gondola system showing three sample rays from the sky reflecting from the nodding primary, coming to the prime focus and entering the dewar containing cold optics and bolometer detectors. Also show are the gondola frame work with the angular momentum wheels on top and two side boxes holding the electronics for the pointing system and the detector signal processing. The first stage of detector electronics is inside the dewar, the next stage in an RF shielded backpack attached to the dewar.

lowing the instrument to sample the shape of the power spectrum over the range $40 < \ell < 1000$.

An additional feature is new detector electronics with AC coupling in order to allow linear scanning in a total power mode, making maps and power spectrum measurements directly. This approach is different than that of making a number of different window functions. The idea is to use a scan or raster scan of the CMB anisotropies on the sky directly rather than obtaining a set of differences at different chop angles. One is thus mapping directly and measuring the power spectrum as the fourier transform of the data. At this stage the instrument is designed to operate in this mode by scanning the primary mirror in a sawtooth pattern rapidly (3 Hz) and more slowly moving the entire gondola in azimuth to cover a larger angle.

Another major change will be going from a single pixel four-frequency photometer to a fourteen-pixel receiver. This will allow taking data at eight times the rate and thus make two-dimensional mapping feasible. The receiver design has been completed and it and the new cold optics being mounted in the new large dewar as show in Figure 6. The bolometers have a spider-web (silicon nitride micromesh) substrate so that cosmic ray transient occurrences will be reduced by more than an order of magnitude. The first flight of the new gondola was September 1995 and we anticipate a flight with the arrary receiver in the summer/fall 1997. We can anticipate that within three years MAXIMA will have made maps and will have measured the anisotropy power spectrum around the location of the first doppler peak. Figure 7 indicates an estimate of the accuracy of the power spectrum determination.

7.1.3. *BOOMERANG*

BOOMERANG is equivalent to the long-duration balloon-borne version of MAXIMA and an intermediate step toward the bolometer space mission, the Planck HFI. Plans call for a northern hemisphere flight in June 1997 followed by a many-day flight circumnavigating Antarctica in the austral summer beginning December 1998. BOOMERANG will move more directly towards mapping a significant region of the sky. The BOOMERANG focal plane contains 8 pixels: four multiband photometers (6, 9 and 14 cm^{-1}) and four monochromatic channels (3 cm^{-1}). The diffraction limited angular resolution is 12' above 6 cm^{-1} and 20' at 3 cm^{-1}. In total power mode, the largest resolution is limited only by the length of a scan.

The high cosmic ray flux over the Antarctic requires detectors which are insensitive to cosmic rays. "Spider web" bolometers have been developed specifically to minimize the effect of cosmic rays on the detector. These bolometers are called composite because the functions of absorbing radiation and measuring the temperature increase are separated. The absorber

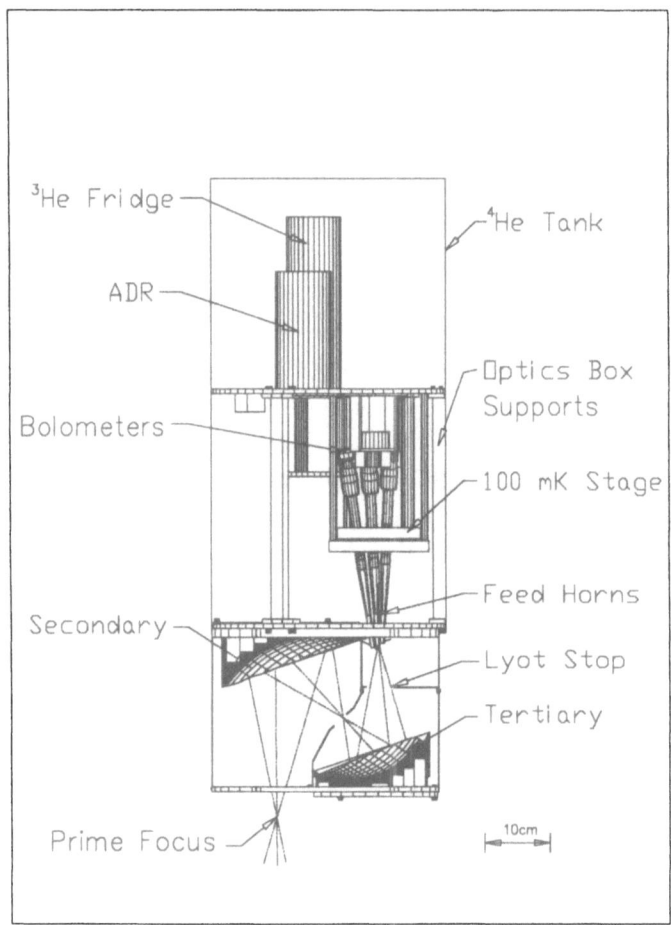

Figure 6. The cold optical system for MAXIMA. Both MAXIMA and BOOMERANG use fast off-axis LHe-cooled secondary and tertiary mirrors to reduce optics emissions. The design incorporates cold black baffles and a cold Lyot stop which controls the illumination on the primary mirror giving smaller offsets and better control of spillover than can be achieved in optical systems that are not re-imaged. The figure also shows the location of the feedhorns and the bolometers as well as the Adiabatic Demagnitization Refridgerator (ADR) and He3 fridges used to maintain the bolometers at 100 mK. The optics are shown in cross-section for clarity.

is the "spider web" and a thermometer is a neutron transmutation doped (NTD 14) thermistor. Spider web bolometers for BOOMERANG will have a Noise Equivalent Power (NEP) of $\sim 1 \times 10^{-17}$ W/Hz$^{1/2}$.

The electrical circuit of BOOMERANG is split in subcircuits which reside at different temperatures. There are bolometers at 300 mK, which are AC-biased at 200 Hz and dual JFET source followers, providing a low impedance line going out of the cryostat. At 300 K, the signal is preamplified, demodulated and filtered through a small bandwidth, thus enabling the detection of signals of the order of nanovolts.

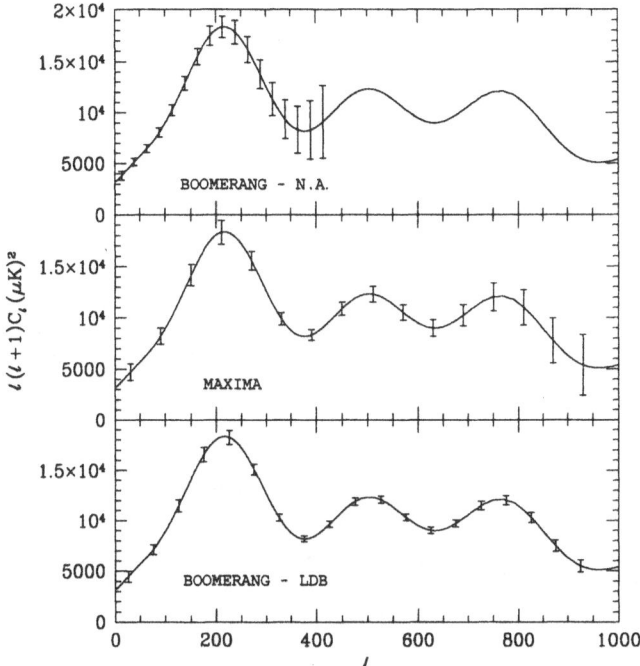

Figure 7. Estimates of the accuracy with which each of the three proposed flights for MAXIMA and BOOMERANG will determine the power spectrum of CMB anisotropy. The sold line in each frame represents the standard cold dark matter ($\Omega = 1$, $\Omega_b = 0.05$, $h = 50$, no reionization) power spectrum. The spacing of the error flags indicates the resolution provided by each experiment; the amplitude of the error flags indicate one standard deviation uncertainty and includes the effects of instrumental noise, sky coverage, and differences strategy.

The BOOMERANG North American flight will produce a map covering a region near the North Celestial Pole. It is about 10% of the full sky. This region will be mapped by sweeping in azimuth between 60 degrees and 90 degrees North latitude while holding the elevation constant; after 12 hours, the rotation of the earth will produce a map of half of a circular region centered on the NCP. The North American flight will produce the first total power map of a significant fraction of the northern sky.

The region to be mapped by the Antarctic flight is centered on the Southern Hole, a region of exceptionally low Galactic dust emission in the southern sky. During the Antarctic summer, the anti-solar direction passes through the Southern Hole; the BOOMERANG scans will be centered on the antipode of the sun's path to minimize sidelobe response and thermal response from the sun.

The Antarctic flight will consist of three scans, each lasting five days: two fast scans of adjoining $30° \times 60°$ fields, which cover 10% of the sky,

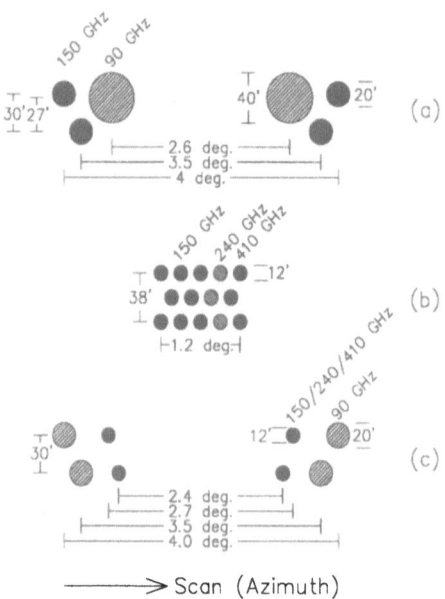

⟶ Scan (Azimuth)

Figure 8. Layout of the three BOOMERANG and MAXIMA focal planes. The focal planes are: (a) BOOMERANG North American flight, (b) MAXIMA, and (c) BOOMERANG LDB (Antarctica). Circles indicate the relative position and FWHM of the beams projected on the sky. Each circle represents a dual-polarization, high efficiency feed, with the exception 12′ beams shown in (c), which represent multi-frequency photometers of the type flown previously on MAX. Each experiment is scanned in azimuth, which is horizontal in the figure.

and a deep integration of a 12° × 12° subset of this region. The fast scans will result in 130 000 pixels (12 arcmin) with a sensitivity of 20 μK/pixel (thermodynamic temperature), obtained by sweeping the telescope at 1 degree/s in azimuth while varying the elevation through 40 degrees over a 24 hour period. The deep integration region will be mapped with a sensitivity of $\sim$ 10 μK/pixel. This scan will test for systematic variations in the data and will serve as a diagnostic for the fast scans.

BOOMERANG will complement the DMR with high sensitivity measurement of CMB power on scales between 12 arcmin and $\sim$ 10°. In Figure 7, the error bars represent the $\pm 1\sigma$ limit which BOOMERANG will determine for a standard cold dark matter model. The BOOMERANG-MAXIMA program will return excellent scientific data and will be a good test of the instrumentation and techniques for Planck.

7.2. MSAM/TOPHAT

MSAM/TOPHAT is a collaboration working on a series of experiments to measure the medium-scale anisotropy of the cosmic microwave background radiation. The collaboration includes researchers at the Bartol Research Institute, Brown University, the University of Chicago, the Danish Space Research Institute, and NASA/Goddard Space Flight Center. The MSAM/TOPHAT program is similar to the MAX/MAXIMA/BOOMERANG program at present.

7.2.1. *MSAM*

MSAM is an acronym for Medium Scale Anisotropy Measurement. A notable difference between MSAM and MAX has been that MSAM used a three-position chop analyzed either as a triple beam or double beam (two chop angles on the sky) observation. MSAM angular resolution is 0.5° between 5 and 23 cm^{-1} (150 and 700 GHz or wavelengths 0.4 to 2.0 mm). MSAM has had three flights (June 1992, May 1994, June 1995) all from Palestine, Texas [16],[17].

MSAM-I: The first phase of the Medium-Scale Anisotropy Measurement (MSAM-I) probed CMB anisotropy at 0.5 degree angular scales between 5 and 23 cm^{-1} (150 and 700 GHz, 0.4 and 2.0 mm). The first flight of this package (June 1992 from Palestine, TX) has resulted in a detection of $0.5 \times 10^{-5} < \delta T/T < 2 \times 10^{-5}$. Fluctuations at these angular scales are believed to be the precursors of the largest structures we observe today. This level of anisotropy is at the lower end of the predicted values from standard Cold Dark Matter theories of structure formation. The interpretation of the results, however, is complicated by the presence of two point sources in the data. This hints at the possibility of a previously unsuspected population of objects which will challenge anisotropy measurements at these sensitivity levels. Such sources may be distinguished by the use of multiple spectral bands bracketing the peak of the CMB, such as those in MSAM-I and MSAM-II. One can also interpret the two extra peaks as CMB fluctuations and then the data are in good agreement with the anisotropy predicted by SCDM.

MSAM-II is the second phase of the Medium-Scale Anisotropy Measurement. An adiabatic demagnetization refrigerator (ADR) is used to cool monolithic silicon bolometers to 100mK in a new radiometer. The new radiometer has expanded frequency coverage in 5 spectral bands between 2.3 and 5 cm^{-1} (70 to 150 GHz or wavelengths 2.0 to 4.3 mm). The instrument is expected to improve the signal-to-noise ratio by about a factor of three over the previous results.

MSAM observations are along a ring surrounding the north celestial pole with a 20′ beam on the sky giving sensitivity to the medium angular-scale power spectrum of the radiation.

7.2.2. *TOPHAT*

TOPHAT is conceived as a long-duration balloon-borne experiment with the detectors located on the top of the balloon rather than in a gondola hanging below the balloon. The extended observation time (~2 weeks) made possible by LDB will permit a substantial fraction of the flight to be dedicated to studying and characterizing systematics in-flight while still maintaining high sensitivity to CMB anisotropy. TOPHAT will observe in five spectral bands between 5 and 21 cm^{-1} (150 and 630 GHz or wavelengths between 0.5 to 2.0 mm). The current plans call for the measurement of 40 points on the sky, each with an rms sensitivity of $\delta T_{rms} \approx 1~\mu K$ or $\delta T_{rms}/T_{CMB} \approx 3 \times 10^{-7}$ including removal of the Galactic foreground dust emission.

7.3. ACE/BEAST

As a follow up to their South Pole HEMT observations the Santa Barbara group has proposed ACE (Advanced Cosmic Explorer). It is a large, light-weight (200 kg), system aimed at making flights lasting 90 days or more. They plan to utilize advanced HEMTs, active refrigerators, and a 2-m diameter mirror to cover the frequency range 25 to 90 GHz. In three such flights such a system could map 75% of the sky to an angular resolution of 10 arcminutes at a level of about 20 μK. This project is still in the early phase but is indicative of what with sufficient funding one might achieve by the year 2000.

7.4. GROUND-BASED INSTRUMENTS

Ground-based instruments have made a significant contribution to CMB anisotropy observations. They have been successful as a result of the observers' clever strategies to minimize and reduce the effect of the atmosphere. These strategies have included going to high, dry sites such as the South Pole and Teide peak on Tenerife and using triple-beam chopping or other similar techniques. These techniques are more difficult to use when going to mapping and making observations over an extended portion of the power spectrum. Here again it is possible that significant progress can be made though it is likely to be eventually limited before the science is exhausted.

An exciting exception is the use of aperture synthesis interferometers. The Ryle Telescope images of the Sunyaev-Zeldovich effect in clusters and the CAT (Cambridge Anisotropy Telescope) results have convinced many that interferometers have a bright future in actually mapping anisotropy on small angular scales over selected regions of the sky.

7.4.1. *CAT: Cambridge Anisotropy Telescope; 30′ to 2°*

The CAT [66] is a three-element interferometer which can operate at frequencies between 13 and 17 GHz with a bandwidth of 500 MHz. This frequency range was chosen as a compromise between the effects of atmospheric emission, which increase with frequency, and Galactic synchrotron and bremsstrahlung emission, which decrease with frequency. The most important contaminating signal for the CAT is that from discrete extragalactic radio sources. The observation strategy is to chose fields with minimum source content and then observe the sources with the higher resolution Ryle Telescope at 15.7 GHz.

The CAT has a system temperature of approximately 50 K. Variations in the system temperature are continuously measured using a modulated 1-K noise signal injected into each antenna. The interferometer baselines can be varied from 1 to 5 m, and are scaled to give the same synthesized beam at different frequencies. The antennas have a primary beam FWHM of 2.2 degrees at 15 GHz. The CAT simultaneously records data from the two orthogonal linear polarizations. Its alt-az mount causes the plane of polarization to rotate on the sky as the telescope tracks a given field.

The CAT is situated within a 5-m high earth bank which is lined with aluminium. This shielding reduces the effect of spillover and terrestrial radio interference, but limits observations to elevations above 25 degrees. The control hut is located about 100 m away. Each element of the telescope is a corrugated-conical horn with a parabolic reflector. The horns are mounted on a single turntable which can track in azimuth. Each antenna has an individual elevation drive. Preliminary tests have shown that crosstalk, correlator offsets, and antenna shadowing - particular problems associated with interferometers - do not affect the performance of the CAT at elevations greater than 40 degrees [67], [58]. Results [37] [71] are shown in Figure 2 summarizing anisotropies.

7.4.2. *Interferometers: VSA, CBI, & VCA*

Three major interferometer projects are funded and underway. They are the VSA (Very Small Array, 15′ to 4°) in England, the Caltech interferometer CBI (Cosmic Background Interferometer, 4′ to 20′) and the University of Chicago VCA (Very Compact Array, 15′ to 1.4°). These interferometers are likely to provide a very good first look at the CMB anisotropy power spectrum on angular scales less than about 0.5° ($\ell > 200$).

VSA is a joint project between the Mullard Radio Astronomy Observatory (Cambridge), the Nuffield Radio Astronomy Laboratories (Jodrell Bank, Manchester), and the Instituto de Astrofisical de Canarias (Tenerife). The VSA is similar in design to CAT which was a prototype for VSA. The VSA will have 15 antennas and a 2-GHz bandwidth, analog correlators,

TABLE 3. Characteristics of Next Generation Interferometers

Initials:	VSA	VCA	CBI
Name:	Very Small Array	Very Compact Array	Cosmic Background Imager
Frequency (GHz)	28 - 37	26 - 36	26 - 36
N_a	14, 15	13	13
No. of Channels	1 tunable	10	10
T_{sys} (K)	~25	~25	~25
$\Delta\nu$	1.75	1	1
ℓ range	150-1600	150-750	400-3500
Resolution	15′	15′	4.5′
Site	Tenerife	Anarctica	California, Chile
Point Sources	Ryle & Bonn	A. T.	VLA & 40-m
Correlations	analog	analog	analog
Operational	1999	1999	1999

and other technology operated on CAT. The operating frequency is 26-36 GHz which is set by the atmospheric window and the natural waveguide bands for which high sensitivity HEMT amplifiers have been developed. The increase in frequency from CAT to VSA will also decrease the effect of discrete radio sources and Galactic emission. The atmospheric emission fluctuations will increase so that VSA will be operated on Mt. Teide on the island of Tenerife. The VSA will operate with two sets of horns: one set with a 15-cm aperture giving a 4° field of view and the second with a 30-cm aperture giving a 2° field of view. The baselines and thus resolution will scale proportionally to maintain about a 1 μK sensitivity per resolution element. The VSA will get about 10 independent points of the anisotropy power spectrum with resolution of $\Delta\ell = 100$ at low ℓ and $\Delta\ell = 200$ at high ℓ covering the range $130 < \ell < 1800$.

The CBI and VCA are planning to observe the same portion of sky in the southern hemisphere. The proposed VCA is expected to image about 2500 square degrees around the South Pole region and if that goes well continued operation to cover eventually about 25% of the sky. The VCA interferometer consists of 13 scalar feed horns arranged in a closed packed configuration which fill about 50% of the aperture to provide maximum brightness sensitivity. The horns feed low-noise HEMT amplifiers operating at 26 to 36 GHz with noise temperatures of about 10 K. The estimated sensitivity is 4 to 10 μK in pixels ranging from 0.25° to 1.4° in the 3° field of view. The VCA will be operated from the South Pole Station and is scheduled for installation in the fall of 1998 with first results expected the following spring.

The sensitivity of an interferometer system can be estimated using the following formulae for flux density and temperature:

Flux density

$$\Delta S_{rms} = \frac{2kT_{sys}}{\eta_a A_a \eta_c [n_a (n_a - 1)\Delta\nu\tau]^{1/2}}$$

Temperature

$$\Delta T_{rms} = \frac{\lambda^2 T_{sys}}{\theta_s^2 \eta_a A_a \eta_c [n_a (n_a - 1)\Delta\nu\tau]^{1/2}}$$

Same expressions evaluated with typical numbers:

Flux density

$$\Delta S_{rms} = \frac{6\left(\frac{T_{sys}}{30\ K}\right)}{\left(\frac{\eta_a}{0.6}\right)\left(\frac{d}{20\ cm}\right)^2\left(\frac{\eta_c}{0.9}\right)\left[\frac{n_a}{14}\frac{n_a-1}{13}\frac{\Delta\nu}{10^9}\frac{\tau}{1\ month}\right]^{1/2}}\mathrm{mJy}$$

Temperature

$$\Delta T_{rms} = \frac{6\left(\frac{\lambda}{1\ cm}\right)^2\left(\frac{T_{sys}}{30\ K}\right)}{\left(\frac{\theta_s}{20'}\right)^2\left(\frac{\eta_a}{0.6}\right)\left(\frac{d}{20\ cm}\right)^2\left(\frac{\eta_c}{0.9}\right)\left[\frac{n_a}{14}\frac{n_a-1}{13}\frac{\Delta\nu}{10^9}\frac{\tau}{1\ month}\right]^{1/2}}\mu K$$

One can then evaluate these formulae and compare with Table 3 to estimate the sensitivity and sky area that can be surveyed in a given observing time and see that on small angular scales interferometers are competive with many other experiments.

8. Future Satellite Missions

An accurate, extensive imaging of CMB anisotropies with sub–degree angular resolution would provide decisive answers to several major open questions on structure formation and cosmological scenarios. The observational requirements of such an ambitious objective can be met by a space mission with a far–Earth orbit and instruments based on state–of–the–art technologies.

While balloon-borne and ground-based observations can do a credible job in measuring the CMB anisotropy power spectrum, atmospheric disturbance, emission from the Earth and limited integration time are the main limiting factors which prevent ground–based and balloon–borne experiments from obtaining sufficient sensitivity over very large sky regions, with additional difficulties in reaching accurate foreground removal (see Danese et al. 1995 for a recent discussion). Only a suitably designed space mission can meet the scientific goals sought by cosmologists. On the other hand it should be stressed that experiments from the ground or from balloons are not alternative to a space mission like Planck, but rather complementary.

8.1. MAP

MAP (Microwave Anisotropy Probe) was selected by NASA in 1996 as a MidEX class mission. Its launch is expected to be roughly 2001. The goal of MAP is to measure the relative CMB temperature over the full sky with an angular resolution of $0.3°$, a sensitivity of 20 μK per $0.3°$ square pixel, and with systematic effects limited to 5 μK per pixel. Details about the major aspects of the mission design are given below.

8.1.1. *Galactic Emission Foreground*

Galactic foreground signals are distinguishable from CMB anisotropy by their differing spectra and spatial distributions.

Figure 9 shows the estimated spectra of the Galactic foreground signals and a range of expected cosmological signal intensities. The three physical mechanisms that contribute to the Galactic emission are synchrotron radiation, free-free radiation, and thermal radiation from interstellar dust. Results from CMB and other measurements show that at high Galactic latitudes CMB anisotropy dominates the Galactic signals in the range 30–150 GHz. However, the Galactic foreground will need to be measured and removed from some of the MAP data.

There are two techniques that MAP will use to evaluate and remove the Galactic foreground. The first uses existing Galactic maps at lower (radio) and higher (far-infrared) frequencies as foreground emission templates. Uncertainties in the original data and position-dependent spectral index variations introduce errors with this technique. There is no good free-free emission template because there is no frequency where it dominates the microwave emission. High resolution, large scale maps of H-alpha emission will be a template for the free-free emission when they become available. One surprise has been the apparent correlation between free-free emission and dust infrared emission, sometimes more so than to H-alpha emission. This indicates that we may not yet have a template for free-free emission.

The second technique is to form linear combinations of multi-frequency MAP observations such that signals with specified spectra are cancelled.

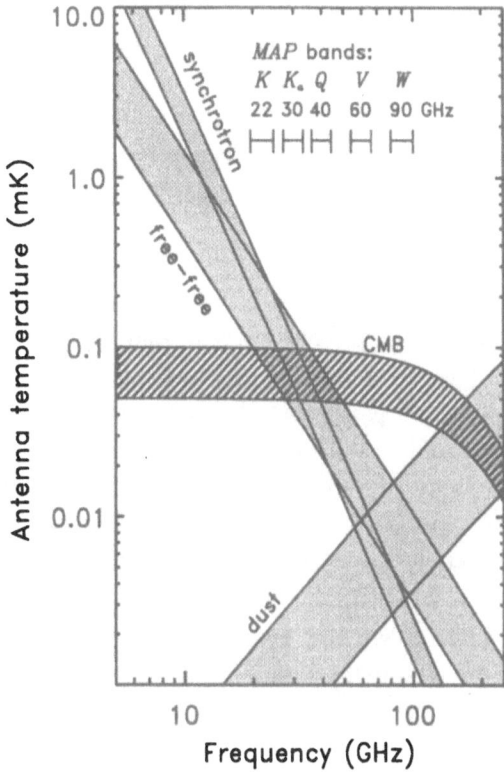

Figure 9. Galactic Foreground Emission estimates plotted as a function of frequency. The width of the band indicates the estimated range for Galactic latitudes varying between $20° < b < 70°$. The proposed MAP frequency bands are indicated.

The linear combination of multi-frequency data makes no assumptions about the foreground signal strength or spatial distribution, but requires knowledge of the spectra of the foregrounds. Both techniques were successfully employed by COBE.

The contamination from extragalactic radio sources is not yet a solved issue. Flat spectrum radio sources observed with a diffraction limited system produce a signal with very nearly the same frequency dependence as the CMB anisotropy making the spectral shape technique ineffective. It will be necessary to compile a list of significant radio sources and excise them from the data or find another approach.

Five frequency bands with comparable sensitivity are desirable to solve for the four signals (synchrotron, free-free, dust, and CMB anisotropy) and the fifth degree of freedom is used to maximize signal-to-noise. The range of frequency coverage is more important than the specific choice of frequencies within the range. The lowest frequency to survey from space should be at the 22 GHz atmospheric water line since frequencies below this can (with difficulty) be accurately measured from the ground. The highest frequency to survey should be about 100 GHz to reduce the dust contribution and minimize the number of competing foreground signals. The choice of frequencies between 22 and 100 GHz can be dictated by the practical consideration of standard waveguide bands. Based on these considerations, MAP has selected the five frequency bands, which are indicated in Figure 9 and in Table 4.

8.1.2. *MAP Mission Goals*

CMB anisotropy information from current and proposed high resolution ($< 0.3°$) measurements over limited sky regions will likely succeed from ground and balloon-based platforms. The priority for the MAP mission is to map the entire sky with $> 0.3°$ angular resolution where the cosmological return is high, and the data cannot be readily obtained in any other way. The MAP optics feature back to back 1.5-meter primary reflectors which lead to an angular resolution of 0.29° in the highest frequency (90 GHz) channel.

The following table gives the angular resolution to be obtained from each of the five MAP frequency bands. The value quoted is the full width at half maximum (FWHM) of the approximately Gaussian central beam lobe, in degrees.

TABLE 4. MAP Angular Resolution

Freq (GHz)	Band Name	Wavelength (mm)	Beam Spec	FWHM Design	No. of Channels	Sensitivity 0.3° by 0.3° pixel
22	K band	13.6	0.90°	0.93°	4	35μK 26μK
30	Ka band	10.0	0.65°	0.68°	4	35μK 32μK
40	Q band	7.5	0.53°	0.47°	8	35μK 27μK
60	V band	5.0	0.39°	0.35°	8	35μK 35μK
90	W band	3.3	0.29°	0.21°	16	35μK 35μK

The MAP specification calls for an equal noise sensitivity per frequency band of 35 μK per 0.3° × 0.3° square pixel. The mission duration required to meet this specification is one year of continuous observation. If Galactic emission is negligible at high latitudes above 40 GHz, as was the case for COBE, the sensitivity achievable by combining the three highest frequency channels is 20 μK per 0.3° × 0.3° pixel.

The corresponding sensitivity to the angular power spectrum, obtained with simple analytic formulae, is illustrated in Figure 10 which shows the

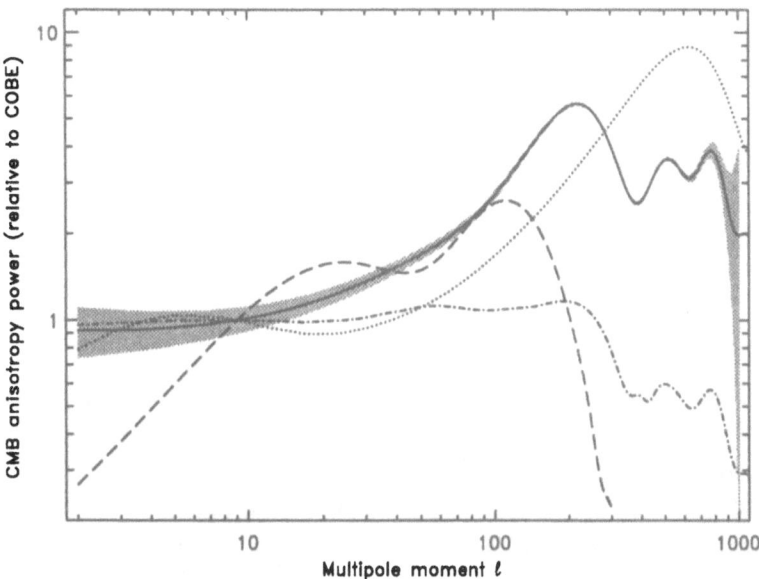

Figure 10. Projected MAP Power Spectrum Sensitivity. This plot shows a simple-minded estimate of the MAP sensitivity in measuring the CMB power spectrum. The gray band straddling the solid (CDM) curve indicates the MAP sensitivity after combining the three highest frequency channels and averaging the spectrum over a 10% band in spherical harmonic order. The curves plotted correspond to a standard CDM model (solid), a highly reionized CDM model (dot-dash), an open CDM model (dotted), and a primordial baryon isocurvature (PBI) model (dashed).

predicted power spectra for a number of competing structure formation models. The gray band straddling the solid (CDM) curve indicates the MAP sensitivity after combining the three highest frequency channels and averaging the spectrum over a 10% band in spherical harmonic order.

8.1.3. *MAP Trajectory and Orbit*

To minimize environmental disturbances and maximize observing efficiency, MAP will observe from a Lissajous orbit about the L2 Sun-Earth Lagrange point 1.5 million km from Earth. The trajectory selected to attain such an orbit consists of 2.5-3.5 lunar phasings loops followed by about a 100 day cruise to L2. No thruster firings are required to enter the L2 orbit.

The L2 Lagrange point offers a virtually ideal location from which to carry out CMB observations. Because of its distance, 1.5 million km from Earth, it affords great protection from the Earth's microwave emission, magnetic fields, and other disturbances. It also provides for a very stable thermal environment and near 100% observing efficiency since the Sun, Earth, and Moon are always behind the instrument's field of view.

The following description indicates the path MAP will follow to L2. The trajectory features 2.5 or 3.5 lunar phasing loops which assist the spacecraft in reaching L2. The cruise time to L2 is approximately 100 days after the lunar phasing loops are completed. The launch window for this trajectory is about 20 minutes/day for 7 consecutive days each month. Once in orbit about L2, the satellite maintains a Lissajous orbit such that the MAP-Earth vector remains between 1° and 10° off the Sun-Earth vector to satisfy communications requirements while avoiding eclipses. Station-keeping maneuvers will be required about 4 times per year to maintain this orbit.

8.1.4. *MAP Instrumentation*

The MAP instrument consists of two back-to-back, off-axis Gregorian tele-scopes that produce two focal planes, A and B, on opposite sides of the spacecraft symmetry axis. A set of 10 corrugated feeds lie in each focal plane and collect the signal power that goes to the amplification electron-ics. The microwave system consists of 10 4-channel differencing assemblies that are designed to eliminate low frequency gain instabilities and amplifier noise in the differential signal.

The reflector design incorporates two back-to-back off-axis Gregorian telescopes with 1.5-m primary reflectors and 0.52 m secondary reflectors. Each primary is an elliptical section of a paraboloid, while the secondaries are nearly elliptical. This arrangement produces two slightly convex focal surfaces on opposite sides of the spacecraft spin (symmetry) axis with plate scales of about 15'/cm. The 99.5% encircled energy spot size diameter is less than 1 cm over a 15 x 15 cm region of the focal plane, and less than 0.33 cm over the central 8 x 8 cm region.

In order to limit diffracted signals to less than 0.5 μK, diffraction shields are employed above, below, and to the sides of each secondary. In addition, the deployable solar panels and multi-layer insulation guarantee that the secondaries remain at least 6° into the shadow from the Sun during observ-ing.

The feed design calls for as small an aperture as possible consistent with a primary edge taper requirement of -25 dB, and a length that places the throat of each differential feed pair in close proximity to the other. The feed aperture diameters scale inversely with frequency, while the primary is equally illuminated at each frequency, leading to a frequency dependent beam size. The feeds are corrugated to produce beams with high symmetry, low-loss, and minimal sidelobes: the extremely low loss HE_{11} hybrid mode dominates. The phase center of each feed is kept as close as possible to its aperture, resulting in a frequency-independent beam for each feed. Since the distance from the focal plane to the spacecraft symmetry axis is nearly the same for all the feeds, the high frequency feeds are extended with low loss corrugated waveguide, while the low frequency feeds are "profiled" to

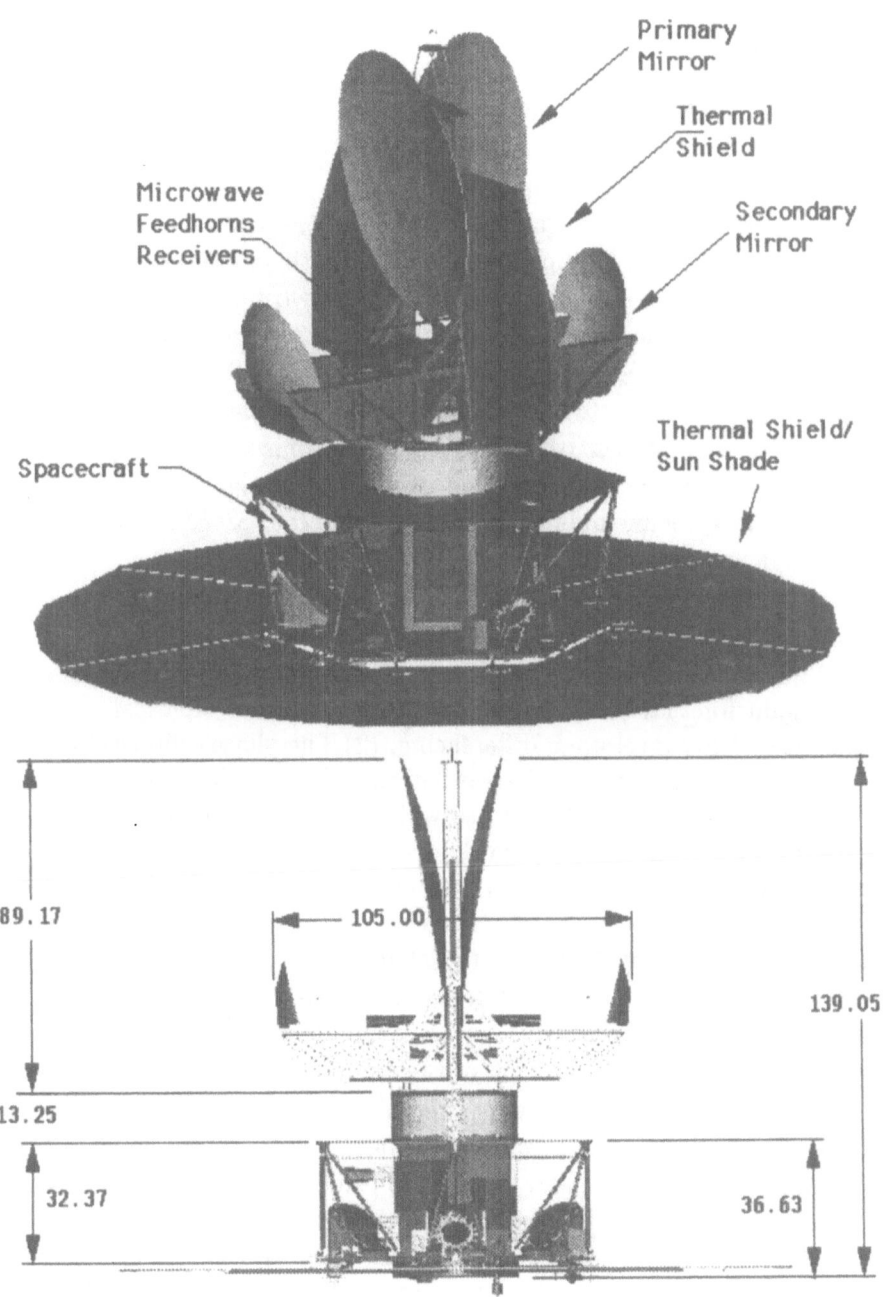

Figure 11. MAP (Microwave Anisotropy Probe) artists conception shown in two views.

reduce their length, while limiting excitation of the TE_{11} mode to less than -30 dB.

The microwave system consists of 10 4-channel differencing assemblies, one for each pair of feeds. One assembly operates at 22 GHz, one at 30 GHz, two at 40 GHz, two at 60 GHz, and four at 90 GHz. The base of an A-side feed in the Focal Plane Assembly (FPA) is attached to a low-loss orthomode transducer (OMT) which separates the signal into two orthogonal polarizations, A and A'. The A side signal is differenced against the orthogonal polarization, B', from the corresponding B-side feed, and vice-versa.

The differencing is accomplished by first combining the two signals A and B' in a hybrid tee to form $(A+B')/\sqrt{2}$ and $(A-B')/\sqrt{2}$, then amplifying each in two cold HEMT amplifiers and sending the phase-matched outputs to the warm receiver box via waveguide. The two signals are amplified in two warm HEMT amplifiers, phase switched between $0°$ and $+90°$ or $-90°$, respectively, at 2.5 kHz, then split back into A and B' in a second hybrid tee. At this point, the two signals are square-law detected, amplified by two line drivers, and sent to the Analog Electronics Unit for synchronous demodulation and digitization. The other pair of signals, A' and B, are differenced in the same manner giving a total of four amplification channels per differencing assembly.

The splitting, phase switching, and subsequent combining of the signals enhances the instrument's performance in two ways:

(1) Since both signals to be differenced are amplified by both amplifier chains, gain fluctuations in either amplifier chain act identically on both signals and thus cancel upon differencing. (2) The phase switches introduce a 180 degree relative phase change between the two signal paths, thereby interchanging which signal is fed to which square law detector. Thus, low frequency noise from the detector diodes is common mode and also cancels, further reducing susceptibility to systematic effects.

8.1.5. *Map Making with Differential Data*

MAP will observe temperature differences between points separated by 135° on the sky. Maps of the relative sky temperature will be produced from the difference data by a modification of the algorithm used by COBE-DMR.

The algorithm MAP will use to reconstruct sky maps from differential data is iterative. It is mathematically equivalent to a least squares fitting of the temperature differences to the map pixel temperatures. However, the scheme has a very intuitive interpretation: for a given pair of differential feeds, A and B, the A feed can be thought of as viewing the sky while the B feed can be thought of as viewing a comparative reference signal, or vice versa. In MAP's case, the comparative signal is a different point in the sky. The actual signal MAP measures is the temperature difference between two points on the sky, $\Delta T = T(A)-T(B)$, where $T(A)$ is the temperature seen by feed A, and likewise for B. If the temperature $T(B)$ is known, one could recover $T(A)$ using $T(A) = \Delta T+T(B)$, but since $T(B)$ is not known, the algorithm makes use of an iterative scheme in which $T(B)$ is estimated from the previous sky map iteration. Thus the temperature in a pixel of a map is given by the average of all observations of that pixel after correcting each observation for the estimated signal seen by the opposite feed.

For this scheme to be successful it is imperative for a given pixel to be observed with many different pixels on its ring of partners 135° away. Thus the method requires a carefully designed scan strategy. The MAP strategy achieves this while simultaneously avoiding close encounters with the Sun, Earth, and Moon. The algorithm has been tested with the MAP scan strategy using an end-to-end mission simulation that incorporates a realistic sky signal, instrument noise, and calibration methods. The results of these simulations are described in detail in an Astrophysical Journal article [87]. After 40 iterations of the algorithm, the artifacts that remain in the map due to the map-making itself have a peak-peak amplitude of less than 0.2 μK, even in the presence of Galactic features with a peak brightness in excess of 60 mK.

8.1.6. *MAP Sky Coverage and Scan Strategy*

MAP will observe the full sky every six months. The MAP scan strategy combines spacecraft spin and precession to achieve the following: 1) The MAP instrument observes more than 30% of the sky each day; 2) The spacecraft spin (and symmetry) axis maintains a fixed angle of 22.5° from the Sun-Earth line to mitigate systematic effects; and 3) Each sky pixel is connected to thousands of other sky pixels to ensure high quality map solutions with negligible noise correlations.

Since a major goal of cosmology is to determine the statistical properties of the Universe, it is clear that the largest possible number of sky samples improves constraints on cosmological models. The measurement of each individual position on the sky is an independent sample of the cosmology of the Universe. Moreover, full sky coverage is absolutely required to accurately determine the low-order spherical harmonic moments. While the largest angular scales were observed by COBE, MAP will remeasure the full sky with higher resolution to:

• Avoid relative calibration errors when two or more experimental results are combined (e.g., COBE and MAP).

  • Provide greater sensitivity to the angular power spectrum.

  • Independently verify the COBE results.

The goals of the MAP scan strategy include the following:

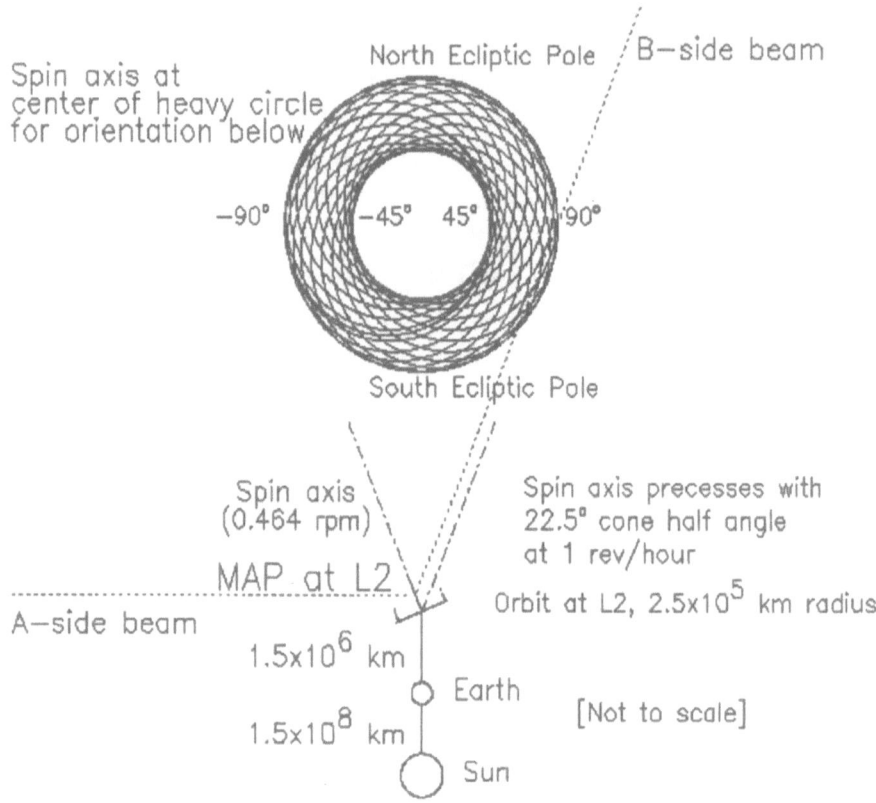

Figure 12. The MAP scan pattern for one hour of observation. The lines show the path for one side of a differential antenna pair. The other pair member follows a similar path, only delayed by 1.1 min. There are four principal time scales for the observations. The phase of the difference signal is switched by 180° at 2.5 KHz. The spacecraft spins around its symmetry axis with a 2.2 min period (bold circle) with cone opening angle of roughly 135°. This pattern precesses about the Earth-Sun line with a period of 60 minutes. Thus, in about 1 hour, over 30% of the sky is covered. Every six months, the whole sky is observed. Note that any pixel is differenced to another pixel in many directions.

- The angular separation between the two observing beams should be "large" in order to maintain sensitivity to signal at large angular scales. This is important for comparing the MAP results to COBE, for properly normalizing the angular power spectrum, and for retaining sensitivity to the dipole which will serve as MAP's primary calibration source.

- Observe a large fraction of the sky every day. This guarantees that sky pixels will be observed at many different times in the mission which provides the capability to monitor instrument stability on many different time scales.

- Maintain a fixed angle between the spacecraft spin axis and the local solar vector. This provides for stable illumination of the spacecraft solar panels which lie normal to the spin axis, and provides a thermally stable environment to mitigate systematic effects.

• Connect each sky pixel to as many other sky pixels as possible to provide high quality map solutions from the differential data, and to render negligible pixel-pixel noise correlations. Since the MAP beam separation is fixed, this implies observing as many pixels on the differential ring of pixels as possible.

The MAP beam separation is 135°. Each beam axis points 67.5° away from the spin and symmetry axis of the spacecraft. The spin axis will precess in a 22.5° angle about the local solar vector. The combined spacecraft spin and precession will cause the observing beams to fill an annulus centered on the local solar vector with inner and outer radii of 45° and 90° respectively. Thus MAP will observe more than 30% of the sky each day and will observe the ecliptic poles every day. The spin period will be 2.2 minutes while the precession period will be 1 hour. As the Earth orbits the Sun, the whole observing annulus revolves with it producing full sky coverage.

The MAP mission is moving ahead quickly. Its spacecraft preliminary design review occurred in January 1997 and instrument review was held in March 1997. At that point many of the major design features were fixed and only smaller modifications will occur. However, MAP has a WWW page http://map.gsfc.nasa.gov which can be consulted for the latest information.

8.2. PLANCK – THE THIRD GENERATION SPACE MISSION

The Planck mission is the result of the merging of two proposals presented in 1993 to the European Space Agency *M3 Call for Mission Ideas*: COBRAS (Cosmic Background Radiation Anisotropy Satellite [54] and SAMBA (Satellite for Measurements of Background Anisotropies [65]). The COBRAS/SAMBA team completed the ESA assessment study in May 1994, and the project continued and completed the Phase A study in May 1996. COBRAS/SAMBA, renamed Planck Surveyor, has been selected to continue within the European Space Agency M3 programme.

The Planck mission is designed for extensive, accurate mapping of the anisotropy of the CMB, with angular sensitivity from sub–degree ($\sim 8'-30'$) scales up to the full sky thus overlapping with the COBE–DMR maps and with signal sensitivity approaching $\Delta T/T \sim 10^{-6}$. Planck will survey the entire sky at frequencies from 30 to 850 GHz ($1\,\text{cm} \leq \lambda \leq 350$ microns). Its 1.3-m passively cooled telescope will be diffraction limited at frequencies below 375 GHz. The primary science goal is a definitive measurement of the structure in the CMB on all angular scales of $10'$ or larger. The case for orbital measurements of CMB anisotropy has been well made. An accurate measurement of CMB anisotropy with angular resolution of 10′ will revolutionize cosmology. A full description of the baseline Planck mission is available from ESTEC at http://astro.estec.esa.nl/SA-general/Projects/ COBRAS/cobras.html. Additional information is available at http://aether.lbl.gov/www/cosa/. This information will be updated regularly and new links added.

Planck has been selected as ESA's next medium-scale mission (M3), this was confirmed in the November 1996 review of the Ariane V demise of the Cluster Mission. Planck is scheduled to fly in 2004, approximately four years after the MAP NASA MidEX mission. Planck is well-designed to follow MAP. Using more sophisticated detector technology, Planck will have 10 times the sensitivity, 2 or 3 (depending on frequency) times the angular resolution, and 6 times the frequency coverage of MAP. This performance will allow Planck to:

1. Measure the power spectrum of the CMB with accuracy limited by cosmic variance over almost the entire range of angular frequency space in which useful cosmological information is expected.

2. Separate Galactic and extragalactic foregrounds from the CMB with high accuracy and confidence. The broad frequency coverage will allow determination of all important foreground components without any prior assumptions about their spectra.

3. Separate secondary anisotropies due to the Sunyaev-Zel'dovich effect from primary anisotropies, and measure the SZ effect with precision in thousands of clusters of galaxies. This information, combined with X-ray data, will yield an independent measurement of the Hubble constant on large scales and probe the peculiar velocity field of clusters to high redshift.

4. Separate polarization of the CMB from that in local foregrounds and measure it with precision on angular scales as small as 7′.

5. Survey the sky at sub-mm (350, 550 and 850 microns) wavelengths that complement the wavelength coverage of SIRTF. The chance for serendiptous discovery in this survey is great.

Planck has two instruments: the Low Frequency Instrument (LFI), based on transistor (HEMT) amplifiers, which covers the frequency range from 30–100 GHz; and the High Frequency Instrument (HFI), based on bolometers, which covers the frequency range 90–850 GHz. The implementation of these technologies fits comfortably within the mass, power, volume and schedule constraints of the ESA M3 opportunity.

8.2.1. *Planck Active Cooling Option*
One area being pursued at the present is the use of actively cooled HEMTs, which requires the HEMT amplifier chain to be broken into a low-temperature portion and a higher-temperature portion. This split reduces the thermal

load on the focal plane allowing passive cooling to a significantly lower temperature (i.e., 65 K) and allowing for the use of active cooling technologies such as a hydrogen sorption cooler.

In addition, there are technical advantages of a combined LFI/HFI focal assembly that uses a hydrogen sorption cooler to cool both the HEMTs and the 20 K shield of the bolometer dewar. The sorption cooler would replace the 20 K Stirling cooler in the baseline design, with its attendant problems in vibration and instrument integration, and reduce the overall mass and power of the instruments. This is an extremely attractive option that will be studied in detail by both the HFI and LFI teams.

The split HEMT design and sorption cooler enable a LFI design with the following features:

• An increase in the sensitivity of the LFI by a factor of roughly five at the highest frequency over that of the Phase A baseline design.

• Division of the HEMT radiometers into a cold focal assembly and a room-temperature assembly. The power dissipated in the focal assembly is more than an order of magnitude lower than assumed in the Phase A study design, allowing radiative cooling of the focal assembly to a temperature of $\leq 65\,\mathrm{K}$ instead of $\sim 100\,\mathrm{K}$.

• Active cooling of the HEMTs in the focal assembly to $\leq 20\,\mathrm{K}$. This reduces potential thermal interactions between the LFI and HFI.

• Elimination of the $\sim 80\,\mathrm{K}$ Stirling cooler in the baseline design.

• The option (introduced above) of using the sorption cooler to cool the 20 K shield around the HFI as well as the HEMTs eliminates the need for the 20 K Stirling cooler in the baseline design. An important benefit of this option is that the vibration associated with Stirling coolers would be eliminated from the focal assembly. The overall mass and power requirements of the instruments on the spacecraft would decrease by roughly 25 kg and 100 W as well.

This design will allow a full identification of the primordial density perturbations which grew to form the large–scale structures observed in the present Universe. The Planck maps will provide decisive answers to several major open questions relevant to the structure formation epoch and will provide powerful tests for the inflationary model as well as several astrophysical issues. Planck will utilize a combination of bolometric and radiometric detection techniques to ensure the sensitivity and wide spectral coverage required for accurate foreground discrimination. An orbit far from Earth has been selected to minimize the unwanted emission from the Earth as a source of contamination.

8.2.2. *Planck Scientific Objectives*

The Planck mission will produce near all-sky maps of the background anisotropies in 8 frequency bands in the range 30–800 GHz, with peak sensitivity $\Delta T/T \sim 10^{-6}$. The maps will provide a detailed description of the background radiation fluctuations. Individual hot and cold regions should be identified above the statistical noise level, at all angular scales from $\lesssim 10'$ up to very large scales, thus providing a high resolution imaging of the last scattering surface.

The Planck maps will provide all multipoles of the temperature anisotropies from $\ell = 1$ (dipole term) up to $\ell \simeq 1500$ (corresponding to $\sim 7'$). It is the information contained in this large number of multipoles that can probe the various proposed scenarios of structure formation and the shape of the primordial fluctuation spectrum (for comparison, the COBE–DMR maps are limited to $\ell \lesssim 20$).

Table 5 compares the ability of various CMB missions to determine cosmological parameters in a model-dependent but self-consistent way. The details of the calculation are not important, with two exceptions. First, $\Omega = 1$ was assumed. If Ω is smaller differences in angular resolution become even more important. Second, it was assumed that confusing foregrounds were completely removed. In practice this will not be the case. The advantage that Planck's wide frequency coverage gives is therefore not reflected in the table. Nevertheless, the power of Planck in general, and the advantages of cooling the HEMTs, are immediately apparent.

The high resolution Planck maps will provide a key test for structure formation mechanisms, based on the statistics of the observed $\Delta T/T$ distribution. The inflationary model predicts Gaussian fluctuations for the statistics of the CMB anisotropies, while alternative models based on the presence of topological defects, such as strings, monopoles, and textures, predict non–Gaussian statistics (e.g. [21]). Due to the different nature of their early history causality constrains primordial perturbations from a source such as inflation and from topological defects to have a different anisotropy power spectra particularly in the region of the "Doppler" peaks [1]. The angular resolution and sensitivity of Planck will allow discrimination between these alternatives with tests of both the power spectrum and statistics.

The high-order multipoles will allow an accurate measure of the spectral index n of the primordial fluctuation spectrum:

$$(\delta\phi)^2 \propto \lambda^{(1-n)} \tag{1}$$

where $\delta\phi$ is the potential fluctuation responsible for the CMB anisotropies, and λ is the scale of the density perturbation. This corresponds to CMB temperature fluctuations $(\Delta T/T)^2 \propto \theta^{(1-n)}$ for angles $\theta > 30' \, \Omega_0^{1/2}$. The

TABLE 5. UNCERTAINTIES IN COSMOLOGICAL PARAME-
TERS

PARAMETER	MAP	Planck HFI[a]	Planck LFI[b]	Planck LFI[c]
$Q_{\mathrm{rms-ps}}/20\mu K$	0.23	0.12	0.18	0.14
h	0.13	0.032	0.12	0.065
$h^2\Omega_b$	0.0072	0.0019	0.0062	0.0036
Ω_Λ	0.67	0.19	0.59	0.33
Ω_ν	0.38	0.12	0.36	0.28
Ω	0.11	0.012	0.068	0.029
n_s	0.12	0.017	0.074	0.029
τ	0.35	0.15	0.26	0.20
N_ν	0.43	0.16	0.40	0.26
$T_0\,\mu K$	0.01	0.01	0.01	0.01
Y	0.01	0.0098	0.01	0.01
T/S	0.47	0.17	0.28	0.19
$Q_{\mathrm{ps}}/0.2\mu K$	0.02	0.000021	0.06	0.0006
$Q_{\mathrm{diffuse}}/20\mu K$	0.19	0.17	0.18	0.18

[a] HFI based on "spider web" bolometers, as given in Phase A study

[b] LFI based on InP HEMTs at 100 K, as given in Phase A study

[c] LFI based on InP HEMTs at 20 K

proposed observations will be able to verify accurately the nearly scale invariant "Harrison–Zel'dovich" spectrum ($n = 1$) predicted by inflation. Any significant deviation from that value would have extremely important consequences for the inflationary paradigm. The COBE–DMR limit on the spectral index after four years of observations ($n = 1.1^{+0.2}_{-0.3}$, 68% CL; [32]) can be constrained ~ 10 times better by the Planck results.

The proposed observations will provide an additional, independent test for the inflationary model. Temperature anisotropies on large angular scales can be generated by gravitational waves (tensor modes, T), in addition to the energy-density perturbation component (scalar modes, S). Most inflationary models predict a well determined, simple relation between the ratio of these two components, T/S, and the spectral index n [23], [53]:

$$n \approx 1 - \frac{1}{7}\frac{T}{S}. \tag{2}$$

The Planck maps may be able to verify this relationship, since the temperature anisotropies from scalar and tensor modes vary with multipoles in different ways.

A good satellite mission will be able not only to test the inflationary concept but also to distinguish between various models and determine inflationary parameters. There is an extensive literature on what can be determined about inflation such as the scalar and tensor power spectra, the energy scale of inflation and so on (see e.g. [76], [42]). Such quality measurements lead also to good observations or constraints for Ω_0, Ω_b, Λ, H_0, etc. Sub–degree anisotropies are sensitive to the ionization history of the Universe. In fact, they can be erased if the intergalactic medium underwent reionization at high redshifts. Moreover, the temperature anisotropies at small angular scales depend on other key cosmological parameters, such as the initial spectrum of irregularities, the baryon density of the Universe, the nature of dark matter, and the geometry of the Universe (see e.g. [22], [41], [38], [70]. The Planck maps will provide constraints on these parameters within the context of specific theoretical models.

Moreover, Planck should measure the Sunyaev–Zel'dovich effect for more than 1000 rich clusters, using the higher resolution bolometric channels. This will allow a rich analysis of clusters. Combined with X–ray observations these measurements can be used to estimate the Hubble constant H_0 as a second independent determination.

8.2.3. *Foreground Emissions*

In order to obtain these scientific goals, the measured temperature fluctuations need to be well understood in terms of the various components that add to the cosmological signal. In fact, in addition to the CMB temperature fluctuations, foreground structures will be present from weak, unresolved extragalactic sources and from radiation of Galactic origin (interstellar dust, free–free and synchrotron radiation).The Planck observations will reach the required control on the foreground components in two ways. First, the large sky coverage ($\geq 90\%$ of the sky) will allow accurate modeling of these components where they are dominant (e.g. Galactic radiation near the Galactic plane). Second, the observations will be performed in a very broad spectral range.

The Planck channels will span the spectral region of minimum foreground intensity (in the range 50–300 GHz), but with enough margin at high and low frequency to monitor "in real–time" the effect of the various foreground components (see e.g. Brandt et al. 1994). By using the Planck spectral information and modeling the spectral dependence of Galactic and extragalactic emissions it will be possible to remove the foreground contributions with high accuracy.

It should be noted that in most channels the ultimate limitation to the cosmological information of high-quality CMB maps is expected to be due to the residual uncertainties in the separation of the foreground components

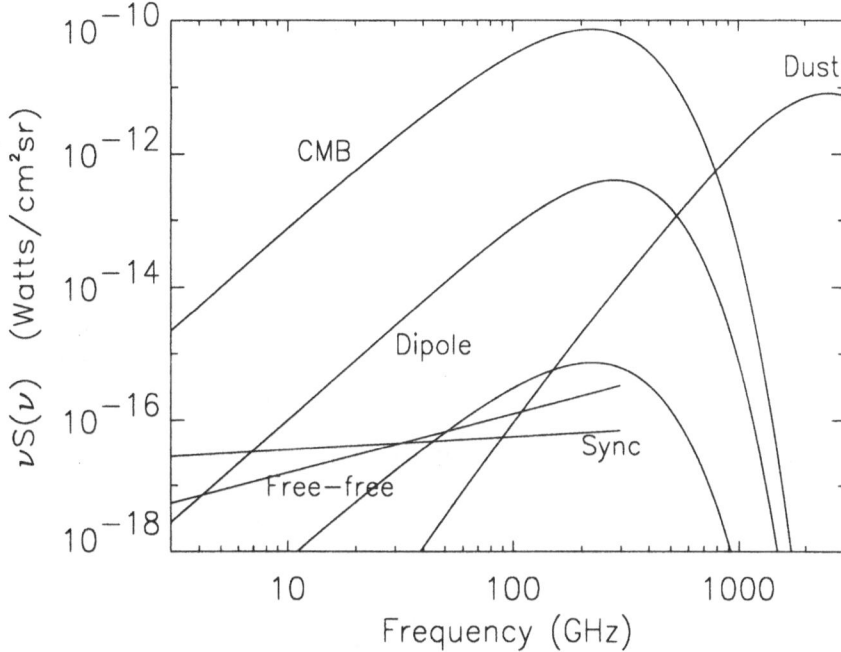

Figure 13. The intensity of the microwave sky from 3 to 3000 GHz near a Galactic latitude of $b = 20°$. The ordinate is the brightness of the sky times the frequency. which means the plot indicates the distribution of power per unit bandwidth. For synchrotron emission $S(\nu) \propto \nu^{-0.7}$; for free-free emission $S(\nu) \propto \nu^{-0.1}$; and for dust emission near 100 GHz $S(\nu) \propto \nu^{3.7}$. The scaling in effective temperature is $T(\nu) \propto \nu^{-2} S(\nu)$. The lowest Planck-like curve is for $T = 27 \mu K$ or anisotropy at the 10^{-5} level.

rather than statistical noise. This explains why the overall design of Planck is highly driven by the need of achieving as large a spectral coverage as possible. Making the observations where the dominant foreground components are different will permit a powerful cross check on residual systematic errors in the CMB temperature fluctuation maps.

The Phase A Study provides a baseline design for the mission, spacecraft, and instruments. Broad frequency coverage is achieved with arrays of HEMT amplifiers (30–100 GHz) and bolometers (100–850 GHz). The amplifiers use GaAs MMICs (monolithic microwave integrated circuits) cooled passively to about 100 K. At the end of the Phase A study the number of detectors and focal plane layout were optimized for the passively cooled LFI configuration. This is summarized in Table 7

During the time since the Phase A study work has continued and it has been learned that converting to InP HEMT amplifiers and actively cooling to 20 K provides an improvement in performance and that changing some operating parameters the instrument might be improved in other

1st Phase A Planck Payload Characteristics							
Telescope	1.5 m Diam. Gregorian; system emissivity $\leq 1\%$						
	Viewing direction offset $\geq 70°$ from spin axis						
Instrument	LFI				HFI		
Center Frequency (GHz)	31.5	53	90	125	140	222	400	714
Wavelength (mm)	9.5	5.7	3.3	2.4	2.1	1.4	0.75	0.42
Bandwidth ($\frac{\Delta\nu}{\nu}$)	0.15	0.15	0.15	0.15	0.4	0.5	0.7	0.6
Detector Technology	HEMT receiver arrays				Bolometers arrays			
Detector Temperature	~ 100 K				0.1 - 0.15 K			
Cooling Requirements	Passive				Cryocooler + Dilution			
Number of Detectors	13	13	13	13	8	11	16	16
Angular Resolution (arcmin)	30	20	15	12	10.5	7.5	4.5	3
Optical Efficiency	1	1	1	1	0.3	0.3	0.3	0.3
$\frac{\Delta T}{T}$ Sensitivity (1σ, $[10^{-6}]$, 90% sky coverage, 2 years)	1.7	2.7	4.1	7.2	0.9	1.0	8.2	10^4
$\frac{\Delta T}{T}$ Sensitivity (1σ, $[10^{-6}]$, 2 % sky coverage, 2 years)	0.6	0.9	1.4	2.4	0.3	0.3	2.7	5000

TABLE 6. Instrumental Parameters for Planck. The most important factors are the frequency coverage, the angular resolution, sky coverage, and sensitivity.

ways. Table 8 shows another possible configuration taking advantage of the developments since the end of Phase A.

The need of accurate characterization of all non-cosmological components, of course, brings the benefit of additional astrophysical information. The very large Planck data base, particularly when combined with the IRAS survey, can provide information on several non–cosmological issues, such as the evolution of starburst galaxies, the distribution of a cold–dust component, or the study of low–mass star formation.

8.2.4. *The Payload*

The Planck model payload consists mainly of a shielded, off-axis Gregorian telescope, with a parabolic primary reflector and a secondary mirror, leading to an integrated instrument focal plane assembly. The payload is part of a spinning spacecraft, with a spin rate of 1 rpm. The focal plane assembly is divided into low-frequency (LFI) and high-frequency (HFI) in-

Final PHASE A INSTRUMENT SUMMARY

Characteristic	LFI				HFI				
Detector technology	HEMT arrays				Bolometer arrays				
Detector temperature	~ 100 K				0.1–0.15 K				
Center freq [GHz]	31.5	53	90	125	143	217	353	545	857
Number of detectors	4	14	26	12	8	12	12	12	12
Angular resolution [']	30	18	12	12	10.3	7.1	4.4	4.4	4.4
Bandwidth [$\Delta\nu/\nu$]	0.15	0.15	0.15	0.15	0.37	0.37	0.37	0.37	0.37
Noise/res. element, in 15 months [μK]	21	20	39	97	3.3	5.5	33	210	11,000

TABLE 7. Proposed "conservative" Planck (COBRAS/SAMBA) instrument configuration at the end of the Phase A study. It was based upon MMIC GaAs and early micromesh bolometer technology and a mix of cryocooler options.

POTENTIAL PLANCK INSTRUMENT SUMMARY

Characteristic	LFI				HFI					
Detector technology	HEMT arrays				Bolometer arrays					
Detector temperature	20 K				0.1–0.15 K					
Center freq [GHz]	30	44	70	100	100	143	217	353	545	857
Number of detectors	4	6	10	30	4	12	12	6	6	6
Angular resolution [']	34	23	16	10	14	10	7.1	4.4	4.4	4.4
Bandwidth [$\Delta\nu/\nu$]	0.20	0.20	0.20	0.20	0.25	0.25	0.25	0.25	0.25	0.25
Noise/res. element, in 15 months [μK]	5	7	10	17	3	4	7	41	240	10,600
Both polarizations?	yes	yes	yes	yes	no?	yes	yes	no?	no	no

TABLE 8. Potential Planck detector configuration based upon active LFI cooling, InP HEMT amplifiers, more advanced micromesh, filter, feedhorn, and polarizer technology.

strumentation according to the technology of the detectors. Both the LFI and the HFI are designed to produce high-sensitivity, multifrequency measurements of the diffuse sky radiation. The LFI will measure in four bands in the frequency range 30–130 GHz (2.3–10 mm wavelength). The HFI will measure in four channels in the range 140–800 GHz (0.4–2.1 mm wavelength). The highest frequency LFI channel and the lowest HFI channel overlap near the minimum foreground region. Both the HFI and LFI teams are reinvestigating the optimal frequency bands. Table 6 summarizes the main characteristics of the Planck payload.

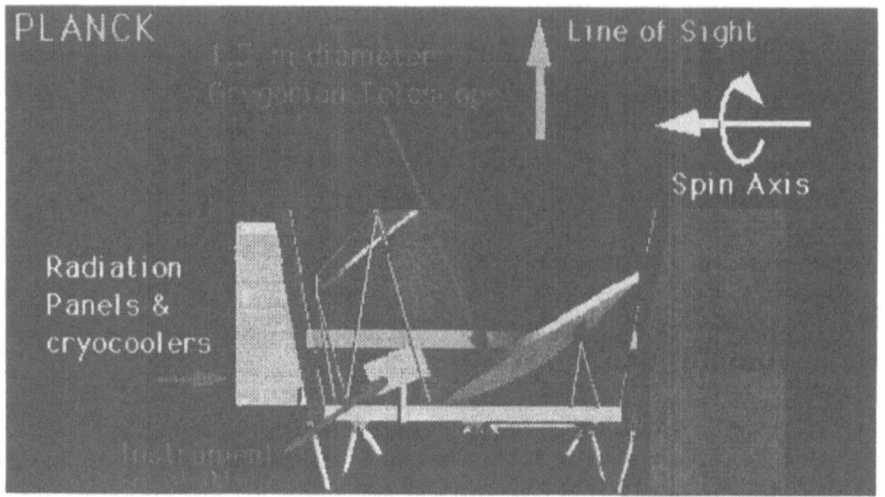

Figure 14. Artist's concept of one possible configuration of the Planck Surveyor optics and focal plane layout.

8.2.5. *The Main Optical System*

A clear field of view is necessary for the optics of a high-sensitivity CMB anisotropy experiment to avoid spurious signals arising from the mirrors or from supports and mechanical mounting. The off-axis Gregorian configuration has a primary parabolic mirror of 1.5 meter, and an elliptic secondary mirror (0.57 m diameter). Stray satellite radiation and other off-axis emissions are minimized by underilluminating the low–emissivity optics. The telescope reimages the sky onto the focal plane instrument located near the payload platform. The telescope optical axis is offset by 70° to 90° from the spin axis. Thus at each spacecraft spin rotation the telescope pointing direction sweeps a large (approaching a great) circle in the sky, according to the sky scan strategy.

Blockage is a particularly important factor since several feeds and detectors are located in the focal plane, and unwanted, local radiation (e.g. from the Earth, the Sun and the Moon) needs to be efficiently rejected. A large, flared shield surrounds the entire telescope and focal plane assembly, to screen the detectors from contaminating sources of radiation. The shield also plays an important role as an element of the passive thermal control of the spacecraft.

8.2.6. *The Focal Plane Assembly*

The necessary wide spectral range requires the use of two different technologies, bolometers and coherent receivers incorporated in a single instrument. Both technologies have shown impressive progress in the last ten years or so, and more is expected in the near future. The thermal requirements of the two types of detectors are widely different. The coherent radiometers (LFI), operating in the low frequency channels, give good performance at and operational temperature of ~ 100 K, which is achievable with passive cooling. Splitting the HEMT chain into cool and warmer portions leads to a passive cooling temperature of 65 K and with active cooling 20 K. The bolometers, on the other hand, require temperatures ≤ 0.15 K in order to reach their extraordinary sensitivity performances. The main characteristics of the LFI and HFI are summarized in Table 6.

The LFI consists of an array of 26 corrugated, conical horns, each exploited in the two orthogonal polarization modes, feeding a set of state–of–the–art, high sensitivity receivers. The receivers will be based on MMIC (Monolithic Microwave Integrated Circuits) technology with HEMT (High Electron Mobility Transistor) ultra–low noise amplifiers (see e.g. Pospieszalski et al. 1993). Since the whole LFI system will be passively cooled, it can be operated for a duration limited only by spacecraft consumables (up to 5 years). If actively cooled with a sorption cooler, it can still operate for 5 years as there is no significant cryogen depletion; however, the HEMT chain must be broken into two sections. The three lowest center frequencies of the LFI were chosen to match the COBE-DMR channels, to facilitate the comparison of the product maps. The exact frequency bands are being reviewed in the upgraded design.

About 50 bolometers will be used in the HFI instrument, which require cooling at ~ 0.1 K. The cooling system combines active coolers reaching 4 K with a dilution refrigeration system working at zero gravity. The refrigeration system will include two pressurized tanks of ^{3}He and ^{4}He for an operational lifetime of 2 years.

8.2.7. *Orbit and Sky Observation Strategy*

One of the main requirements for the Planck mission is the need of a far–Earth orbit. This choice greatly reduces the problem of unwanted radiation from the Earth which is a serious potential contaminant at the high goal sensitivity and angular resolution. The requirements on residual Earth radiation are basically the same for the LFI and the HFI systems. Adopting a low–earth orbit, such as that used by the COBE satellite, the requirement on straylight and sidelobe rejection would be a factor of 10^{13}, which is beyond the capabilities of present microwave and sub–mm systems and test equipment. Two orbits have been considered for Planck: a small orbit around the L5 Lagrangian point of the Earth–Moon system, at a distance of about 400,000 km from both the Earth and the Moon and the L2 Lagrange point of the Earth–Sun system. From the Earth–Moon Lagrange point the required rejection is relaxed by four orders of magnitude, which is achievable with careful, standard optical designs. For the Earth–Sun L2 point the situation for the Earth and Moon is even better and the Sun is basically unchanged but because the Earth, Moon, and Sun are all roughly in the same direction, the spacecraft can be oriented very favorably.

These orbits are also very favorable from the point of view of passive cooling and thermal stability [28]. The spacecraft will be normally operated in the anti-solar direction, with part of the sky observations performed within ±40° from anti–solar.

Other potential missions considered both a heliocentric orbit and the Earth–Sun L2 point. All concerned seemed to have come to the conclusion that the Earth–Sun L2 point is the best choice. Operationally, it is difficult to find a more optimum location.

The main goal of the mission is to observe nearly the whole sky ($\gtrsim 90\%$) with a sensitivity of 10–15 μK within the two year mission lifetime. Deeper observation of a limited ($\sim 2\%$) sky region with low foregrounds could significantly contribute to the cosmological information. Simulations have shown that these observational objectives can be achieved simultaneously in a natural way, using the spinning and orbit motion of the spacecraft, with relatively simple schemes.

9. Interpretation and Future

In five short years the field of CMB anisotropy observations and theory has made great strides. Until April 1992 all plots of CMB anisotropy showed only upper limits, except for the $\ell = 1$ dipole. Now we are beginning to trace out the shape of the power spectrum and to make maps of the anisotropies. This observational program promises to deliver a wealth of new information to cosmology and to connect it to other fields. The COBE DMR has now released the full four-year data set. We can expect little in the way of improvement compared to the final DMR results from future experiments on the large angular scales but scientific interest has moved to covering the full spectrum and learning what the medium and small angular scales will tell us. Already we are seeing plots showing the CMB anisotropy spectrum related to and overlaid on the primordial density perturbation power spectrum and attempts to reconstruct the inflaton potential. These are the first steps in a new period of growth.

The last table gives an example of the level of sensitivity that might be achieved by the many experiments underway, planned, and approved. Nearly every group has data under analysis and is also at work on developing new experiments. Some of these are the natural extensions of the ongoing experiments. Some groups are considering novel approaches. Real long-term progress depends on avoiding the potential foregrounds: fluctuations of the atmosphere, a source of noise that largely overwhelms recent ad-

	1997	BOOM/MAX	MAP*	Planck*
Ω	0.01 - 2	6%	18%	1%
Ω_b	$0.01h^{-2}$	30%	10%	0.7%
$\Lambda(\Omega_\Lambda)$	< 0.65	±0.10	±0.43	±0.05
Ω_ν	< 2	±0.25	±0.08	±0.03
t_0	12-18 Gyr	—	—	—
H_0	30-80 km/s/Mpc	10%	20%	2%
σ_8	0.5-0.6	30%	30%	10%
Q	$20 \pm 2\,\mu K^\dagger$	"	"	"
n_s	1.0 ± 0.5	30%	5%	1%
τ	0.01 - 1	±0.5	±0.2	±0.15
T_0	$2.73 \pm 0.01^\dagger$	—	—	—
Y	0.2-0.25	10%	10%	7%
T/S	0.0 - 1	±1.6	±0.38	±0.09

TABLE 9. Projected Parameter Errors: Assumes variation around Standard CDM. (*Bond et al., 1997.) Note that parameters are not all independent, e.g., $H_0 t_0 = f(\Omega, \Lambda)$. The symbol † indicates current precise results from CMB observations (COBE).

vances in detector technology, and Galactic and extragalactic signals. This requires instruments having sufficient information (usually only through multifrequency observations) and observing frequencies to separate out the various components. It also means going above the varying atmosphere. Collaborations are working on long-duration ballooning instruments. Ultimately, as COBE has shown, going to space really allows one to overcome the atmospheric problem and to get data in a very stable and shielded environment. The two selected satellite mission are actively being developed. We can anticipate a steady and significant advance in observations. With the new data that are appearing, can be expected, and ultimately will come from the Planck mission we can look forward to a very significant improvement in our knowledge of cosmology.

Acknowledgements

This work was supported in part by the Director, Office of Energy Research, Office of High Energy and Nuclear Physics, Division of High Energy Physics of the U.S. Department of Energy under contract No. DE-AC03-76SF00098.

236

References

1. Albrecht, A., Coulsen, D., Ferreira, P., & Magueijo, J., 1995 Phys. Rev. Lett. 76, 1413. astro-ph 9505030.
2. Alsop, D. C., et al. 1992, ApJ, 395, 317
3. Alpher, R.A. & R.C. Herman 1948 Physics Today, Vol. 41, No. 8, p. 24
4. Banday, A. et al. 1996a, ApJ, 468, L85. astro-ph/9601064
5. Banday, A. et al. 1996b, ApJ, 464, L1, astro-ph/9601066
6. Bennett, C. L., et al. 1996, ApJ, 464, L1, astro-ph/9601067
7. Bennett, C. L., et al. 1994, ApJ, 436, 423
8. Bennett, C. L., et al. 1993, ApJ, 414, L77
9. Bennett, C. L., et al. 1992, ApJ, 396, L7
10. Bennett, C. L., et al. 1991, ApJ, 391, 466
11. Bock, J. et al. Proceedings of "Submillimeter and Far-Infrared Space Instrumentation," 30th ESLAB Symposium, 24-26 Sept. 1996, ESTEC, Noordwijk, The Netherlands.
12. J.R. Bond astrop-ph/9407044 (1994)
13. J.R. Bond, G. Efstathiou, & M. Tegmark astrop-ph/9702100 (1997)
14. Brandt, W.N., Lawrence, C.R., Readhead, ACS, Pakianathan, & Fiola, T.M., 1994, Ap.J. 424, 1
15. Clapp, A. C., Devlin, M. J., Gundersen, J. O., Hagmann, C. A., Hristov, V. V., Lange, A. E., Lim, M., Lubin, P. M., Mauskopf, P. D., Meinhold, P. R., Richards, P. L., Smoot, G. F., Tanaka, S. T., Timbie, P. T. & Wuensche, C. A. 1994, ApJL, 433, 57-60. astro-ph/9404072
16. Cheng, E. S., Cottingham, D. A., Fixsen, D. J., Inman, C. A., Kowitt, M. S., Meyer, S. S., Page, L. A., Puchalla, J. L., Ruhl, J. & Silverberg, R. F. 1996, ApJ, 456, L71 astro-ph/9508087.
17. Cheng et al. astro-ph/9705041
18. Clapp, A. C., Devlin, M. J., Gundersen, J. O., Hagmann, C. A., Hristov, V. V., Lange, A. E., Lim, M., Lubin, P. M., Mauskopf, P. D., Meinhold, P. R., Richards, P. L., Smoot, G. F., Tanaka, S. T., Timbie, P. T. & Wuensche, C. A. 1994, ApJL, 433:57-60 astro-ph/9404072
19. Cayon, L., & Smoot, G.F., 1995 Astrophysical Journal, 452:487 astro-ph/9504072.
20. Costa, A. de Oliveira-, & Smoot, G.F., 1995, Astrophysical Journal, 448:477. astro-ph/9412003
21. Coulson, D., Ferreira, P., Graham, P., & Turok, N., 1994, Nature, 368, 27
22. Crittenden, R., Bond, J.R., Davis, R.L., Efstathiou, G., & Steinhardt, P., 1993, Phys. Rev. Lett., 71, 324
23. Davis, R.L., Hodges, H.M., Smoot, G.F., Steinhardt, P.J., Turner, M.S.; 1992, PRL, 69, 1856. erratum 70:1733
24. Devlin, M. J., Clapp, A. C., Gundersen, J. O., Hagmann, C. A., Hristov, V. V., Lange, A. E., Lim, M., Lubin, P. M., Mauskopf, P. D., Meinhold, P. R., Richards, P. L., Smoot, G. F., Tanaka, S. T., Timbie, P. T. & Wuensche, C. A. 1994, ApJL, **430**, L1. astro-ph/940403
25. R.H. Dicke, P.J.E. Peebles, P.G. Roll, and D.T. Wilkinson1965 Ap.J. 142, 414.
26. A.G. Doroshkevich, V.N. Lukash & I.D. Novikov 1973 The Isotropization of Homogeneous Cosmological Models *Zh. Eksper. Teor. Fiz* **64**, 739-746.
27. A.G. Doroshkevich, V.N. Lukash & I.D. Novikov 1974 Primordial Radiation in a Homogeneous but Anisotropic Universe *Astron. Zh.* **51**, 554-560.
28. Farquhar, R.W & Dunham D.W., 1990; *Observatories in Earth Orbit & Beyond*, Y.Kondo Ed., Kluwer, p.391.
29. Fischer, M. L., et al. 1992, ApJ, 388, 242
30. Fixsen, D. J., et al. 1996, ApJ, accepted, astro-ph/9605054
31. Ganga, K., et al. 1993, ApJ, 410, L57
32. Górski K.M. et al., 1996, ApJ, 464, L11. astro-ph/9601063

237

33. Harrison, E. R. 1970, Phys. Rev. D, 1, 2726
34. Hauser, M. G. & Peebles, P. J. E. 1973, ApJ, 185, 757
35. Hinshaw, G., et al. 1996a, ApJ Letters, 464, L17. astro-ph/9601058
36. Hinshaw, G., et al. 1996b, ApJ Letters, 464, L25. astro-ph/9601061
37. Hobson M.P., Lasenby A.N. & Jones M., 1995, MNRAS, 275, 863.
38. Hu, W., Scott, D., & Silk, J., 1994, ApJ., 430, L5
39. Hu, W., D. Spergel, & M. White 1996 PRD 55, 3288 astro-ph/9605193
40. Jungman, G., Kamionkowski, M., Kosowsky, A., & Spergel, D. N. 1995, PRL 76, 1007. astro-ph/9512139
41. M. Kamionkowski, D.N. Spergel, & N. Sugyiama 1994 ApJ 426, L1.
42. Knox, L. 1995, Phys. Rev. D 1995 52, 4307. astro-ph/9504054
43. Kogut, A., et al. 1992, ApJ, 401, 1
44. Kogut, A., Banday, A. J., Bennett, C. L., Hinshaw, G., Loewenstein, K., Lubin, P., Smoot, G. F., & Wright, E. L. 1994, ApJ, 433, 435
45. Kogut, A., et al. 1996a, ApJ, 460, 1. astro-ph/9601066
46. Kogut, A., et al. 1996b, ApJ, 464, L5. astro-ph/9601060
47. Kogut, A., et al. 1996c, ApJ, 464, L29. astro-ph/9601062
48. Lee, A. T. et al. 1996, Preprint, To appear in Appl. Phys. Lett.
49. M. A. Lim, A. C. Clapp, M. J. Devlin, N. Figueiredo, J. O. Gundersen, S. Hanany, V. V. Hristov, A. E. Lange, P. M. Lubin, P. R. Meinhold, P. L. Richards, J. W. Staren, G. F. Smoot, S. T. Tanaka astro-ph/9605142
50. Lineweaver, C., et al. 1994 Ap. J., 436, 452. astro-ph/9403021
51. Lineweaver, C., et al. 1995, ApJ, 448, 482.
52. Lineweaver, C., et al. 1996, ApJ, 470, 38. astro-ph/9601151
53. Little & Lyth 1992 PLB, 291, 391.
54. N. Mandolesi, G.F. Smoot, M. Bersanelli, C. Cesarsky, M. Lachieze-Rey, L. Danese, N. Vittorio, P. De Bernardis, G. Dall'Oglio, G. Sironi, P. Crane, M. Janssen, B. Partridge, J. Beckman, R. Rebolo, J. L. Puget, E. Bussoletti, G. Raffelt, R. Davies, P. Encrenaz, V. Natale, G. Tofani, P. Merluzzi, L. Toffolatti, R. Scaramella, E. Martinez-Gonzales, D. Saez, A. Lasenby, G. Efstathiou; 1993, *COBRAS*, proposal submitted to ESA M3.
55. Mandolesi, N., Bersanelli, M., Cesarsky, C., Danese, L., Efstathiou, g, Griffin, M., Lamarre, J.M., Norgaard-Nielson, H.U., Pace, O., Puget, J.L., Raisanen, A., Smoot, G.F., Tauber, J., & Volonte, S. 1995, Planetary & Space Sciences 43, 1459.
56. Mather, J. C., et al. 1994, ApJ, 420, 439
57. Meinhold, P., et al. 1993, ApJ, 409, L1
58. O'Sullivan C. et al., 1995, MNRAS, submitted.
59. Ganga, K, Page, L. Cheng, E., Meyer, S. 1994 astro-ph/9404009
60. L. Page 1997 astro-ph/9703054
61. Peebles, P. J. E. 1973, ApJ, 185, 413
62. Peebles, P. J. E. & Yu, J. T. 1970, ApJ, 162, 815
63. P.J.E. Peebles, 1993, "Principles of Physical Cosmology," Princeton U. Press, p. 168
64. A.A. Penzias and R. Wilson 1965, Ap. J. 142, 419.
65. Puget, J.L., Ade, P., Benoit, A., De Bernardis, P., Bouchet, F., Cesarsky, C., Desert, F.X., Gispert, R., Griffin, M., Lachieze-Rey, M., Lamarre, J.M., De Marcillac, P., Masi, S., Melchiorri, F., Pajot, F., Rowan-Robinson, M., Serra, G., Torre, J.P., Vigroux, L., White, S.; 1993, *SAMBA*, proposal submitted to ESA M3.
66. Robson M., Yassin G., Woan G., Wilson D.M.A., Scott P.F., Lasenby A.N., Kenderdine S., Duffett-Smith P.J., 1993 A&A, 277, 314.
67. Robson M., O'Sullivan C.M.M., Scott P.F., Duffett-Smith P.J., 1994, A&A, 286, 1028.
68. Robson M., 1994, PhD thesis, (University of Cambridge).
69. R. K. Sachs & A.M. Wolfe 1967, Ap.J. 147, 73.
70. D. Scott. J. Silk, and M. White 1995, Science, 268, 829
71. Scott. et al. ApJ 461, L1.

238

72. J. Silk 1967, Nature, 215, 1155-1156.
73. Smoot, G.F., et al. 1990, ApJ, 360, 685
74. Smoot, G. F., et al. 1992, ApJ, 396, L1
75. Smoot, G. F., Tenorio, L., Banday, A.J., Kogut, A., Wright, E.L., Hinshaw, G., & Bennett, C.L. 1994, ApJ, 437, 1
76. Steinhardt, P.J., 1995, Proc. of Snowmass Workshop,
77. S. T. Tanaka, A. C. Clapp, M. J. Devlin, N. Figueiredo, J. O. Gundersen, S. Hanany, V. V. Hristov, A. E. Lange, M. A. Lim, P. M. Lubin, P. R. Meinhold, P. L. Richards, G. F. Smoot, J. Staren 1996, ApJ 468, L81 astro-ph/9512067
78. S. T. Tanaka, et al. in prep.
79. Tegmark, M. & Bunn, E. F. 1995, ApJ, 455, 1
80. M. Tegmark, A. N. Taylor, & A.F. Heavens astro-ph/9603021 (1996)
81. M. Tegmark, A. de Oliveira-Costa, D.J. Devlin, C.B. Netterfield, L. Page & E.J. Wollack astro-ph/9608019 (1996)
82. White, M. & E.F. Bunn 1995, ApJ 443, L53 astro-ph/9510088
83. M. White, D. Scott, and J. Silk, Ann. Rev. Astron. & Astrophys. **32**, 329 (1994)
84. White, M. 1996, Phys Rev D, 53, 3011 astro-ph/9601158
85. Wright, E. L., et al. 1992, ApJ, 396, L13
86. Wright, E. L., et al. 1996, ApJ Letters, 464, L21. astro-ph/9601059
87. Wright, E. L., F. Hinshaw & C.L. Bennett 1996, ApJ, 458, L53
88. Zeldovich, Ya B. 1972, MNRAS, 160, 1

CMB Anisotropy Experiment References

89. ACE and BEAST. These two new projects are aimed at using HEMTs between 26 and 100 GHz on both super-pressure and conventional long-duration balloon platforms. The finest angular resolution will be near 1/5°.
90. APACHE. This experiment will observe from Dome-C in the Antarctic. Web site http://tonno.tesre.bo.cnr.it/ valenzia/APACHE/apache.html contains more information.
91. ARGO. A balloon-borne bolometer based experiment. Results are reported in de Bernardis, et al. 1994, Ap.J. 422:L33.
92. ATCA: Australia Telescope Compact Array. An interferometer operating at 8.7 GHz with a 2' resolution produced a map that was analyzed for anisotropy. The results are reported in Subrahmanyan R., Ekers, R. D., Sinclair, M. & Silk, J. 1993, MNRAS 263:416.
93. BAM: Balloon Anisotropy Measurement. This uses a differential Fourier transform spectrometer to measure the spectrum of the anisotropy between 90 and 300 GHz. Recent results are reported in astro-ph/9609108. More information may be obtained from http://cmbr.physics.ubc.edu.
94. Bartol. This is a bolometer-based experiment designed to look at 2° angular scales. It observed from the Canary Islands. Results are reported in Piccirillo et al., astro-ph/9609186.
95. BOOMERanG is a collaboration between the Caltech, Berkeley, Santa Barbara (Ruhl) and Rome groups. It will use bolometers to measure the anisotropy in the CMB between 90 and 410 GHz. The ultimate goal is a circumpolar Antarctic flight.
96. CBI: Cosmic Background Imager. This is an interferometer that plans to produce maps of the microwave sky near 30 GHz.
97. HACME/SP. This uses HEMTs on the ACME telescope. Observations were made from the South Pole. Recent results are reported in Gundersen, J. et al. 1995, Ap.J. 443:L57.
98. IAB. A bolometer-based experiment carried out at spectrum of the anisotropy between 90 and 300 GHz. Recent results are reported in astro-ph/9609108. More information may be obtained from http://cmbr.physics.ubc.edu.
99. Bartol. This is a bolometer-based experiment designed to look at 2° angular scales. It observed from the Canary Islands. Results are reported in Piccirillo et al., astro-ph/9609186.

100.   BOOMERanG is a collaboration between the Caltech, Berkeley, Santa Barbara (Ruhl) and Rome groups. It will use bolometers to measure the anisotropy in the CMB between 90 and 410 GHz. The ultimate goal is a circumpolar Antarctic flight.

101.   CAT. This is the Cambridge Anisotropy Telescope. It operates near 15 GHz and produces images of the microwave background. Early results are reported in Scott et al. 1996, Ap.J. 461:L1.

102.   CBI: Cosmic Background Imager. This is an interferometer that plans to produce maps of the microwave sky near 30 GHz.

103.   COBE is the COsmic Background Explorer. The three experiments aboard the satellite are the Differential Microwave Radiometers (30-90 GHz, DMR), the Far-InfraRed Absolute Spectrophotometer (60-630 GHz, FIRAS), and the Diffuse InfraRed Background Experiment (1.2-240 μm, DIRBE) All the experiments produce maps of the sky.

104.   FIRS. The Far InfraRed Survey. This is an experiment that started at MIT but has since moved to Princeton, University of Chicago and NASA/GSFC. It is a bolometer-based balloon-borne radiometer. It confirmed the initial COBE/DMR discovery.

105.   HACME/SP. This uses HEMTs on the ACME telescope. Observations were made from the South Pole. Recent results are reported in Gundersen, J. et al. 1995, Ap.J. 443:L57.

106.   IAB. A bolometer-based experiment carried out at the Italian Antarctic Base. Results are reported in Piccirillo, L. & Calisse, P. 1993, Ap.J. 413:529.

107.   MAT. This is the Mobile Anisotropy Telescope. It is similar to QMAP but is designed to operate from the ground in Chile.

108.   MAX was a collaboration between UCSB and Berkeley. It is a balloon-borne bolometer-based radiometer spanning roughly between 90 and 420 GHz. Recent results are reported in Lim et al. 1996, Ap. J. 469:L69. It flew on the ACME telescope.

109.   MAXIMA is a collaboration between Berkeley, Italy, and CalTech. It is the next generation of MAX. Web site http://physics7.berkeley.edu/group/cmb/gen.html contains more information.

110.   MSAM. There are a number of versions of MSAM. All use bolometers of various sorts and fly on balloons. The MSAM collaboration includes NASA/GSFC, Bartol Research Institute, Brown University, and the University of Chicago.

111.   OVRO. The Owen's Valley Radio Observatoty telescopes operate with various receivers between 15 and 30 GHz. The 40 m dish has a $2'$ beam, and the 5.5 m has a $7.3'$ beam. The experiments are aimed primarily at small angular scales.

112.   PYTHON. A multi-pixel bolometer- and HEMT- based experiment operated from the ground at the South Pole. The experiment has run in a number of configurations. Recent results are reported in Ruhl, J., et al. 1995, Ap.J., 453:L1.

113.   QMAP. This is a balloon-borne experiment that uses a combination of HEMTs and SIS detectors. The angular resolution is $1/5°$. This experiment is designed to produce "true" maps of the sky.

114.   SASK. These experiments are based on HEMT amplifiers operating between 26 and 46 GHz. They were performed in Saskatoon, Saskatchewan CA. Three years of observations have gone into the final data set.

115.   SuZIE is a bolometer-based experiment that observes from the ground. It is primarily intended to measure the SZ effect at high frequencies though it will also give information on the anisotropy at small scales.

116.   TOPHAT is a collaboration between Bartol Research Institute, Brown University, DSRI, NASA/GSFC, and the University of Chicago. The group plans to observe with an extremely light-weight bolometer-based payload mounted on top of a scientific balloon that circumnavigates the Antarctic. For more information see http://cobi.gsfc.nasa.gov/msam-tophat.html.

117.   Tenerife. Ground-based differential radiometers with 10-33 GHz receivers. The resolution is about $6°$. The experiment observes from the Observatorio del Teide in Tenerife, Spain. It has operated for many years. Recent results are discussed in Han-

cock et al. 1994, Nature, 367, 333.

118. VCA: Very Compact Array. This is an interferometer being developed at the University of Chicago. It will produce maps of the CMB at 30 GHz and be sensitive to larger angular scales than CBI.

119. VLA. This is work done near 5 GHz, on arcminute and smaller angular scales. It uses the Very Large Array. Recent results are reported in Fomalont et al. 1993, Ap.J. 404:8-20.

120. VSA: Very Small Array. This is a 30 GHz interferometer; the next generation of CAT. Web site http://www.mrao.com.ac.uk/telescopes/cat/vsa.html contains more information.

121. White Dish. This experiment uses and on-axis Cassegrain telescope and a 90 GHz single-mode bolometer. It observed at the South Pole and is sensitive to small angular scales. Results are reported in Tucker et al., 1993, Ap.J. 419:L45.

THE CMBR SPECTRUM

A Theoretical Introduction

ALBERT STEBBINS

NASA/Fermilab Astrophysics Group
Box 500, Batavia, IL 60510, USA

1. Introduction

The Cosmic Microwave Background Radiation (CMBR) provides a strong observational foundation for the standard cosmological scenario, the Big Bang theory. It is difficult to understand how to produce a 2.7°K blackbody spectrum except in the context of the Big Bang scenario. The near blackbody spectrum of the CMBR along with its near isotropy provides compelling evidence for a period of fairly quiescent Friedman-Robertson-Walker expansion for many expansion times before recombination. The past decade has seen huge advances in the measurement of the CMBR, with COBE's definitive discovery of anisotropies and measurement of a near perfect blackbody spectrum. The small deviations from isotropy have and will continue to tell us a great deal about the inhomogeneities in our Universe, and small deviations from a blackbody spectrum can also tell us about the energetics in our Universe. Such deviations have already been discovered in the direction of clusters of galaxies, although the mean CMBR spectrum is, so far, indistinguishable from a blackbody spectrum.

Here we give a introduction to the observed spectrum of the CMBR and discuss what can be learned about it. Particular attention will be given to how Compton scattering can distort the spectrum of the CMBR. This is left toward the end though. Unfortunately the author has no expertise in the area of how these measurements are made, but Smoot has covered this area in his lectures. An incomplete bibliography of relevant papers is also provided. Some old but still highly useful reviews of the physics behind the spectra are by Danese and DeZotti [20], and Sunyaev and Zel'dovich [69]. Theoretically not much has changed in this field in over 25 years. Much of the interesting work was done by Zel'dovich and Sunyaev in 1969 [78].

C. H. Lineweaver et al. (eds.), The Cosmic Microwave Background, 241–270.

2. Executive Summary

The Universe today is fairly diffuse and cold; however the Universe is observed to be expanding, and in the past we may deduce that the Universe was more dense and, because of $p\,dV$ work, the matter in the Universe would also have been hotter. Extrapolating the expansion back to very early epochs, the Universe would have been very hot and very dense and the Universe must have been expanding very rapidly in order to have grown as large as it is observed to be. Hence the Hot Big Bang. When the matter in the Universe is hot and dense the thermal equilibration time becomes very short. Thus, we expect everything to rapidly approach thermal equilibrium and we therefore expect the photons in the Universe to have a thermal (blackbody) spectrum at early times. It is easily shown that expansion of the Universe (or traversal through any gravitational field) leaves a blackbody spectrum a blackbody spectrum although the temperature may change. This temperature change is known as the redshift and can sometimes be thought of as either a Doppler shift or a gravitational redshift. Formally speaking the two may be thought of as the same phenomena and there is often no physical sense in trying to separate them.

Thus, as a first approximation we expect the photons in the Universe to have a blackbody spectrum. The fact that the cosmic microwave background radiation (CMBR) has nearly a blackbody spectrum is strong evidence for the Hot Big Bang hypothesis. There is simply not enough matter around today to thermalize so many photons (there are $\gtrsim 10^9$ photons for every atom) and in any case most of the matter in our Universe is much hotter that 3 K. The reason we might expect a deviation from blackbody is because some of the matter in the Universe has gone out of thermal equilibrium with the photons and may either heat or cool the photons. This can be done by non-equilibrium scattering or absorption of existing photons or by non-equilibrium emission of new photons. Clearly most of the matter in the Universe is not today in thermal equilibrium with the CMBR and the spectrum offers us a probe of this. However there are so many more CMBR photons in the Universe than there are protons or electrons that it is difficult for the matter to significantly distort the spectrum of the CMBR. Thus the fact that the observed CMBR spectrum is so close to a blackbody should come as no surprise.

3. Measures of Temperature

The *brightness* or *specific intensity* of light, I_ν, is defined as the incident energy per unit area, per unit solid angle, per unit frequency, per unit time.

It may be written

$$I_\nu = \frac{2h\nu^3}{c^2} n_\gamma \qquad (1)$$

where ν is the frequency and $n_\gamma(\nu)$ is the quantum-mechanical occupation number, i.e., the number of photons (in each polarization state) per unit phase space volume measured in units of h^3. Here h is Planck's constant, and we assume the light is not (linearly) polarized so that there are an equal number of photons in each polarization state. A *blackbody* or *Planck spectrum* has

$$n_\gamma^{\mathrm{BB}} = \frac{1}{\exp\left(\frac{h\nu}{kT}\right) - 1} \qquad (2)$$

where T is the the temperature. The high-frequency ($h\nu \gg kT$) limit of the Planck spectrum is known as *Wien's law*:

$$I_\nu^{\mathrm{W}} = \frac{2h\nu^3}{c^2} \exp\left(-\frac{h\nu}{kT}\right) \qquad (3)$$

while the low frequency ($h\nu \ll kT$) limit of the Planck spectrum is known as the *Rayleigh-Jeans law*:

$$I_\nu^{\mathrm{RJ}} = \frac{2\nu^2 kT}{c^2} . \qquad (4)$$

Note that the intensity is proportional to the temperature in this case. One may invert the Planck spectrum and characterize the intensity by the *thermodynamic temperature* or *brightness temperature*:

$$T_b = \frac{h\nu}{k \ln\left|1 + \frac{2h\nu^3}{c^2 I_\nu}\right|} \qquad (5)$$

Occasionally radio astronomers may define the brightness temperature by its Rayleigh Jeans limit:

$$T_b^{\mathrm{RJ}} = \frac{c^2 I_\nu}{k \, 2\nu^2} . \qquad (6)$$

In the radio region this is an excellent approximation to the thermodynamic temperature and is simply related to the intensity, and is therefore closer to what is actually measured.

Here we are interested in small deviations from a blackbody spectrum, i.e., we have some temperature T_γ which is a good fit to T_b at many frequencies and want to express the actual spectrum in terms of small deviations from a blackbody with this temperature, in particular in terms of the deviations in intensity from the blackbody spectrum, ΔI_ν. For small deviations the deviation in brightness temperature is

$$\Delta T_b \equiv T_b - T_\gamma = \frac{(e^x - 1)^2}{x^2 e^x} \frac{c^2 \Delta I_\nu}{k \, 2\nu^2} \qquad x \equiv \frac{h\nu}{kT_\gamma} \, . \tag{7}$$

Experimentally it is often easier to measure differences rather than absolute numbers: differences in intensity in different directions on the sky, or between the sky and internal calibrators. Note that in the Wien region differences in brightness temperature are greater than in the Rayleigh-Jeans region for the same difference in intensity. In what follows we will tend to plot spectral distortions in terms of differences in brightness temperature versus the dimensionless frequency, x. These "derived" quantities are probably more appealing to a theorist than an observer since they are further removed from what is actually measured.

4. Measured Mean Spectrum of the CMBR

Over the years there have been many measurements of the CMBR. There have been many claims that the spectrum deviated significantly from a blackbody, especially in the Wien region; however recent measurements with FIRAS (Far-InfraRed Absolute Spectrophotometer) on the COBE satellite have shown definitively that the spectrum is very close to a blackbody. Contemporary and quite competitive with the first FIRAS measurements was a rocket experiment [33] which also found a blackbody. The most recent FIRAS results have appeared in Ref [29] which are plotted in Figs 1 & 2. In the 2nd figure we have converted to a more theoretical representation by plotting versus $x = \frac{h\nu}{kT_\gamma}$ and converting to fractional changes in temperature. The reason that this transformation is useful is that we can predict the shape of the deviations from blackbody in terms of x, which we cannot in terms of ν since we have no *a priori* knowledge of T_γ. To transform to an x variable one must decide on a fiducial temperature. We have used the best-fit temperature, $T_\gamma = 2.728$ K, taken from the most recent results of FIRAS [29]: $T_\gamma = 2.728 \pm 0.002$ K. Note that the uncertainty in T_γ is larger than the error bars on most of the individual points in the plots. The reason that the uncertainty in $T_b - T_\gamma$ can be smaller than the uncertainty in T_γ is that the experiment measures the difference in brightness between a blackbody and the sky. The ± 0.002 K represents the uncertainty in the temperature of the reference blackbody, while the accuracy to which this reference is thought to be a blackbody is much better than this. Note that when fitting for a distortion from a blackbody, one must fit for both the amplitude of the distortion and for T_γ simultaneously.

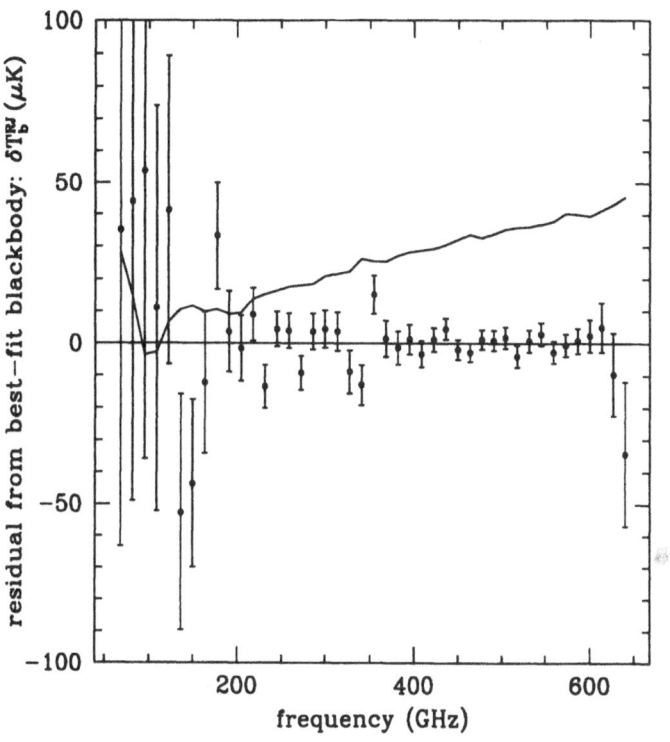

Figure 1. Plotted are the residuals in Rayleigh-Jeans temperature from the best fit blackbody as a function of frequency as stated by Fixsen *et al.* (1996). The error bars are 1-σ. The solid line is the subtracted Galaxy model at the Galactic poles. We see that these measurements are running up against a fundamental limitation of Galactic contamination. Given the large correction that has been made for Galaxy, one should treat the quoted error bar with caution.

While FIRAS certainly revolutionized the field, and does make obsolete most other short wavelength measurements of the CMBR spectrum, it only looked at the frequency range $68-640$ GHz. The bolometric techniques used by FIRAS only work at high frequencies and therefore the spectrum at low frequency was not touched by FIRAS. There has been ongoing measurements of the absolute CMBR flux in the Rayleigh-Jeans region for 30 years and we list some results from the last 15 years in Table 1. One can see that measurement did not stop after COBE. As we shall see, some of the spectral distortions we are looking for are most visible in the Rayleigh-Jeans regime. We have selected some of the most sensitive of measurements to plot in Fig 3. The uncertainties vary widely with frequency and are orders of magnitude larger than those of FIRAS. Several authors have noted that these low frequency measurements tend to indicate a temperature lower

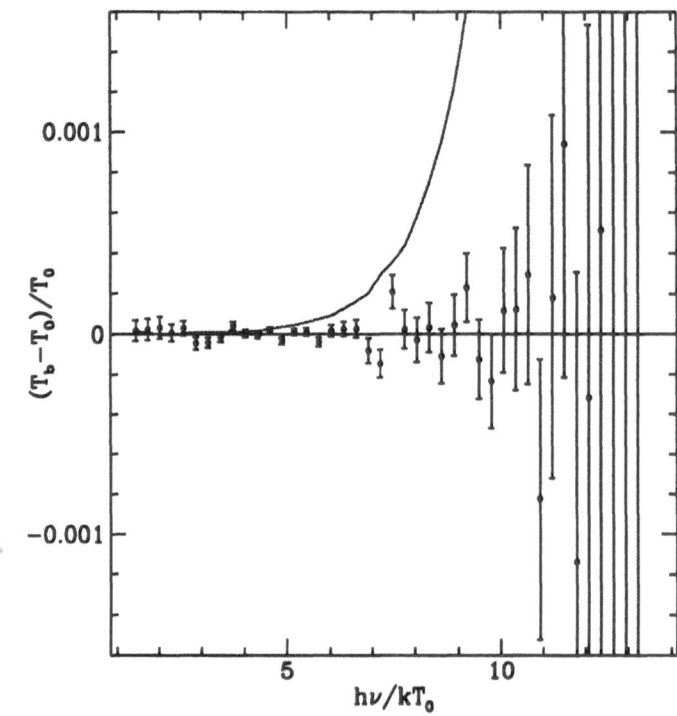

Figure 2. Same as Fig 1 except here we plot the fractional deviation in brightness temperature vs. the dimensionless frequency $x = \frac{h\nu}{kT}$. To do this we have used the best-fit photon temperature $T_\gamma = 2.728$ K. Plotting things in this way accentuates the deviations at high frequencies.

than that obtained by FIRAS, suggesting that there may be a deviation from a blackbody spectrum at low frequencies.

Measurements of the absolute CMBR spectrum, at the present level of sensitivity, face significant problems of Galactic contamination at both long and short wavelengths. Synchrotron radiation contaminates the long-wavelength spectrum while the short wavelength region is contaminated by dust emission. Since we cannot expect to observe the CMBR from outside of the Galaxy, this is a fundamental limitation. Many of the results plotted here include significant corrections for this contamination. While there is a limit to how well one can subtract off the Galaxy, we can look forward to improvements in Galaxy modeling using results from anisotropy experiments which will have increasingly better sensitivity, sky coverage and angular resolution. While anisotropy experiments cannot generally make absolute measurements of intensity, they can help to map out the Galaxy.

TABLE 1. Listed are measurements, made over the past 15 years, of the absolute CMBR brightness temperature at a variety of wavelengths. The results often include significant corrections for Galactic emission. Millimeter wavelengths are omitted as they have been superseded by the results of FIRAS ($> 68\,\mathrm{GHz}$ - see Fixsen *et al.* 1996). The ADS code given refers to the paper where these results are presented or reviewed and may be used to find the papers and abstracts online at the NASA Astrophysics Data System and mirror sites: *adsabs.harvard.edu, cdsads.u-strasbg.fr, d01.mtk.nao.ac.jp*. These codes are of the form *year journal volume page*.

Frequency (GHz)	Wavelength (cm)	T_{CMBR} (Kelvin)	1st Author	ADS Bibliographic Code
1.47	20.4	$2.26\,^{+0.19}_{-0.19}$	Bensadoun	1993ApJ...409....1B
90.	0.22	$2.60\,^{+0.09}_{-0.09}$	Bersanelli	1989ApJ...339..632B
2.0	15.	$2.55\,^{+0.14}_{-0.14}$	"	1994ApJ...424..517B
3.7	8.1	$2.59\,^{+0.13}_{-0.13}$	De Amici	1988ApJ...329..556D
3.8	7.9	$2.64\,^{+0.07}_{-0.07}$	"	1990ApJ...359..219D
3.8	7.9	$2.64\,^{+0.06}_{-0.06}$	"	1991ApJ...381..341D
25.	1.2	$2.783\,^{+0.025}_{-0.025}$	Johnson	1987ApJ...313L...1J
7.5	4.0	$2.60\,^{+0.07}_{-0.07}$	Kogut	1990ApJ...355..102K
1.410	21.26	$2.11\,^{+0.38}_{-0.38}$	Levin	1988ApJ...334...14L
7.5	4.	$2.64\,^{+0.06}_{-0.06}$	"	1992ApJ...396....3L
4.75	6.3	$2.70\,^{+0.07}_{-0.07}$	Mandolesi	1986ApJ...310..561M
2.5	12.	$2.79\,^{+0.15}_{-0.15}$	Sironi	1986ApJ...311..418S
0.600	50.	$3.0\,^{+1.2}_{-1.2}$	"	1990ApJ...357..301S
2.5	12.	$2.50\,^{+0.34}_{-0.34}$	"	1991ApJ...378..550S
0.82	36.6	$2.7\,^{+1.6}_{-1.6}$	"	"
2.5	12.0	$2.78\,^{+0.3}_{-0.3}$	Smoot	1987ApJ...317L..45S
33.0	0.909	$2.81\,^{+0.2}_{-0.2}$	"	"
1.41	21.2	$2.22\,^{+0.55}_{-0.55}$	"	"
3.66	8.2	$2.59\,^{+0.14}_{-0.14}$	"	"
10.	3.0	$2.61\,^{+0.06}_{-0.06}$	"	"
90.	0.33	$2.60\,^{+0.10}_{-0.10}$	"	"
1.4	21.	$2.65\,^{+0.33}_{-0.30}$	Staggs	1993PhDT.........6S
10.7	2.80	$2.730\,^{+0.014}_{-0.014}$	"	1996ApJ...473L...1S
90.	0.33	$2.57\,^{+0.12}_{-0.12}$	Witebsky	1986ApJ...310..145W

5. Spectral Distortions of the CMBR

While one should not be surprised that the CMBR has close to a blackbody spectrum, there are various mechanisms which should cause deviations from a thermal spectrum. Now we discuss a few of them.

248

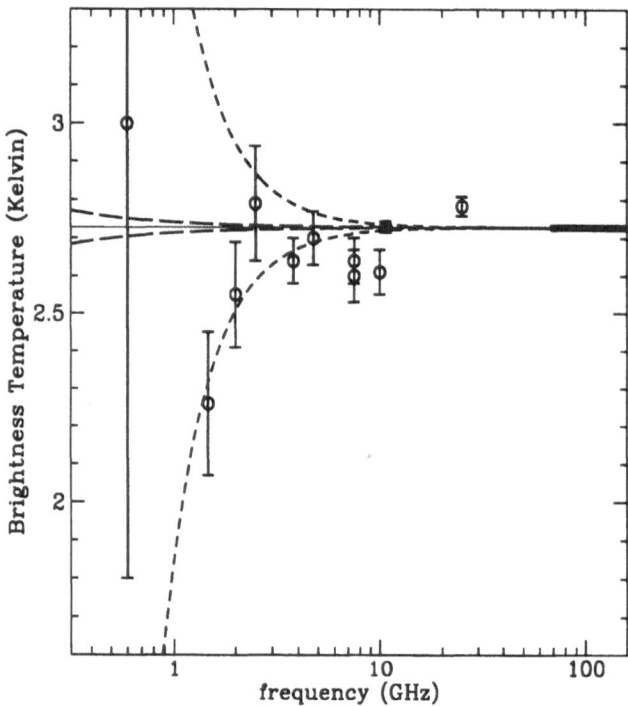

Figure 3. Plotted are a selection of the low frequency measurements of the CMBR brightness temperature listed in Table 1. From left to right, the points are from Sironi *et al.* (1990), Bensadoun *et al.* (1993), Bersanelli *et al.* (1994), Sironi & Bonelli (1986), De Amici *et al.* (1991), Mandolesi *et al.* (1986), Kogut *et al.* (1990), Levin *et al.* (1992), Smoot *et al.* (1987), Staggs (1996), Johnson & Wilkinson (1987) and were chosen because of the small errorbars. The black band at the right indicates the FIRAS data (Fixsen *et al.* 1996), while the horizontal straight line represents a temperature 2.728 K given by the FIRAS best-fit blackbody spectrum. The long-dashed curves represent chemical potential distortions with amplitude $\mu = \pm 9 \times 10^{-5}$, while the short-dashed curves give free-free distortions with amplitude $Y_{ff} = \pm 10^{-4}$. These are both idealized curves and one may expect (model dependent) corrections long-ward of 1 GHz (see Burigana, DeZotti, and Danese 1995).

5.1. ANISOTROPIES

The most common way in which the CMBR spectral distortion occurs is when the photons have a blackbody spectrum in each direction but the temperatures characterizing these spectra are different in different directions. This direction dependent temperature difference is called anisotropy. The anisotropy can either be caused by a Doppler/gravitational effect or because the gas emitting the photons really did have different temperatures.

The first anisotropy discovered was the dipole anisotropy, i.e., the tem-

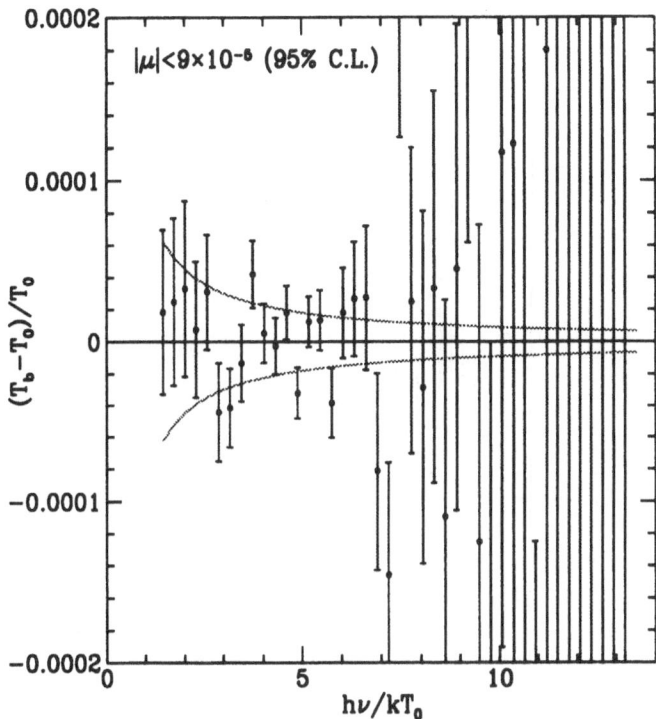

Figure 4. Superimposed on the FIRAS data of Fig 2 is the largest chemical potential distortion allowed by the data (Fixsen *et al.* 1996): $\mu = \pm 9 \times 10^{-5}$. The falling positive curve is the far more plausible positive chemical potential distortion and the negative rising curve is a negative chemical potential distortion.

perature varies like the cosine of the angle from some point on the celestial sphere. It is usually attributed to the Sun moving at 371 km/s. Note that to a first approximation, when averaging over the sky, the dipole does not contribute to the mean spectrum of the CMBR.

One way to check that measured anisotropies are what they are supposed to be is to measure the spectrum. For a small anisotropy the change in flux from the mean spectrum should be proportional to the derivative of the flux of a blackbody with respect to temperature. FIRAS has done just that for the dipole [29] and found just what was expected. Most modern anisotropy experiments use many frequency channels in order to check the spectrum of the anisotropy, or more specifically to be able to subtract off contamination of the measurements by effects other than the anisotropy.

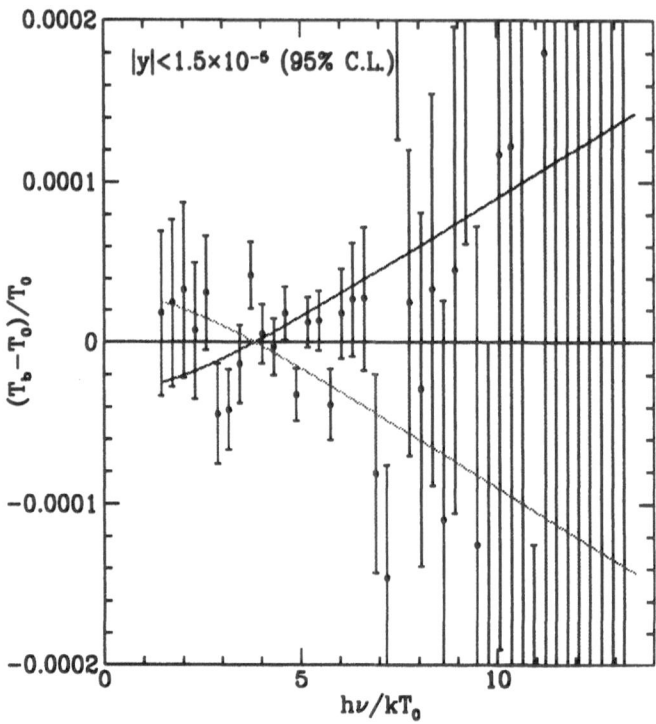

Figure 5. Superimposed on the FIRAS data of Fig 2 is the largest y-distortion allowed by the data (Fixsen *et al.* 1996): $y = \pm 1.5 \times 10^{-5}$. The rising curve is the more plausible positive y-distortion and the falling curve is a negative y-distortion.

5.2. CHEMICAL POTENTIAL DISTORTIONS

There are three processes which are important from thermalizing the CMBR spectrum in the early Universe: *Compton scattering, double Compton scattering,* and *free-free scattering* (also known as *bremsstrahlung*). Compton scattering is a much more rapid process, but since it conserves the number of photons, it can only thermalize the energy distribution of the photons and not the number of photons. All of these processes become more efficient as one goes to earlier and earlier epochs, and eventually photon non-conserving processes start to become important.

There is an epoch between $z = 10^5$ and $z = 2 \times 10^6$ during which Compton scattering is efficient in thermalizing the energy distribution while other processes are not capable of thermalizing the photon number. During this epoch, if the energy-to-photon ratio is perturbed from that required for a blackbody spectrum, the spectrum will instead approach a Bose-Einstein distribution

$$n_l^{BE} = \frac{1}{\exp\left(\dfrac{h\nu}{kT_\gamma} + \mu\right) - 1} \qquad (8)$$

where T_γ and μ (the *dimensionless chemical potential*) are determined by the total energy available and the total number of photons available. If one starts out with a thermal distribution of photons at temperature T_γ and injects a fractional increase in the energy density, $\frac{\Delta U}{U}$, without significantly increasing the number of photons, one obtains

$$T_b \approx T_\gamma \left(1 - \mu\left[0.456 - \frac{1}{x}\right]\right) \qquad \frac{\Delta U}{U} = 0.714\mu \qquad \mu \ll 1, \qquad (9)$$

Since one must fit the observations to both T_γ and μ, one really can only measure the $\frac{1}{x}$ term. Double Compton scattering and free-free scattering become increasingly more efficient at lower frequencies and there are usually corrections to this formula at small frequencies $x \ll 1$ [9, 10]. These corrections are not liable to be important for FIRAS measurements.

This distortion to the spectrum is greatest at small x, however the FIRAS measurements at high frequencies are so accurate that they yield much better constraints on μ than do the low frequency experiments. Comparing with the FIRAS data, one finds $|\mu| < 9 \times 10^{-5}$ at the 2σ level [29]. We compare the maximal allowed distortion to the low & high frequency data in Figs 3 & 4, respectively.

Thus we find the extremely stringent constraint at a very early epoch

$$\frac{\Delta U}{U} < 6 \times 10^{-5} \qquad 10^5 < z < 2 \times 10^6. \qquad (10)$$

Of course this is not to say that one expects large energy injection at these epochs.

Note that for $z > 10^6$ the CMBR spectrum is not telling us much about the energetics of the Universe. However one can use Big Bang Nucleosynthesis to probe the total energy of the Universe up to $z \sim 10^{10}$.

5.3. Y DISTORTIONS

If energy is injected into the Universe after $z \sim 10^4$, Compton scattering is unable to thermalize the distribution. The fact that we observe very little deviation from a blackbody spectrum tells us that not much energy could have been injected compared with the thermal energy of the CMBR. If a small amount of energy is injected then one may solve for the linear perturbation from a blackbody spectrum under the action of Compton scattering, as was done by Zel'dovich and Sunyaev [78]. One finds that the perturbation in the photon occupation number is

$$\Delta n = y \frac{x e^x}{(e^x - 1)^2} \left(x \frac{e^x + 1}{e^x - 1} - 4 \right) \tag{11}$$

where the "y-parameter" is

$$y = \int dt\, \sigma_T\, c\, n_e\, \frac{k(T_e - T_\gamma)}{m_e c^2}, \tag{12}$$

n_e is the number density of free-electrons, and σ_T is the Thomson cross-section. If one could manage to cool gas below the radiation temperature one could produce a distortion with negative y, but typically this distortion is produced by ionized gas which is much hotter than the photons. In the early Universe when the density of electrons is large, even a small heating of the gas over the photon temperature may lead to a significant distortion. This distortion is generally referred to as a y-distortion when applied to the mean CMBR spectrum, but is usually called the Sunyaev-Zel'dovich distortion when referring to an anisotropy in the spectrum because there is more or less hot gas in one direction than another. Large amounts of hot ionized gas exist in clusters of galaxies and the "S-Z effect" has been observed in the directions of several clusters. There is no evidence for a y of the mean CMBR spectrum, although with sensitive enough measurements we should see the hot gas we know is out there. Fixsen et $al.$ [29] have placed a limit of $|y| < 1.5 \times 10^{-5}$ from the FIRAS data. We compare the maximal distortions to the FIRAS data in Fig 5. Note that a positive y-distortion produces a negative change in T_b at low frequencies and positive change in T_b at high frequencies, just what one expects if one was heating a fixed number of photons. This negative ΔT_b is sometimes called the "S-Z decrement", for the S-Z effect was first looked for at radio wavelengths.

5.4. DISTORTION FROM FREE-FREE

Another important process for the CMBR spectrum is free-free scattering, which is the scattering of a free electron off of a charged nucleon by either emitting or absorbing a photon; in most cases of interest, emitting rather than absorbing. This is the same process which produces the X-rays observed from hot cluster gas operating at microwave and radio frequencies. The effect of free-free scattering on the CMBR spectrum is most likely to be seen at the very longest wavelengths measured. In this Rayleigh-Jeans limit, the distortion it produces may be approximated by [3]

$$\frac{\Delta T_b}{T_\gamma} \approx \frac{Y_{ff}}{x^2} \qquad Y_{ff} = \int dt\, \frac{T_e - T_\gamma}{T_e}\, \kappa\, dt \tag{13}$$

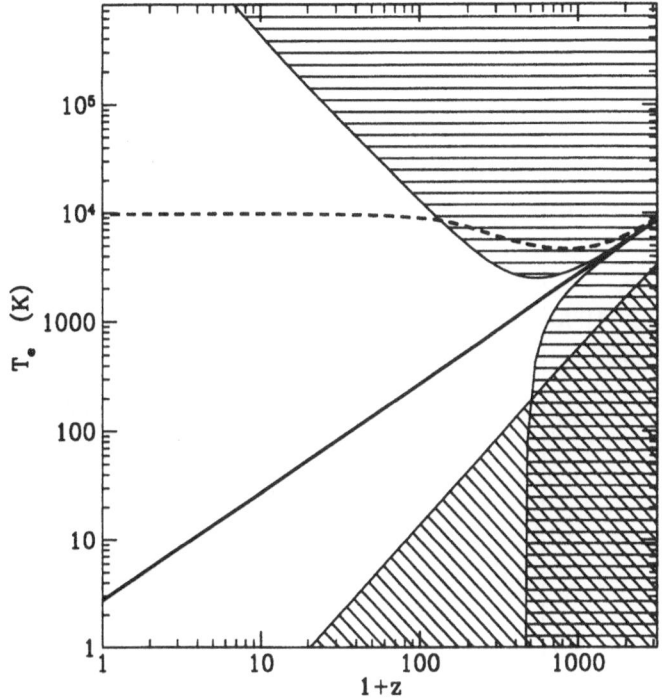

Figure 6. Plotted is the constraint on the temperature of a fully ionized universe as a function of redshift. The horizontally hatched region is excluded since $|y| < 1.5 \times 10^{-5}$ while the diagonally hatched region would be excluded if $Y_{ff} < 10^{-4}$. The solid line indicates where the gas temperature equals the photon temperature and the dashed line gives the temperature as a function of redshift for a model where the gas is ionized by very massive stars (VMOs - see Stebbins & Silk 1986). The cosmological parameters used are $H_0 = 65$km/sec/Mpc, $\Omega_0 = 0.4$ and $\Omega_b = 0.10$.

where

$$\kappa \equiv \frac{8\pi e^6 h^2 n_e^2 g}{3m_e (kT_\gamma)^3 \sqrt{6\pi m_e kT_e}} \tag{14}$$

and the Gaunt factor, $g \sim 2$ in most cases of interest. Note the $T_e^{-\frac{1}{2}}$ factor in κ, which means that the effect is suppressed for higher temperature gas. The $1/x^2$ dependence of this distortion means that it is the low frequency measurements which will constrain its amplitude. In Fig 3 we plot free-free distortions with $Y_{ff} = \pm 10^{-4}$. We see that this size distortion is close to what is being constrained by these measurements, although no proper statistical analysis has been done. This size free-free distortion would produce no significant effect in the FIRAS data, although if one went far enough into the Wien tail one would find large distortions from free-free emission.

254

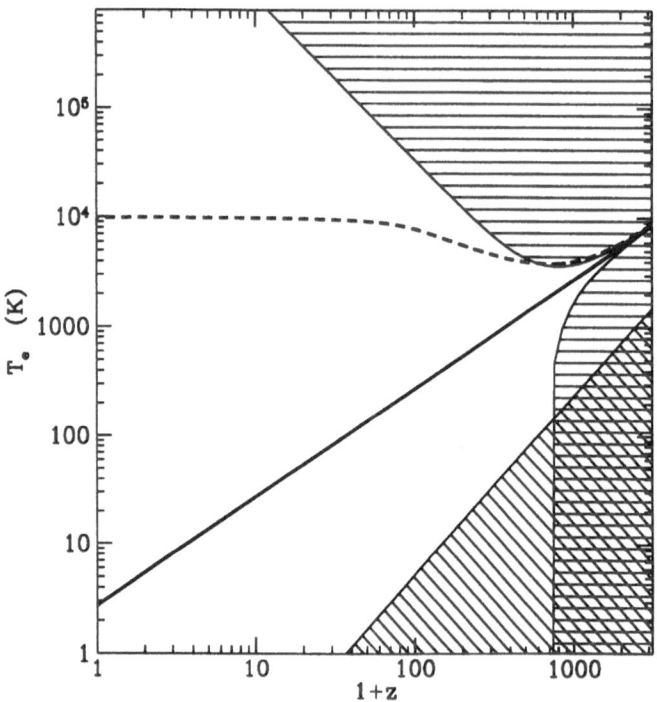

Figure 7. The same as the previous figure except with $\Omega_0 = 1$ and $\Omega_b = 0.06$. The constraints are less severe for larger Ω_0 and smaller Ω_b.

In Figs 6 & 7 we have used constraints on y and Y_{ff} to put constraints on the temperature and epoch of a reionized universe, assuming presently favored cosmological parameters. We see that it is really the y-distortion which is most important, the free-distortion only being detectable on the off chance that the gas was ionized and cold. Even though the limits on y are quite small, we see that there is not too much of a constraint of ionization after $z \sim 100$. The constraints could be made stronger if one assumes a larger baryon density or a smaller total density.

6. Physical Processes

Now we will take a closer look at the physical processes which could cause a distortion of the spectrum. Here we will only discuss Compton scattering although free-free emission and double-Compton deserve an equally thorough treatment.

6.1. COMPTON SCATTERING

6.1.1. *Collisional Boltzmann Equation*

One may describe the state of the primeval gas of photons and electrons in terms of the the density of particles in phase space, i.e., momentum and position space. Here we are not interested in the polarization state of electrons and photons so we average over the two polarization states. [1] It is convenient to measure the phase space density in units of $h = 2\pi\hbar$ which gives the the quantum mechanical occupation numbers, n_γ and n_e, for photons and electrons, respectively. The evolution of n_γ can be described by the collisional Boltzmann equation which has the form

$$\frac{Dn_\gamma(\mathbf{p}_\gamma)}{Dt} = C(\mathbf{p}_\gamma), \qquad (15)$$

where $C(\mathbf{p}_\gamma)$ is the scattering term which describes the interactions with other particles. Here $\frac{D}{Dt}$ is a convective derivative along the photon's trajectory in phase space, while the right-hand-side gives the collision integral. If there were no collisions then the Boltzmann equation states that the occupation number remains constant along photon trajectories. [2] Included in this convective derivative are all the effects of gravity on the photons, which include many of the effects which produce anisotropy. We will not discuss these effects further as they are covered extensively in Bunn's lectures.

The collision integral for Compton scattering of unpolarized particles after averaging over the polarization state of scattered particles is of the form[3]

$$C^C(\mathbf{p}_\gamma) = \frac{2}{(2\pi\hbar)^3} \int d^3\mathbf{p}_e \int d^2\hat{\mathbf{n}}'\, c\,(1 - \hat{\mathbf{n}}\cdot\vec{\beta}) \frac{d^2\sigma}{d^2\hat{\mathbf{n}}'}$$

$$\times \left[(1 - n_e(\mathbf{p}_e))\, n_e(\mathbf{p}_e')\,(1 + n_\gamma(\mathbf{p}_\gamma))\, n_\gamma(\mathbf{p}_\gamma') \right.$$

$$\left. - (1 - n_e(\mathbf{p}_e'))\, n_e(\mathbf{p}_e)\,(1 + n_\gamma(\mathbf{p}_\gamma'))\, n_\gamma(\mathbf{p}_\gamma) \right] \qquad (16)$$

where we have (or will) use the notation

$$\mathbf{p}_\gamma = \frac{\epsilon}{c}\hat{\mathbf{n}} \qquad \mathbf{p}_\gamma' = \frac{\epsilon'}{c}\hat{\mathbf{n}}' \qquad |\hat{\mathbf{n}}| = |\hat{\mathbf{n}}'| = 1 \qquad E = \sqrt{(m_e c^2)^2 + (c\mathbf{p}_e)^2}$$

[1] Compton scattering in an inhomogeneous medium will produce some polarization of the photons, which can be measured, and also effects the anisotropy at the several percent level. See Melchiorri and Vittorio this volume.

[2] This is true for a phase space defined by a position, x^i, and its canonically conjugate momentum, p_i; $n(\mathbf{p})$ measures the particle density with volume measure: $d^3x^i d^3p_i$. In general relativity there is both the covariant momentum, p_i, and contravariant momentum, p^i. If one measures the density of particles per unit $d^3x^i d^3p^i$, the Boltzmann equation as expressed above does not apply!

[3] This form is determined by the *principle of detailed balance* which results from the time-reversal symmetry of the S-matrix (of classical or quantum mechanics) [44].

$$\mathbf{p_e} = (m_e c)\gamma\vec{\beta} \qquad \beta = |\vec{\beta}| \qquad \gamma = \frac{1}{\sqrt{1 - \beta^2}} . \tag{17}$$

In Eq 16 the values of $\mathbf{p'_e}$ and ϵ' are determined by energy-momentum conservation. The 2nd term in square brackets describes the scattering $\mathbf{p_\gamma} + \mathbf{p_e} \rightarrow \mathbf{p'_\gamma} + \mathbf{p'_e}$, while the 1st term results from the inverse process, $\mathbf{p'_\gamma} + \mathbf{p'_e} \rightarrow \mathbf{p_\gamma} + \mathbf{p_e}$. The $1 + n_\gamma$ factor represents the increased scattering rate due to the stimulated emission of the bosonic photons, while the $1 - n_\gamma$ is the Pauli blocking factor giving the exclusion principle for fermionic electrons. The factor of 2 in the prefactor counts the two polarization states of the incoming electrons. The factor $c(1 - \hat{n}\cdot\vec{\beta})$ in Eq 16 is a measure of the relative velocity between the ingoing electron and photon.[4] Of course, $\frac{d^2\sigma}{d^2\hat{n}'}$ is the differential Compton cross-section[5]

Note that this form of the collision integral guarantees that a thermal distribution is a fixed point. Substituting a Fermi-Dirac distribution for the electrons and a Bose-Einstein distribution for the photon, i.e.,

$$n_e(E) = \frac{1}{\exp(\frac{E}{k_B T} + \mu_e) + 1} \qquad n_\gamma(\epsilon) = \frac{1}{\exp(\frac{\epsilon}{k_B T} + \mu_\gamma) + 1} \tag{18}$$

will cause the integrand of the collision integral to be zero so long as the temperature is the same for both. Here μ_e, and μ_γ are (dimensionless) chemical potentials given by the total electron and photon density, each of which is conserved by Compton scattering. We expect such a thermal distribution to be a stable fixed point since it is the highest entropy state and entropy increases according to Boltzmann's H-theorem [44]. In the contexts we are interested in, the density of electrons is sufficiently low that $\mu_e \gg 1$ and Fermi-blocking is unimportant, so we may set $1 - n_e \rightarrow 1$. In this limit the equilibrium distribution for the electrons becomes a simple Boltzmann distribution, i.e.,

$$n_e(E) = \exp(-\frac{E}{k_B T} - \mu_e) . \tag{19}$$

Henceforth we will ignore Fermi-blocking.

We are not really interested in the scattered electrons, so we may "integrate out" the electron distribution function. The idea is that we know

[4] This relative velocity factor is really determined by the definition of the cross-section. The factor is equal to $\sqrt{|\mathbf{v}_1 - \mathbf{v}_2|^2 - \frac{1}{c^2}|\mathbf{v}_1 \times \mathbf{v}_2|^2}$ which reduces to the above expression when one of the particles is massless. If both incoming particles are non-relativistic then it reduces to the "usual" definition of relative-velocity: $|\mathbf{v}_1 - \mathbf{v}_2|$.

[5] The outgoing particle momenta $\mathbf{p_e}$ and $\mathbf{p_\gamma}$ are described by six numbers, however four are fixed by energy and momentum conservation. The differential cross-section is a function of the remaining two parameters, which in this case we have taken to be the outgoing photon direction, $\hat{n}'$. Any two parameters would do!

the electron distribution function *a priori* - which is often is a true since Coulomb scattering is usually very effective in thermalizing the electron momenta. Thus we may rewrite the collision integral as

$$C^C(\mathbf{p}_\gamma) = \int d^2\hat{\mathbf{p}}'_\gamma \left[\frac{\epsilon^2}{\epsilon'^2} S(\mathbf{p}'_\gamma, \mathbf{p}_\gamma)\,(1 + n_\gamma(\mathbf{p}_\gamma))\,n_\gamma(\mathbf{p}'_\gamma) \right.$$

$$\left. - S(\mathbf{p}_\gamma, \mathbf{p}'_\gamma)\,(1 + n_\gamma(\mathbf{p}'_\gamma))\,n_\gamma(\mathbf{p}_\gamma) \right] \qquad (20)$$

where

$$S(\mathbf{p}_\gamma, \mathbf{p}'_\gamma) = \frac{2}{(2\pi\hbar)^3} \int d^3\mathbf{p}_e n_e(\mathbf{p}_e)\,(1 - \vec{\beta}\cdot\hat{\mathbf{n}})\,\frac{d^2\sigma}{d^2\hat{\mathbf{n}}'}(\mathbf{p}_e, \mathbf{p}_\gamma \hat{\mathbf{n}}')\,\frac{\delta(\epsilon' - \epsilon(1 + \bar{\Delta}))}{\epsilon'^2}$$

$$(21)$$

and $\bar{\Delta}(\mathbf{p}_e, \mathbf{p}_\gamma, \hat{\mathbf{n}}')$ gives the fractional change in the energy determined by energy-momentum conservation, i.e., it is the solution to the equation

$$\sqrt{(m_e c^2)^2 + |c\mathbf{p}_e|^2} + c|\mathbf{p}_\gamma|$$
$$= \sqrt{(m_e c^2)^2 + |c\mathbf{p}_e + c\mathbf{p}_\gamma - \epsilon(1 + \bar{\Delta})\hat{\mathbf{n}}'|^2} + c|\mathbf{p}_e|\,(1 + \bar{\Delta}). \qquad (22)$$

A unique solution always exists with $\bar{\Delta} \in [-1, \infty)$.

We know that the CMBR is very nearly isotropic today, and it is reasonable to assume that the background radiation was always isotropic. Since we are interested in changes in the spectrum and not anisotropy, we may also average the collision integral over $\hat{\mathbf{n}}$ to find the mean change in the spectrum. Performing the two averages $\hat{\mathbf{n}}$ and $\hat{\mathbf{n}}'$ the collision integral becomes

$$C^C(\epsilon, \Delta) = \int d\Delta \left[\frac{1}{(1+\Delta)^3}\overline{S}(\frac{\epsilon}{1+\Delta}, \Delta)\,(1 + n_\gamma(\epsilon))\,n_\gamma(\frac{\epsilon}{1+\Delta}) \right.$$

$$\left. - \overline{S}(\epsilon, \Delta)\,(1 + n_\gamma(\epsilon(1 + \Delta)))\,n_\gamma(\epsilon) \right]. \qquad (23)$$

where

$$\overline{S}(\epsilon, \Delta) = \frac{\epsilon^3(1+\Delta)^2}{4\pi} \int d^2\hat{\mathbf{n}} \int d^2\hat{\mathbf{n}}'\, S(\frac{\epsilon}{c}\hat{\mathbf{n}}, \frac{\epsilon}{c}(1+\Delta)\hat{\mathbf{n}}'). \qquad (24)$$

To obtain Eq 23 we have used a little trick of changing the variable of integration for inverse scattering from Δ to $\frac{1}{1+\Delta} - 1$, and renamed this new dummy variable Δ. If one looks closely at Eq 23 one can see that the total photon number is preserved by scattering, no matter what the form of $\overline{S}(\epsilon, \Delta)$.

6.1.2. *Fokker-Planck Equation*

One important property of cosmological Compton scattering is that, at the low redshifts we are interested in, the background radiation photons have much lower (total) energy and are moving much faster than the electron they are scattering off of. One is bouncing a very light object (the photon) off of a much more slowly moving heavy object (the electron) and energy and momentum conservation dictate that the energy of the light object is nearly unchanged by the scattering (consider bouncing a ping-pong ball off of a bowling ball). The electrons are not infinitely massive nor are they completely stationary so that the photon energy will change slightly in each collision. All this will be reflected in the fact that $\overline{S}(\epsilon, \Delta)$, when considered as a function of Δ, will be a very narrow function sharply peaked around $\Delta = 0$ with width much less than unity. In contrast, the Δ-dependence of $n_\gamma(\epsilon(1 + \Delta))$ and $\overline{S}(\frac{\epsilon}{1+\Delta},)$ are much smoother functions in the sense that they do not vary much over the region in Δ where $\overline{S}(\epsilon, \Delta)$ is significantly non-zero. Thus it should be a good approximation to Taylor expand the integrand of Eq 23 in Δ about $\Delta = 0$, but excluding the rapid dependence through the 2nd argument of $\overline{S}$ and truncating at a given order. This is a kind of Fokker-Planck equation[6]. If we expand to 2nd order in Δ, the Boltzmann equation becomes a partial differential equation (see Eq 8 of Ref [1])

$$\frac{Dn_\gamma}{D\tau} = \frac{1}{\epsilon^2} \frac{\partial}{\partial \epsilon} \left[\epsilon^3 \left(\frac{1}{2} \epsilon \overline{\overline{\Delta^2}} \frac{\partial n_\gamma}{\partial \epsilon} + \left(-\overline{\overline{\Delta}} + 2\overline{\overline{\Delta^2}} + \frac{1}{2} \epsilon \overline{\overline{\Delta^{2\prime}}} \right) (1 + n_\gamma) n_\gamma \right) \right] \tag{25}$$

where

$$\overline{\overline{\Delta^n}} = \frac{1}{N_e \sigma_T} \int_{-1}^\infty d\Delta \, \Delta^n \, \overline{S}(\epsilon, \Delta) \qquad \overline{\overline{\Delta^{n\prime}}} = \frac{1}{N_e \sigma_T} \int_{-1}^\infty d\Delta \, \Delta^n \, \frac{\partial \overline{S}(\epsilon, \Delta)}{\partial \epsilon}, \tag{26}$$

and we have used the electron density, N_e, introduced the Thomson cross-section[7], σ_T, and defined the Thomson optical depth, τ:

$$N_e = \frac{2}{(2\pi\hbar)^3} \int d^3\mathbf{p}_e \, n_e(\mathbf{p}_e) \qquad \sigma_T = \frac{8\pi}{3} \left(\frac{e^2}{m_e c^2} \right)^2 \qquad d\tau = N_e c \, \sigma_T dt \, . \tag{27}$$

[6] Fokker and Planck actually considered the case where the momentum is only slightly changed in each scattering and proposed Taylor expanding to 2nd order in the small change in momentum. For Compton scattering the direction of the photon will change significantly, so the momentum change is not small, but the energy change is, and expanding in the small fractional energy change, Δ, is an obvious generalization. It is useful to consider expanding to higher order than 2nd.

[7] Thomson scattering is the non-relativistic and classical limit of Compton scattering as first described by J.J. Thomson.

This optical depth gives the expected number of Compton scatterings of low energy photons off non-relativistic electrons.

The form of the equation is reminiscent of a diffusion equation which is a good description of the physics; the small changes in photon energy at each scattering cause the photons to diffuse in energy space. The $\frac{\partial n_\gamma}{\partial \epsilon}$ term causes a net drift toward increasing energies while the $(1+n_\gamma)n_\gamma$ causes a net drift to lower energies (if its coefficient is positive). We expect these drifts and diffusion to sum to zero in thermal equilibrium, i.e., when n_γ has a Bose-Einstein distribution (Eq 18), the electrons have a Boltzmann distribution (Eq 19), and the two share a common temperature. This consideration alone suggests that for a thermal electron distribution with temperature T_e, we should expect

$$
\frac{-2\overline{\overline{\Delta}} + 4\overline{\overline{\Delta^2}} + \epsilon\,\overline{\overline{\Delta^{2\prime}}}}{\overline{\overline{\Delta^2}}} = \frac{\epsilon}{kT_e} .
\tag{28}
$$

Another feature of Eq 25 is the differential operator $\frac{1}{\epsilon^2}\frac{\partial}{\partial \epsilon}$ in front, which guarantees conservation of photon number. This will persist to all order in the Δ-expansion. In fact one can pretty much guess the 2nd order Fokker-Planck equation without knowing much about the Compton cross-section. We will take a more constructive approach below.

6.1.3. Compton Cross-Section

To compute the Compton collision integral, or its Fokker-Planck approximations, one needs to use the Compton cross-section. The relativistic (Klein-Nishina) differential Compton cross-section in an arbitrary rest-frame is [1]

$$
\frac{d^2\sigma}{d^2\hat{\mathbf{n}}'} = \frac{3\sigma_T}{16\pi}\frac{1-\beta^2}{[1-\hat{\mathbf{n}}'\cdot\vec{\beta}+\alpha\gamma^{-1}(1-\hat{\mathbf{n}}\cdot\hat{\mathbf{n}}')]^2}
$$
$$
\times\left[1+\left(1-\frac{(1-\beta^2)(1-\hat{\mathbf{n}}\cdot\hat{\mathbf{n}}')}{(1-\hat{\mathbf{n}}\cdot\vec{\beta})\,(1-\hat{\mathbf{n}}'\cdot\vec{\beta})}\right)^2\right.
$$
$$
\left.+\frac{\alpha^2(1-\beta^2)\,(1-\hat{\mathbf{n}}\cdot\hat{\mathbf{n}}')^2}{(1-\hat{\mathbf{n}}'\cdot\vec{\beta})\,[1-\hat{\mathbf{n}}'\cdot\vec{\beta}+\alpha\gamma^{-1}(1-\hat{\mathbf{n}}\cdot\hat{\mathbf{n}}')]}\right]
\tag{29}
$$

where

$$
\alpha = \frac{\epsilon}{m_e c^2} .
\tag{30}
$$

For many astrophysical applications, and especially those related to the CMBR, there are two small numbers which enter this cross-section. Firstly, α is very small for the microwave photons we see observe today, roughly 10^{-9}. Clearly as we go to higher redshifts the background photons become

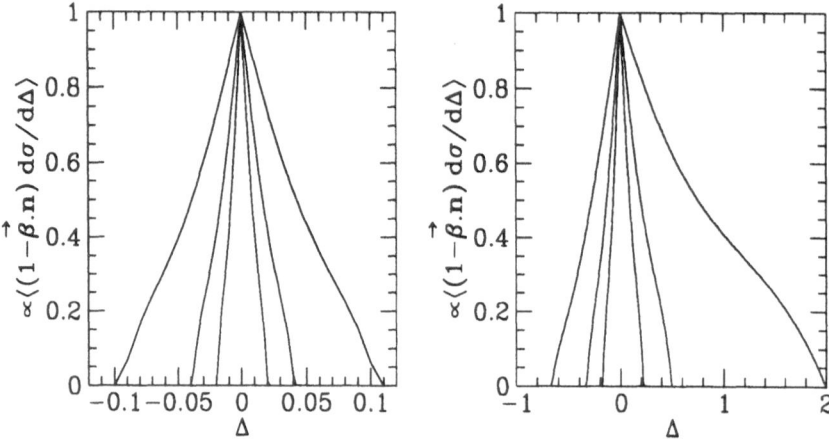

Figure 8. Plotted is the distribution of fractional energy changes, Δ, experienced by low energy photons scattering off an isotropic distribution of electrons with velocity βc. The left panel shows the distribution for $\beta = 0.01$, 0.02, and 0.05; while the right panel shows the distribution for $\beta = 0.1$, 0.2, and 0.5. For graphical clarity we have adjusted the heights of the curves to have unit amplitude at $\Delta = 0$. The maximum and minimum values for Δ are dictated by energy and momentum conservation. The distribution is narrow and symmetric for small β and becomes wider and more skew for larger velocity electrons. This positive skewness gives the heating of the photons by the electrons. The Fokker-Planck equation approximates the photon distribution function by the first few terms in its Taylor series about $\Delta = 0$ when convolving with these distributions. This is liable to be a good approximation for scattering off low velocity electrons since the Δ-distribution is sharply peaked around $\Delta = 0$.

more energetic, but α remains small in the redshift range relevant to the CMBR spectrum $z \lesssim 10^7$. The 2nd small number is β since we are almost always interested in non-relativistic electrons. If one is interested in a thermal electron velocity distribution then a small β expansion is equivalent to a small $\frac{kT_e}{m_e c^2}$ expansion. In most applications, $\alpha \ll \frac{kT_e}{m_e c^2}$, so we will concentrate more on the higher order terms in T_e and not α.

To proceed it is probably easiest to follow the methodology of Barbosa [1], where one expands the cross-section in α but not β. For a thermal electron distribution one can compute the moments, $\overline{\Delta^n}$, in the Fokker-Planck expansion analytically, and only at the end should one Taylor expand the result in T_e about $T_e = 0$. One may rewrite Eq 24 as

$$\overline{S}(\epsilon, \Delta) = N_e \left\langle (1 - \vec{\beta} \cdot \hat{\mathbf{n}}) \frac{d\sigma}{d\Delta} \right\rangle_{\hat{\mathbf{n}}, \hat{\mathbf{n}}'} \tag{31}$$

where $\frac{d\sigma}{d\Delta}$ gives the differential cross-section with respect to the fractional change in photon energy. So for example, expanding everything to zeroth order in α (which we denote by the superscript $^{(0)}$) we find

$$\left\langle (1 - \vec{\beta} \cdot \hat{n}) \frac{d\sigma^{(0)}}{d\Delta} \right\rangle_{\hat{n},\hat{n}'} = \sigma_T \, \overline{F}(\Delta, \beta \, \mathrm{sgn}(\Delta)) \qquad (32)$$

where

$$\overline{F}(\Delta, b) = \mathrm{sgn}(\Delta) \times \mathcal{H}(1 - \frac{(1 - b)\Delta}{2b})$$

$$\times \left[\frac{3(1 - b^2)^2(3 - b^2)(2 + \Delta)}{16b^6} \ln \frac{(1 - b)(1 + \Delta)}{1 + b} \right.$$

$$+ \frac{3(1 - b^2)(2b - (1 - b)\Delta)}{32b^6(1 + \Delta)} \Big(4(3 - 3b^2 + b^4)$$

$$+ 2(6 + b - 6b^2 - b^3 + 2b^4)\Delta$$

$$\left. + (1 - b^2)(1 + b)\Delta^2 \Big) \right], \qquad (33)$$

and $\mathcal{H}()$ is the Lorentz-Heaviside function which is unity for positive arguments and zero otherwise. We see that this function is only non-zero for

$$\Delta \in \left[-\frac{2\beta}{1 + \beta}, \frac{2\beta}{1 - \beta} \right] \qquad (34)$$

and, as promised, for small β is sharply peaked around $\Delta = 0$. We plot this function for various values of β in Fig 8.

6.1.4. *Moments of Δ*

With this general expression for $\frac{d\sigma^{(0)}}{d\Delta}$ given above, one can compute to 0th order in α, the Δ-moments which are the coefficients in the Fokker-Planck equation (some of these may be found in Ref [1]):

$$\overline{\Delta^0}^{(0)} = 1$$

$$\overline{\Delta^1}^{(0)} = \frac{4}{3}\gamma^2\beta^2 \qquad = 4\left(\frac{kT_e}{m_e c^2}\right) + 10\left(\frac{kT_e}{m_e c^2}\right)^2 + \mathcal{O}\left[\left(\frac{kT_e}{m_e c^2}\right)^3\right]$$

$$\overline{\Delta^2}^{(0)} = \frac{2}{15}\gamma^4\beta^2(5 + 16\beta^2) = 2\left(\frac{kT_e}{m_e c^2}\right) + 47\left(\frac{kT_e}{m_e c^2}\right)^2 + \mathcal{O}\left[\left(\frac{kT_e}{m_e c^2}\right)^3\right]$$

$$\overline{\Delta^3}^{(0)} = \frac{4}{25}\gamma^6\beta^4(21 + 23\beta^2) = \frac{252}{5}\left(\frac{kT_e}{m_e c^2}\right)^2 + \mathcal{O}\left[\left(\frac{kT_e}{m_e c^2}\right)^3\right]$$

$$\overline{\Delta^4}^{(0)} = \frac{4}{525}\gamma^8\beta^4(147 + 1554\beta^2 + 859\beta^4) = \frac{84}{5}\left(\frac{kT_e}{m_e c^2}\right)^2 + \mathcal{O}\left[\left(\frac{kT_e}{m_e c^2}\right)^3\right]$$

$$(35)$$

and we also find that $\overline{\overline{\Delta^{n\prime}}}^{(0)} = 0$ since $\frac{d\sigma^{(0)}}{d\Delta}$ has no dependence on ϵ. The fact that $\overline{\overline{\Delta^0}}^{(0)} = 1$ tells us that to 0th order in α and all orders in β, the scattering rate per unit volume is $cN_e\sigma_T$.[8] The coefficients in the Fokker-Planck equations are determined by the average of the electron velocities indicated, and these expressions hold whether or not the electrons are in thermal equilibrium. For a thermal distribution these velocity moments can be computed exactly in terms of modified Bessel functions [1], however we have found it convenient to expand these functions to the appropriate order in temperature. It seems that a Taylor series to a given order in Δ is less accurate than the same order Taylor series expansion in T_e. To keep track of the various terms in the expansion let us devise the notation

$$\mathcal{O}(n, m) = \mathcal{O}\left(\left(\frac{kT_e}{m_e c^2} \right)^n \left(\frac{\epsilon}{m_e c^2} \right)^m \right) \tag{36}$$

There are no terms $\sim \mathcal{O}(0, 0)$. One finds that

$$\overline{\overline{\Delta^{2n-1}}}^{(m)} \sim \overline{\overline{\Delta^{2n}}}^{(m)} \sim \mathcal{O}(n, m) . \tag{37}$$

So to include all the terms of order $\sim \mathcal{O}(n, m)$ one must make a Fokker-Planck expansion to order $2n$ in Δ. It is probably not worthwhile to go to high order in these expansions, since one can circumvent this expansion by doing the collision integral. Nevertheless the first few terms give useful analytical expressions.

6.1.5. *The Kompaneets Equation and Relativistic Corrections*
The lowest order non-zero Fokker-Planck equation, given by the expansion of Eq 25, is the Kompaneets equation

$$\frac{\partial n_\gamma}{\partial \tau} = \frac{1}{\epsilon^2} \frac{\partial}{\partial \epsilon} \left[\epsilon^3 \left(\frac{kT_e}{m_e c^2} \epsilon \frac{\partial n_\gamma}{\partial \epsilon} + \frac{\epsilon}{m_e c^2} (1 + n_\gamma) n_\gamma \right) \right],$$

$$\underset{\mathcal{O}(1,0)}{|} \qquad\qquad \underset{\mathcal{O}(0,1)}{|} \tag{38}$$

where the order of the two terms are indicated. This equation was first published by Kompaneets [41] in 1957 and probably developed earlier as

---

[8] The total (Klein-Nishina) cross-section starts to fall below the Thomson cross-section when the center-of-mass photon energy rises to close to $m_e c^2$, i.e., when $\gamma\alpha \gtrsim 1$. A careful look at Eq 29 will show that by setting $\alpha = 0$ in this equation we are ignoring terms of order $\alpha\gamma$. For microwave photons this approximation should be good for computing the total cross-section as long as $\gamma \lesssim 10^9$ i.e., for anything less energetic than $\sim 500\mathrm{TeV}$ electrons. In contrast, to compute the small effects on the spectrum from Compton scattering one should include 1st order terms in α whenever $\alpha \gtrsim \beta^2$.

part of the Soviet thermonuclear weapons program. For hotter gas one can add terms $\mathcal{O}(2,0)$, which yields an extended Kompaneets equation [67]

$$
\frac{\partial n_\gamma}{\partial \tau} = \frac{1}{\epsilon^2} \frac{\partial}{\partial \epsilon} \left[\epsilon^3 \left(\frac{kT_e}{m_e c^2} \left(1 + \frac{5}{2} \frac{kT_e}{m_e c^2} \right) \epsilon \frac{\partial n_\gamma}{\partial \epsilon} \right. \right.
$$

$$
+ \frac{7}{10} \left(\frac{kT_e}{m_e c^2} \right)^2 \left(6\epsilon^2 \frac{\partial^2 n_\gamma}{\partial \epsilon^2} + \epsilon^3 \frac{\partial^3 n_\gamma}{\partial \epsilon^3} \right)
$$

$$
\left. \left. + \frac{\epsilon}{m_e c^2} (1 + n_\gamma) n_\gamma \right) \right] \tag{39}
$$

Further terms in this expansion will be derived in Ref [67], although it is not clear how useful they will be since extensive numerical work has been done with the more accurate collision integral (e.g. Ref [53]).

6.1.6. The Generalized Sunyaev-Zel'dovich Effect

The idea of the Sunyaev Zel'dovich distortion is that one starts out with a background radiation which is close to a blackbody spectrum, just what we expect to be produced by the early Universe, and it is *slightly* distorted by the action of hot ionized gas through the Compton scattering process we have just described. In this small distortion limit we need just substitute a blackbody spectrum, n^{BB} of Eq 2, into the right-hand-side of the Kompaneets equation. Let us generalize this idea a bit by instead considering the more general Fokker-Planck expansion which is an expansion in T_e and α. In this small distortion limit the different terms will add linearly to the total distortion which we may write as a sum

$$
\Delta n_\gamma = \sum_{n \geq 0} \sum_{m \geq 0} Y_C^{(n,m)} \Delta n_{SZ}^{(n,m)}(x) \qquad x = \frac{\epsilon}{kT_\gamma}, \tag{40}
$$

where

$$
Y_C^{(n,m)} = \int d\tau \left(\frac{kT_e}{m_e c^2} \right)^n \left(\frac{kT_\gamma}{m_e c^2} \right)^m, \tag{41}
$$

and the superscript $^{(n,m)}$ correspond to the $\mathcal{O}(n,m)$ contributions to the Fokker-Planck expansion. Substituting $n^{BB}(x)$ into the various terms of Eq 39 we find that

$$
\Delta n_{SZ}^{(0,0)}(x) = 0
$$

$$
\Delta n_{SZ}^{(1,0)}(x) = \frac{xe^x}{(e^x - 1)^2} \left(x \frac{e^x + 1}{e^x - 1} - 4 \right)
$$

$$
\Delta n_{SZ}^{(0,1)}(x) = -\frac{xe^x}{(e^x - 1)^2} \left(x \frac{e^x + 1}{e^x - 1} - 4 \right)
$$

$$
\Delta n_{SZ}^{(2,0)}(x) = \frac{xe^x}{(e^x - 1)^2} \left(-10 + \frac{47}{2} x \frac{e^x + 1}{e^x - 1} - \frac{42}{5} x^2 \frac{e^{2x} + 4e^x + 1}{(e^x - 1)^2} + \right.
$$

264

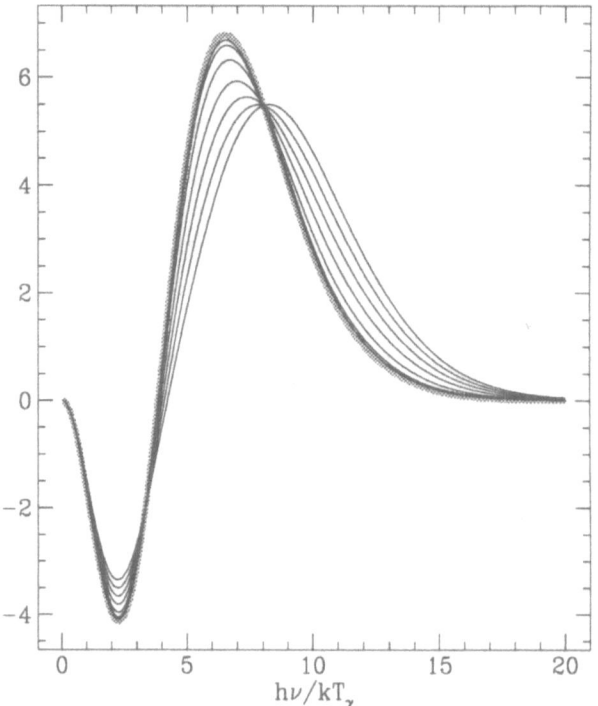

Figure 9. Plotted is the small deviation in intensity from a blackbody divided by the classical y-parameter caused when blackbody photons pass through a hot gas of electrons. This is computed using the extension of the y-distortion given in the text. The gray band is centered on the classical y-distortion which applies when $kT_e \ll m_e c^2$. The black lines are for electron temperatures of 1, 2, 5, 10, 15, 20, and 25 keV. We have of course assumed $T_e \gg T_\gamma$. The curves intersect at the zeros of the function $\Delta n_{\text{SZ}}^{(2,0)}$.

$$+ \frac{7}{10} x^3 \frac{(e^x + 1)(e^{2x} + 10e^x + 1)}{(e^x - 1)^3}\Bigg) . \qquad (42)$$

One does expect that, to each order in energy, a blackbody spectrum is a stable solution when the electron and photon temperatures are equal, so one should expect the sum rule

$$\sum_{n=0}^{N} \Delta n_{\text{SZ}}^{(n, N-n)}(x) = 0, \qquad (43)$$

and this does seem to be true for $N = 0$ and and $N = 1$.

The classical Sunyaev-Zel'dovich y-distortion contains only the $\mathcal{O}(1,0)$ and $\mathcal{O}(0,1)$ terms and may be written

$$\Delta n = y \frac{xe^x}{(e^x - 1)^2} \left(x \frac{e^x + 1}{e^x - 1} - 4 \right), \tag{44}$$

where

$$y = Y_C^{(1,0)} - Y_C^{(0,1)} = \int d\tau \frac{k(T_e - T_\gamma)}{m_e c^2}. \tag{45}$$

This is the y-distortion plotted in Fig 5 and used in Eqs 11 & 12. To see how much this classical formula errs, we compare the different expressions for a range of electron temperatures in Fig 9. We see that the $\mathcal{O}(2,0)$ corrections become significant when $T_e \gtrsim 5$ keV. This 2nd order distortion agrees well with the computation of the collision integral by Rephaeli [53], so higher order corrections do not seem to be important for $T_e \lesssim 15$ keV.

7. The Future

In the past decade we have witnessed astounding advances in the measurement of the CMBR spectrum. After decades of tantalizing evidence of deviations from a blackbody spectrum we find that the spectrum is amazingly close to a perfect blackbody. No longer is it possible to consider a universe with a very hot inter-galactic medium, or that hydrodynamic forces could have played a large role in forming the *large* scale structure. There is also little room for non-equilibrium energetic events in the early Universe at redshifts $< 10^7$. In a way this is most unfortunate. The thermal equilibrium state contains the least information - all remnant of events in the Universe before $z \sim 10^7$ have been thermalized to nothing, or more precisely to one number: the temperature. At the moment we really don't know how to interpret this number, other than to make a rough comparison to the number of baryons which is observationally rather ill-determined. Perhaps some day we will have cosmogenic theories which will predict the baryon-to-photon ratio with great accuracy.

Observationally we are approaching a brick wall which is the Galaxy. At the present level of sensitivity Galactic contamination from dust and synchrotron radiation is an important contaminant at all wavelengths. Galaxy modeling which makes use of a spectral and spatial structure of the Galaxy observed at a variety of wavelengths will improve as sensitivities improve, however there will be a limit to how accurately one can subtract off the Galaxy even given perfect data. We won't be making observations outside of the Galactic Plane any time soon!

Yet there is still a lot of room for improvement on the decimeter- and meter-scale anisotropies. Also there is this tantalizing evidence for *negative* spectral deviations ....[9].

Things are not bleak. In fact the study of spectral distortions of the CMBR is a rapidly growing field. Multi-frequency observations are beginning to be the norm for CMBR anisotropy experiments, and with the CMBR satellites we can expect literally millions of measurements of the CMBR spectrum in different directions on the sky. Admittedly there is a big difference between absolute measurements of the CMBR flux and differential measurements which are required for anisotropy since the anisotropy spectral measurements are modulo any DC spectral distortion. However it is just this sort of measurement which will make improved Galaxy subtraction possible. The spectral information obtained will tell us mostly about the Galaxy and extragalactic radio sources, however with millions of measurements one can always hope for something a little more interesting. Along these lines there is the cluster S-Z effect which is a rapidly growing field. With increased sensitivity we can look forward to S-Z selected cluster catalogs, measurements of radial cluster velocities through the kinematic S-Z effect, and these studies can work their way down to galaxy groups and even large scale structure filaments of hot gas. As we have seen we can even hope to measure the gas temperature from the spectrum if it is hot enough. In the future we can expect the spectrum and anisotropy measurements to become increasingly intertwined.

8. Bibliography

What follows is not a list of articles cited in this work, although it includes all articles cited, but rather an (incomplete) bibliography of published works related to the CMBR spectrum, including title, listed alphabetically by the name of the first author. Many of these papers are of only historic interest: theories have been ruled out and observations superseded. I hope some readers will find it a useful reference.[10]

References

1. Barbosa, D.D. (1982) *"A Note on Compton Scattering"* Ap. J. **254** 301-308.
2. Barbosa, D., Bartlett, J.G., Blanchard, A., and Oukbir, J. (1996) *"The Sunyaev-Zel'dovich effect and the value of Ω_0"* A&A **314** 13-17.

---

[9]Some people never learn.

[10]The compilation method was somewhat haphazard and the author apologizes to the authors of the many important works which are not listed. Incompleteness is probably fairly large for papers written in the past decade, and entire subject areas (e.g. decaying particles) have been omitted. Conference proceedings and other articles in books have been excluded.

3. Bartlett, J.G. and Stebbins, A. (1991) *"Did the Universe Recombine?"*, Ap.J., **371**, 8.

4. Bensadoun, M., Bersanelli, M., De Amici, G., Kogut, A., Levin, S.M., Limon, M., Smoot, G.F., and Witebsky, C. (1993) *"Measurements of the cosmic microwave background temperature at 1.47 GHz"* Ap.J. **409** 1-13.

5. Bernstein, G.M., Fischer, M.L., Richards, P.L., Peterson, J.B., and Timusk, T (1990) *"A measurement of the spectrum of the cosmic background radiation from 1 to 3 millimeter wavelength"* Ap. J. **326** 107-113.

6. Bersanelli, M., Witebsky, C., Bensadoun, M., De Amici, G., Kogut, A., Levin, S.M., and Smoot, G.F. (1989) *"Measurements of the cosmic microwave background radiation temperature at 90 GHz"* Ap.J. **339** 632-637.

7. Bersanelli, M., Bensadoun, M., De Amici, G., Levin, S., Limon, M., Smoot, G.F., and Vinje, W. (1994) *"Absolute measurement of the cosmic microwave background at 2 GHz"* Ap.J. **424** 517-529.

8. Bontz, J., Price, R.H., and Haughn, M.P. (1981) *"Implications of the Deviations in the Spectrum of the Cosmic Background Radiation"* Ap. J. **246** 592-611.

9. Burigana, C., Danese, L., and De Zotti, G. (1991) *"Formation and evolution of early distortions of the microwave background spectrum - A numerical study"* A&A **246** 49-58.

10. Burigana, C., Danese, L., and De Zotti, G. (1995) *"Analytical description of spectral distortions of the cosmic microwave background "* A&A **303** 323.

11. Chan, K.L. and Jones, B.J.T. (1975a) *"Distortion of 3°K Background Radiation Spectrum: Observational Constraints on the Early Universe"* Ap. J. **195** 1-11.

12. Chan, K.L. and Jones, B.J.T. (1975b) *"Distortion of the Microwave Background Radiation Spectrum in the Submillimeter Wavelength Region"* Ap. J. **198** 245-248.

13. Chan, K.L. and Jones, B.J.T. (1975c) *"The Evolution of the Cosmic Radiation Spectrum Under the Influence of Turbulent Heating. I. Theory"* Ap. J. **200** 454-460.

14. Chan, K.L. and Jones, B.J.T. (1975d) *"The Evolution of the Cosmic Radiation Spectrum Under the Influence of Turbulent Heating. I. Numerical Calculations and Applications"* Ap. J. **200** 461-470.

15. Chang, J.S. and Cooper, G. (1970) *"A Practical Scheme for Fokker-Planck Equations"* J. Comp. Phys. **6** 1-16.

16. Colafrancesco, S., Mazzotta, P., Rephaeli, and Vittorio, N. (1997) *"Intracluster Comptonization of the CMB: Mean Spectral Distortion and Cluster Number Counts"* astro-ph9703121.

17. Crane, P., Hegyi, D.J., Mandolesi, N., and Danks, A.C. (1986) *"Cosmic background radiation temperature from CN absorption"* Ap.J. **309** 822-827.

18. Crane, P., Hegyi, D.J., Kutner, M.L., and Mandolesi, N. (1989) *"Cosmic background radiation temperature at 2.64 millimeters"* Ap.J. **346** 136-142.

19. Daly, R.A. (1991) *"Spectral distortions of the microwave background radiation resulting from the damping of pressure waves"* Ap. J. **371** 14-28.

20. Danese, L. and De Zotti, G. (1977) *"The Relic Radiation Spectrum and the Thermal History of the Universe"* Rev. Nou. Cim. **7** 277-362.

21. Danese L. and De Zotti G. (1981) *"Dipole Anisotropy and Distortions of the Spectrum of the Cosmic Microwave Background Radiation"* A&A **94** L33-L34.

22. Danese, L. and De Zotti, G. (1982) *"Double Compton and the Spectrum of the Microwave Background"* A&A **107** 39-42.

23. De Amici, G., Witebsky, C., Smoot, G., and Friedman, S. (1984) *"Measurement of the Cosmic Background Radiation at 3.3 and 9.1 mm"* Phys. Rev **D29** 2673.

24. De Amici, G., Smoot, G.F., Aymon, J., Bersanelli, M., Kogut, A., Levin, S.M., and Witebsky, C. (1988) *"Measurement of the intensity of the cosmic background radiation at 3.7 GHz"* Ap.J. **329** 556-566.

25. De Amici, G., Bensadoun, M., Bersanelli, M., Kogut, A., Levin, S. Smoot, G.F., and Witebsky, C. (1990) *"The temperature of the cosmic background radiation - Results from the 1987 and 1988 measurements at 3.8 GHz"* Ap.J. **359** 219-227.

26. De Amici, G., Limon, M., Smoot, G.F., Bersanelli, M., Kogut, A., and Levin, S. (1991) *"The temperature of the cosmic microwave background radiation at 3.8 GHz - Results of a measurement from the South Pole site"* Ap.J. **381** 341-347.

27. Field G.. (1972) *"Intergalactic Matter"* Ann. Rev. Astro. Ap. **10** 227.

28. Field, G. and Perrenod, S. (1977) *"Constraints on a Dense Hot Intergalactic Medium"* Ap. J. **215** 717.

29. Fixsen, D.J., Cheng, E.S., Gales, J.M., Mather, J.C., Shafer, R.A., and Wright, E.L. (1996) *"The Cosmic Microwave Background Spectrum from the Full COBE FIRAS Data Set"* Ap. J. **473** 576-587.

30. Friedman, S., Smoot, G., De Amici, G., and Witebsky, C. (1984) *"Measurement of the Cosmic Background Radiation at 3.0 cm"* Phys. Rev. **D29** 2677.

31. Ginzburg, V. and Ozernoi, L. (1966) *"The Temperature of Intergalactic Gas"* Sov. Astro. - AJ **9** 726.

32. Gush, H. (1981) *"Rocket Measurement of the Cosmic Background Submillimeter Spectrum"* Ap. J. **218** 592.

33. Gush, H.P., Halpern, M., & Wishnow, E.H. (1990) *"Rocket measurement of the cosmic-background-radiation mm-wave spectrum"* Phys. Rev. Lett. **65** 537-540.

34. Hawkins, I. and Wright, E.L. (1988) *"Needling the Universe"* Ap. J. **324** 46-59.

35. Howell, T. and Shakeshaft J. (1967) *"Spectrum of the 3° K Cosmic Microwave Radiation"* Nature **216** 753.

36. Illarionov, A.F. and Sunyaev, R.A. (1975) *"Comptonization, characteristic radiation spectra, and thermal balance of low-density plasma"* Sov. Astr. **18** 413-419.

37. Johnson, D.G., and Wilkinson, D.T. (1987) *"A 1 percent measurement of the temperature of the cosmic microwave radiation at lambda = 1.2 centimeters"* Ap.J. **313** L1-L4.

38. Kogut, A., Bersanelli, M., De Amici, G., Friedman, S.D., Griffith, M., Grossan,B., Levin, S., Smoot, G.F., and Witebsky, C. (1988) *"The temperature of the cosmic microwave background radiation at a frequency of 10 GHz"* Ap.J. **325** 1-15.

39. Kogut, A., Bersanelli, M., De Amici, G., Friedman, S.D., Griffith, M., Grossan, B., Levin, S., Smoot, G.F., and Witebsky, C. (1988) *"The Temperature of the Cosmic Microwave Background Radiation at a Frequency of 10 GHz: Erratum"* Ap.J. **332** 1092.

40. Kogut, A., Bensadoun, M., De Amici,G., Levin, S., Smoot, G.F., and Witebsky, C. (1990) *"A measurement of the temperature of the cosmic microwave background at a frequency of 7.5 GHz"* Ap.J. **355** 102-113.

41. Kompaneets, A. (1957) *"The Establishment of Thermal Equilibrium between Quanta and Electrons"* Sov. Phys. - JETP **4** 730-737.

42. Levin, S.M., Witebsky, C., Bensadoun, M., Bersanelli, M., De Amici, G., Kogut, A., and Smoot, G.F. (1988) *"A measurement of the cosmic microwave background radiation temperature at 1.410 GHz"* Ap.J. **334** 14-21.

43. Levin, S., Bensadoun, M., Bersanelli, M., De Amici, G., Kogut, A., Limon, M., and Smoot, G. (1992) *"A measurement of the cosmic microwave background temperature at 7.5 GHz"* Ap.J. **396** 3-9.

44. Lifshitz, E.M., and Pitaevskii, L.P. (1981) *Physical Kinetics* Pergamon Press, Oxford.

45. Lyubarsky, Y.E. and Sunyaev, R.A. (1983) *"The spectral feature in the microwave background radiation spectrum due to energy release in the early Universe"* Ann. Rev. Astro. and Ap. **123** 171-183.

46. Mandolesi, N., Calzolari, P., Cortiglioni, S., Morigi, G., Danese, L., and De Zotti, G. (1986) *"Measurements of the cosmic background radiation temperature at 6.3 centimeters"* Ap.J. **310.** 561-567.

47. Matsumoto, T., Hayakawa, S., Matuo, H., Murakami, H., Sato S., Lange, A.E., and Richards P.L (1988) *"The submillimeter spectrum of the cosmic background radiation"* Ap. J. **329** 567-571.

48. Meyer, D.M. and Jura, M. (1984) *"The microwave background temperature at 2.64*

and 1.32 millimeters" Ap.J.Lett. **276** L1-L3.

49. Meyer, D.M., Jura, M. (1985) *"A precise measurement of the cosmic microwave background temperature from optical observations of interstellar CN"* Ap.J. **297** 119-132.

50. Palazzi, E., Mandolesi, N., Crane, P., Hegyi, D.J., and Blades, J.C. (1990) *"The cosmic background radiation temperature at 1.3 mm"* No. Cim. C **13** 537-540.

51. Palazzi, E., Mandolesi, N., Crane, P., Kutner, M.L., Blades, J.C., and Hegyi, D.J. (1990) *"A precise measurement of the cosmic background radiation at 1.32 millimeters"* Ap.J. **357** 14-22.

52. Peterson, J.B., Richards, P.L., and Timusk, T. (1985) *"Spectrum of the Cosmic Background Radiation at Millimeter Wavelengths"* Phys. Rev. Lett. **55** 332.

53. Rephaeli, Y. (1995) *"Cosmic Microwave Background Comptonization by Hot IntraCluster Gas"* Ap. J. **445** 33-36.

54. Rephaeli, Y. and Yankovitch, D. (1997) *"Relativistic Corrections in the Determination of H_0 from X-ray and Sunyaev-Zel'dovich Measurements"* Ap. J. Lett. **481** L55-L58.

55. Sironi G. and Ferrari A. (1984) *"Measurement of the Cosmic Background Radiation at 12 cm"* Phys. Rev. **D29** 2686.

56. Sironi, G. and Bonelli, G. (1986) *"The temperature of the diffuse background radiation at 12 centimeter wavelength"* Ap.J. **311** 418-424.

57. Sironi, G., Limon, M., Marcellino, G., Bonelli, G., Bersanelli, M., Conti, G., and Reif, K. (1990) *"The absolute temperature of the sky and the temperature of the cosmic background radiation at 600 MHz"* Ap.J. **357** 301-308.

58. Sironi, G., Bonelli, G., and Limon, M. (1991) *"The brightness temperature of the south celestial pole and the temperature of the cosmic background radiation measured at 36.6 and 12 centimeter wavelength"* Ap.J. **378** 550-556.

59. Sironi, G., Bonelli, G., and Limon, M. (1992) *"Measurements of the absolute temperature of the relic radiation observations at 0.6, 0.82 and 2.5 GHz from Alpe Gera and the South Pole"* No. Cim. C **15** 983-991.

60. Smoot, G.F., De Amici, G., Friedman, S.D., Witebsky, C., Mandolesi, N., Partridge, R.B., Sironi, G., Danese, L., and De Zotti, G. (1983) *"Low-frequency measurement of the spectrum of the cosmic background radiation"* Phys.Rev.Lett. **51** 1099-1102.

61. Smoot, G. et al. (1984) *"Low-Frequency Measurement of the Cosmic Background Radiation Spectrum"* Ap. J. Lett. **291** L23.

62. Smoot, G.F., De Amici, G., Friedman, S.D., Witebsky, C., Sironi, G., Bonelli, G., Mandolesi, N., Cortiglioni, S., Morigi, G., and Partridge, R.B. (1985) *"Low-frequency measurements of the cosmic background radiation spectrum"* Ap.J.Lett. **291** L23-L27.

63. Smoot, G.F., Bensadoun, M., Bersanelli, M., De Amici, G., Kogut, A., Levin, S., and Witebsky, C. (1987) *"Long-wavelength measurements of the cosmic microwave background radiation spectrum"* Ap.J.Lett. **317** L45-L49.

64. Smoot, G., Levin, S.M., Witebsky, C., De Amici, G., and Rephaeli,Y. (1988) *"An analysis of recent measurements of the temperature of the cosmic microwave background radiation"* Ap.J. **331** 653-659.

65. Staggs, S. (1993) *"An absolute measurement of the cosmic background radiation temperature at 1.4 GHz"* Ph.D.Thesis, Princeton.

66. Staggs, S.T., Jarosik, N.C., Meyer, S.S., and Wilkinson, D.T. (1996) *"An Absolute Measurement of the Cosmic Microwave Background Radiation Temperature at 10.7 GHz"* Ap.J.Lett. **473** L1.

67. Stebbins, A. (1997) *in preparation.*

68. Stebbins, A. and Silk, J. (1986) *"The Universe Between $z = 10$ and $z = 1000$: Spectral Constraints on Reheating"* Ap.J. **300**, 1-19.

69. Sunyaev, R.A. and Zel'dovich Y.B. (1980) *"Microwave Background Radiation as a Probe of the Contemporary Structure and History of the Universe"* Ann. Rev. Astro. and Ap. **18** 537-560.

70. Sunyaev, R.A. and Zel'dovich Y.B. (1970) *"The Interaction of Matter and Radiation*

in a Hot Model of the Universe II" Ap. and Spac. Sci. **7** 20-30.

71. Weiss R. (1980) *"Measurements of the Cosmic Background Radiation"* Ann. Rev. Astro. Ap. **18** 489.

72. Weymann, R. (1966) *"The Energy Spectrum of Radiation in the Expanding Universe"* Ap. J. **145** 560-571.

73. Weymann, R. (1967) *"Possible Thermal Histories of Intergalactic Gas"* Ap. J. **147** 887.

74. Witebsky, C., Smoot, G., De Amici, G., and Friedman, S.D. (1986) *"New measurements of the cosmic background radiation temperature at 3.3 millimeter wavelength"* Ap.J. **310** 145-159.

75. Woody, D. and Richards, P. (1979) *"Spectrum of the Cosmic Background Radiation"* Phys. Rev. Lett. **42** 14.

76. Wright, E.L. (1979) *"Distortion of the Microwave Background by a Hot Intergalactic Medium"* Ap. J. **232** 348-351.

77. Zel'dovich Y.B., Illarionov, A.F., and Sunyaev, R.A. (1969) *"?"* Sov. Phys. - JETP **35** 643.

78. Zel'dovich, Y.B. and Sunyaev, R.A. (1969) *"The Interaction of Matter and Radiation in a Hot Model Universe"* Ap. and Spac. Sci. **4** 301-316.

THE CMB SPECTRUM

Cosmic Microwave Background

GEORGE F. SMOOT
Lawrence Berkeley National Lab & Physics Department
University of California
Berkeley CA 94720

1. Introduction

The observed cosmic microwave background (CMB) radiation provides strong evidence for the hot Big Bang. The success of primordial nucleosynthesis calculations ("Big Bang nucleosynthesis") requires a cosmic background radiation (CBR) characterized by a temperature $kT \sim 1\,\mathrm{MeV}$ at a redshift of $z \simeq 10^9$. In their pioneering work, Gamow, Alpher, and Herman [2] realized this and predicted the existence of a faint residual relic of the primordial radiation, with a present temperature of a few degrees. The observed CMB is interpreted as the current manifestation of the hypothesized CBR.

The CMB was serendipitously discovered by Penzias and Wilson [49] in 1964. Its spectrum is well characterized by a 2.73 ± 0.01 K blackbody (Planckian) spectrum over more than three decades in frequency (see Figure 1). A non-interacting Planckian distribution of temperature T_i at redshift z_i transforms with the universal expansion to another Planckian distribution at redshift z_r with temperature $T_r/(1 + z_r) = T_i/(1 + z_i)$. Hence thermal equilibrium, once established (e.g., at the nucleosynthesis epoch), is preserved by the expansion, in spite of the fact that photons decoupled from matter at early times. Because there are about 10^9 photons per nucleon, the transition from the ionized primordial plasma to neutral atoms at $z \sim 1000$ does not significantly alter the CBR spectrum [48].

2. Theoretical Spectral Distortions

The remarkable precision with which the CMB spectrum is fitted by a Planckian distribution provides limits on possible energy releases in the early Universe, at roughly the fractional level of 10^{-4} of the CBR energy,

C. H. Lineweaver et al. (eds.), The Cosmic Microwave Background, 271–308.

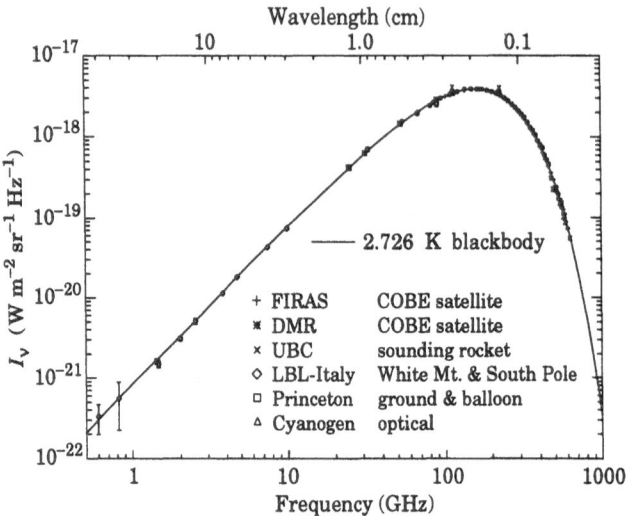

Figure 1. Precise measurements of the CMB spectrum. The line represents a 2.73 K blackbody, which describes the spectrum very well, especially around the peak of intensity. The spectrum is less well constrained at frequencies of 3 GHz and below (10 cm and longer wavelengths). See Table 1 for references.

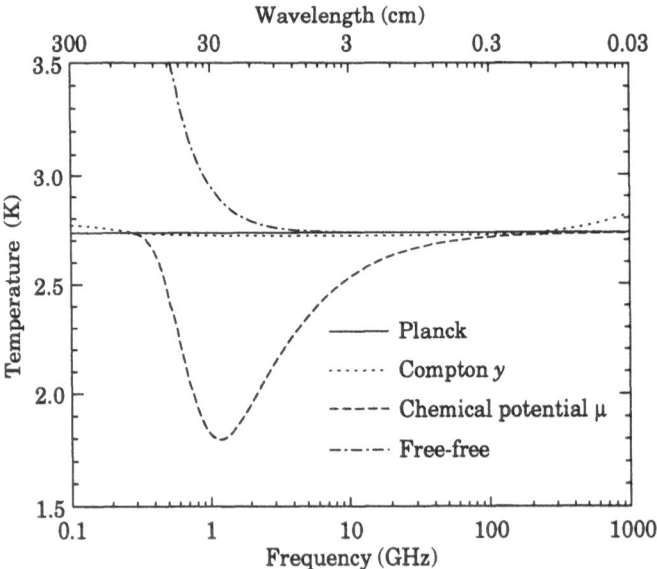

Figure 2. The shapes of expected, but so far unobserved, CMB distortions, resulting from energy-releasing processes at different epochs.

for redshifts $\lesssim 10^7$ (corresponding to epochs $\gtrsim 1$ year). The following three important classes of spectral distortions (see Figure 2) generally correspond to energy releases at different epochs. The distortion results from interactions with a hot electron gas at temperature T_e.

2.1. COMPTON DISTORTION

Late energy release $(z \sim 10^5)$. Compton scattering $(\gamma e \to \gamma' e')$ of the CBR photons by a hot electron gas creates spectral distortions by transferring energy from the electrons to the photons. Compton scattering cannot achieve thermal equilibrium for $y < 1$, where

$$y = \int_0^z \frac{kT_e(z') - kT_\gamma(z')}{m_e c^2} \, \sigma_T \, n_e(z') \, c \, \frac{dt}{dz'} \, dz' \, , \qquad (1)$$

is the integral of the number of interactions, $\sigma_T \, n_e(z) \, c \, dt$, times the mean-fractional photon-energy change per collision [59]. For $T_e \gg T_\gamma$ y is also proportional to the integral of the electron pressure $n_e k T_e$ along the line of sight. For standard thermal histories $y < 1$ for epochs later than $z \simeq 10^5$.

The resulting CMB distortion is approximately a temperature decrement

$$\Delta T_{\rm RJ} = -2y \, T_\gamma \qquad (2)$$

in the Rayleigh-Jeans $(h\nu/kT_\gamma \ll 1)$ portion of the spectrum, and a rapid rise in temperature in the Wien $(h\nu/kT_\gamma \gg 1)$ region, i.e., photons are shifted from low to high frequencies. The magnitude of the distortion is related to the total energy transfer [59] ΔE by

$$\Delta E / E_{\rm CBR} = e^{4y} - 1 \simeq 4y \, . \qquad (3)$$

A prime candidate for producing a Comptonized spectrum is a hot intergalactic medium. A hot $(T_e > 10^5 \, {\rm K})$ medium in clusters of galaxies also produces a partially Comptonized spectrum as seen through the cluster, known as the Sunyaev-Zel'dovich effect. Based upon X-ray data, the predicted large angular scale total combined effect of the hot intracluster medium should produce $y < 10^{-6}$ [6].

2.2. BOSE-EINSTEIN OR CHEMICAL POTENTIAL DISTORTION

Early energy release $(z \sim 10^5 - 10^7)$. After many Compton scatterings $(y > 1)$, the photons and electrons will reach statistical (not thermodynamic) equilibrium, because Compton scattering conserves photon number. This equilibrium is described by the Bose-Einstein distribution with non-zero chemical potential:

$$n = \frac{1}{e^{x + \mu_0} - 1} \, , \qquad (4)$$

where $x \equiv h\nu/kT$ and $\mu_0 \simeq 1.4 \Delta E/E_{\rm CBR}$, with μ_0 being the dimensionless chemical potential that is required to conserve photon number.

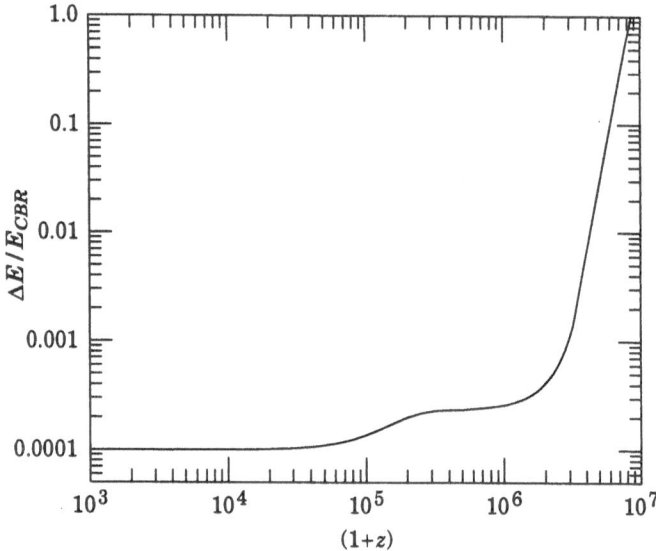

Figure 3. Upper Limits (95% CL) on fractional energy ($\Delta E/E_{\mathrm{CBR}}$) releases as set by lack of CMB spectral distortions resulting from processes at different epochs. These can be translated into constraints on the mass, lifetime and photon branching ratio of unstable relic particles, with some additional dependence on cosmological parameters such as Ω_b [21], [55].

The collisions of electrons with nuclei in the plasma produce free-free (thermal bremsstrahlung) radiation: $eZ \rightarrow eZ\gamma$. Free-free emission thermalizes the spectrum to the plasma temperature at long wavelengths. Including this effect, the chemical potential becomes frequency-dependent,

$$\mu(x) = \mu_0 e^{-2x_b/x} , \qquad (5)$$

where x_b is the dimensionless transition frequency at which Compton scattering of photons to higher frequencies is balanced by free-free creation of new photons. The resulting spectrum has a sharp drop in brightness temperature at centimeter wavelengths [5]. The minimum wavelength is determined by Ω_b.

The equilibrium Bose-Einstein distribution results from the oldest non-equilibrium processes ($10^5 < z < 10^7$), such as the decay of relic particles or primordial inhomogeneities. Note that free-free emission (thermal bremsstrahlung) and radiative-Compton scattering effectively erase any distortions [13], [58], [16] to a Planckian spectrum for epochs earlier than $z \sim 10^7$.

2.3. FREE-FREE DISTORTION

Very late energy release ($z \ll 10^3$). Free-free emission from recent reionization ($z < 10^3$) and from a warm intergalactic medium can create rather than erase spectral distortion in the late Universe. The distortion arises because of the lack of Comptonization at recent epochs. The effect on the present-day CMB spectrum is described by

$$\Delta T_{ff} = T_\gamma\, Y_{ff}/x^2, \tag{6}$$

where T_γ is the undistorted photon temperature, x is the dimensionless frequency, and Y_{ff}/x^2 is the optical depth to free-free emission:

$$Y_{ff} = \int_0^z \frac{T_e(z') - T_\gamma(z')}{T_e(z')} \frac{8\pi e^6 h^2 n_e^2\, g}{3 m_e (kT_\gamma)^3 \sqrt{6\pi\, m_e\, kT_e}} \frac{dt}{dz'} dz' . \tag{7}$$

Here h is Planck's constant, n_e is the electron density and g is the Gaunt factor [3].

3. Spectrum Observations

Beginning with the original discovery of the CMB by Penzias and Wilson there was a rush of observations in the period 1965 through 1967. For the most part, interest shifted to the short-wavelength (high-frequency) portion of the spectrum to observe the peak and Wien turn down to show that the spectrum was thermal. In the early 1980's effort was renewed for observing the low-frequency (Rayleigh-Jeans) region primarily by a USA-Italian collaboration consisting of my group (LBNL-Berkeley), Bruce Partridge (Haverford), Reno Mandolesi's group (Bologna), and Giorgio Sironi's group (Milano) with theoretical support from Luigi Danese and Gianfranco DeZotti (Padua). We determined that scientific goals and technology had advanced to the point that the long wavelength (> 1 cm) region should and could be measured more accurately.

The general experimental concept is to observe the total power coming from the sky and compare that to a well-known reference source using specially designed radio receivers called radiometers. The references we used were carefully designed blackbodies with total temperature of about 3.8 K, which was very close to the total sky signal so that the gain calibration of our radiometers was not a critical component of the observation. The sky signal is the sum of many terms

$$T_{sky} = T_{CMB} + T_{atmosphere} + T_{Galaxy} + T_{Sun/Moon} + T_{instrum} + T_{terrestial}. \tag{8}$$

The total sky signal is dominated by the CMB, $T_{CMB} = 2.73$ K, and the atmospheric emission, $T_{atmosphere} \approx 1$ K. The other terms are generally much smaller giving a total sky signal close to 3.8 K of total power which closely matches the absolute reference load. At very low frequencies the Galactic signal tends to be greater but the atmospheric component decreases.

The signal from the atmosphere is determined by scanning the radiometer beam through various zenith angles and thus varying the air mass observed and the data are then constrained by continuity and comparison to an atmospheric model.

Our groups carried out a series of measurements from the high, dry sites at White Mountain University of California Research Station (3800 meters) which is in the rain shadow of the Sierras and from the NSF South Pole Station. Table 1 gives a summary of the low-frequency observations and observations excluding those from *COBE* FIRAS and the UBC rocket-borne experiment.

3.1. INTERSTELLAR MOLECULES AND ATOMIC SYSTEMS

Observations of interstellar molecules and atomic systems, most especially cyanogen (CN), provide a probe of the CMB temperature in narrow wavelength bands at remote locations. CN has proved most used and most precise as the observations are made at optical wavelengths and use well-developed technology. They are by their nature indirect observations in that what is measured is the relative population of various energy levels and the ratio can be used to estimate T_{CMB}. Thus one is observing the total excitation of the system and accounting must be done to subtract the contribution of other potential sources. Fortunately, fue to cold, non-dense clouds, the contribution from sources other than the CMB tends to be quite small, typically on the order of 0.1 K compared to 2.73 K.

CN molecules exist abundantly in interstellar clouds. If a cloud lies along the line of sight from a bright optical source, the CN produces narrow absorption-line features on the source spectrum. These absorbtion lines were detected in 1941 (more than half a century ago and 23 years before the CMB was discovered by Penzias and Wilson) by Adams [1], McKellar [38], and others and were noted as a puzzle. McKellar estimated a value for the excitation temperature of the CN of a few degrees. Immediately after the CMB discovery many people began observations of CN systems for information about the CMB temperature. Those efforts have continued to the present and the most recent results are shown in Table 2 along with some results for other molecules. Note that there is a discrepancy of results on CN between [45] and [54].

3.2. *COBE* FIRAS

It is the measurements of the FIRAS (Far-InfraRed Absolute Spectrophotometer) on the *COBE* satellite that have shown definitively that the CMB spectrum is very nearly Planckian. The FIRAS instrument is a twin-input,

TABLE 1. Measurements of T_{CMB}

Frequency (GHz)	Wavelength (cm)	Temperature (K)	Location (calibration)	Reference
0.408	73.5	3.7 ± 1.2	Ground (LN)	Howell & Shakeshaft 1967, Nature, 216, 753
0.6	50	3.0 ± 1.2	Ground (Term)	Sironi et al. 1990, Ap.J., 357, 301
0.610	49.1	3.7 ± 1.2	Ground (LN)	Howell & Shakeshaft 1967, Nature, 216, 7
0.635	47.2	3.0 ± 0.5	Ground (LN)	Stankevich et al 1970, Australian J. Phys, 23, 529
0.820	36.6	2.7 ± 1.6	Ground (Term)	Sironi et al. 1991, Ap.J., 378, 550
1	30	2.5 ± 0.3	Ground (LN)	Pelyushenko & Stankevich 1969, Sov. Astron., 13, 223
1.4	21.3	2.11 ± 0.38	Ground (CLC)	Levin et al. 1988, Ap.J., 334, 14
1.42	21.2	3.2 ± 1.0	Ground (Term)	Penzias and Wilson 1967, AJ, 72, 315
1.43	21	$2.65^{+0.33}_{-0.30}$	Ground (LN)	Staggs et al. 1996, ApJ, 458, 407
1.44	20.9	2.5 ± 0.3	Ground (LN)	Pelyushenko & Stankevich 1969, Sov. Astron., 13, 223
1.45	20.7	2.8 ± 0.6	Ground (Term)	Howell & Shakeshaft 1966, Nature, 210, 1318
1.47	20.4	2.27 ± 0.19	Ground (CLC)	Bensadoun et al. 1993, Ap. J., 409, 1
2	15	2.5 ± 0.3	Ground (LN)	Pelyushenko & Stankevich 1969, Sov. Astron., 13, 223
2.3	13.1	2.66 ± 0.77	Ground (Term)	Otoshi et al. 1975, IEEE Trans Inst & Meas, 24, 174
2.5	12	2.71 ± 0.21	Ground (CLC)	Sironi et al. 1991, Ap. J., 378, 550
3.8	7.9	2.64 ± 0.06	Ground (CLC)	De Amici et al. 1991, Ap.J., 381, 341
4.08	7.35	3.5 ± 1.0	Ground (Term)	Penzias & Wilson 1965, Ap.J., 142, 419
4.75	6.3	2.70 ± 0.07	Ground (CLC)	Mandolesi et al. 1986, Ap.J., 310, 561
7.5	4.0	2.60 ± 0.07	Ground (CLC)	Kogut et al. 1988, Ap.J., 355, 102
7.5	4.0	2.64 ± 0.06	Ground (CLC)	Levin et al. 1992, Ap.J., 396, 3
9.4	3.2	3.0 ± 0.5	Ground (Term)	Roll & Wilkinson 1966, Phys. Rev. Lett., 16, 405
9.4	3.2	$2.69^{+0.16}_{-0.21}$	Ground (CLC)	Stokes et al. 1967, Phys. Rev. Lett., 19, 1199
10	3.0	2.62 ± 0.06	Ground (CLC)	Kogut et al. 1990, Ap.J., 355, 102
10.7	2.8	2.730 ± 0.014	Balloon (LHe)	Staggs et al. 1996, ApJ, 458, 407
19.0	1.58	$2.78^{+0.12}_{-0.17}$	Ground (CLC)	Stokes et al. 1967, Phys. Rev. Lett., 19, 1199
20	1.5	2.0 ± 0.4	Ground (CLC)	Welch et al. 1967, Phys. Rev. Lett, 18, 1068
24.8	1.2	2.783 ± 0.025	Balloon	Johnson & Wilkinson 1987, Ap.J. Lett, 313, L1
32.5	0.924	3.16 ± 0.26	Ground (CLC)	Ewing et al. 1967, Phys. Rev. Lett, 19, 1251
33.0	0.909	2.81 ± 0.12	Ground (CLC)	De Amici et al. 1985, Ap.J., 298, 710
35.0	0.856	$2.56^{+0.17}_{-0.22}$	Ground (CLC)	Wilkinson 1967, Phys. Rev. Lett., 19, 1195
37	0.82	2.9 ± 0.7	Ground (LN)	Puzanov et al. 1968, Sov. Astr., 11, 905
83.8	0.358	2.4 ± 0.7	Ground (LN)	Kislyakov et al. 1971, Sov. Ast., 15, 29
90	0.33	$2.46^{+0.40}_{-0.44}$	Ground (CLC)	Boynton et al. 1968, Phys. Rev. Lett., 21, 462
90	0.33	2.61 ± 0.25	Ground (CLC)	Millea et al. 1971, Phys. Rev. Lett., 26, 919
90	0.33	2.48 ± 0.54	Plane (Term)	Boynton & Stokes 1974, Nature, 247, 528
90	0.33	2.60 ± 0.09	Ground (CLC)	Bersanelli et al. 1989, Ap.J., 339, 632
90.3	0.332	2.97 ± 1.0	Balloon	Bernstein et al. 1990, Ap.J., 362, 107
113.6	0.264	2.70 ± 0.04	CN (z Per)	Meyer & Jura 1985, Ap.J., 297, 119
113.6	0.264	2.74 ± 0.05	CN (z Oph)	Crane et al. 1986, Ap.J., 309, 12
113.6	0.264	2.76 ± 0.07	CN (HD 21483)	Meyer et al. 1989, Ap.J. Lett, 343, L1
113.6	0.264	$2.796^{+0.014}_{-0.039}$	CN (z Oph)	Crane et al. 1989, Ap.J., 346, 136
113.6	0.264	2.75 ± 0.04	CN (z Per)	Kaiser & Wright 1990, Ap.J. Lett, 356, L1
113.6	0.264	2.834 ± 0.085	CN (HD 154368)	Palazzi et al. 1990, Ap.J., 357, 14
113.6	0.264	2.807 ± 0.025	CN (16 stars)	Palazzi et al. 1992, Ap.J., 398, 53
154.8	0.194	3.02 ± 1.0	Balloon	Bernstein et al. 1990, Ap.J., 362, 107
195.0	0.154	2.91 ± 1.0	Balloon	Bernstein et al. 1990, Ap.J., 362, 107
227.3	0.132	2.76 ± 0.20	CN (z Per)	Meyer & Jura 1985, Ap.J., 297, 119
227.3	0.132	$2.75^{+0.24}_{-0.29}$	CN (z Oph)	Crane et al. 1986, Ap.J., 309, 822
227.3	0.132	2.83 ± 0.09	CN (HD 21483)	Meyer et al. 1989, Ap.J. Lett, 343, L1
227.3	0.132	2.832 ± 0.072	CN (HD 154368)	Palazzi et al. 1990, Ap.J., 357, 14
266.4	0.113	2.88 ± 1.0	Balloon	Bernstein et al. 1990, Ap.J., 362, 107

TABLE 2. Recent Molecular Measurements of T_{CMB}

Reference (Year)	Molecule	Wavelength (mm)	Temperature (K)
Meyer & Jura (1985)	CN	2.64	2.70 ± 0.04
Meyer et al. (1989)	CN	2.64	2.75 ± 0.03
Meyer et al. (1989)	CN	2.64	2.77 ± 0.07
Meyer et al. (1989)	CN	2.64	2.75 ± 0.08
Meyer & Jura (1985)	CN	1.32	2.76 ± 0.20
Meyer et al. (1989)	CN	1.32	2.83 ± 0.09
Meyer et al. (1989)	CN	1.32	2.85 ± 0.16
Crane et al. (1986)	CN	2.64	2.74 ± 0.05
Crane et al. (1989)	CN	2.64	$2.796^{+0.014}_{-0.039}$
Palazzi et al.(1990)	CN	1.32	2.83 ± 0.07
Palazzi et al.(1990)	CN	1.32	2.82 ± 0.11
Kaiser & Wright (1990)	CN	2.64	2.75 ± 0.04
Palazzi et al.(1992)	CN	2.64	2.817 ± 0.02
Roth et al. (1993)	CN	2.64	$2.729^{+0.023}_{-0.031}$
Roth & Meyer (1995)	CN	2.64	"
Thaddeus (1972)	CH	0.76	< 5.23
Thaddeus (1972)	CH^+	0.36	< 7.35
Kogut et al. (1990)	H_2CO	2.1	3.2 ± 0.9
Weighted mean $\pm 1\sigma$			2.76 ± 0.03

twin-output polarizing Michelson interferometer that achieves high precision by making a differential rather than an absolute measurement.

One input is connected to view the sky through a large, low side-lobe sky horn. The other input is connected to an internal calibrator at all times. The internal calibrator is nearly a blackbody (96-98% emissivity) over the full wavelength range and is very stable. The calibrator temperature is adjusted to give nearly null interferometer output.

The sky horn can be filled by the external calibrator by swinging it on its pivot. The external calibrator is a re-entrant absorbing cone. The combined external calibrator and sky horn cavity is a very good blackbody with emissivity measured to be greater than 99.99% and calculated to be greater than 99.999%. The external calibrator temperature is commandable and was adjusted around null defined by the sky signal to provide an absolute and relative calibration. This operation is possible since one does not have to be concerned with windows or freezing of the atmosphere on the instrument and calibrator or with serious thermal loading.

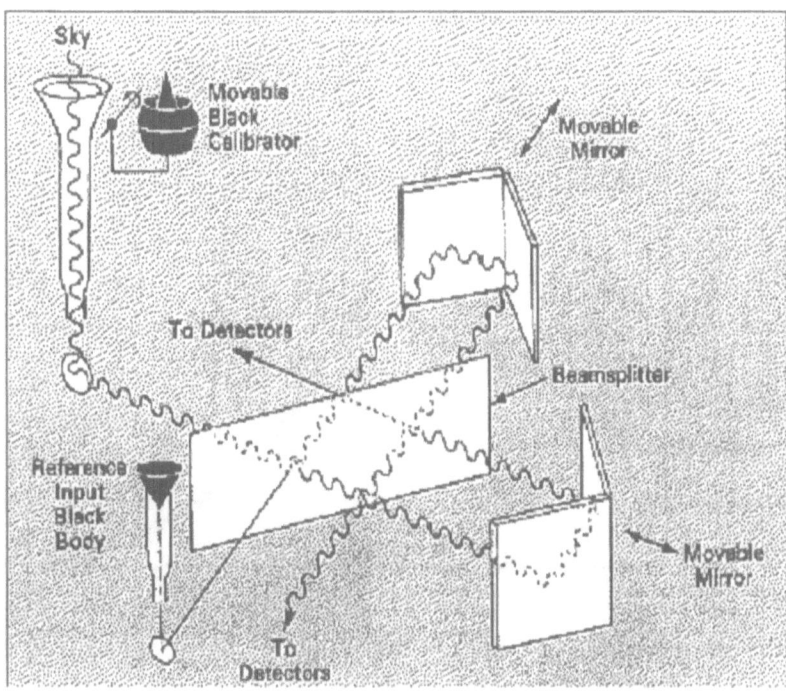

Figure 4. Schematic of the FIRAS instrument. There are two symmetric inputs: the power from the sky and from an internal reference blackbody. A high precision blackbody can be substituted for the sky signal input. FIRAS is a polarizing Michelson interferometer with two dihedral mirrors.

Comparison of the signal from the sky with the signal from the external calibrator with temperature adjusted to match gives an accurate and precise measurement of deviations of the sky spectrum from a blackbody. When these small deviations are added to the calculated Planck spectrum, the FIRAS observed spectrum is produced. See Figure 5 for the measured spectrum and a 2.728 K Planckian. The temperature of the external calibrator, when the output matches the sky viewing output, is the sky temperature. A number of small corrections must be made, e.g., to the GRT (germanium resistance thermometers) readings, cosmic ray hits, extra signal from interstellar dust or the experiment. Another method is to use the wavelength of the peak of the brightness spectrum determined by the length scale set by the dimensions of the interferometer and which is accurately checked and calibrated by the molecular lines observed in the Galactic emission by the interferometer. A third approach is to use the dipole spectrum (see Section 10) to set the temperature scale.

Since the RMS deviation of the spectral intensity I_ν from a blackbody

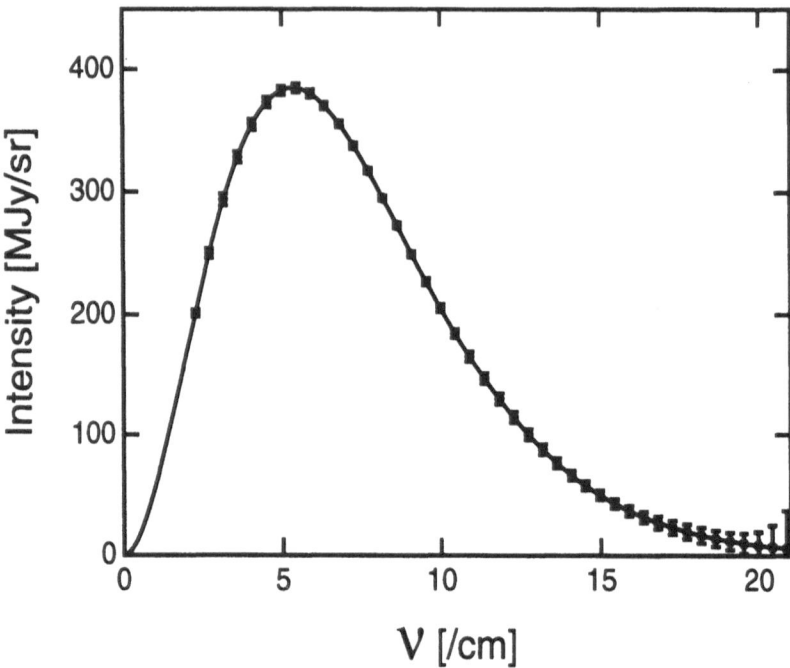

Figure 5. The spectrum of the CMB measured by FIRAS. The error bars shown are $\pm 400\sigma$. The solid line is a 2.728 K blackbody.

TABLE 3. *COBE* Measurement of T_{CMB}

Method	Temperature
GRT at sky match	2.730 ± 0.001 K
Peak of $\partial B_\nu / \partial T$	2.726 ± 0.001 K
FIRAS Dipole Spectrum	2.717 ± 0.007 K
DMR Annual Dipole	2.725 ± 0.015 K
Weighted mean $\pm 1\sigma$	2.728 ± 0.002 K

B_ν is 5×10^{-5} of the peak B_ν amplitude, the Planck function must be subtracted before plotting, for residuals to be seen (e.g., see Figure 6). In fact the data are fitted to a form

$$I_\nu = B_\nu(T_0) + \Delta T \frac{\partial B_\nu}{\partial T} + gG(\nu) \qquad (9)$$

where $G(\nu)$ is the observed spectrum of the Galactic emission and the parameters ΔT and g are adjustable to allow for a temperature correction and an unknown amount of residual Galactic emission in the darkest parts of the sky.

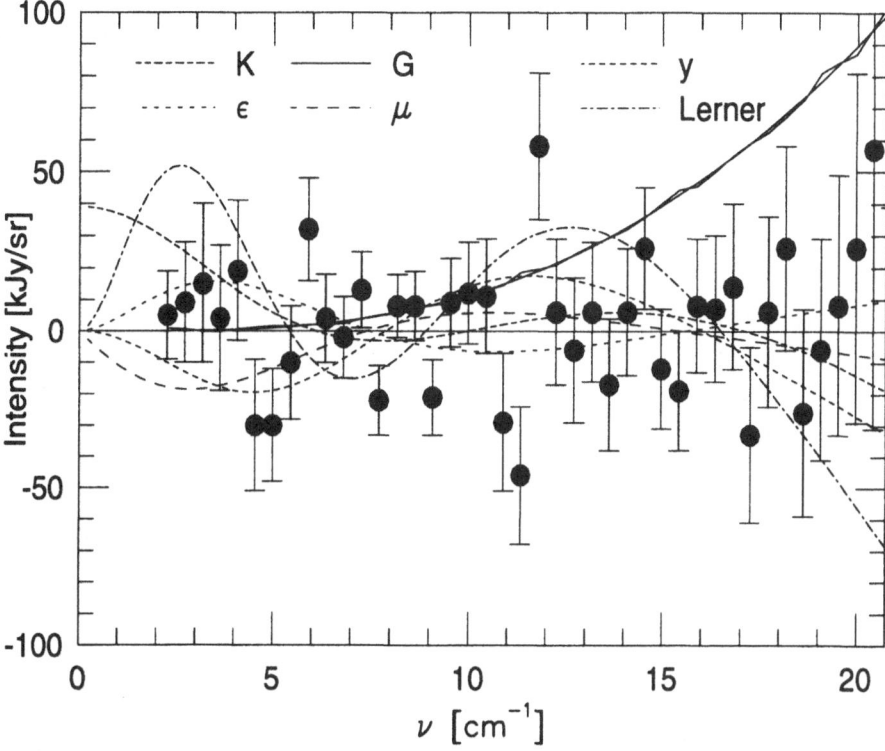

Figure 6. The residuals after subtracting a 2.728 K blackbody and the best-fitted Galactic emission plotted at their 95% confidence interval limit amplitudes. The distortions are shown in the same manner as the data: they are the residuals from a least squares fit to a temperature change and the best-fitted Galactic emission amplitude. The solid line (G) shows the observed Galactic emission and the smooth model fitting it. The curve K shows a constant intensity, I_ν, distortion. The curve labeled ϵ shows an emissivity different from unity. The curve labeled μ shows a Bose-Einstein (chemical potential) distortion. The curve labeled y shows a Compton distortion. The curve labeled "Lerner" is a fit to the 1994 FIRAS data combining the ϵ and y distortions which represented an effort to have a 'plasma' model that explained the data; but it is a poor fit to the improved data.

Using the FIRAS measured spectrum or deviations, one can fit for distortions and find the results in Table 4. The first two distortions are the Compton and chemical potential distortions discussed above. The next is allowing for an emissivity different than unity. It is clear that the CMB is extremely close to the blackbody thermal spectrum. The last ΔI_ν allows for an offset either from the sky or the instrument.

FIRAS also measures the spectrum of the dipole anisotropy which is shown here but is discussed in Section 10.

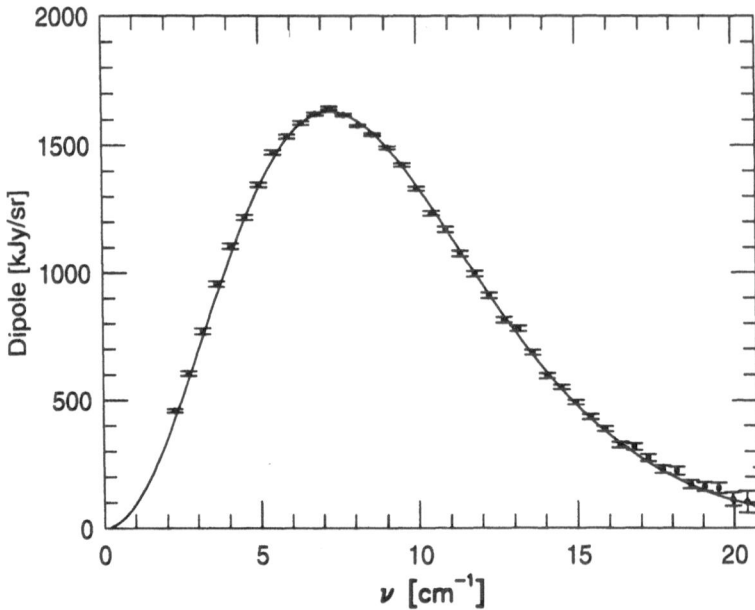

Figure 7. The spectrum of the CMB dipole as measured by FIRAS. The solid line is the derivative of a T = 2.728 K Planck function.

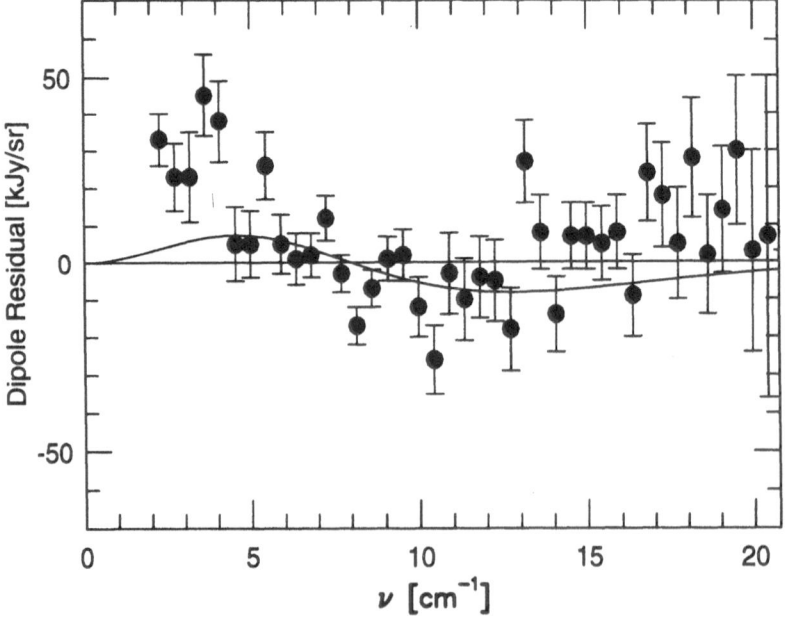

Figure 8. The CMB dipole spectrum residuals measured by FIRAS after subtracting the derivative of the T = 2.728 K Planck function. The curve shows the best fit letting both the dipole amplitude and CMB temperature vary.

TABLE 4. FIRAS Spectral Distortion Limits

Distortion	Best Fit $\pm\sigma$	95% CL Limit
y	$(-1 \pm 6) \times 10^{-6}$	1.5×10^{-5}
μ	$(-1 \pm 4) \times 10^{-5}$	9×10^{-5}
$\epsilon - 1$	$(1 \pm 5) \times 10^{-5}$	11×10^{-5}
ΔI_ν	(9 ± 15) kJy/sr	39 kJy/sr

4. Spectrum Summary

The CMB spectrum is consistent with a blackbody spectrum over more than three decades of frequency around the peak. A least-squares fit to all CMB measurements yields:

$$T_\gamma = 2.73 \pm 0.01 \text{ K}$$
$$n_\gamma = (2\zeta(3)/\pi^2)T_\gamma^3 \simeq 413 \text{ cm}^{-3}$$
$$\rho_\gamma = (\pi^2/15)T_\gamma^4 \simeq 4.68 \times 10^{-34} \text{ g cm}^{-3} = 0.262 \text{ eV cm}^{-3}$$
$$|y| < 1.5 \times 10^{-5} \quad (95\% \text{ CL})$$
$$|\mu_0| < 9 \times 10^{-5} \quad (95\% \text{ CL})$$
$$|Y_{ff}| < 1.9 \times 10^{-5} \quad (95\% \text{ CL})$$

The distortion parameter limits quoted here correspond to limits [37], [4] on energetic processes $\Delta E/E_{\text{CBR}} < 2 \times 10^{-4}$ occurring between redshifts 10^3 and 5×10^6 (see Figure 3). The best-fit temperature from the COBE FIRAS experiment is $T_\gamma = 2.728 \pm 0.002$ K [19].

5. Spectrum Interpretation and Discussion

5.1. SIGNIFICANCE OF CMB BEING PLANCKIAN

Possibly the strongest arguments for the Big Bang model are the CMB's existence and particularly its Planckian nature. This means that the CMB is both very cold and highly thermalized. Since there are roughly 10^9 photons to each baryon in the Universe, it is very difficult to produce the CMB in astrophysical processes such as the absorption and re-emission of starlight by cold dust (even iron needles) or the absorption or emission by plasmas.

All alternative models and modifications to the simplest Big Bang model produce distortions to the CMB spectrum that have a y component. It is interesting to note that any deviation from a perfectly homogeneous, isotropic, and isochronous universe causes a spectral distortion. This is a result of the fact that the sum of two blackbody spectra of different temperatures does not result in a blackbody spectrum. In the form discussed above a y distortion is simply the convolution of Planckian spectra.

Thus for example, although the energy content of the CMB is comparable to that in starlight and it is possible that dust absorption, processing, and re-emission could shift the radiation frequency to this range, it is extremely unlikely that the sum of all this radiation would just match a Planckian. If somehow the dust were optically thick on cosmological scales, it is still not possible that the sum of redshifted emission from each shell would add to a Planckian for all observers. Full arguments for dust or plasma filled universes must make use of additional observations but in general there is an inconsistency with being able to see distant extragalactic sources at many wavelengths, the observed CMB spectrum, and the Copernican Principle.

Likewise, this means that for all angular scales less than the FIRAS beam size of 7°, rms anisotropies cannot exceed about $\Delta T/T < 10^{-3}$, otherwise the superposition of temperatures would produce a $y > 10^{-5}$.

5.2. KNOWLEDGE OF T_{CMB}

The CMB temperature, T_{CMB}, is now known to a precision of 1%. This makes it the best known cosmological parameter. If we assume that the CMB spectrum is blackbody, we can calculate the number of photons in the CMB:

$$n_\gamma = (2\zeta(3)/\pi^2)T_\gamma^3 = 413(T_{CMB}/2.728 \text{ K})^3 \text{ cm}^{-3} \qquad (10)$$

It is a small change to include simple distortions provided we know their value. We can also compute the present energy density in CMB photons

$$\rho_\gamma = (\pi^2/15)T_\gamma^4 \simeq 4.68 \times 10^{-34} \text{ g cm}^{-3} = 0.262 \text{ eV cm}^{-3}. \qquad (11)$$

Since the temperature scales as $T_0 = T_i(1+z_i)$, we can calculate the photon number density, $n_{\gamma i} = (1+z_i)^3 n_\gamma$, and energy density, $\rho_{\gamma i} = (1+z_i)^4 \rho_\gamma$, for any epoch i with redshift z_i.

In the early Universe the CBR (cosmic background radiation), which is the cosmologically redshifted present day CMB radiation, dominated over the matter energy density and thus was critical to the development of the Universe. In addition most cosmological models and calculations, such as Big Bang Nucleosynthesis (BBN), are done in terms of the CBR temperature or density. In particular matter density is usually expressed in terms of the ratio either to the critical density or to the CBR density. For example, BBN gives the number density of baryons, n_b, as

$$n_b/n_\gamma = 2.3 \times 10^{-8}(\Omega_b h^2) = 5 \times 10^{-10} h^2. \qquad (12)$$

There is also the effect of the CBR/CMB on high energy cosmic rays which depends primarily on the energy density and less so on the spectrum. But the CMB implies a strong cut off of high energy protons at roughly 10^{21} eV due to the photoproduction of pions. Likewise, the existence of the CMB causes a cut off for high energy photons (and electrons/positrons) due to electron-positron pair production (Compton scattering).

5.3. LIMITS ON PROCESSES IN THE EARLY UNIVERSE

There are many possible sources of energy release or augmentation from processes occurring in the early Universe, including decay of primeval turbulence, elementary particles, cosmic strings, or black holes. The growth of black holes, quasars, galaxies, clusters, and superclusters might also convert energy from other forms.

5.3.1. *Early Generation of Stars and Reionization*

Wright et al.(1994) also give limits on hydrogen burning following the decoupling. These results depend on using geometrical arguments (a csc $|b|$ fit) to estimate the maximum amount of extragalactic energy that could have a spectrum similar to that of our own Galactic dust. We found a limit that is a factor of about 3 smaller than the polar brightness of the Milky Way. A better understanding of the Galactic dust would help produce a tighter limit on these extragalactic signals.

Consider first population III stars liberating energy that is converted by dust into far infrared light (using an optical depth of 0.02 per Hubble radius), and assume that $\Omega_b h^2 = 0.015$. In that case less than 0.6% of the hydrogen could have been burned after $z = 80$. As a second example, consider evolving infrared galaxies as observed by the IRAS. For reasonable assumptions, we found that less than 0.8% of the hydrogen could have been burned in evolving IR galaxies.

We also obtained limits on the heating and reionization of the intergalactic medium. It does not take very much energy to reionize the intergalactic medium, relative to the CBR energy, because there are so few baryons relative to CBR photons. Even the strict FIRAS limits permit a single reionization event to occur as recently as $z = 5$. More detailed calculations by Daly (1993) show that the energy required to keep the intergalactic medium ionized over long periods of time is much more substantial and quite strict limits can be obtained. If the current y limits were about a factor of 5 more strict, then it would be possible to test the ionization state of the IGM all the way back to the decoupling.

If the IGM were hot and dense enough to emit the diffuse X-ray background light, it should distort the spectrum of the CMB by inverse Compton scattering. This is a special case of the Comptonization process, with small optical depth and possibly relativistic particles. Calculations show that a smooth hot IGM could have produced less than 10^{-4} of the X-ray background, and that the electrons that do produce the X-ray background must also have a filling factor of less than 10^{-4}.

5.3.2. *Limits on Primordial Anisotropy*

Primordial perturbations will undergo energy dissipation via Silk damping which is more effective at short wavelengths where there are more oscillations. Limits on energy release are also limits on the primordial perturbation power spectrum. Hu, Scott, and Silk (1994) find an upper limit on the power spectrum index of about $n = 1.55$. It is interesting that these calculations give tighter limits than existing direct measurements, even though the spectrum is only an upper limit. These results are dependent on assuming that a power law is the correct form for the fluctuations over 7 orders of magnitude of scale. There is little possibility of observational evidence to confirm this assumption over such a wide range, since small scale fluctuations have long since been replaced by non-linear phenomena.

5.3.3. *Limits on Gravitational Energy from LSS Formation*

Together, free-free and Comptonized spectra can be used to detect the onset of nuclear fusion by the first collapsed objects. Ultraviolet radiation from the first collapsed objects is expected to photoionize the intergalactic medium. Since these objects form by non-linear collapse of rare high-density peaks in the primordial density distribution, the redshift at which they form is a sensitive probe of the statistical distribution of density peaks and the matter content of the Universe. Various models [61], [36] of structure formation predict significant ionization at redshifts ranging from $10 < z < 150$, depending on the matter content and power spectrum of density perturbations, with a "typical" value $z_{ion} \approx 50$.

5.3.4. *Limits on Particle Decay*

Exotic particle decay provides another source for non-zero chemical potential. Particle physics provides a number of dark matter candidates, including massive neutrinos, photinos, axions, or other weakly interacting massive particles (WIMPs). In most of these models, the current dark matter consists of the lightest stable member of a family of related particles, produced by pair creation in the early Universe. Decay of the heavier, unstable members to a photon or charged particle branch will distort the CMB spectrum provided the particle lifetime is greater than a year. Rare decays of quasi-stable particles (e.g., a small branching ratio for massive neutrino decay $\nu_{heavy} \rightarrow \nu_{light} + \gamma$) provide a continuous energy input, also distorting the CMB spectrum. The size and wavelength of the CMB distortion are dependent upon the decay mass difference, branching ratio, and lifetime. Stringent limits on the energy released by exotic particle decay provides an important input to high-energy theories including supersymmetry and neutrino physics [16].

5.3.5. *Limits on Antimatter-Matter Mixing*

In baryon-symmetric cosmologies matter-antimatter annihilations give rise to excessive distortions of the CMB spectrum [28].

5.3.6. *Limits on Primordial Black Hole Evaporation*

Only a very small fraction, $f = M_{planck}/M$, of matter can have formed black holes in the mass range $10^{11} \leq M \leq 10^{13}$ gm, otherwise their evaporation in the epoch preceding recombination would have resulted in excessive distortions. For smaller black holes the limit is much weaker, since for $M < 10^{11}$ gm, evaporation would have taken place during the epoch when the photon spectrum would be completely thermalized. The constraints follow from the condition that no more than all the entropy in the Universe can come from black hole evaporation so that $f < 10^9 M_{planck}/M$.

5.3.7. *Limits on Superconducting Cosmic Strings and Explosive Formation*

If they are to play an important role in large-scale structure formation, superconducting cosmic strings would be significant energy sources, keeping the Universe ionized well past standard recombination. As a result, the energy input distorts the spectrum of the CMB through the Sunyaev-Zel'dovich effect. The Compton-y parameter attains a maximum value in the range of $(1-5) \times 10^{-3}$ [43]. This is well above the observed value.

Explosive models of large-scale structure formation must create distortions in the CMB spectrum from the energy released in the shock waves. The limits on the Compton-y parameter rule out explosive models for structure on scales > 15 Mpc [33].

5.3.8. *Limits on the Variation of Fundamental Constants*

Noerdlinger [42] pointed out that the intensity of the Rayleigh-Jeans portion of the CMB spectrum gives the present values of kT, independently of the value of the Planck constant, h, while the wavelength at which the spectrum peaks gives kT in combination with h. That the two temperatures agree within errors implies that the variation of h must not have exceeded a few per cent since recombination ($z \sim 1000$). Likewise a wide variety of G-varying cosmologies predict that the CMB spectrum will follow the standard Planckian formula multiplied by an epoch-dependent factor, which in turn is related to $G(t)$ [41]. The agreement between the brightness temperature in the Rayleigh-Jeans region and the temperature determined by the peak location constrains the possible variation in the gravitational constant G. Likewise one can obtain limits on the variation in the cosmological constant (energy density of the vacuum) [50]. The shape of the spectrum also constrains the number of large spatial dimensions (taking into account the possibility of fractal dimensions) to very nearly three (± 0.02).

6. Future Observations and Results

Since FIRAS has done such a splendid job of measuring the spectrum for the bulk of the CMB energy and since at long wavelengths Galactic emission is such a serious foreground, it is at first difficult to imagine the motivation necessary to gather the resources for significant improvement. However, there are scientific motivations for improved measurements and there are experiments that one can envisage that may make worthwhile improvements in the observations of the CMB spectrum.

6.1. INTERSTELLAR/EXTRAGALACTIC MOLECULES AND ATOMS

The use of interstellar molecules, such as CN (cyanogen), offer a probe of the CBR at a remote location. There are two distinct potential scientific gains from such observations. The first is demonstrating that the CBR is universal, a thing that observations of the Sunyaev-Zeldovich effect also establishes a little more indirectly. The second is that the CBR temperature scales as $(1 + z)$ with redshift. A number of indications that this might be the case exist but I would not consider them to yet be definitive (i.e., strong enough to rule out a model like the Big Bang). The best direct upper limit is a measurement [55] of the background temperature in high-redshift primordial clouds from an experiment aimed at measuring the primordial deuterium abundance. The claimed direct measurement [56] is based upon measuring the relative populations of hyperfine states in neutral carbon atoms observed in a gas cloud at a redshift $z = 1.776$, which indicate a thermodynamic temperature of 7.4 ± 0.8 K, which is consistent with the Big Bang prediction $T(z) = (1 + z)2.73$ K which is 7.58 K.

Another recent measurement by Ge et al. [22] has measured C I again in a gas cloud at a redshift $z = 1.79$ with a result of $T(z = 1.97) = 7.9 \pm 1.0$ K at 0.61 mm. Scaling by $1 + z$, one finds the Big Bang predicted value is 8.1 K which is again consistent. With accumulating observations and understanding of excitation mechanisms these measurements provide a definite tightening of allowed regions for alternative cosmologies.

ARCADE Schematic

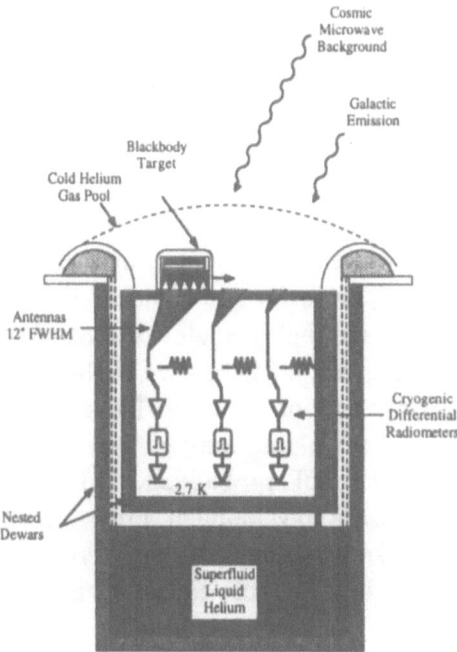

Figure 9. A schematic view of the ARCADE balloon-borne instrument

6.2. ARCADE

We consider that our long-wavelength ground-based observations have come near the fundamental limits set by the atmosphere and the Galactic foreground. Only a very great effort or a large space-based mission is likely to generate more than a very modest improvement. At the very longest wavelengths a much better understanding of the Galactic emission is required in order to make more than just a modest improvement.

However, at intermediate wavelengths - those in the centimeter (1-6 cm), it is possible to improve the spectrum measurements significantly by balloon-borne or satellite-based experiments. It takes a large effort and very precise measurements, including careful control of systematics and very good absolute calibration, to actually improve the various limits or measurements of distortion parameters such as μ and Y_{ff}. But it is possible.

ARCADE (Absolute Radiometer for Cosmology, Astrophysics, and Diffuse Emission) is a balloon-borne instrument designed to make measurements of the intermediate wavelength spectrum. A conceptual schematic drawing of the instrument is shown in Figure 9. The instrument lives in a big bucket dewar, with a second dewar nested inside to allow the aperture plane to remain cold even through it is nearly flush with the mouth of the outer dewar. Fountain-effect pumps squirt superfluid liquid helium into a reservoir under the aperture plane assembly, where it boils to keep the top plate cold (dumping the radiative heat load from the IR lines in the atmosphere). Pinholes in the aperture plane vent the boiloff gas; a set of helium-cooled flares provide a bowl filled with a "puddle" of cold helium gas. Provided the gas is colder than 20K, it is denser than ambient-temperature nitrogen and sits quietly as a transparent blocking layer between the cold optics and the warm atmosphere. The antennas are tipped 30 degrees with respect to the dewar symmetry axis, so that the dewar can remain upright (most of the time) while the antennas scan a circle 30 degrees in radius centered on the zenith. The dewar tips occasionally to scan various atmospheric columns, (i.e., different zenith angles to look through various amounts of atmosphere), but this will be disruptive to the absolute target performance, so this happens only part of the time. The anticipated measurement sensitivity is 1 mK from a balloon, limited by the ability to estimate/measure emission from the atmosphere, balloon, flight train, and Earth. ARCADE is basically a hardware development project for the eventual space mission. The design is kept such that the instrument can come off the balloon gondola and be put in a Spartan rocket with minimal changes.

6.3. DIMES

The Diffuse Microwave Emission Survey (DIMES) has been selected for a mission concept study for NASA's New Mission Concepts for Astrophysics program [32]. DIMES will measure the frequency spectrum of the CMB and diffuse Galactic foregrounds at centimeter wavelengths to 0.1% precision (0.1 mK), and will map the angular distribution to 20 μK per 6° field of view. It consists of a set of narrow-band cryogenic radiometers, each of which compares the signal from the sky to a full-aperture blackbody calibration target. All frequency channels compare the sky to the *same* blackbody target, with common offset and calibration, so that deviations from a blackbody spectral *shape* may be determined with maximum precision. Measurements of the CMB spectrum complement CMB anisotropy experiments and provide information on the early Universe unobtainable in any other way; even a null detection will place important constraints on the matter content, structure, and evolution of the Universe. Centimeter-wavelength measurements of the diffuse Galactic emission fill in a crucial wavelength range and test models of the heat sources, energy balance, and composition of the interstellar medium.

The FIRAS measurement at sub-mm wavelengths shows no evidence for Compton heating from a hot IGM. Since the Compton parameter $y \propto n_e T_e$, the IGM at high redshift must not be very hot ($T_e \sim 10^5$ K) or reionization

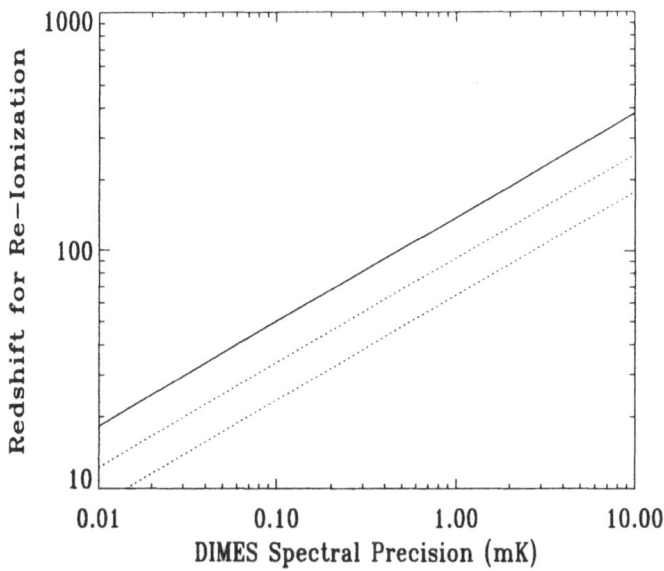

Figure 10. Upper limits to the redshift z_{ion} at which the intergalactic medium (IGM) becomes reionized, as a function of the DIMES spectral precision. The cosmologically interesting region $z_{ion} < 100$ requires precision 0.1 mK or better.

must occur relatively recently ($z_{ion} < 10$). DIMES provides a definitive test of these alternatives. Since the free-free distortion $Y_{ff} \propto n_e^2/\sqrt{T_e}$, lowering the electron temperature *increases* the spectral distortion [3]. Figure 10 shows the limit to z_{ion} that could be established from the combined DIMES and FIRAS spectra, as a function of the DIMES sensitivity. A spectral measurement at centimeter wavelengths with 0.1 mK precision can detect the free-free signature from the ionized IGM, allowing direct detection of the onset of hydrogen burning.

DIMES also provides a sensitive test for early energy releases, such as the decay of exotic heavy particles or metric perturbations from GUT and Planck-era physics.

DIMES will provide a substantial increase in sensitivity for non-zero chemical potential (Figure 11). Such a distortion arises naturally in several models. The *COBE* anisotropy data are well-described [24] by a Gaussian primordial density field with power spectrum $P(k) \propto k^n$ per comoving wave number k, with power-law index $n = 1.2 \pm 0.3$. Short-wavelength fluctuations which enter the horizon while the Universe is radiation-dominated oscillate as acoustic waves of constant amplitude and are damped by photon diffusion, transferring energy from the acoustic waves to the CMB spectrum and creating a non-zero chemical potential [10], [27]. The energy transferred, and hence the magnitude of the present distortion to the CMB spectrum, depends on the amplitude of the perturbations as they enter the

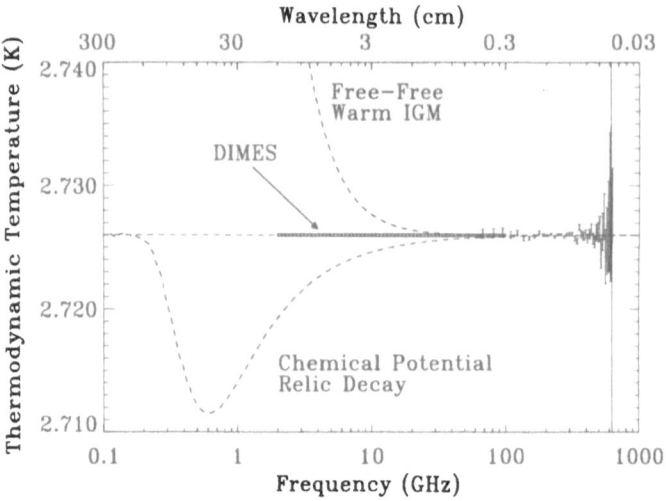

Figure 11. Current 95% confidence upper limits to distorted CMB spectra. The FIRAS data and DIMES 0.1 mK error box are also shown; error bars from existing cm-wavelength measurements are larger than the figure height. An absence of distortions at millimeter and sub-mm wavelengths does *not* imply correspondingly small distortions at centimeter wavelengths.

horizon through the power-law index n. Models with "tilted" spectra $n > 1$ produce observable distortions.

6.3.1. *Galactic Astrophysics*

Measurements of the diffuse sky intensity at centimeter wavelengths also provide valuable information on astrophysical processes within our Galaxy. Figure 12 shows the relative intensity of cosmic and Galactic emission at high Galactic latitudes. Diffuse Galactic emission at centimeter wavelengths is dominated by three components: synchrotron radiation from cosmic-ray electrons, electron-ion bremsstrahlung (free-free emission) from the warm ionized interstellar medium (WIM), and thermal radiation from interstellar dust. Despite surveys carried out over many years, relatively little is known about the physical conditions responsible for these diffuse emissions. Precise measurements of the diffuse sky intensity over a large fraction of the sky, calibrated to a common standard, will provide answers to outstanding questions on physical conditions in the interstellar medium (ISM):

• What is the heating mechanism in the ISM? Is the diffuse gas heated by photoionization from the stellar disk, shocks, Galactic fountain flows, or decaying halo dark matter?

• How are cosmic rays accelerated? Is the energy spectrum of local cosmic-ray electrons representative of the Galaxy as a whole?

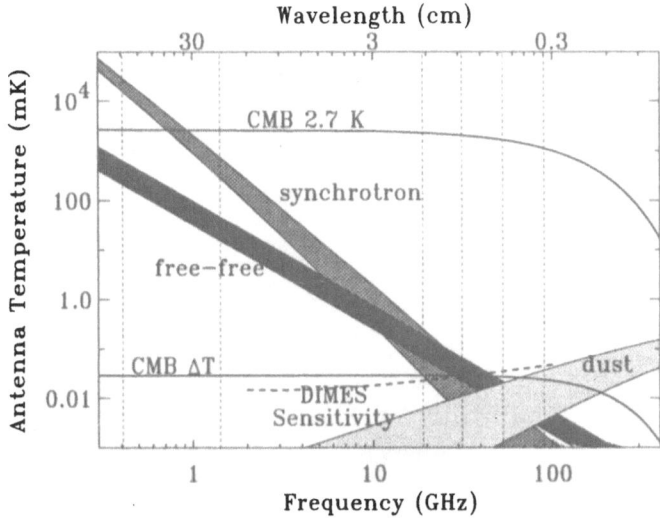

Figure 12. CMB and Galactic emission spectra. The shaded regions indicate the signal range at high latitude ($|b| > 30°$) and include the effects of spatial structure and uncertainties in the spectral index of the Galactic emission components. Solid lines indicate the mean CMB spectrum and rms amplitude of anisotropy. Vertical dashed lines indicate existing sky surveys. The DIMES sensitivity for a 6-month mission is shown.

• What is the shape, constitution, and size distribution of interstellar dust? Is there a distinct "cold" component in the cirrus?

The Galactic radio foregrounds may be separated from the CMB by their frequency dependence and spatial morphology. DIMES will map radio free-free emission from the warm ionized interstellar medium. The ratio of radio free-free emission to Hα emission will map the temperature of the WIM to 20% precision, probing the heating mechanism in the diffuse ionized gas. DIMES will have sufficient sensitivity to map the high-latitude synchrotron emission, probing the magnetic field and electron energy spectrum throughout the Galaxy. Cross-correlation with the DIRBE far-infrared dust maps will fix the spectral index of the high-latitude cirrus to determine whether the dust has enhanced microwave emissivity.

6.3.2. *Instrument Description*

Figure 13 shows a schematic of the DIMES instrument. It consists of a set of narrow-band cryogenic radiometers ($\Delta\nu/\nu \sim 10\%$) with central frequencies chosen to cover the gap between full-sky surveys at radio frequencies ($\nu < 2$ GHz) and the *COBE* millimeter and sub-mm measurements. Each radiometer measures the difference in power between a beam-defining antenna (FWHM $\sim 6°$) and a temperature-controlled internal reference load. An independently controlled blackbody target is located on the aperture

DIMES

DIFFUSE MICROWAVE EMISSION SURVEY

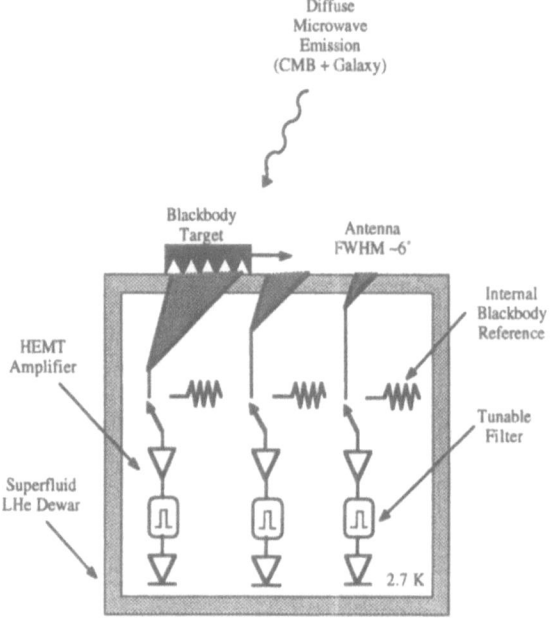

Figure 13. Schematic drawing of DIMES instrument.

plane, so that each antenna alternately views the sky or a known black-body. The target temperature will be adjusted to null the sky-antenna signal difference in the longest wavelength channel. With temperature held constant, the target will then move to cover the short-wavelength antennas: DIMES will measure small spectral shifts about a precise blackbody, greatly reducing dependence on instrument calibration and stability. The target, antennas, and radiometer front-end amplifiers are maintained near thermal equilibrium with the CMB, greatly reducing thermal gradients and drifts.

DIMES uses multiple levels of differences to reduce the effects of offset, drifts, and instrumental signatures. To reduce gain instability or drifts, each receiver is rapidly switched between a cryogenic antenna and a temperature-controlled internal reference load. To eliminate the instrumental signature, each antenna alternately views the sky or a full-aperture target with emissivity $\epsilon > 0.9999$. To maximize sensitivity to spectral shape, all frequency channels view the *same* target in progression, so that deviations from a blackbody spectrum may be determined much more precisely than the absolute blackbody temperature.

DIMES will remove the residual instrument signature by comparing the sky to an external full-aperture blackbody target. The precision achieved will likely be dominated by the thermal stability of the target. While the use of a single external target rejects common-mode uncertainties in the absolute target temperature, thermal gradients within the target or variations of target temperature with time will appear as artifacts in the derived spectra and sky maps. Thermal gradients within the external target are reduced by using a passive multiply-buffered design in which a blackbody absorber (Eccosorb CR-112, an iron-loaded epoxy) is mounted on a series of thermally conductive plates with conductance G_1 separated by thermal insulators of conductance G_2. Thermal control is achieved by heating the outermost buffer plate, which is in weak thermal contact with a superfluid helium reservoir. Radial thermal gradients at each stage are reduced by the ratio G_2/G_1 between the buffer plates. Typical materials (Fiberglass and copper) achieve a ratio $G_2/G_1 < 10^{-3}$; a two-stage design should achieve net thermal gradients well below 0.1 mK. No heat is applied directly to the absorber, and a conductive copper layer surrounds the absorber on all sides except the front: the Eccosorb lies at the end of an open thermal circuit, eliminating thermal gradients from heat flow.

DIMES will not be limited by raw sensitivity. HEMT amplifiers cooled to 2.7 K easily achieve *rms* noise 1 mK Hz$^{-1/2}$, reaching 0.1 mK sensitivity in 100 seconds of integration. The DIMES spectra are derived from comparison of the sky to the external blackbody target. The largest systematic uncertainties arise from thermal drifts or gradients within the target and emission from warm objects outside the DIMES dewar (e.g., the Earth). Thermometers buried in the microwave absorber monitor thermal gradients and drifts to precision 0.05 mK. Emission from the Earth must be rejected at the -70 dB level to avoid contributing more than 0.1 mK to the total sky signal. DIMES will achieve this rejection using corrugated antennas with 6° beam and good sidelobe response; two sets of shields between the aperture plane and the Earth provide further attenuation of thermal radiation from the Earth. *COBE* achieved -70 dB attenuation with a 7° beam and a single shield [31], so the DIMES requirement should be attainable.

DIMES will eliminate atmospheric emission completely by observing from low Earth orbit. We are currently investigating the possibility of utilizing the Spartan-400 carrier, which will provide free-flyer capability to Shuttle orbits for 700 kg instruments for a nominal mission of 6 to 9 months.

7. Monopole Spectrum Summary

The previous discussion reviews the observations, results, and future possiblities of the spectrum of the total CMB power. In the next sections we consider the expected signal for a Planckian spectrum for the monopole, dipole and higher order anisotropies and how spectral distortions would appear in the frequency spectrum of various anisotropies.

8. Planckian Radiation Formula

The specific intensity, I_ν, of light is defined as the incident energy per unit area, per unit solid angle, per unit frequency.

$$I_\nu = \frac{2h\nu^3}{c^2}n \tag{13}$$

where h is Planck's constant, ν is the frequency, c is the speed of light, and n is the photon occupation number per mode. The intensity or spectral brightness of a blackbody is a function of only one parameter, the temperature

$$B_\nu(T) = \frac{2h\nu^3}{c^2}\frac{1}{e^x - 1} \tag{14}$$

where $x = h\nu/kT$. In the Rayleigh-Jeans region $x \ll 1$ and thus

$$B_\nu(T) = \frac{2\nu^2}{c^2}kT. \tag{15}$$

The generalization of Equation (15) to any x defines the antenna temperature of a blackbody

$$B_\nu(T) = \frac{2\nu^2}{c^2}kT_{ant}(\nu). \tag{16}$$

Rewriting Equation (16) yields the relation between antenna and thermodynamic temperature

$$T_{ant}(\nu) = \frac{h\nu/k}{e^x - 1} = T\frac{x}{e^x - 1}. \tag{17}$$

In the Rayleigh-Jeans portion of a blackbody spectrum the antenna temperature and the thermodynamic temperature are equal ($T_{ant} = T$). Taking the derivative of Equation (17) one obtains the relation between antenna and thermodynamic temperature *differences*

$$\frac{dT_{ant}}{dT} = \frac{x^2 e^x}{(e^x - 1)^2} \tag{18}$$

where here $x = h\nu/kT_o$. The temperature difference conversion depends on a knowledge of T_o while Equation (16) does not. For example plugging 31.5, 53 and 90 GHz into Equation (18) with $T_o = 2.73$ K, we get the conversion factors 1.026, 1.074, 1.227 respectively.

9. Dipole Formulae

Observers with velocity $\vec{\beta} = \vec{v}/c$ through a Planckian radiation field of temperature T_o will measure directionally dependent temperatures,

$$T_{obs}(\theta) = T_o \frac{(1 - \beta^2)^{1/2}}{(1 - \beta\mu)} \tag{19}$$

where $\mu = cos\theta$ and θ is the angle between $\vec{\beta}$ and the direction of observation as measured in the observer's frame [47].We expand this through order β^3 to show that the dipole is the largest member of a family of kinetic anisotropies,

$$\frac{\Delta T}{T_o} = \beta\mu + \frac{\beta^2}{2}(2\mu^2 - 1) + \frac{\beta^3}{4}(4\mu^3 - 2\mu) + O(\beta^4). \tag{20}$$

or

$$\frac{\Delta T}{T_o} = \beta cos\theta + \frac{\beta^2}{2}cos2\theta + \frac{\beta^3}{4}(4\mu^3 - 2\mu) + O(\beta^4). \tag{21}$$

The antenna temperatures of the CMB and the normalizing quadrupole amplitude are plotted in Figure 12.

In the more general case of non-Planckian spectra I_ν we can define an equivalent antenna temperature by

$$I_\nu = \frac{2\nu^2}{c^2} kT_{ant}, \tag{22}$$

which when combined with Equation (18) yields

$$\frac{\Delta I}{I_o} = \frac{\Delta T_{ant}}{T_{ant}} = \frac{\Delta T}{T_o} \frac{xe^x}{(e^x - 1)}, \tag{23}$$

where I_o is an isotropic but not necessarily Planckian radiation field as seen by an observer in the rest frame of the field.

10. The Dipole Anisotropy and Distortions of the CMB Spectrum

The generalization of Equation (21) for motion through an isotropic but not necessarily Planckian radiation field of intensity $I_o(\nu)$ yields an observed intensity anisotropy,

$$\frac{\Delta I}{I_o}(\nu, \theta) = \frac{I_{obs}(\nu, \theta) - I_o(\nu)}{I_o(\nu)}. \tag{24}$$

where ν is the frequency in the observer's frame and I_o is the intensity in the rest frame of the radiation. The result to third order in β is [34]

$$\frac{\Delta I}{I_o} = \beta\mu(3 - \alpha_1) + \frac{\beta^2}{2}\left[2\mu^2(6 - 3\alpha_1 + \frac{\alpha_2}{2}) - (3 - \alpha_1)\right]$$
$$+ \frac{\beta^3}{4}\left[4\mu^3(10 - 6\alpha_1 + \frac{3}{2}\alpha_2 - \frac{1}{6}\alpha_3) - 2\mu(9 - 5\alpha_1 + \alpha_2)\right] \tag{25}$$

where $\alpha_n = \frac{\nu^n}{I(\nu)}\frac{\partial^n I(\nu)}{\partial \nu^n}$. A pedagogical check of this formula can be made by noticing that for a Planckian spectrum $\Delta I/I_o = \Delta T_{ant}/T_{ant} = \frac{xe^x}{(e^x-1)}\Delta T/T_o$, where T_{ant} is antenna temperature and $x = h\nu/kT_o$. In the Rayleigh-Jeans limit, $\alpha_1 = \alpha_2 = 2$, $\alpha_3 = 0$ and one obtains $\Delta I/I_o = \Delta T/T_o$. An analogous simplification does not occur in the Wien limit because of the ν dependence of the derivatives α_n. Another check is that an $I \propto \nu^3$ non-Planckian spectrum yields no kinetic anisotropy since I/ν^3 is a Lorentz invariant. For this case, $\alpha_1 = 3$ and $\alpha_2 = \alpha_3 = 6$.

Summary The frequency dependence of the dipole anisotropy provides a means to determine the CMB temperature and to detect CMB spectral distortions. In particular accurate measurements of the CMB dipole anisotropy at multiple wavelengths may help in limiting or detecting small spectral distortions. On the other hand accurate spectral measurements are needed for a precise interpretation of the observed anisotropy. It is important to make measurements at as many wavelengths as possible.

10.1. INTRODUCTION TO DIPOLE ANISOTROPY SPECTRUM

The dipole anisotropy has been measured well at many wavelengths, particularly by the *COBE* DMR and FIRAS instruments. Prior to that, several experiments also measured the dipole anisotropy amplitude and direction.

The most obvious interpretation of the dipole anisotropy is in terms of the peculiar velocity of the Solar System; on the other hand it might result from a combination of very long wavelength primordial perturbations [64] [63] [46]. We can certainly expect that on the order of 1% of the dipole anisotropy is due to primordial anisotropies based upon a simple extrapolation of the observed anisotropy power spectrum.

Assuming that the observed dipole anisotropy results primarily from the Doppler shift due to the peculiar motion of the Solar System, small spectral distortions must give rise via the Compton-Getting effect to a characteristic frequency dependence of the dipole amplitude arising from the shape of the spectral distortions.

10.2. THE COMPTON-GETTING EFFECT

The Compton-Getting effect is, in its original formulation [7], the 24-hour variation in the cosmic ray intensity due to the peculiar velocity of the Earth. This effect is easily generalized as it is a straight consequence of the Lorentz invariance of the distribution functions of the particles and photons in phase space (see [21] for a comprehensive discussion).

An observer with velocity $\mathbf{v}$ ($\beta = v/c$) with respect to the reference frame in which the photon distribution function $n(\nu)$ is isotropic to at least first order in β will measure a difference between the intensity received in the direction of motion and that received in a direction perpendicular to its motion proportional to

$$\frac{\Delta n}{n} = \frac{d \ln n}{d \ln \nu}\beta. \tag{26}$$

Thus measurements of the dipole anisotropy of the CMB intensity yield information on the slope of the spectrum.

To first order in β the dipole anisotropy of the CMB intensity is

$$\begin{aligned} T_d &= \frac{h\nu}{k}\left[\frac{1}{\ln(1+1/n(\nu))} - \frac{1}{\ln(1+1/n(\nu[1+\beta]))}\right] \\ &\approx -\frac{h\nu}{k(1+n)}\ln^{-2}(1+\frac{1}{n})\frac{d \ln n}{d \ln \nu}\beta \end{aligned} \tag{27}$$

In the case of a Planckian spectrum $[n = (exp(x) - 1)^{-1}; x = h\nu/kT]$ the temperature anisotropy is independent of frequency and

$$\frac{T_d}{T} \approx \frac{v}{c} = \beta. \tag{28}$$

Deviations from a Planckian spectrum, however, lead to a dependence of the dipole anisotropy amplitude, T_d, specific to the shape of the distortion. Define δ, the first order fractional change in the dipole anisotropy amplitude, to be

$$\delta \equiv \frac{\Delta T_d}{T_{do}} \approx [\frac{T_d}{T} - \beta]\beta^{-1}. \tag{29}$$

Now we can calculate and plot the fractional change in dipole amplitude δ from predicted potential distortions.

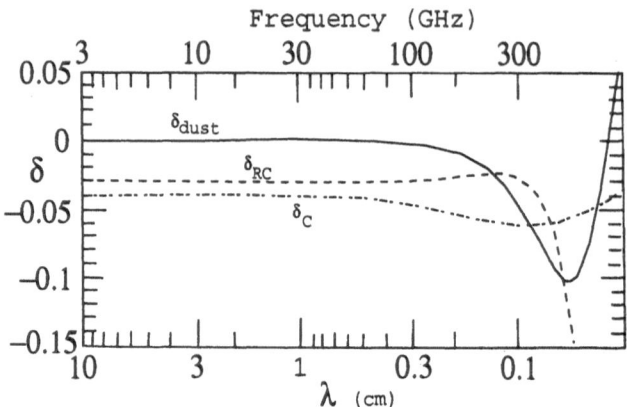

Figure 14. Predicted fractional variation of the dipole amplitude ($\delta = [(T_d/T) - \beta]/\beta$) for models with non-relativistic Comptonization (C - dot-dashed line), relativisitc Comptonization (RC - dashed line), and cold dust (dust - solid line) producing a sub-mm excess.

10.3. APPLICATION TO POTENTIAL DISTORTIONS

The three types of spectral distortions normally discussed are: Comptonization distortion, Bose-Einstein distortion, and free-free distortion. In addition it is sometimes pointed out that there are some very low-level distortions expected from the final stages of recombination. Finally, it is possible that there is a generic distortion caused by effects which have not been anticipated, calculated, or otherwise expected. We can make estimates of these also.

10.3.1. *Comptonization Distortion*
The first order approximation to the photon occupation number for a Comptonized spectrum is

$$n_c = n_p \left[1 + ux exp(x) n_p \left(\frac{x}{tanh(x/2)} - 4 \right) \right] \tag{30}$$

where the parameter $u = k(T_e - T_\gamma)/m_e c^2$ is a measure of the amount of extra energy injected into the radiation field. Figure 14 shows the dipole deviation, δ spectra predicted for such distortions.

10.3.2. *Bose-Einstein Distortion*
In a Bose-Einstein or chemical potential distortion the photon occupation number n is

$$n = \frac{1}{e^{x+\mu_0} - 1}, \tag{31}$$

where $x \equiv h\nu/kT$ and μ_0 is the dimensionless chemical potential. The chemical potential is predicted to be frequency-dependent

$$\mu(x) = \mu_0 e^{-2x_b/x}, \tag{32}$$

where x_b is the transition frequency at which Compton scattering of photons to higher frequencies is balanced by free-free creation of new photons. The resulting spectrum has a sharp drop in brightness temperature at centimeter wavelengths [5] with a minimum at $\lambda_{min} \simeq 4.5\, \Omega_b^{-5/8}$ cm. Thus the minimum wavelength is determined by Ω_b.

We can use this expression for the photon occupation number in the formula for the dipole anisotropy amplitude and find the fractional variation in the dipole anisotropy, δ, for the various possible values of energy release $\mu_0 \simeq 1.4 \Delta E/E_{CBR}$ and other cosmological parameters, i.e., Ω_b.

To first order the deviation is proportional to μ

$$\frac{T_d}{T} \approx \beta \frac{x^2}{(x+\mu)^2}(1 + \mu \frac{2x_b}{x^2}), \tag{33}$$

where the second term in the parenthesis is generally small so that

$$\delta = -\frac{2\mu x + \mu^2}{(x+\mu)^2} = -2\mu \frac{x + \mu/2}{(x+\mu)^2}. \tag{34}$$

10.3.3. Free-Free Distortion

Thermal bremsstrahulung from an ionized intergalactic medium distorts the observed CMB spectrum changing the temperature by an amount

$$\Delta T_{ff} = T_\gamma\, Y_{ff}/x^2, \tag{35}$$

where T_γ is the undistorted photon temperature, x is the dimensionless frequency, and Y_{ff}/x^2 is the optical depth to free-free emission. The predicted distortion is shown in Figure 16.

10.3.4. Recombination Line Distortion

Since there are on the order of 10^9 CMB photons per baryon, recombination does not have a large effect on the CMB spectrum. However, there are small features ($10^{-4} - 10^{-8}$) that result from atomic lines. The best known and calculated are the Lyman-α lines for hydrogen which appear deep in the Wien region [48]. The result is a step at the high-frequency side of the Lyman-α resonance (divided by the redshift $z \lesssim 1100$ of recombination). There is a slight smearing due to the natural line width set by atomic

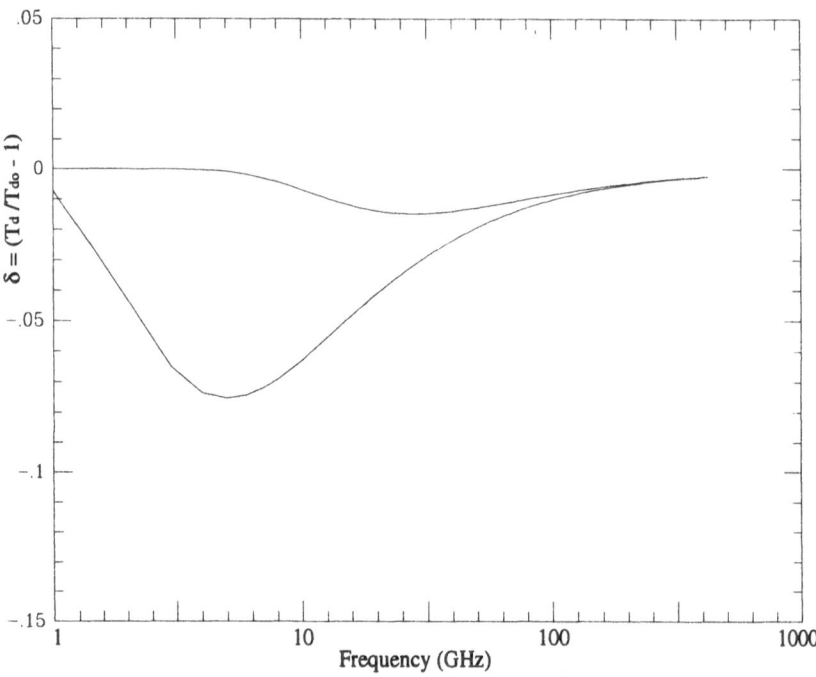

Figure 15. Fractional deviation in the dipole anisotropy amplitude due to the Doppler shift for two cases of a Bose-Einstein (chemical potential) distortion.

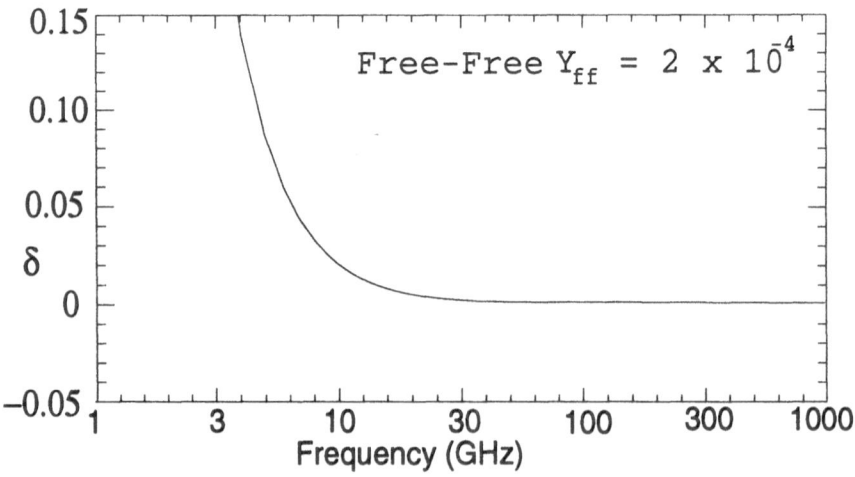

Figure 16. Fractional deviation in the dipole anisotropy amplitude from a free-free emission distortion with $Y_{ff} = 2 \times 10^{-4}$.

parameters and the thermal motion. The dominant effect is the cosmological expansion redshift which pulls photons from the low frequency side and deposits them on the high frequency side. The final result is a slight step down at the highest frequency at which the resonance was effective.

There are hydrogen resonances at lower frequencies, not only bound-free but also bound-bound transitions, that are manifest in the radio and mm wavelength range. It appears difficult to detect these with an absolute measurement even using a frequency switching system without a very substantial effort. It is likely that using a narrow-bandwidth or spectral receiver observing the dipole anisotropy is a more effective way to observe such a line. The calibration of either such system requires a great deal of care.

10.3.5. *Unanticipated Distortions*

It is always possible that there are spectral distortions that do not fall in the categories discussed above. In particular, it is quite possible that astrophysical or particle decay/interaction effects could alter the photon occupation number at long wavelengths not yet measured precisely.

Although the precise *COBE* measurements carry implications for possible distortions at longer wavelengths, the absence of distortions near the peak CMB intensity does *not* imply correspondingly small distortions at longer wavelengths. Distortions as large as 5% could exist at wavelengths of several centimeters or longer without violating existing observations.

11. SZ Measurements as a Probe of Spectral Distortions

The Sunyaev-Zeldovich effect [60] in the direction of rich clusters of galaxies provides another probe of the CMB spectral shape by means of differential measurements ([23], [17], [51], [66], [52]). The change in the CMB brightness temperature or intensity is essentially a second order Doppler effect. The amplitude of the effect is proportional to the second derivative of the intensity at the frequency of observation:

$$\frac{\partial I}{\partial y_C} = \frac{\partial^2 I}{\partial (lnx)^2} - 3\frac{\partial I}{\partial \ln x} , \qquad (36)$$

where y_C is the Comptonization parameter of the cluster and $x = h\nu/kT_0$. If the intensity ($I \propto x^\alpha$) is locally a power law with exponent α, then $\partial I/\partial y_C \propto \alpha(\alpha - 3)$. In the Rayleigh-Jeans region, $\alpha = 2$; it then decreases with increasing frequency and becomes negative in the Wien region. The Sunyaev-Zeldovich effect changes signs around the CMB spectrum peak. The spectrum of the SZ effect is sensitive to the detailed shape of the original CMB spectrum. Figure 17 shows examples of the predicted effect. Two

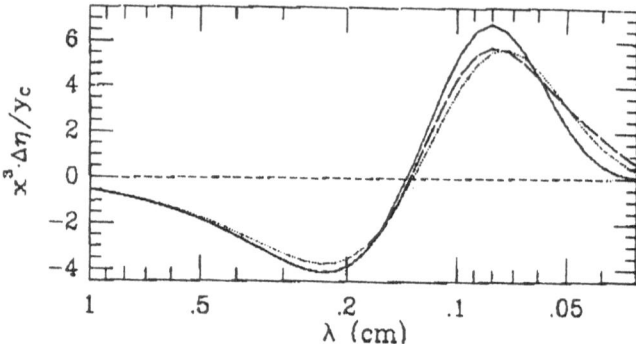

Figure 17. The predicted Sunyaev-Zeldovich effect corresponding to a Planck spectrum (solid line) and two spectral distortions: non-relativisitc Comptonization (dotted line) and a cold dust emission (long dashes) creating a sub-mm excess. For the ordinate $x = h\nu/kT$; $x^3 \Delta n$ is a quantity proportional to the change of the CMB intensity, y_C is the Comptonization parameter for the cluster.

things, in addition to observational noise and errors, act to confuse. The first confusion is that the shape of the SZ effect is slightly modified by the temperature distribution of the hot electrons in the galaxy cluster [52]. The second confusion is any local cluster or foreground emission contributions to the observed intensity. Fortunately, foreground emissions will not have a dipole pattern or SZ effect and measurements of this kind can be used to separate out extragalactic contributions to the observed flux.

12. CMB Anisotropy Frequency Spectrum

Given the precise observations of the monopole and dipole frequency spectrum, we can confidently predict the frequency spectrum of higher order CMB anisotropies. The frequency spectrum should be the same as that for the dipole anisotropy (except for the special case of the thermal SZ effect). This is a fundamental assumption underlying techniques for separating the observations of the microwave sky into its CMB and foreground components.

We can ask, based on the COBE data, how well is this assumption verified. It turns out that FIRAS alone does not have sufficient resolution to measure the higher order anisotropy frequency spectrum. That is FIRAS can readily measure the dipole frequency spectrum but is not able to measure that of the quadrupole, octopole, etc. on its own. However, if the FIRAS observations are cross-correlated with the DMR observations, then one can make an estimate of the anisotropy frequency spectrum (see Figure 18) [20]. This technique can and has been used with external experiments such as FIRS, Tenerife, Saskatoon and will be used with future observa-

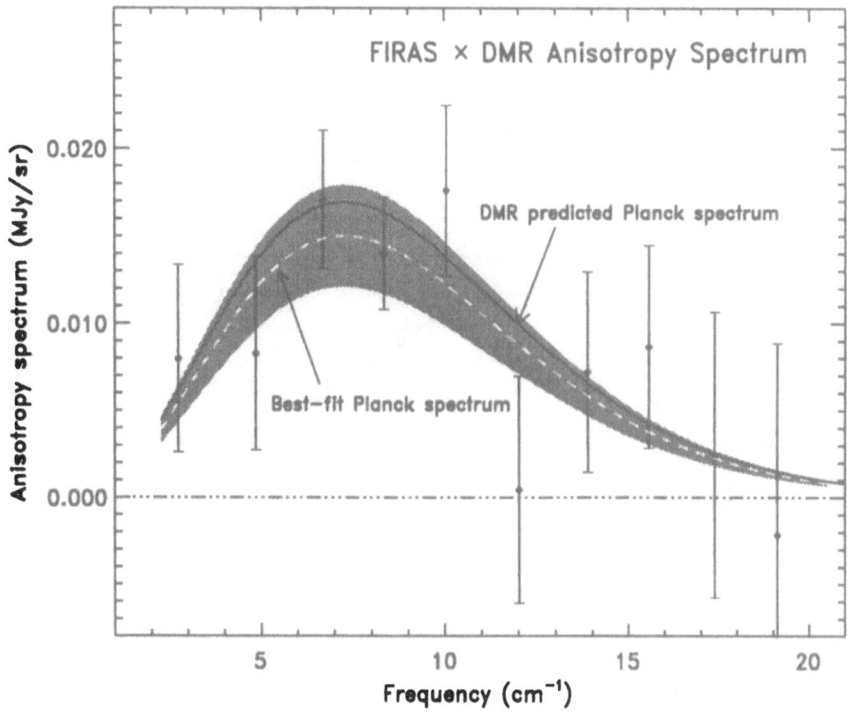

Figure 18. The frequency spectrum of primordial CMB distortions determined by cross-correlation of the COBE DMR and FIRAS data sets. The points with errors are the results of correlating the DMR CMB anisotropy map with the FIRAS map generated by removing the monopole, dipole, and an estimate of the Galactic dust emission. The thin line is the predicted spectrum based upon the DMR data alone and the assumption of a precisely Planckian CMB spectrum. The band with centered line is the best-fitted Planckian CMB spectrum to the FIRAS-DMR cross-correlation.

tions; but these other observations are currently much more limited than FIRAS in frequency sampling.

The observations of the CMB thermal spectrum and the frequency spectra of anisotropies point to a precisely thermal spectrum for the CMB. Figure 19 [20] shows the three levels of frequency spectra: monopole, dipole, higher order anisotropies left after Galactic dust emission is removed.

13. Acknowledgments

This work was supported in part by the Director, Office of Energy Research, Office of High Energy and Nuclear Physics, Division of High Energy Physics of the U.S. Department of Energy under contract No. DE-AC03-76SF00098.

306

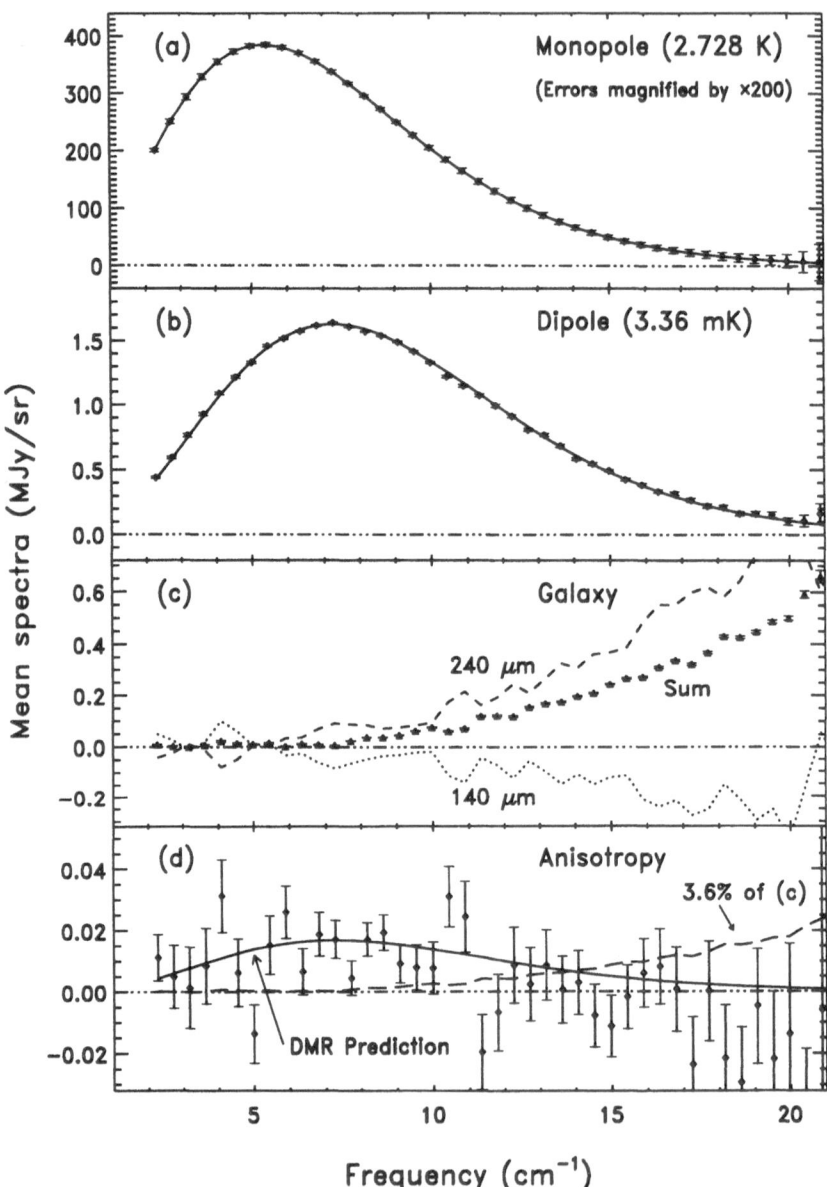

Figure 19. Frequency Spectrum of CMB features: (a) The CMB monopole frequency spectrum as measured by FIRAS and a line for a Planckian. (b) The CMB dipole frequency spectrum as measured by FIRAS and a line indicating the anisotropy frequency spectrum expected for a Planckian. (c) The frequency spectrum of Galactic dust emission removed from the FIRAS spectrum. (d) The CMB higher-order anisotropy frequency spectrum as measured by FIRAS and a line indicating the anisotropy frequency spectrum expected for a Planckian.

References

1. W.S. Adams Ap. J., 93, 11 (1941)
2. R.A. Alpher and R.C. Herman, Physics Today, Vol. 41, No. 8, p. 24 (1988)
3. J.G. Bartlett and A. Stebbins, Ap. J., 371, 8 (1991)
4. M. Bersanelli et al., Ap. J., 424, 517 (1994)
5. C. Burigana, L. Danese, and G.F. De Zotti, Astron. & Astrophysics, 246, 49 (1991)
6. M.T. Ceballos and X. Barcons, MNRAS, **271**, 817 (1994)
7. Compton, A.H., Getting, I.A., Phys. Rev., 47, 817 (1935)
8. P. Crane, D.J. Hegyi, N. Mandolesi, & A.C. Danks, Ap. J., 309, 822 (1986)
9. P. Crane, D.J. Hegyi, M.L. Kutner, & N. Mandolesi, Ap. J., 346, 136 (1989)
10. R. Daly, Ap. J., 371, 14 (1992)
11. Daly, R.A., Ap. J., 406, 47 (1993)
12. Danese, L. & De Zotti, G., Astron. & Astrophysics, 84, 364 (1981)
13. L. Danese and G.F. De Zotti, Astron. & Astrophysics, 107, 39 (1982)
14. G. De Zotti, ppnp, 17, 117 (1987)
15. R.H. Dicke, P.J.E. Peebles, P.G. Roll, and D.T. Wilkinson, Ap. J., 142, 414 (1965)
16. John Ellis, G.B. Gelmini, Jorge L. Lopez, D.V. Nanopoulos, Subir Sarkar, Nucl. Phys. B, 373, 399 (1992)
17. R. Fabbri, F. Melchiorri, & V. Natale, Astropysics & Space Science, 59, 223 (1978)
18. R. Fabbri, Astropysics & Space Science 77, 529 (1981)
19. D.J. Fixsen et al., Ap. J., 473, 576 (1996)
20. Fixsen D.J., Hinshaw, G., Bennett, C.L., & Mather, J.C., Ap. J., in press (1997), astro-ph/9704176.
21. Forman, M.A., Planet. Space Sci. 18, 25 (1970)
22. Ge et al., Ap. J., 474, 67 (1997)
23. R.J. Gould & Y. Rephaeli, Ap. J.219, 12 (1978)
24. Górski K.M., Banday, A.J., Bennett, C.L., Hinshaw, G., Kogut, A., Smoot, G.F., & Wright, E.L., Ap. J., 464, L11 (1996)
25. Gush et al., Phys. Rev. Lett, 65, 537 (1990)
26. W. Hu and J. Silk, Phy. Rev.170, 2661 (1993)
27. W. Hu, D. Scott, & J. Silk, Ap. J.430, L5 (1994)
28. B.J.T. Jones & G. Steigman, MNRAS, 183, 585 (1978)
29. A. Kogut, S.J. Petuchowski, C.L. Bennett, & G.F. Smoot, Ap. J., 348, L45 (1990)
30. A. Kogut et al., Ap. J.419, 1 (1993)
31. A. Kogut et al., Ap. J.470, 1? (1996)
32. A. Kogut, Moriond CMB Conference Proceedings (1997)
33. J.J. Levin, K. Freese & D.N. Spergel, Ap. J.389, 464 (1992)
34. C. Lineweaver, G. F. Smoot, L. Tenorio, & A. Kogut, Astrophysical Letters & Communications 32, 173 (1995)
35. C. Lineweaver et al., Ap. J., 470, 38 (1996) (astro-ph/9601151)
36. Liddle, A.R. & Lyth, D.H., MNRAS, 273, 1177 (1995)
37. J.C. Mather et al., Ap. J., 420, 439 (1994)
38. A. Mckellar, Publ. Dominion Astrophys. Observatory, 1, 251 (1941)
39. D.M. Meyer & M. Jura, Ap. J., 297, 119 (1985)
40. D.M. Meyer, K.C. Roth, & I. Hawkins, Ap. J., 343, L1 (1989)
41. J.V. Narlikar & N.C. Rana, Phys. Lett. 77A, 219 (1980)
42. P.D. Noerdlinger, Phy. Rev. Let. 30, 761 (1973)
43. J.P. Ostriker & C. Thompson, Ap. J., 323, L97 (1987)
44. E. Palazzi et al., Ap. J., 357, 14 (1990)
45. E. Palazzi, N. Mandolesi, & P. Crane, Ap. J., 398, 53 (1992)
46. Paczynski & Piran, Ap. J., 364, 314 (1990)
47. P.J.E. Peebles & D.T. Wilkinson, Phys. Rev., 174, 2168 (1968)
48. P.J.E. Peebles, "Principles of Physical Cosmology," Princeton U. Press, p. 168 (1993)

49. A.A. Penzias and R. Wilson, Ap. J., 142, 419 (1965)
50. M. D. Pollock, MNRAS, 193, 825 (1973)
51. Y. Rephaeli, Ap. J., 241, 858 (1980)
52. Y. Rephaeli, Ann. Rev. Astron. & Astrophysics, 33:541-579 (1994)
53. K.C. Roth, D.M. Meyer, & I. Hawkins, Ap. J., 420, L67 (1993)
54. K.C. Roth & D.M. Meyer, Ap. J., 441, 129 (1995)
55. A. Songaila et al., Nature, 368, 599 (1994)
56. A. Songaila et al., Nature, 371, 43 (1994)
57. Smoot, G.F., Gorenstein, M.V., Muller, R.A., Phys. Rev. Let., 39, 898 (1977)
58. Sarkar, S., & Cooper, A.M., Phys. Lett. 148B, 347 (1984)
59. R.A. Sunyaev and Ya.B. Zel'dovich, Ann. Rev. Astron. & Astrophysics, 18, 537 (1980)
60. R.A. Sunyaev and Ya.B. Zel'dovich, Comm. Astrophys. Space Physics, 4, 173 (1972)
61. Tegmark, M., Silk, J., & Blanchard, A., Ap. J., 420, 484 (1994)
62. P. Thaddeus, Ann. Rev. Astron. & Astrophysics, 10, 305 (1972)
63. M. Turner, Phys. Rev. D. 44, 3737 (1991)
64. Wilson, M.L. & Silk, J., Ap. J., 243, 14 (1981)
65. E.L. Wright, Ap. J., 232, 348 (1979)
66. E.L. Wright, Ap. J., 232, 348 (1979)
67. E.L. Wright et al., Ap. J., 420, 450 (1994)

CMB Spectrum References

LBL-Italy: G. F. Smoot et al., Phy. Rev., 151,,1099,(1983)
M. Bensadoun et al., Ap. J., 409, 1 (1993)
M. Bersanelli et al., Ap. J., (1994)
M. Bersanelli et al., Astro Lett and Communications 32, 7 (1996)
De Amici, G. et al., Ap. J., 381, 341 (1991)
N. Mandolesi et al., Ap. J., 310, 561 (1986)
G. Sironi, G. Bonelli, and M. Limon, Ap. J., 378, 550 (1991)
FIRAS: J. C. Mather et al., Ap. J., 432, L15 (1993)
D. Fixsen et al., Ap. J., 420, 445 (1994)
D. Fixsen et al., Ap. J., 473, 576, (1996)
DMR: Kogut et al., Ap. J., 419, 1 (1993)
A. Kogut et al., Ap. J., submitted (1996)
Princeton: Staggs, S. et al., Astrophys. Lett & Comm, 32, 99 (1995)
D. G. Johnson and D. T. Wilkinson , Ap. J., 313, L1 (1987)
UBC: H. P. Gush, M. Halperin, and E. H. Wishnow Phy. Rev. Let., 65, 537 (1990)
Cyanogen: D. M. Meyer et al., Ap. J., 297, 119 (1985)
E. Palazzi, et al., Ap. J., 357, 14 (1990)
Staggs Staggs et al. 1996, Ap. J., 473, L1
Staggs et al. 1996, Ap. J., 458, 407

INFLATION AND THE COSMIC BACKGROUND RADIATION

What every cosmologist needs to know

MICHAEL S. TURNER
Enrico Fermi Institute
5640 South Ellis Avenue
The University of Chicago
Chicago, IL 60637-1433 USA

Theoretical Astrophysics
Fermi National Accelerator Laboratory
Box 500
Batavia, IL 60510-0500 USA

Abstract. The Cosmic Background Radiation (CBR) is the cleanest and most powerful test of inflation. Because there is no standard model of inflation it also has the potential to reveal much about the underlying model of inflation. In this Chapter I carefully review the predictions of inflation, emphasizing their implications for CBR anisotropy, discuss reconstruction of the inflationary potential, and summarize the observational status of inflation.

1. Inflation is a Bold and Expansive Paradigm

The hot big-bang cosmology is very successful. It provides a physical description of the Universe from about 10^{-2} sec onward [1]. However, it raises fundamental questions about initial conditions: the origin of the smoothness and flatness of our Hubble volume, the small (one part in 10^5) density inhomogeneities needed to seed all the structure seen in the Universe today, and the tiny baryon asymmetry that results in the existence of matter today.

Inflation explains how a region of size much, much greater than our Hubble volume could have become smooth and flat [2] as well as the origin of the density inhomogeneities needed to seed structure [3]. With regard to the smoothness and flatness, inflation is a temporary fix: it does not

C. H. Lineweaver et al. (eds.), The Cosmic Microwave Background, 309–344.

guarantee that the observable Universe in the exponentially distant future will be isotropic and homogeneous [4].

Models of inflation are based upon well-defined, albeit speculative physics – usually the semi-classical evolution of a weakly coupled scalar field. The physics is speculative because a) there is no evidence for the existence of even a single fundamental scalar field and b) the energy scale associated with inflation is typically much greater than 1 TeV and in most models around 10^{14} GeV.

I believe that it is fair to say that inflation has revolutionized the way cosmologists view the Universe. It leads to the current working hypothesis for an extension of the standard cosmology: the inflation/cold dark matter paradigm. This paradigm has the potential to extend the standard cosmology back to times as early as 10^{-32} sec and address almost all the pressing questions in cosmology. The key elements of this paradigm are: flat universe, non-baryonic dark matter in the form of slowly moving elementary particles (cold dark matter), and nearly scale-invariant, adiabatic density perturbations. As I will emphasize, the inflation/cold dark matter paradigm is highly testable and a flood of observations are doing so. At the outside, within the next decade this paradigm will have been falsified or more firmly established.

There are even grander implications of inflation, albeit very difficult to test [5]. Cosmologists have long used the Copernican principle to argue that the entire Universe must be smooth because of the smoothness of our Hubble volume. In the post-inflation view, our Hubble volume is smooth because it is a small part of a region that underwent inflation, and thus it need not reflect the large-scale features of the Universe as a whole. On the largest scales the structure of the Universe is likely to be very rich: different regions may have undergone different amounts of inflation, beginning at different times; some regions may not have undergone inflation and may have collapsed to black holes; other regions may be governed by different realizations of the laws of physics because they evolved into different vacuum states of equivalent energy. It is likely that most of the volume of the Universe is still undergoing inflation and that inflationary patches are being constantly produced (eternal inflation). In this case, "the age of the Universe" is a meaningless concept: our expansion age merely measures the time back to the end of our inflationary event.

If inflation is correct, it will be a major advance in our understanding of the origin and evolution of our Hubble volume and it will open a new window on physics beyond the standard model of particle physics. It is possible that inflation is just plain wrong, and over the years other explanations will be put forth to address the dilemma of the initial data. For example, Penrose has suggested the smoothness and flatness of the Universe has to do with the nature of initial singularities [6]. It has, however, been shown that any microphysical solution to the horizon and flatness problems must involve the two key elements of inflation – superluminal expansion and entropy production – suggesting to me that inflation or something closely related is likely to be the correct explanation [7].

2. There is No Standard Model of Inflation

It would be nice if there were a standard model of inflation, but there isn't. Because inflation involves physics beyond the standard model of particle physics and is probably too tied to fundamental physics at energies of $\mathcal{O}(10^{14}\,\text{GeV})$ this is not surprising. What is important, is that inflationary models make three robust predictions (see next section) which allow the paradigm to be decisively tested. Moreover, cosmological measurements should also be able to discriminate between different models (see final section).

The two required elements of any inflationary model are: superluminal expansion (i.e., accelerated expansion, $\ddot{R} > 0$) and massive entropy production [7]. They are usually achieved by means of the classical evolution of a scalar field rolling down its potential-energy curve. During the first part of its evolution, the field rolls so slowly that its potential energy density is nearly constant; this drives a nearly exponential expansion (superluminal expansion). During the late part of its evolution, the scalar field rapidly oscillates about the minimum of its potential and the decay of these oscillations eventually leads to the production of particles and the reheating of the Universe (entropy production). The entropy produced is the heat that today is the Cosmic Background Radiation (CBR). Because of the massive entropy produced, any initial baryon asymmetry is diluted to a level much, much less than the observed $\mathcal{O}(10^{-10})$ baryon number per photon and baryogenesis after inflation is mandatory. The basic mechanics of inflation are well understood and summarized in Ref. [8].

All models of inflation have one feature in common: the scalar field responsible for inflation has a very flat potential-energy curve and is very weakly coupled. This typically implies a dimensionless coupling of the order of 10^{-14}. Such a small number, like other small numbers in physics (e.g., the ratio of the weak to Planck scales $\approx 10^{-17}$ or the ratio of the mass of the electron to the W/Z boson masses $\approx 10^{-5}$), runs counter to one's belief that a truly fundamental theory should have no tiny parameters, and cries out for an explanation.[1] In some models, the small number in the

---

[1] It is sometimes stated that inflation is unnatural because of the small coupling of the scalar field responsible for inflation; while the small coupling certainly begs explanation, inflationary models are not unnatural in the technical sense as the small number is always

inflationary potential is related to other small numbers in particle physics: for example, the ratio of the electron mass to the weak scale or the ratio of the unification scale to the Planck scale. Explaining the origin of the small number associated with inflation is both a challenge and an opportunity.

Models of inflation range from the very simple (e.g., chaotic inflation [9]) to those that attempt to be part of a grander scheme (e.g., models that make contact with speculations about physics at very high energies – grand unification [10], supersymmetry [11, 12, 13], preonic physics [14], or supergravity [15].) Some have attempted to link inflation with superstring theory [16]; others have focussed on the naturalness issue, trying to explain the small dimensionless number associated with inflation [17].

While the scale of the vacuum energy that drives inflation is typically of the order of $(10^{14}\,\text{GeV})^4$, a model of inflation at the electroweak scale, vacuum energy $\approx (1\,\text{TeV})^4$, has been proposed [18]. Multiple epochs of inflation are also possible [19]. Inflation has been considered in the context of alternative theories of gravity. The most successful is first-order inflation [20, 21], where gravity is described by Jordan – Brans – Dicke theory (or a similar theory of gravity) and inflation is triggered by a strongly first-order phase transition (e.g., GUT symmetry breaking) of the kind originally envisioned by Guth [2].

There are certainly details of inflation that are both model-dependent and not completely understood. For example, the basics of reheating were laid out early on [22]; however, important details are still under study today [23]. While the physics issues such as reheating and model building are important and interesting, they do not affect the basic predictions of inflation that are crucial to its testing. In the end, observations may give the best guidance about models and even physics issues.

3. Inflation Makes Three Robust Predictions

Inflation theorists are very inventive and there are probably no set of predictions that are common to all models of inflation. However, a theory without definite predictions is not testable – and is hardly a theory at all (Mach's principle provides an interesting case in point). The philosopher of science Karl Popper argued that the status of a scientific theory is tied to its vulnerability – strong theories constantly subject themselves to falsification. I believe that inflation is a strong theory in the sense of Popper and that it makes three predictions which allow it to be falsified. They are:

1. **Flat universe.** This is perhaps the most fundamental prediction of inflation. Through the Friedmann equation it implies that the total

stabilized against the effect of quantum corrections.

energy density is always equal to the critical energy density; it does not however predict the form (or forms) that the critical density takes on today or at any earlier or later epoch.

2. **Nearly scale-invariant spectrum of gaussian density perturbations.** These density perturbations (scalar metric perturbations) arise from quantum-mechanical fluctuations in the field that drives inflation [3]; they begin on very tiny scales (of the order of 10^{-23} cm) and are stretched to astrophysical size by the tremendous growth of the scale factor during inflation (factor of e^{60} or greater). Scale invariant refers to the fact that the fluctuations in the gravitational potential are independent of length scale; or equivalently that the horizon-crossing amplitudes of the density perturbations are independent of length scale. While the shape of the spectrum of density perturbations is common to all models, the overall amplitude is model dependent. Achieving density perturbations that are consistent with the observed anisotropy of the CBR and large enough to produce the structure seen in the Universe today requires a horizon crossing amplitude of around $(\delta\rho/\rho)_H \approx 2 \times 10^{-5}$. This is the most severe constraint on inflationary models and leads to the small dimensionless number associated with inflation.

3. **Nearly scale-invariant spectrum of gravitational waves.** These gravitational waves (tensor metric perturbations) arise during inflation from quantum-mechanical fluctuations in the metric itself and today have wavelengths from $\mathcal{O}(1 \text{ km})$ to the size of the present Hubble radius and beyond [24]. Scale invariant here refers to the fact that gravitational waves of all wavelength cross the horizon with the same dimensionless strain amplitude. Once again, the overall amplitude is model dependent (proportional to the inflationary vacuum energy). The uniformity of the CBR provides a cosmological upper bound to the overall amplitude, but unlike density perturbations, there is no cosmological lower bound to the amplitude of gravity-wave perturbations.

There are other interesting consequences of inflation that are not generic. For example, in models of first-order inflation, where reheating occurs through the nucleation and collision of vacuum bubbles, there is an additional, much larger amplitude, but narrow-frequency-band spectrum of gravitational waves ($\Omega_{GW}h^2 \sim 10^{-7}$) [25]. Large-scale primeval magnetic fields of interesting size can be seeded during inflation [26]. It is also possible to produce topological defects during or near the end of inflation [27] or isocurvature perturbations in a matter component (axions [28], baryons [29], or something else [30]). Such auxiliary predictions are interesting, but are not part of the core predictions that can be used to falsify inflation. On the other hand, they could prove very helpful in establishing inflation.

4. Can Inflation Lead to an Open Universe?

Yes, BUT!!

Whether or not flatness is a generic prediction of inflation has been the topic of much debate recently. I believe that flatness should be taken as a firm prediction of inflation and I will explain why. If there is one episode of inflation, solving the "horizon" problem and solving the "flatness" problem (maintaining Ω very close to unity until the present epoch) are linked geometrically by the simple expression [31]

$$
|\Omega_0 - 1| \lesssim \left(\frac{H_0^{-1}}{d_{\text{Patch}}} \right)^2 \tag{1}
$$

where d_{Patch} is the present size of the inflationary patch that our Hubble volume resides within, which is assumed to have size H^{-1} at the beginning of inflation. If we make no assumption about the smoothness of the Universe on superHubble scales before inflation or about our location within our inflationary region, solving the horizon problem requires $H_0^{-1} \ll d_{\text{Patch}}$ and this implies $|\Omega_0 - 1| \ll 1$.

Open inflation requires that this linkage be evaded and that the amount of inflation be tuned to a specific value. The number of e-folds of inflation N is determined by the shape of the scalar-field potential,

$$
N = \frac{8\pi}{m_{\text{Pl}}^2} \int_{\phi_i}^{\phi_f} \frac{V(\phi)d\phi}{V'} . \tag{2}
$$

The value of N required to achieve a given value of Ω today depends upon the reheating temperature after inflation, the value of Ω before inflation, and the temperature today,

$$
N = \frac{1}{2} \ln \left[\frac{|\Omega_i - 1|}{|\Omega_0 - 1|} \right] + \ln \left[\frac{T_{\text{RH}}T_0}{m_{\text{Pl}}H_0} \right] . \tag{3}
$$

The amount of inflation needed is linked to both the initial state and the epoch of our existence and invites one to invoke the anthropic principle. I see this as a major step backward and counter to the spirit of the inflationary program.

In any case, the simplest way to evade Eq. (1) is to assume that the smooth patch that inflates has an initial size that is ten or even one hundred times larger than the Hubble radius. The more elegant way is to assume two epochs of inflation, the first ending with the nucleation of a bubble and second tied to the slow roll of a scalar field [32]. The open universe resides within the bubble nucleated by the first episode of inflation (which looks like an open universe [33]) and is reheated by the second, slow-roll epoch of inflation.

Open, double inflation in the context of eternal inflation can trade tuning for a distribution of values of Ω_0. My hunch is that the distribution is likely to be very strongly peaked, either around $\Omega_0 = 0$ or $\Omega_0 = 1$, rather than uniform. The recent work of Vilenkin and Winitzki suggests that this is the case [34].

The scientific question of the flatness of the Universe will be answered, probably within the next five years by using the fine-scale anisotropy of the CBR. If the Universe is found to be flat, I will score it as an important victory for inflation. If the flatness prediction is falsified I will consider it a major defeat. If the Universe is found not to be flat, but other tests of inflation prove successful (e.g., CBR anisotropy and/or gravitational waves), I will be willing to take another look at open inflation.

5. Inflation Implies Particle Dark Matter and Maybe More

While inflation predicts a flat, critical-density universe, it sheds no light on the form that the critical density should take. Cosmological observations have narrowed the possibilities. Denote the fraction of critical density contributed by all forms of energy density by Ω_0; inflation predicts $\Omega_0 = 1$. Big-bang nucleosynthesis constrains the baryon density to be well below the critical density: $0.007h^{-2} \leq \Omega_B \leq 0.024h^{-2} < 0.10$ (for $h > 0.5$) [35], which implies that most of the critical density must be in a form other than baryons. When the primeval deuterium abundance is pinned down by definitive determinations of D/H in high-redshift hydrogen clouds, the baryon density will be pegged to a precision of around 10% or so. Tytler and his collaborators have made a very strong case for a primeval deuterium abundance of D/H $\simeq 2.5 \times 10^{-5}$ [36], which implies that $\Omega_B h^2 \simeq 0.024$ or $\Omega_B \simeq 0.05(0.7/h)^2$.

It is also known that: most of the matter is dark (luminous matter contributes less than 1% of the critical density, $\Omega_{LUM} \simeq 0.003h^{-1}$) and the fraction of critical density in matter that clusters exceeds 30%, $\Omega_M > 0.3$ [37]. Thus, it follows that at least 25% of the critical density should be in the form of non-baryonic matter in the form of particles, $\Omega_{nbparticles} = \Omega_M - \Omega_B \gtrsim 0.25$. Particle physics has provided three very good candidates whose relic abundance (if they exist) should be of the order of the critical density: an axion of mass around 10^{-5} eV [38]; a neutralino of mass between 10 GeV and 500 GeV [39]; and a neutrino of mass of the order of 10 eV [40].

All are well motivated: the axion is a prediction of the most promising solution to the strong-CP problem; the neutralino is predicted by supersymmetric extensions of the standard model; and essentially all attempts to unify the forces and particles of Nature lead to the prediction that neutrinos have small, but nonzero, masses. In fact, these three particle dark matter candidates are so well motivated that we should probably take seriously the possibility that more than one might contribute significantly to the matter density today.

Finally, it is possible that there is another, even more exotic component, which is smoothly distributed and contributes up to 70% of the critical density, $\Omega_X = \Omega_0 - \Omega_M \lesssim 0.7$. The fact that evidence for $\Omega_M \sim 1$ is still lacking and that a case is mounting for $\Omega_M \sim 0.3$ [37], suggests that inflationists should consider the possibility of a smooth component seriously. Candidates for such include vacuum energy, tangled strings, and rolling scalar fields [41].

While Occam's Razor argues against a smorgasbord, Nature might enjoy a more interesting meal, and inflation gives no guidance.

6. Large-scale Structure from Quantum Fluctuations

This is perhaps the most striking prediction of inflation, and I believe, the motivation for Stephen Hawking's description of the COBE DMR discovery of CBR anisotropy [42] as "the most important discovery of all time." I believe Hawking said this thinking the COBE discovery might prove to be the first evidence that the density perturbations that seeded all structure in the Universe arose from quantum fluctuations during inflation.

Recall, scale invariant refers to density perturbations that cross the horizon with the same amplitude, independent of length scale. Different scales cross the horizon at different times, so the spectrum of density perturbations today is not independent of scale.

Gaussian means that the density contrast, $\delta\rho(\mathbf{x}, t)/\bar{\rho}$, is a gaussian random field, described fully by its two-point correlation function, or equivalently by the power spectrum, which is the Fourier transform of the correlation function and the square of the Fourier transform of the density contrast.

Both scale invariant and gaussian are generic predictions as they are linked to central features of inflation. Because the scalar field that drives inflation is very weakly coupled, it behaves like a free field and its fluctuations are gaussian [3]. The density perturbations are proportional to the scalar-field fluctuations and hence they too should be gaussian. The deviation of the fluctuations from scale invariance is related to the steepness of the scalar potential; since the scalar field responsible for inflation must take the 60 or so Hubble times to evolve to the minimum of its potential in order to solve the horizon/flatness problems its potential cannot be too steep.

The relationship between the inflationary potential and the power spectrum of density perturbations today $(P(k) \equiv \langle|\delta_k|^2\rangle)$ is given by

$$P(k) = \frac{1024\pi^3}{75} \frac{k}{H_0^4} \frac{V_*^3}{m_{\rm Pl}^6 V_*'^2} \left(\frac{k}{k_*}\right)^{n-1} T^2(k)$$

$$n - 1 = -\frac{1}{8\pi}\left(\frac{m_{\rm Pl}V_*'}{V_*}\right)^2 + \frac{m_{\rm Pl}}{4\pi}\left(\frac{m_{\rm Pl}V_*'}{V_*}\right)'$$

$$\frac{dn}{d\ln k} = -\frac{1}{32\pi^2}\left(\frac{m_{\rm Pl}^3 V_*'''}{V_*}\right)\left(\frac{m_{\rm Pl}V_*'}{V_*}\right)$$

$$+\frac{1}{8\pi^2}\left(\frac{m_{\rm Pl}^2 V_*''}{V_*}\right)\left(\frac{m_{\rm Pl}V_*'}{V_*}\right)^2 - \frac{3}{32\pi^2}\left(m_{\rm Pl}\frac{V_*'}{V_*}\right)^4$$

$$T(q) = \frac{\ln\left(1 + 2.34q\right)/2.34q}{\left[1 + 3.89q + (16.1q)^2 + (5.46q)^3 + (6.71q)^4\right]^{1/4}}, \tag{4}$$

where $V(\phi)$ is the inflationary potential, prime denotes $d/d\phi$, V_* is the value of the scalar potential when the scale k_* crossed outside the horizon during inflation, $T(k)$ is the transfer function which accounts for the evolution of the mode k from horizon crossing until the present, $q = k/h\Gamma$, and $\Gamma \simeq \Omega_M h$ is the "shape" parameter [43]. It is very convenient to chose $k_* = H_0$ so that V_* is the value of the inflationary potential when the scale that fixes the CBR quadrupole crossed outside the horizon. These expressions are given to lowest-order in the deviation from scale invariance, and assume a matter-dominated universe today; the next-order corrections are given in Ref. [44] and the analogous expressions which include the possibility of a cosmological constant are given in Ref. [45].

There are several important things to take note of:

1. The overall amplitude of the power spectrum depends upon the combination $V_*^3/V_*'^2$.
2. The quantity $n-1$, which measures the deviation from scale invariance, is generally not equal to zero [46].
3. The deviation from scale invariance depends upon the steepness of the potential $(m_{\rm Pl}V_*'/V_*)$ and the change in the steepness [47, 48].
4. Typically n is less than 1, and for many models is in the range 0.94 to 0.96 [47] (e.g., chaotic inflation [9] and new inflation [49]).
5. There are inflationary models where n is larger than 1 (e.g., hybrid inflation [50]) or as small as 0.7 (e.g., power-law inflation and natural inflation).
6. Generally, the spectrum of perturbations is nearly, but not exactly, a power law: $dn/d\ln k$ is typically of the order of -10^{-3}; only for power-law inflation is the spectrum an exact power law [51].

318

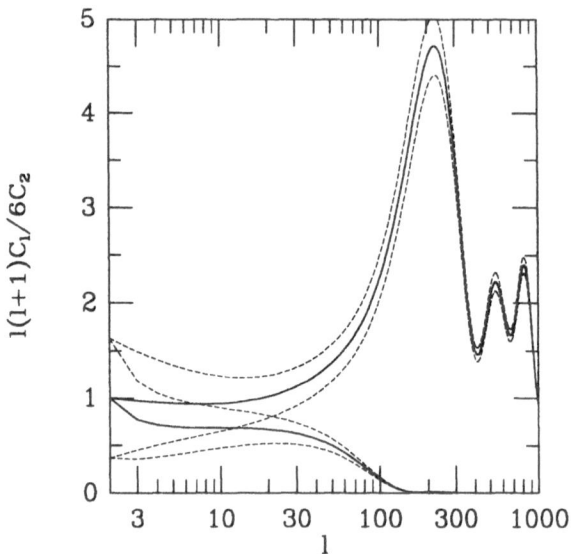

Figure 1. Angular power spectra ($C_l \equiv \langle |a_{lm}|^2 \rangle$) of CBR anisotropy for gravity waves (lower curves) and density perturbations (upper curves), normalized to the quadrupole anisotropy; broken lines indicate sampling variance. Temperature fluctuations measured on angular scale θ are approximately, $(\delta T/T)_\theta \sim \sqrt{l(l+1)C_l/2\pi}$ with $l \sim 200°/\theta$ (courtesy of M. White and U. Seljak).

7. The shape of the transfer function, which determines the level of inhomogeneity on small scales when the power spectrum on large scales is normalized by COBE, depends upon $\Gamma = \Omega_M h$ (and to a lesser extent upon the baryon density [52]).

Density perturbations give rise to CBR anisotropy which can be computed very precisely [53]. CBR anisotropy probes the power spectrum at early times ($z \simeq 1100$), when the perturbations were still in the linear regime and astrophysical effects were minimal. Thus, it is one of the most important tests and powerful probes of inflation. Expanding the CBR temperature on the sky in spherical harmonics,

$$\frac{\delta T(\theta, \phi)}{T} = \sum_{lm} a_{lm} Y_{lm}(\theta, \phi), \qquad (5)$$

the anisotropy is fully characterized by its angular power spectrum $C_l \equiv \langle |a_{lm}|^2 \rangle$, shown in Fig. 1. The ensemble average for the variance of the multipoles, $\langle |a_{lm}|^2 \rangle$, is related to the power spectrum (as described in Ref. [53]); note, isotropy in the mean implies $\langle a_{lm} \rangle = 0$. The variance of multipole l is dominated by modes of wavenumbers $k \simeq lH_0/2$. CBR anisotropy on large-angular scales ($l \ll 100$) arises almost solely from the Sachs-Wolfe effect and to good approximation can be computed analytically [8, 54]

$$C_l = \frac{H_0^4}{2\pi} \int_0^\infty (u/u_{\text{EQ}}) |j_l(u)|^2 P(u/u_{\text{EQ}}) du/u, \tag{6}$$

where $u = k\tau_0$ and $u_{\text{EQ}} = k_{\text{EQ}}\tau_0$. The variance of the quadrupole anisotropy provides a handy means of normalizing the power spectrum

$$S \equiv \frac{5\langle |a_{2m}|^2 \rangle}{4\pi} \simeq 2.2 \frac{V_*/m_{\text{Pl}}^4}{(m_{\text{Pl}} V_*'/V_*)^2}. \tag{7}$$

The detection of CBR anisotropy by the COBE DMR was a major advance as it allowed the spectrum of density perturbations to be normalized on very large scales ($k \sim H_0$). For precisely scale-invariant density perturbations and no gravitational-wave contribution to CBR anisotropy the normalization procedure is easy to describe: S is equated to the COBE determination of the variance of the quadrupole anisotropy, $Q_{\text{COBE}} = (17\mu\,\text{K}/2.2728\,\text{K})^2 \simeq 3.8 \times 10^{-11}$ [55], which then implies

$$\frac{V_*/m_{\text{Pl}}^4}{(m_{\text{Pl}} V_*'/V_*)^2} = 1.8 \times 10^{-11} \tag{8}$$

Bunn et al. [56] have done a careful analysis of the COBE four-year data which takes account of the fact that the COBE normalization for S depends upon n as well as the possible contribution of gravitational waves to CBR anisotropy. (The "pivot point" for the COBE data is $l \sim 15$; that is, the COBE determinations of C_{15} and n are almost uncorrelated.) This leads to the more accurate normalization

$$\frac{V_*/m_{\text{Pl}}^4}{(m_{\text{Pl}} V_*'/V_*)^2} = 1.7 \times 10^{-11} \frac{\exp[-2.02(n-1)]}{\sqrt{1 + \frac{2}{3}\frac{T}{S}}} \tag{9}$$

where T is the tensor contribution to the variance of the CBR quadrupole. The 1σ error is 15%. (Bunn et al. have also generalized this result to allow for the possibility of a cosmological constant [56].)

The Bunn et al. normalization can also be expressed in terms of the horizon-crossing amplitude for the comoving scale $k = H_0$:

$$\delta_H(k = H_0) \equiv \left[\frac{k^{3/2}|\delta_k|}{\sqrt{2\pi^2}}\right]_{k=H_0} = 1.9 \times 10^{-5} \frac{\exp[-1.01(n-1)]}{\sqrt{1 + \frac{2}{3}\frac{T}{S}}}. \tag{10}$$

7. Gravitational Waves: The Smokin' Gun

The inflation-produced gravitational waves are the smokin' gun signature of inflation and crucial to learning about the inflationary potential. Both a flat universe [57] and scale-invariant density perturbations (so called Harrison–Zel'dovich spectrum [58]) were discussed as features of any "sensible cosmology" long before inflation. The nearly scale-invariant spectrum of gravitational waves which arises from quantum mechanical fluctuations excited in the space-time metric during inflation is a very important prediction of inflation that sets it apart from just any other sensible cosmology. Detecting these gravitational waves will be very challenging.

Unlike the scalar perturbations, which must have an amplitude of around 10^{-5} to seed structure formation, there is no astrophysical clue as to the amplitude of the tensor perturbations. They can be characterized by their power spectrum today [59]

$$P_T(k) \equiv \langle |h_k|^2 \rangle = \frac{8}{3\pi} \frac{V_*}{m_{\rm Pl}^4} \left(\frac{k}{k_*} \right)^{n_T - 3} T_T^2(k)$$

$$n_T = -\frac{1}{8\pi} \left(\frac{m_{\rm Pl} V_*'}{V_*} \right)^2$$

$$\frac{dn_T}{d \ln k} = \frac{1}{32\pi^2} \left(\frac{m_{\rm Pl}^2 V''}{V} \right) \left(\frac{m_{\rm Pl} V'}{V} \right)^2 - \frac{1}{32\pi^2} \left(\frac{m_{\rm Pl} V'}{V} \right)^4 = -n_T[(n-1) - n_T]$$

$$T_T(k) \simeq \left[1 + \frac{4}{3} \frac{k}{k_{\rm EQ}} + \frac{5}{2} \left(\frac{k}{k_{\rm EQ}} \right)^2 \right]^{1/2}, \tag{11}$$

where $T_T(k)$ is the transfer function for gravity waves and describes the evolution of mode k from horizon crossing until the present, $k_{\rm EQ} = 6.22 \times 10^{-2}\,{\rm Mpc}^{-1} (\Omega_M h^2 / \sqrt{g_*/3.36})$ is the scale that crossed the horizon at matter-radiation equality, Ω_M is the fraction of critical density in matter, and g_* counts the effective number of relativistic degrees of freedom (3.36 for photons and three light neutrino species). The quantity $k^{3/2} |h_k| / \sqrt{2\pi^2}$ corresponds to the dimensionless strain (metric perturbation) on length scale $\lambda = 2\pi/k$.

Like density perturbations, gravity waves lead to CBR anisotropy which can be fully described by an angular power spectrum. The gravity-wave angular power spectrum is related to the power spectrum and can be computed very accurately [60]; it is shown in Fig. 1. The following analytical expression is accurate to about 10%,

$$C_l = 36\pi^2 \frac{\Gamma(l+3)}{\Gamma(l-1)} \int_0^\infty F_l(u)(u/u_{\rm EQ})^3 P_T(u/u_{\rm EQ})\, du/u$$

$$F_l(u) = -\int_{(\tau_{\rm LS}/\tau_0)u}^{u} dy \left(\frac{j_2(y)}{y} \right) \left(\frac{j_l(u-y)}{(u-y)^2} \right) \tag{12}$$

$$\tag{13}$$

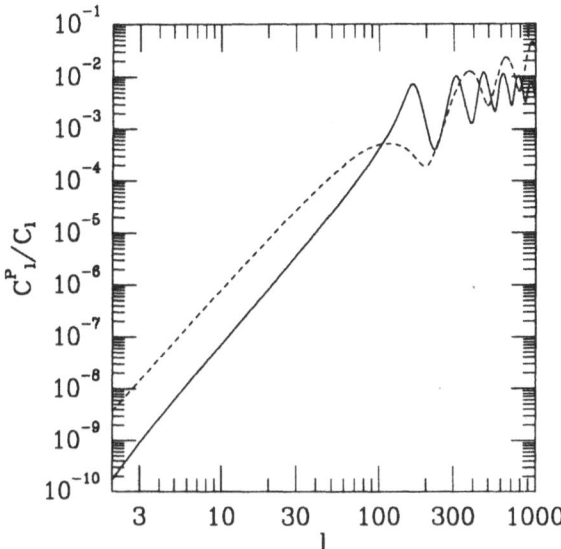

Figure 2. Polarization angular power spectra for gravity waves (broken) and density perturbations (solid). The polarization of the CBR anisotropy is roughly $\sqrt{C_l^P/C_l}$ (courtesy of M. White and U. Seljak).

where $u = k\tau_0$, $\tau_0 = 2H_0^{-1}$ is the conformal age of the Universe today, and $\tau_{LS} = \tau_0/\sqrt{(1+z_{LS})}$ is the conformal age at last scattering. The tensor contribution to the variance of the CBR quadrupole is a convenient normalization for the spectrum:

$$T \equiv \frac{5\langle|a_{2m}|^2\rangle}{4\pi} = 0.61(V_*/m_{\text{Pl}}^4). \tag{14}$$

The predicted variance of the CBR quadrupole anisotropy is $T + S$.

There are several important things to take note of:

1. The contribution of gravity waves to the variance of the CBR quadrupole is proportional to the value of the vacuum energy that drives inflation, and if T can be determined, the energy scale of inflation can be determined.

2. Using the COBE four-year results as an upper bound, $T < Q_{\text{COBE}}$, it follows that, $V_* < 6 \times 10^{-11} m_{\text{Pl}}^4$, or equivalently, $V_*^{1/4} < 3.4 \times 10^{16}$ GeV. This indicates that inflation, if it occurred, involved energies much smaller than the Planck scale. (To be more precise, inflation could have begun at a much higher energy scale, but the portion of inflation relevant for us, i.e., the last 60 or so e-folds, occurred at an energy scale much smaller than the Planck energy. In chaotic inflation [9], inflation is supposed to begin at the Planck energy density.)

3. The four potential observables, $S, T, n-1$ and n_T, depend upon three properties of the inflationary potential, V_*, V_*' and V_*''. Thus, the po-

tential and its first two derivatives can be expressed in terms of the four observables with an additional consistency relation, $T/S = -7n_T$, which is an important test of inflation [61]. If S, T, and $n - 1$ can be determined, the potential and its first two derivatives can be "reconstructed"; in addition, if n_T can also be measured, the consistency of inflation can be tested [62].

4. The amplitude of the gravity-wave spectrum and its "tilt" (deviation from scale invariance) are related: the larger the amplitude, the greater the tilt. Moreover, this means the spectrum of gravity waves can be described by a single parameter.

5. The tensor tilt, deviation of n_T from 0, and the scalar tilt, deviation of $n - 1$ from zero, are in general not equal; they differ by the rate of change of the steepness. The only model where they are identical is power-law inflation.

6. The variation of the tensor index with scale, $dn_T/d\ln k$, is typically $\mathcal{O}(10^{-3})$.

There are two basic approaches to detecting the tensor perturbations: CBR anisotropy and/or polarization and direct detection of gravity waves. The first approach probes the spectrum at very long wavelengths, $\lambda \sim H_0^{-1}/100 - H_0^{-1} \sim 10^{26}\,\text{cm} - 10^{28}\,\text{cm}$, while the second probes much shorter wavelengths, $\lambda \sim 10^8\,\text{cm} - 10^{14}\,\text{cm}$. If some information (detections/upper limits) could be obtained at both wavelengths, both T and n_T could be measured or at least constrained.

While the scalar and tensor angular power spectra are very different (see Fig. 1), sampling variance sets a fundamental limit to how well the two can be separated from measurements of the one CBR sky we have access to. For multipole l, $2l + 1$ multipole amplitudes can be used to determine the variance; "the variance of the variance" is

$$\frac{\langle (C_l^{\text{estimate}} - C_l)^2 \rangle^{1/2}}{C_l} = \sqrt{\frac{2}{2l + 1}}, \tag{15}$$

where C_l^{estimate} is the estimate based upon CBR measurements; sampling variance is shown in Fig. 1. Using anisotropy alone, sampling variance implies that the tensor contribution can be reliably separated only if $T/S \geq 0.1$ [63]. The tensor and scalar perturbations lead to different levels of polarization of the anisotropy; see Fig. 2. The sampling-variance limit based upon CBR polarization is about a factor of five smaller [63], but requires that polarization on large-angular scales be measured at less than a fraction of a percent. Recently, it has been pointed out that scalar and tensor perturbations excite different patterns of polarization [64], which could allow sampling variance to be evaded. In any case, the potential of polarization

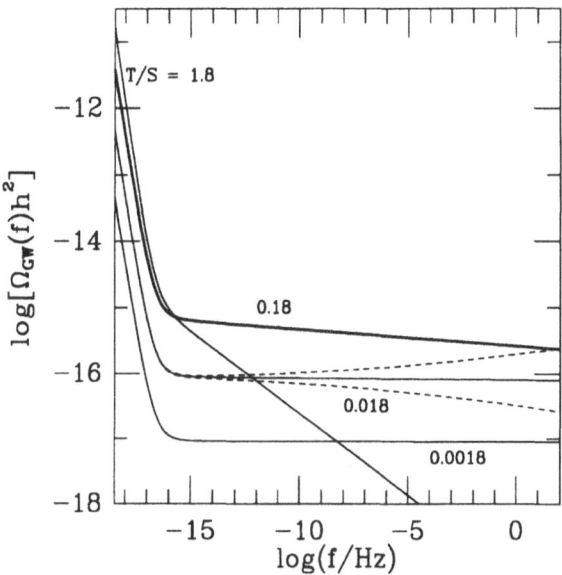

Figure 3. Spectral energy density in gravity waves produced by inflation; for $T/S = 0.018$, $dn_T/d\ln k = -10^{-3}$, 0, 10^{-3}. $T/S = 0.18$ (heavy curve) maximizes the energy density at $f = 10^{-4}$ Hz. Curves are from Eq. (16) using $H_0 = 60\,\mathrm{km\,s^{-1}\,Mpc^{-1}}$, $\Omega_M = 1$, and $g_* = 3.36$ (from Ref. [65]).

remains to be seen: the signal is small (maximum polarization is a few percent of the anisotropy); CBR polarization has yet to be detected; and the severity of polarization foregrounds are yet to be determined.

Earth-based laser interferometers which operate in the 10 Hz to kHz range are being built in the US (LIGO) and in Europe (VIRGO). A space-based interferometer is being planned by ESA (LISA) and ideas for a smaller mission are being discussed in the US (OMEGA). Space-based interferometers could operate at frequencies as low as 10^{-4} Hz.

It is straightforward to go from the power spectrum to the fraction of critical density contributed by gravity waves per log frequency interval

$$\Omega_{\mathrm{GW}}(f) \equiv \frac{1}{\rho_{\mathrm{crit}}} \frac{d\rho_{\mathrm{GW}}}{d\ln k} = \frac{\Omega_M^2\, V_*/m_{\mathrm{Pl}}^4}{(k/H_0)^{2-n_T}} \left[1 + \frac{4}{3}\frac{k}{k_{\mathrm{EQ}}} + \frac{5}{2}\left(\frac{k}{k_{\mathrm{EQ}}}\right)^2\right], \quad (16)$$

where $f = k/2\pi$ and $\Omega_{\mathrm{GW}}(f)$ is shown in Fig. 3. The long plateau, frequencies greater than $f_{\mathrm{EQ}} = k_{\mathrm{EQ}}/2\pi \sim 10^{-15}$ Hz, reflects the scale invariance of the gravitational-wave spectrum. The rise for smaller frequencies, as f^2, traces to the fact that the longest wavelength modes crossed the horizon during the matter-dominated epoch. The energy density in gravitational waves can also be expressed in terms of the *rms* metric perturbation or strain, $h_{rms}^2(k) \equiv k^3|h_k|^2/2\pi^2$,

$$\Omega_{GW}(f) = \frac{2\pi^2}{3}\left(\frac{f}{H_0}\right)^2 h_{rms}^2(k) = 6.3h^{-2} \times 10^{-7}\,(f/\text{Hz})^2 \qquad (17)$$

Using the relationship between the tensor spectral index n_T and the amplitude T, and the COBE determination of the variance of the CBR quadrupole anisotropy, Eq. (16) can be rewritten in terms of n_T (or T/S) alone [65]. Doing so, it then follows that on the "long plateau" ($f \gg 10^{-15}$ Hz)

$$\Omega_{GW}(f)h^2 = 5.1 \times 10^{-15}\,(g_*/3.36)\,\frac{n_T}{n_T - 1/7}$$

$$\times \exp[n_T N + \frac{1}{2}N^2(dn_T/d\ln k)], \qquad (18)$$

where $N \equiv \ln(k/H_0) \simeq 33 + \ln(f/10^{-4}\text{Hz}) + \ln(0.6/h)$ and I have also allowed for the possible variation of the tensor index n_T which is typically -10^{-3}.

The importance of the amplitude – tilt relationship can be seen in Fig. 3. Sadly, tilt goes in the direction of pushing Ω_{GW} down as the amplitude T/S is increased. There is a bright side: $\Omega_{GW}(f \sim \text{Hz})$ is maximized for $T/S \simeq 0.18$, which is close to the value predicted by chaotic inflation and exceeds the sampling-variance limit to the detection of tensor perturbations using CBR anisotropy alone. While there are essentially no inflationary models where $|n - 1| \ll 0.1$, there are many models where $-n_T \ll 0.1$ (e.g., natural inflation and new inflation). Because of the amplitude – tilt relationship, the gravity-wave background in these models is very small.

The range of T/S accessible to a gravity-wave detector operating at $f = 10^{-4}$ Hz (appropriate for LISA) and $f = 100$ Hz (appropriate for LIGO and VIRGO) is shown as a function of the detector energy sensitivity in Fig. 4 [65]. A sensitivity of $\Omega_{GW}(f)h^2 \sim 10^{-15}$ is needed for a serious search for inflation-produced gravity waves. With its initial strain detectors, the earth-based LIGO should be able to detect a stochastic background of gravity waves with 90% confidence provided $\Omega_{GW}(f \sim 100\,\text{Hz})h^2 \geq 2.8 \times 10^{-6}$; with advanced strain detectors this should improve to 2.8×10^{-11} [66, 67]. Unfortunately, this misses the mark by four orders of magnitude. (If one were to ignore the relation between tilt and amplitude and assume $n_T \equiv 0$, LIGO would only miss by three orders of magnitude [67].)

Because the energy density in gravity waves is proportional to strain squared times frequency squared, a detector operating at lower frequency has better energy-density sensitivity for fixed strain sensitivity. Earth-based detectors cannot operate at lower frequencies because of seismic noise, but

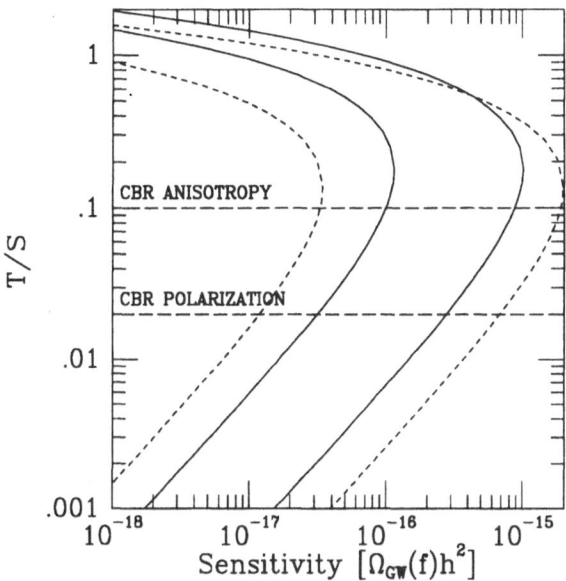

Figure 4. The range of T/S probed (interval interior to parabola) as a function of energy sensitivity for $f = 10^{-4}$ Hz (solid curves) and $f = 100$ Hz (broken curves). The "pessimistic" (left) parabola assumes $dn_T/d\ln k = -10^{-3}$ and the "optimistic" (right) parabola assumes $dn_T/d\ln k = 10^{-3}$. Also shown are the sampling-variance limiting sensitivities for CBR anisotropy and polarization (from Ref. [65]).

space-based detectors can. The design study for the space-based LISA indicates an energy sensitivity of around $\Omega_{\rm GW}(f)h^2 \sim 10^{-13}$ at $f = 10^{-4}$ Hz [68], which is more promising, but still misses by two orders of magnitude. (There is also a worrisome background of coalescing white-dwarf binaries, which could dominate inflation at frequencies greater than around 10^{-4} Hz [69].)

Detection of the inflation-produced gravity waves presents a very great challenge. But the possible payoffs are commensurately large: a smokin' gun test of inflation; a determination of the energy scale of inflation (through T); and a consistency test of inflation (through n_T and T/S).

8. CDM: A Testable, Ten Parameter Theory

CBR anisotropy has been detected on angular scales ranging from 100° to a fraction of degree at the level of about $\delta T/T \sim 10^{-5}$ (see Fig. 5). This provides strong support for the notion that structure formed by the gravitational amplification of small primeval density inhomogeneities. One of the pressing problems in cosmology is the formulation of a detailed and coherent picture of structure formation. The two key elements in any such theory are the quantity and composition of matter in the Universe and the nature of the perturbations that seed the formation of structure. Inflation makes definite predictions about both, and with the rapidly growing number of high-quality observations that bear on the issue, structure formation has become an important testing ground for inflation.

Recall, inflation and astrophysical data indicate the following about the quantity and composition of matter in the Universe: baryons contribute a small fraction of the critical density, $\Omega_B = (0.007 - 0.024)h^{-2}$; particle dark matter plus baryons contribute at least 30% of the critical density; and there may be a smooth component that brings the total to the critical density. The inflationary prediction concerning the seed perturbations is sharper: almost scale-invariant gaussian density perturbations, whose horizon-crossing amplitude is determined by COBE to be about 2×10^{-5}.

Particle dark matter can be classified by its velocity dispersion at the epoch of matter – radiation equality: cold, $v_{rms} \ll 1$; hot, $v_{rms} \sim 1$; and warm, v_{rms} not too much smaller than 1. Neutrinos and neutralinos were once in thermal equilibrium and their velocity dispersion is set by the temperature at matter – radiation equality ($T_{EQ} \simeq 6h^2\,\text{eV}$) and is inversely proportional to their mass. Neutrinos are light and therefore hot; neutralinos are heavy and therefore cold. Axions are cold in spite of their small mass because they were never in thermal equilibrium and were produced by a coherent, rather than thermal, process [38]. Around the time of matter – radiation equality, density perturbations on small scales can be damped by the free-streaming of dark matter particles from regions of high density to regions of low density; for neutrinos this is a significant effect and perturbations on scales less than about 30 Mpc are strongly damped. For cold dark matter the free-streaming scale is much less than 1 Mpc and most likely uninteresting. For warm dark matter the free-streaming scale is around 1 Mpc (essentially by definition) and has interesting consequences.

For hot dark matter structure must form "top down" – superclusters form and fragment into galaxies. More than a decade ago this possibility was studied and was found to be wanting [70]. Put simply, there is every evidence that structure formed from the bottom up: the bulk of the galaxies formed at redshifts from two to four and superclusters are only forming today. Warm dark matter is problematic because subgalactic-sized objects must form from the fragmentation of galaxies; the abundance of high-redshift (up to redshifts of almost five) hydrogen clouds is difficult at best to reconcile with this fact [71]. That leaves cold dark matter as the unique "prediction" of inflation. As we shall see, this prediction has been very successful.

Here are the essential features of CDM [72, 73]: (1) it is hierarchical, with smaller things forming first and larger things forming (slightly) later; (2) because the amplitude of density perturbations on very small scales varies slowly with scale, $k^{2/3}|\delta_k| \propto \log k$ for $k \gg k_{EQ}$, structure formation

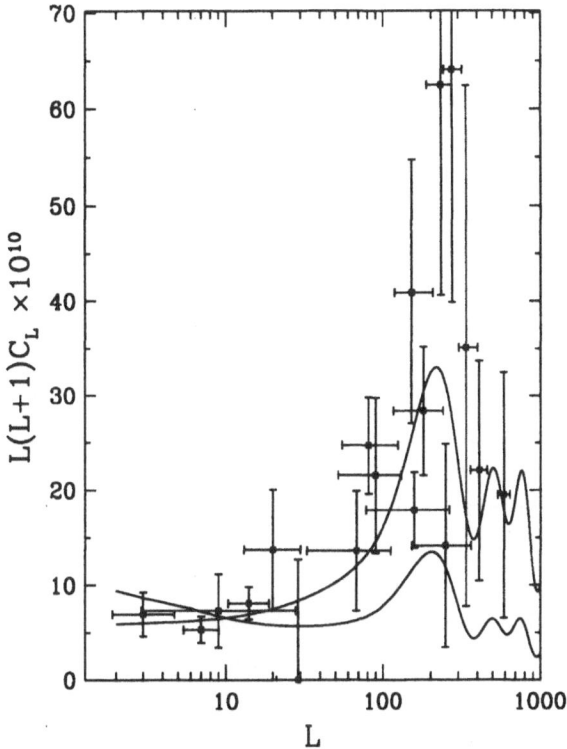

Figure 5. Summary of CBR anisotropy measurements and predictions for two CDM models. Plotted are the squares of the measured multipole amplitudes $(C_l = \langle |a_{lm}|^2 \rangle)$ versus multipole number l. The relative temperature difference on angular scale θ is given roughly by $\sqrt{l(l+1)C_l/2\pi}$ with $l \sim 200°/\theta$. The theoretical curves are standard CDM (upper curve) and CDM with $n = 0.7$ and $h = 0.5$ (lower curve) (from Ref. [74]).

is not strongly hierarchical; (3) in COBE normalized CDM the first stars (in globular-cluster size objects) form at redshifts of ten or so, galaxies begin forming at redshifts of five (with the bulk forming between $z \sim 4$ and $z \sim 2$), clusters begin forming at redshifts around one, and superclusters are just becoming gravitationally bound today; (4) CDM particles form the cosmic infrastructure on all scales – in galaxies they are the dark halos and in clusters they are the dark matter that pervades the cluster.

When the CDM scenario emerged more than a decade ago many referred to it as a no-parameter theory because it was so specific compared to previous models for the formation of structure. This was an overstatement as there are cosmological quantities that must be specified. However, the data available did not require precise knowledge of these quantities to begin testing the CDM paradigm.

Broadly speaking the parameters can be organized into two groups [74]. First are the cosmological parameters: the Hubble constant h; the density of ordinary matter, $\Omega_B h^2$; the amplitude of the scalar perturbations, S, the scalar power-law index, n, and the rate at which it varies, $dn/d\ln k$; the amplitude of the tensor perturbations, T/S, and tensor power-law index, n_T. The inflationary parameters fall into this category because there is no standard model of inflation; on the other hand, once determined they can be used to discriminate among models of inflation.

The second group specifies the composition of invisible matter in the Universe: radiation, dark matter, and vacuum energy. Radiation refers to relativistic particles: the photons in the CBR, three massless neutrino species (assuming none of the neutrino species has a mass), and possibly other undetected relativistic particles (some particle-physics theories predict the existence of additional massless particle species). At present relativistic particles contribute almost nothing to the energy density in the Universe, $\Omega_R \simeq 4.2 \times 10^{-5} h^{-2}$; early on – when the Universe was smaller than about 10^{-5} of its present size – they dominated the energy content; the level of radiation today is important as it determines when the transition from radiation domination to matter domination took place and thereby the shape of the transfer function (through Γ).

Dark matter could include other particle relics besides CDM. For example, each neutrino species has a number density of $113\,\mathrm{cm}^{-3}$, and a neutrino species of mass $5\,\mathrm{eV}$ would account for about 20% of the critical density ($\Omega_\nu = m_\nu/90h^2\,\mathrm{eV}$). Predictions for neutrino masses range from $10^{-12}\,\mathrm{eV}$ to several MeV, and there is some experimental evidence that at least one of the neutrino species has a small mass [75]. Finally, there is the cosmological constant. Introduced and then abandoned by Einstein to prevent the expansion of the Universe, and resurrected by Bondi, Gold and Hoyle in 1948 to address an age crisis, it is still with us. In the modern context it corresponds to an energy density associated with the quantum vacuum. At present, there is no reliable calculation of the value that the cosmological constant should take [76], and so its existence must be regarded as a logical possibility, with its value to be determined by observations. (As mentioned earlier, there are even more exotic possibilities for the smooth component [41].)

The original no-parameter CDM model, often referred to as standard CDM [72], is characterized by simple choices for the cosmological and the invisible matter parameters: precisely scale-invariant density perturbations ($n = 1$), $h = 0.5$, $\Omega_B = 0.05$, $\Omega_{\mathrm{CDM}} = 0.95$; no radiation beyond the photons and the three massless neutrinos; no dark matter beyond CDM; and zero cosmological constant. Standard CDM served its purpose well as the DOS 1.0 of cosmology: it focussed attention on a specific CDM model.

While inflation models predict that the shape of the spectrum is approximately scale-invariant, the overall amplitude depends on the particular inflationary model. For standard CDM the overall amplitude was fixed by comparing the predicted level of inhomogeneity with that seen today in the distribution of bright galaxies. Galaxy-number fluctuations in spheres of radius $8h^{-1}$ Mpc are unity:

$$\sigma_r^2 \equiv \int_0^\infty \frac{dk}{k} \frac{k^3 P(k)}{2\pi^2} \left(\frac{3j_1(kr)}{kr}\right)^2 = 1 \qquad (19)$$

for $r = 8\,h^{-1}$Mpc. This normalization ($\sigma_8 = 1$) corresponds to the assumption that light, in the form of bright galaxies, traces mass. Choosing σ_8 to be less than one means that light is more clustered than mass and is a biased tracer of mass. There is some evidence that bright galaxies are somewhat more clumped than mass with biasing factor $b \equiv 1/\sigma_8 \simeq 1 - 2$; e.g., the number fluctuations of IRAS galaxies on the $8h^{-1}$ Mpc scale is less than one:, $\sigma_8(\text{IRAS}) = 0.69 \pm 0.04$ [77], implying that infrared selected galaxies are less clustered than optically bright galaxies.

As discussed earlier, COBE changed the normalization procedure; given the values of the CDM parameters and normalizing to COBE σ_8 can be computed. Further, an independent means of determining σ_8, based upon the abundance of rich clusters, has been developed; comparing this value to that computed from the COBE normalized spectrum now provides a check/constraint. But I am getting ahead of myself.

Is a ten parameter theory testable? With sufficient high-quality data the answer is yes. The standard model of particle physics has at least nineteen parameters; not only has it been tested, but most of the parameters have been determined, many to better than 1% precision. In the next two sections I hope to make the case that inflation/cold dark matter can be tested with the same decisiveness. Much of my case will rely upon CBR anisotropy; if 2000 or so multipoles can be measured to a precision close to that dictated by sampling variance, I believe this ambitious goal is achievable.

9. Status of Inflation: So Far, So Good

The testing of inflation necessarily focuses on its three robust predictions and their consequences.

9.1. FLATNESS

The first prediction is a flat universe with non-baryonic dark matter. There is strong evidence coming from a number of directions that Ω_M is at least 0.3 [37]. This makes non-baryonic dark matter inescapable since the big-bang nucleosynthesis upper bound is $\Omega_B < 0.024h^{-2} < 0.10$ (for $h > 0.5$) and is half way (on a logarithmic scale) to the simplest realization of a flat universe, $\Omega_M = 1$; see Fig. 6. In the case that $\Omega_M \approx 0.3$ it is possible that the bulk of the closure density resides in a smooth component.

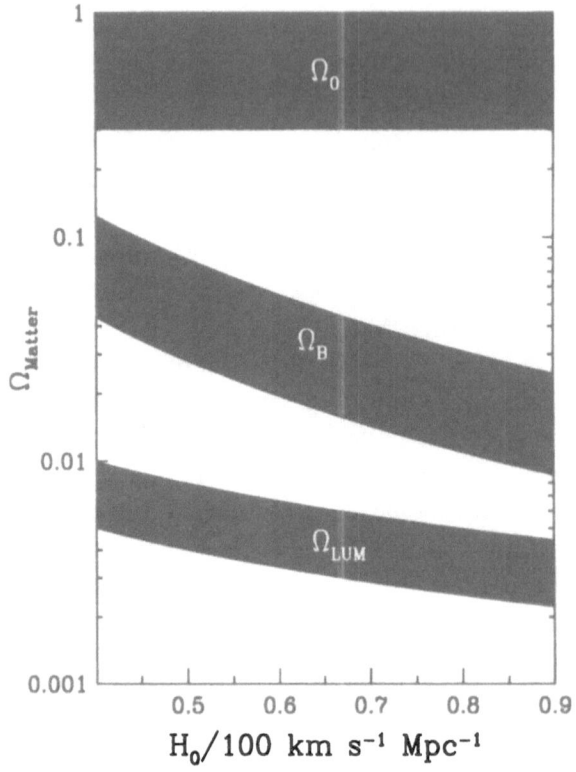

Figure 6. Summary of knowledge of Ω_M. The lowest band is luminous matter, in the form of bright stars and associated material; the middle band is the big-bang nucleosynthesis determination of the density of baryons; the upper region is the estimate based upon the peculiar velocities of galaxies and other dynamical methods [37]. The gaps between the bands illustrate the two dark matter problems: most of the ordinary matter is dark and most of the matter is non-baryonic (from Ref. [74]).

Testing the flatness prediction has an even brighter future. The position of the first acoustic (Doppler) peak is sensitive to Ω_0, $l_{\text{peak}} \simeq 220/\sqrt{\Omega_0}$. The current data, while certainly not definitive, put a smile on my face: Hancock infers $\Omega_0 = 0.7^{+1.0}_{-0.4}$ [78]. It is likely that even before MAP is launched in 2000, ground-based and balloon-based measurements will determine the position of the first acoustic peak well enough to peg Ω_0 to 10%.

Measurements of the deceleration of the Universe using the magnitude – redshift diagram for SNeIa constrain a nearly orthogonal combination, $\Omega_M - \Omega_\Lambda$; together, they can determine both Ω_M and Ω_Λ. (One should keep in mind that there are more exotic possibilities for the smooth component [41, 80].) Assuming a flat universe and using their first seven SNeIa, Perlmutter et al. [79] derive the bound $\Omega_\Lambda < 0.51$ (95%); or without the assumption of flatness, $-0.4 < \Omega_M - \Omega_\Lambda < 2.7$ (95%). Perlmutter's group, the Supernova Cosmology Project, now has a total of more than fifty SNeIa at redshifts $z \sim 0.4 - 0.8$, and the High-Redshift Supernova Team has a similar number; more definitive results should be coming soon.

The combination of the age and Hubble constant can, in principle, determine or at least constrain Ω_M and Ω_Λ. At the moment, the uncertainties preclude any firm conclusions. Taken at face value, the data seem to favor a cosmological constant (if the Universe is flat); see Fig. 7.

A key consequence of the flatness prediction is the existence of non-baryonic dark matter. The cold dark matter scenario won't be fully tested until CDM particles are detected. A large-scale search for halo axions with sufficient sensitivity to detect them (if they are there) is now underway [81], and soon, CDMS, the Stanford – Berkeley – Case Western – UCSB bolometric detector, will begin searching for halo neutralinos with sufficient sensitivity to detect them for some models of low-energy supersymmetry [82]. SuperKamiokande and MACRO can search indirectly for neutralinos that annihilate in the Sun and produce high-energy neutrinos, and Ting's AMS, which will be flown on the shuttle a year from now, will be able to search for positrons and antiprotons produced by neutralino annihilations in the halo.

9.2. GRAVITY WAVES

The search for inflation-produced gravitational waves was summarized in Section 7.

9.3. DENSITY PERTURBATIONS/COLD DARK MATTER

Figure 5 summarizes the status of testing the second robust prediction of inflation through CBR anisotropy: the measurements are generally consistent with the inflationary prediction. The power-law index is constrained to be 1.1 ± 0.2, which is well within the range predicted by inflation, and when COBE is used to normalize the spectrum, there are CDM models that are consistent with all the other observations.

There is much data that can be used to test the cold dark matter scenario. To focus the discussion, I will consider four "families" of models, distinguished by their invisible matter content: standard invisible matter content (CDM); extra radiation (τCDM); small hot dark matter component (νCDM); and cosmological constant (ΛCDM). There are of course other possibilities: extra radiation + cosmological constant, or a more exotic smooth component, which has been analyzed in Ref. [80]. Within each family, the five cosmological parameters (h, $\Omega_B h^2$, n, T/S and n_T) must be specified. Once specified, the power spectrum can be COBE-normalized and the expected level of inhomogeneity in the Universe today computed.

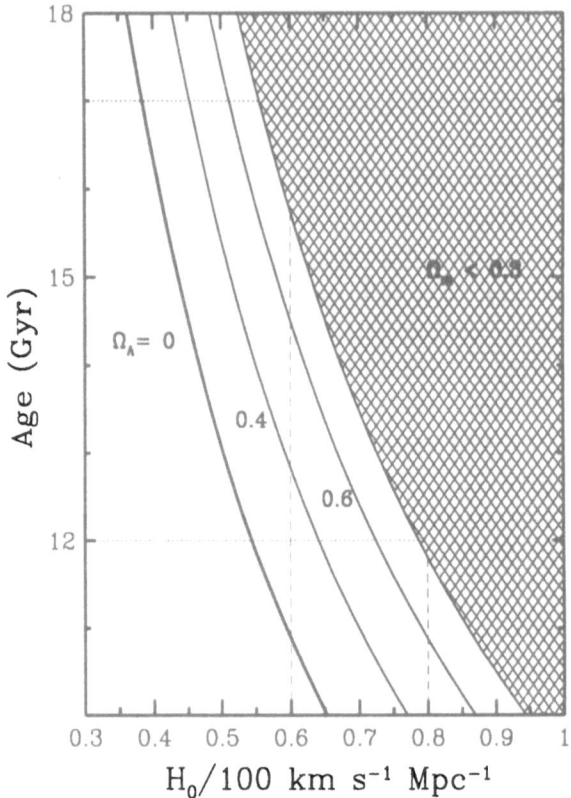

Figure 7. The relationship between age and H_0 for flat-universe models with $\Omega_M = 1 - \Omega_\Lambda$. The cross-hatched region is ruled out because $\Omega_M < 0.3$. The dotted lines indicate the favored range for H_0 and for the age of the Universe (based upon the ages of the oldest stars) (from Ref. [74]).

In assessing the viability of CDM models I will summarize work done in collaboration with Dodelson and Gates [74]; others have done similar work [84]. We began with three robust observational constraints on the power spectrum: the shape of the power spectrum; the power on cluster scales; and the early formation of objects. The first constraint, the shape of the power spectrum on scales from a few Mpc to a few 100 Mpc (see Fig. 8), comes from redshift surveys of the distribution of bright galaxies today [83]. In the absence of an understanding of the relationship between the distribution of light, which is what these surveys determine, and of mass, the bias factor is left as a free parameter. Models whose power spectra deviates from the measured power spectrum (as compiled in Ref. [83]) by more than two sigma (value of χ^2 whose likelihood is less than 5%) were rejected. (Very roughly, this constrains the shape parameter: $\Gamma = \Omega_M h = 0.25 \pm 0.1$.)

The abundance of x-ray emitting clusters is sensitive to the level of inhomogeneity on scales around $8h^{-1}$ Mpc and thus provides a good means

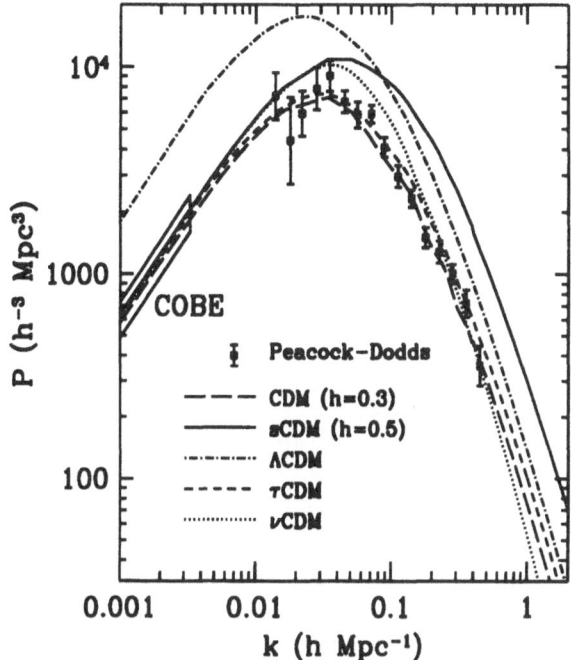

Figure 8.   Measurements of the power spectrum, $P(k) = |\delta_k|^2$, and the predictions of different COBE-normalized CDM models. (COBE constrains the power spectrum at wavenumbers k around $2h \times 10^{-3}$ Mpc as indicated by rectangle.) The points are from several redshift surveys as compiled by [83]; the models are: ΛCDM with $\Omega_\Lambda = 0.6$ and $h = 0.65$; standard CDM (sCDM), CDM with $h = 0.35$; τCDM (with the energy equivalent of 12 massless neutrino species) and νCDM with $\Omega_\nu = 0.2$ (unspecified parameters have their standard CDM values). The offset between a model and the points indicates the level of biasing. Note, ΛCDM does not pass through the COBE rectangle because a cosmological constant alters the relation between the power spectrum and CBR anisotropy (from Ref. [74]).

of inferring the value of σ_8. Following [85] we used $0.5 \leq \sigma_8 \leq 0.8$ for models with $\Omega_M = 1$ and let this range scale with $\Omega^{-0.56}$ for models with a cosmological constant $(\Omega_\Lambda = 1 - \Omega_M)$ [86].

The formation of objects at high redshift (early structure formation) probes the power spectrum on small scales. At redshifts of two to four, hydrogen clouds, detected by their absorption features in the spectra of high-redshift quasars $(z \sim 4-5)$, contribute a fraction of the critical density, $\Omega_{\text{clouds}} \simeq (0.001 \pm 0.0002)h^{-1}$ [87]. Insisting that the predicted level of inhomogeneity is sufficient to account for this leads to a lower limit to the power on small scales $(\lambda \sim 0.2h^{-1}\,\text{Mpc})$.

Figure 9 summarizes the viable models. Models with standard invisible-matter content must lie in a region that runs diagonally from smaller Hubble constant and larger n to larger Hubble constant and smaller n. That is, higher values of the Hubble constant require more tilt. As is well appreciated

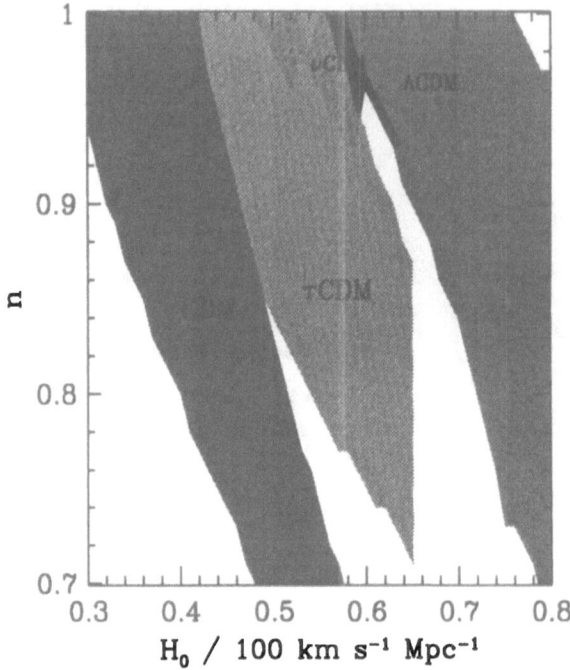

Figure 9. Acceptable values of the cosmological parameters n and h for CDM models with standard invisible-matter content (CDM), with 20% hot dark matter (νCDM), with additional relativistic particles (the energy equivalent of 12 massless neutrino species, denoted τCDM), and with a cosmological constant that accounts for 60% of the critical density (ΛCDM). A model is considered viable if it passes the three tests for *any* value of $\Omega_B h^2$ between 0.01 and 0.02 and any level of gravitational waves. The τCDM models have been truncated at a Hubble constant of $65\,\mathrm{km\,s^{-1}\,Mpc^{-1}}$ because a larger value would result in a universe that is younger than 10 Gyr (from Ref. [74]).

[88, 89], standard CDM is outside of the allowed range – so much for DOS 1.0, onto Windows 95! Current measurements of CBR anisotropy on the degree scale, as well as the COBE four-year anisotropy data, preclude n less than about 0.7, which implies that the largest H_0 consistent with the simplest CDM models is slightly less than $60\,\mathrm{km\,s^{-1}\,Mpc^{-1}}$.

If the invisible-matter content is nonstandard, higher values of H_0 can be accommodated. With tilt and hot dark matter, H_0 as large as $65\,\mathrm{km\,s^{-1}\,Mpc^{-1}}$ is consistent with the constraints. The introduction of a cosmological constant permits H_0 as large as $80\,\mathrm{km\,s^{-1}\,Mpc^{-1}}$, and additional radiation allows a Hubble constant as large as the age constraint permits (we assumed $t_0 \geq 10\,\mathrm{Gyr}$, which requires $H_0 \leq 65\,\mathrm{km\,s^{-1}\,Mpc^{-1}}$).

In passing I mention that a similar analysis has been carried out for open-inflation models and they do not fare nearly as well [90]. The only viable models have $n > 1.1$ or $\Omega_0 > 0.5$. Cold dark matter seems to be weighing in on the side of a flat universe.

A host of other observations test CDM. Some are more controversial and/or open to interpretation. They tend to distinguish the cosmological-constant family of models from the other three families. This is because models with standard invisible matter, extra radiation, or a hot dark matter component are all matter dominated today and have the same kinematic properties – age for a given Hubble constant and distance to an object of given redshift. The introduction of a cosmological constant leads to an older Universe and greater distance to an object at fixed redshift.

Together, the Hubble constant and age of the Universe, have great leverage. Determinations of the Hubble constant based upon a variety of techniques (Type Ia and II supernovae, IR Tully-Fisher and fundamental-plane methods) have converged on a value between $60\,\mathrm{km\,s^{-1}\,Mpc^{-1}}$ and $80\,\mathrm{km\,s^{-1}\,Mpc^{-1}}$ [91]. This corresponds to an expansion age of less than 11 Gyr for a flat, matter-dominated model; for ΛCDM, the expansion age can be significantly higher, as large as 16 Gyr for $\Omega_\Lambda = 0.6$ (see Fig. 7). On the other hand, the ages of the oldest globular clusters indicate that the Universe is between 13 Gyr and 17 Gyr old; further, these age determinations, together with the those for the oldest white dwarfs and the long-lived radioactive elements, provide an ironclad case for a universe that is at least 10 Gyr old [92, 93, 94]. At face value, the age/Hubble constant combination favor ΛCDM. But again, I want to stress that, within the uncertainties in both the age and Hubble constant, all of the models are viable.

Clusters are large enough that the baryon fraction should reflect its "universal value," $\Omega_B/\Omega_M = (0.007 - 0.024)h^{-2}/(1 - \Omega_\Lambda)$. Most of the (observed) baryons in clusters are in the hot, intracluster x-ray emitting gas. From x-ray measurements of the flux and temperature of the gas, baryon fractions in the range $(0.04 - 0.10)h^{-3/2}$ have been determined [95, 96, 97]; further, a recent detailed analysis and comparison to numerical models of clusters in CDM indicates an even smaller scatter, $(0.07 \pm 0.007)h^{-3/2}$ [98]. From the cluster baryon fraction and Ω_B, Ω_M can be inferred: $\Omega_M = (0.25 \pm 0.15)h^{-1/2}$, which for the lowest Hubble constant consistent with current determinations ($h = 0.6$) implies $\Omega_M = 0.32 \pm 0.2$. Unless one of the assumptions underlying this analysis is wrong, ΛCDM is strongly favored.

E. Turner emphasized the frequency of gravitational lensing of distant QSOs as an important cosmological test [99]. The underlying principle is simple: the comoving distance to fixed redshift depends upon the cosmology – it is largest for a cosmological constant, and in a matter-dominated universe decreases with increasing Ω_M – and the probability of lensing increases with comoving distance (more lenses along the line of sight). For a flat universe, Kochanek has obtained the upper limit $\Omega_\Lambda < 0.65$ (95%), and for a matter-dominated universe $0.25 < \Omega_M < 2$ (95%) [100]. Neither result is decisive.

ΛCDM is consistent with all the observations discussed here as well as others; see Fig. 10. For this reason, it is the strawman CDM model [101]. The parameters for this best fit model are: $\Omega_\Lambda \sim 0.5-0.65$ and $h \sim 0.6-0.7$. One should keep in mind that the introduction of a cosmological constant is a big step, one which twice before proved to be a misstep, and raises a fundamental question – the origin of the implied vacuum energy, about $10^{-8}\,\mathrm{eV}^4$ [76].

ΛCDM's hold on the title of best-fit CDM model is by no means unshakeable: should the Hubble constant turn out to be less than $60\,\mathrm{km\,s^{-1}\,Mpc^{-1}}$ and should a flaw be found in the cluster-baryon-fraction argument, the other models become very viable and are theoretically more attractive. Bartlett et al. have pointed out that if the Hubble constant is around $30\,\mathrm{km\,s^{-1}\,Mpc^{-1}}$, then CDM with $n \approx 1$ is consistent with all the measurements discussed above [102]. Lineweaver et al. have analyzed the existing CBR anisotropy data and conclude that it favors a Hubble constant of around this value [103]. The rub is squaring this "determination" of H_0 with the multitude of other determinations that indicate a value almost twice as large. Appealing to a difference between the local value of the expansion rate and the global value is of little help – in the context of CDM, the difference, which arises due to cosmic and sampling variance, is expected to be only 10% or so [104].

In finishing this status report I would like to emphasize three things. First, inflation is currently consistent with all the observational data, which is no mean feat. Second, cold dark matter is consistent with a large body of high-quality cosmological data, ranging from measurements of CBR anisotropy to our growing understanding of the evolution of galaxies and clusters of galaxies. This too is no mean feat; at the moment, the only other potentially viable paradigm for structure formation is topological defects + non-baryonic dark matter. These models, when COBE normalized, appear to be in great jeopardy as they predict very little power on small scales (high bias). Finally, the quantity of high-quality data that bear on inflation and cold dark matter is growing rapidly, leading me to believe that inflation/cold dark matter will be decisively tested soon.

10. The Future: Precision Testing and More

Inflation is a bold attempt to build upon the success of the standard cosmology and extend our understanding of the Universe to times as early as $10^{-32}\,\mathrm{sec}$ after the bang. Its three robust predictions – flat universe, nearly scale-invariant spectrum of density perturbations, and nearly scale-invariant spectrum of gravitationally waves – are the keys to its testing. In addition, much can be learned about the specific, underlying model of infla-

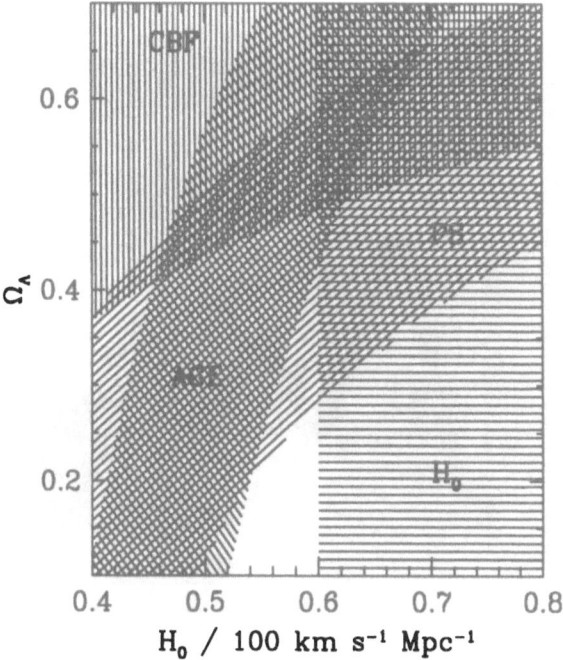

Figure 10. Summary of constraints projected onto the $H_0 - \Omega_M$ plane: (CBF) comes from combining the BBN limit to the baryon density with x-ray observations of clusters; (PS) arises from the power spectrum; (AGE) is based on age determinations of the Universe; (H_0) indicates the range currently favored for the Hubble constant. The constraint $\Omega_\Lambda < 0.66$ has been implicitly taken into account since the Ω_Λ axis extends only to 0.7. The darkest region indicates the parameters allowed by all constraints.

tion if other measurements are made (e.g., the small anticipated deviation from scale invariance and the level of gravitational waves).

Cold dark matter, which is an important means of testing inflation, is a ten-parameter theory, h, $\Omega_B h^2$, S, n, $dn/d\ln k$, T/S, n_T, Ω_ν, g_*, and Ω_Λ. While this is a daunting number of parameters, especially for a cosmological theory, there is good reason to believe that within ten years the data will overdetermine these parameters. Crucial to achieving this goal are the high-precision, high-resolution measurements of CBR anisotropy that will be made over the next decade by earth-based, balloon-borne and satellite-borne instruments (see Fig. 11). As a reminder of the power of high-quality data, the standard model of particle physics has nineteen parameters; precision measurements at Fermilab, SLAC, CERN and other accelerator laboratories, as well as nonaccelerator experiments, have both sharply tested the theory as well as accurately determining the parameters.

Within five years we should be well on our way to precisely testing inflation and cold dark matter. In the next few years ground-based and balloon-borne anisotropy experiments (e.g., VSA, VCA, Boomerang, TOPHAT,

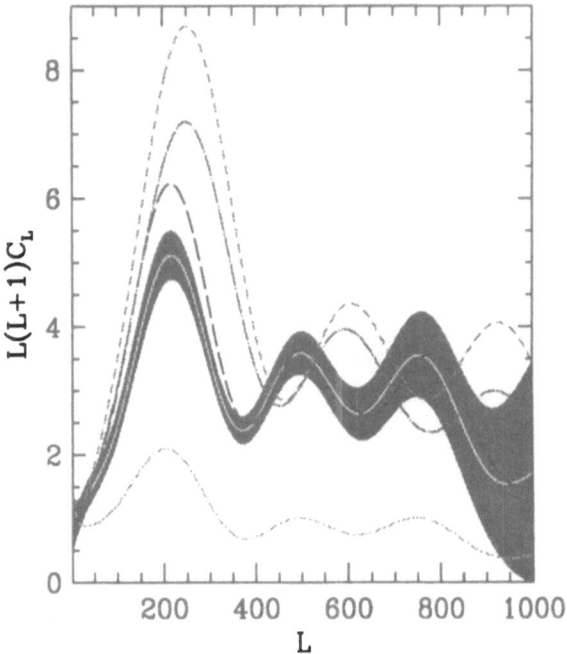

Figure 11. Predicted angular power spectra of CBR anisotropy for several viable CDM models and the anticipated uncertainty (per multipole) from a CBR satellite experiment similar to MAP. From top to bottom the CDM models are: CDM with $h = 0.35$, τCDM with the energy equivalent of 12 massless neutrino species, ΛCDM with $h = 0.65$ and $\Omega_\Lambda = 0.6$, νCDM with $\Omega_\nu = 0.2$, and CDM with $n = 0.7$ (unspecified parameters have their standard CDM values) (from Ref. [74]).

QMAX, and others) should be able to determine the approximate position of the Doppler peak and thereby Ω_0 to an accuracy of around 10%. Because flatness is a fundamental prediction of inflation, perhaps the most fundamental, this is a landmark test. On the same timescale, the Supernova Cosmology Project and the High-Redshift Supernova Team will use the SNeIa magnitude – redshift diagram to determine the deceleration of the Universe. While the Doppler peak determines $\Omega_M + \Omega_\Lambda$, the SNeIa measurement determines an almost orthogonal quantity, $\Omega_M - \Omega_\Lambda$; together, they can determine Ω_M and Ω_Λ.

In the same time frame measurements of the Hubble constant will play an important role. Since the Universe is at least 10 Gyr old, a definitive determination that the Hubble constant is 65 km s^{-1} Mpc^{-1} or greater would rule out all models but ΛCDM; on the other hand, a determination that the Hubble constant is below 55 km s^{-1} Mpc^{-1} would undermine much of the motivation for ΛCDM. The Hubble Space Telescope calibration of secondary distance indicators such as SNeIa with Cepheids in nearby galaxies and the maturation of physics-based methods such as gravitational time delay and Zel'dovich – Sunyaev are helping to pin down H_0 more accurately

[105].

There are other important tests that will be made on a longer time scale. The level of inhomogeneity in the Universe today is determined mainly from redshift surveys, the largest of which contains of order $30,000$ galaxies. Two larger surveys are underway. The Sloan Digital Sky Survey will cover 25% of the sky and obtain positions for two hundred million galaxies and redshifts for a million galaxies [106]. The Anglo – Australian Two-degree Field is targeting hundreds of two-degree patches on the sky and will obtain 250,000 redshifts [107]. These two projects will determine the power spectrum much more precisely and on scales large enough ($500h^{-1}$ Mpc) to connect with measurements from CBR anisotropy on angular scales of up to five degrees, allowing bias to be probed.

The most fundamental element of cold dark matter – the existence of the CDM particles – will be tested over the next decade. Experiments with sufficient sensitivity to detect the CDM particles that hold our own galaxy together if they are in the form of axions of mass 10^{-6} eV $- 10^{-4}$ eV [81] or neutralinos of mass tens of GeV [82] are now underway. Evidence for the existence of the neutralino could also come from particle accelerators searching for other supersymmetric particles [108]. In addition, a variety of experiments sensitive to neutrino masses are operating or are planned: accelerator-based neutrino oscillation experiments at Fermilab, CERN, and Los Alamos; solar-neutrino detectors in Japan, Canada, Germany, Russia and Italy; and experiments at e^+e^- colliders (LEP at CERN and CESR at Cornell) to the study the tau neutrino.

The most telling test of inflation and cold dark matter will come with the two new space missions that have recently been approved – MAP to be launched in 2000 by NASA and Planck to be launched by ESA in 2005. Each will map CBR anisotropy over the entire sky with more than thirty times the angular resolution of COBE (resolution of $0.2°$ for MAP and $0.1°$ for Planck). MAP should determine the angular power spectrum out to multipole number 1000, and Planck out to multipole number 2000, each to a precision close to that dictated by sampling variance alone. Theoretical studies [109] indicate that the results of Planck should be able to determine n to a precision of less than one percent, Ω_Λ to a few percent, Ω_0 to less than one percent, Ω_ν to enough precision to test νCDM [110], the baryon density to less than ten percent, and even the Hubble constant to one percent.

Inflation and cold dark matter are a bold attempt to extend our knowledge of the Universe to within 10^{-32} sec of the bang. The number of observations that are testing the cold dark matter theory is growing fast, and prospects for not only testing the theory, but also discriminating among the different CDM models and models of inflation are excellent. If cold dark matter is shown to be correct, an important aspect of the standard cosmology – the origin and evolution of structure – will have been resolved and a window to the early moments of the Universe and physics at very high energies will have been opened.

While the window has not been opened yet, I would like to end with one example of what one could hope to learn. As discussed earlier, S, $n-1$, T/S and n_T are related to the inflationary potential and its first two derivatives. If one can measure the power-law index of the scalar perturbations and the amplitudes of the scalar and tensor perturbations, one can recover the value of the potential and its first two derivatives around the point on the potential where inflation took place [62]:

$$V_* = 1.65T\, m_{\mathrm{Pl}}{}^4, \tag{20}$$

$$V_*' = \pm\sqrt{\frac{8\pi}{7}\frac{T}{S}}\, V_*/m_{\mathrm{Pl}}, \tag{21}$$

$$V_*'' = 4\pi\left[(n-1) + \frac{3}{7}\frac{T}{S}\right]V_*/m_{\mathrm{Pl}}{}^2, \tag{22}$$

where the sign of V' is indeterminate (under $\phi \leftrightarrow -\phi$ the sign changes). If the tensor spectral index can also be measured the relation, $T/S = -7n_T$, can be used to test the consistency of inflation. Reconstruction of the inflationary scalar potential would shed light both on inflation as well as physics at energies of the order of $10^{14}\,\mathrm{GeV}$.

Acknowledgments. Much of the material in these lectures derives from collaborative work with Scott Dodelson, Evalyn Gates, Lloyd Knox, Andrew Liddle, and Martin White. I thank my collaborators for teaching me so much. This work was supported by the DoE (at Chicago and Fermilab) and by the NASA (at Fermilab by grant NAG 5-2788).

References

1. See e.g., P.J.E. Peebles, *Physical Cosmology* (Princeton University Press, Princeton, 1993).
2. A.H. Guth, *Phys. Rev. D* **23**, 347 (1981).
3. A. H. Guth and S.-Y. Pi, *Phys. Rev. Lett.* **49**, 1110 (1982); S. W. Hawking, *Phys. Lett. B* **115**, 295 (1982); A. A. Starobinskii, *ibid* **117**, 175 (1982); J. M. Bardeen, P. J. Steinhardt, and M. S. Turner, *Phys. Rev. D* **28**, 697 (1983).
4. M.S. Turner and L.M. Widrow, *Phys. Rev. Lett.* **57**, 2237 (1986); L. Jensen and J. Stein-Schabes, *Phys. Rev. D* **35**, 1146 (1987); A.A. Starobinskii, *JETP Lett.* **37**, 66 (1983).
5. See e.g., A.D. Linde, *Inflation and Quantum Cosmology* (Academic Press, San Diego, CA, 1990); or A. Vilenkin, *Phys. Rev. D* **52**, 3365 (1995).
6. R. Penrose, in *General Relativity: An Einstein Centenary Survey*, edited by S.W. Hawking and W. Israel (Cambridge University Press, Cambridge), p. 581.

7. Y. Hu, M.S. Turner, and E.J. Weinberg, *Phys. Rev. D* **49**, 3830 (1994); A.R. Liddle, *ibid* **51**, 5347 (1995).
8. See e.g., E.W. Kolb and M.S. Turner, *The Early Universe* (Addison-Wesley, Redwood City, 1990), Ch. 8.
9. A.D. Linde, *Phys. Lett. B* **129**, 177 (1983).
10. Q. Shafi and A. Vilenkin, *Phys. Rev. Lett.* **52**, 691 (1984); S.-Y. Pi, *ibid* **52**, 1725 (1984).
11. R. Holman, P. Ramond, and G.G. Ross, *Phys. Lett. B* **137**, 343 (1984).
12. K. Olive, *Phys. Repts.* **190**, 309 (1990).
13. H. Murayama et al., *Phys. Rev. D(RC)* **50**, R2356 (1994).
14. M. Cvetic, T. Hubsch, J. Pati, and H. Stremnitzer, *Phys. Rev. D* **40**, 1311 (1990).
15. E.J. Copeland et al., *Phys. Rev. D* **49**, 6410 (1994).
16. See e.g., M. Gasperini and G. Veneziano, *Phys. Rev. D* **50**, 2519 (1994); R. Brustein and G. Veneziano, *Phys. Lett. B* **329**, 429 (1994); T. Banks et al., *Phys. Rev. D* **52**, 3548 (1995).
17. K. Freese, J.A. Frieman, and A. Olinto, *Phys. Rev. Lett.* **65**, 3233 (1990); L. Randall, M. Soljacic, and A. Guth, *Nucl. Phys. B* **472**, 377 (1996).
18. L. Knox and M.S. Turner, *Phys. Rev. Lett.* **70**, 371 (1993).
19. J. Silk and M.S. Turner, *Phys. Rev. D* **35**, 419 (1986); L.A. Kofman, A.D. Linde, and J. Einsato, *Nature* **326**, 48 (1987).
20. D. La and P.J. Steinhardt, *Phys. Rev. Lett.* **62**, 376 (1991).
21. E.W. Kolb, *Physica Scripta* **T36**, 199 (1991).
22. A. Albrecht et al, *Phys. Rev. Lett.* **48**, 1437 (1982); L. Abbott and M. Wise, *Phys. Lett. B* **117**, 29 (1992); A.D. Linde and A.D. Dolgov, *ibid* **116**, 329 (1982).
23. See e.g., L. Kofman et al., *Phys. Rev. Lett.* **73**, 3195 (1994); **76**, 1011 (1996); E.W. Kolb et al., *ibid* **77**, 4290 (1996); S. Khebnikov and I. Tkachev, *ibid* **77**, 219 (1996); D. Boyanovsky et al, hep-ph/9701304.
24. V.A. Rubakov, M. Sazhin, and A. Veryaskin, *Phys. Lett. B* **115**, 189 (1982); R. Fabbri and M. Pollock, *ibid* **125**, 445 (1983); A.A. Starobinskii *Sov. Astron. Lett.* **9**, 302 (1983); L. Abbott and M. Wise, *Nucl. Phys. B* **244**, 541 (1984).
25. M.S. Turner and F. Wilczek, *Phys. Rev. Lett.* **65**, 3080 (1990); A. Kosowsky, M.S. Turner, and R. Watkins, *ibid* **69**, 2026 (1992).
26. M.S. Turner and L.M. Widrow, *Phys. Rev. D* **37**, 2743 (1988); B. Ratra, *Astrophys. J.* **391**, L1 (1992).
27. See e.g., A. Vilenkin, astro-ph/9610125.
28. See e.g., A.D. Linde, *Phys. Lett. B* **158**, 375 (1985); D. Seckel and M.S. Turner, *Phys. Rev. D* **32**, 3178 (1985).
29. M.S. Turner, A. Cohen, and D. Kaplan, *Phys. Lett. B* **216**, 20 (1989).
30. P.J.E. Peebles, (*Astrophys. J.*, in press).
31. M.S. Turner, *Phys. Rev. D* **44**, 3737 (1991).
32. M. Bucher A.S. Goldhaber, and N. Turok, *Phys. Rev. D* **52**, 3314 (1995); A.D. Linde and A. Mezhlumian, *ibid* **52**, 6789 (1995); J.R. Gott, *Nature* **295**, 304 (1992); A. Green and A. Liddle, *Phys. Rev. D* **55**, 609 (1997)
33. J.R. Gott, *Nature* **295**, 304 (1992); A.H. Guth and E.J. Weinberg, *Nucl. Phys. B* **212**, 321 (1983).
34. A. Vilenkin and S. Winitzki, *Phys. Rev. D* **55**, 548 (1997).
35. C. Copi, D.N. Schramm, and M.S. Turner, *Science* **267**, 192 (1995).
36. D. Tytler, X.-M. Fan and S. Burles, *Nature* **381**, 207 (1996); D. Tytler, S. Burles and D. Kirkman, astro-ph/9612121.
37. Y. Sigad et al, in preparation (1997); J. Willick et al, astro-ph/9612240 A. Dekel, D. Burstein, and S. White, astro-ph/9611108. M. Strauss and J. Willick, *Phys. Repts.* **261**, 271 (1995); A. Dekel, *Ann. Rev. Astron. As-*

342

trophys. **32**, 319 (1994); N. Bahcall, L.M. Lubin and V. Dorman, *Astrophys. J.* **447**, L81 (1995); A. Dekel and M.J. Rees, *Astrophys. J.* **422**, L1 (1994); R.G. Carlberg, H.K.C. Yee, and E. Ellingson, *Astrophys. J.*, 478, 462, astro-ph/9512087 and astro-ph/9704060.

38. See e.g., M.S. Turner, *Phys. Rept.* **197**, 67 (1990); G.G. Raffelt, *ibid* **198**, 1 (1990).

39. See e.g., G. Jungman, M. Kamionkowski, and K. Griest, *Phys. Rept.* **267**, 195 (1996).

40. See e.g., G.G. Ross, *Grand Unified Theories* (Addison-Wesley, Redwood City, 1985), or B. Kayser, F. Gibrat-Debu, and F. Perrier, *The Physics of Massive Neutrinos* (World Scientific, Singapore, 1989).

41. A. Vilenkin, *Phys. Rev. Lett.* **53**, 1016 (1984); M.S. Turner, G. Steigman, and L. Krauss, *Phys. Rev. Lett.* **52**, 2090 (1984); R.L. Davis, *Phys. Rev. D* **35**, 3705 (1987); M. Kamionkowski and N. Toumbas, *Phys. Rev. Lett.* **77**, 587 (1996); M. Ozer and M.O. Taha, *Nucl. Phys.* **B287** 776 (1987); K. Freese et al., *Nucl. Phys.* **B287** 797 (1987); B. Ratra and P.J.E. Peebles, *Phys. Rev. D* **37**, 3406 (1988); L.F. Bloomfield-Torres and I. Waga, astro-ph/9504101; K. Coble, S. Dodelson, and J. Frieman, *Phys. Rev. D* **55**, 1851 (1996).

42. G.F. Smoot et al., *Astrophys. J.* **396**, L1 (1992).

43. J.M. Bardeen, J.R. Bond, N. Kaiser, and A.S. Szalay, *Astrophys. J.*, **304**, 15 (1986).

44. A.R. Liddle and M.S. Turner, *Phys. Rev. D* **54**, 2980 (1996).

45. M.S. Turner and M. White, *Phys. Rev. D* **53**, 6822 (1996).

46. P.J. Steinhardt and M.S. Turner, *Phys. Rev. D* **29**, 2162 (1984).

47. M.S. Turner, *Phys. Rev. D* **48**, 3502 (1993).

48. D.H. Lyth and A.R. Liddle, *Phys. Lett. B* **291**, 391 (1992).

49. A.D. Linde, *Phys. Lett. B* **108**, 389 (1982); A. Albrecht and P.J. Steinhardt, *Phys. Rev. Lett.* **48**, 1220 (1982).

50. D.H. Lyth and E. Stewart, *Phys. Rev. D* **54**, 7186 (1996); A. Linde, *ibid* **49**, 748 (1994).

51. A. Kosowsky and M.S. Turner, *Phys. Rev. D* **52**, R1739 (1995).

52. J. Bartlett et al., *Science* **267**, 980 (1995); P.T.P. Viana et al., *Mon. Not. R. astron. Soc.* **283**, 107 (1996).

53. See e.g., W. Hu and N. Sugiyama, *Phys. Rev. D* **51**, 2599 (1991). For a very fast and accurate numerical routine, see U. Seljak and M. Zaldarriaga, *Astrophys. J.*, 469, 437, astro-ph/9603033.

54. R.K. Sachs and A.M. Wolfe, *Astrophys. J.* **147**, 73 (1967).

55. C.L. Bennett et al, *Astrophys. J.* **464**, L1 (1996).

56. E.F. Bunn et al., *Phys. Rev. D* **54**, 5917 (1997); also see e.g., K. Gorski et al, astro-ph/9608054.

57. R.H. Dicke and P.J.E. Peebles, in *General Relativity: An Einstein Centenary Survey*, edited by S.W. Hawking and W. Israel (Cambridge University Press, Cambridge), p. 504.

58. E.R. Harrison, *Phys. Rev. D* **1**, 2726 (1970); Ya.B. Zel'dovich, *Mon. Not. R. astr. Soc.* **160**, 1p (1972).

59. See e.g., M.S. Turner, J.E. Lidsey and M. White, *Phys. Rev. D* **48**, 4613 (1993).

60. M.S. Turner and Y. Wang, *Phys. Rev. D* **53**, 5727 (1996); B. Allen and S. Koranda, *ibid* **52**, 1902 (1995); S. Dodelson et al, *Phys. Rev. Lett.* **72**, 3444 (1994); R. Crittenden et al, *ibid* **71**, 324 (1993).

61. R. Davis et al., *Phys. Rev. Lett.* **69**, 1856 (1992); F. Lucchin, S. Mattarese, and S. Mollerach, *Astrophys. J.* **401**, L49 (1992); D. Salopek, *Phys. Rev. Lett.* **69**, 3602 (1992); A. Liddle and D. Lyth, *Phys. Lett. B* **291**, 391 (1992); J.E. Lidsey and P. Coles, *Mon. Not. R. astron. Soc.* **258**, 57p (1992); T. Souradeep and V. Sahni, *Mod. Phys. Lett. A* **7**, 3541 (1992).

62. See e.g., E.J. Copeland, E.W. Kolb, A.R. Liddle, and J.E. Lidsey, *Phys. Rev. Lett.* **71**, 219 (1993); *Phys. Rev. D* **48**, 2529 (1993); M.S. Turner, *ibid*, 3502 (1993); *ibid* **48**, 5539 (1993); S. Dodelson, W.H. Kinney and E.W. Kolb, astro-ph/9702166.

63. L. Knox and M.S. Turner, *Phys. Rev. Lett.* **73**, 3347 (1994).

64. U. Seljak, *Astrophys. J.*, **482**, 6, astro-ph/9608131; M. Kamionkowski, A. Kosowsky, and A. Stebbins, astro-ph/9609132.

65. M.S. Turner, *Phys. Rev. D* **55**, R435 (1997).

66. See e.g., A. Abramovici et al., *Science* **256**, 325 (1992) and in *Particle and Nuclear Astrophysics and Cosmology in the Next Millennium*, eds. E.W. Kolb and R.D. Peccei (World Scientific, Singapore, 1995), p. 398; N. Christensen, *Phys. Rev. D* **46**, 5250 (1992); E. Flanagan, *ibid* **48**, 2389 (1993).

67. B. Allen, qr-gc/9604033.

68. Pre-phase A Design Study for LISA.

69. K.S. Thorne, in *300 Years of Gravitation*, eds. S.W. Hawking and W. Israel (Cambridge Univ. Press, Cambridge, 1987), p.330.

70. S.D.M. White, C. Frenk and M. Davis, *Astrophys. J.* **274**, L1 (1983).

71. S. Colombi, S. Dodelson and L. Widrow, *Astrophys. J.* **458**, 1 (1996).

72. G. Blumenthal et al., *Nature* **311**, 517 (1984).

73. See e.g., J.P. Ostriker and R. Cen, *Astrophys. J.* **464**, 270, astro-ph/9601021; Y. Zhang et al, astro-ph/9611224; T. Abel et al, astro-ph/9608040;

74. S. Dodelson, E. Gates and M.S. Turner, *Science* **274**, 69 (1996).

75. For example, S. Parke, *Phys. Rev. Lett.* **74**, 839 (1995); C. Athanassopoulos et al, *ibid* **75**, 2650 (1995); J.E. Hill, *ibid*, 2654 (1995); K.S. Hirata et al, *Phys. Lett. B* **280**, 146 (1992); Y. Fukuda et al, *ibid* **335**, 237 (1994); R. Becker-Szendy et al, *Phys. Rev. D* **46**, 3720 (1992).

76. S. Weinberg, *Rev. Mod. Phys.* **61**, 1 (1989).

77. K. Fisher et al., *Mon. Not. R. astron. Soc.* **266**, 50 (1994).

78. S. Hancock and G. Rocha, astro-ph/9612016.

79. S.J. Perlmutter et al., *Astrophys. J.* **483**, 565, astro-ph/9608192.

80. M.S. Turner and M. White, astro-ph/9701138.

81. C. Hagmann et al, *Nucl. Phys. B (Proc. Suppl.)* **51**, 209 (1996).

82. T. Shutt et al, in *Proceedings of the XVIIIth Texas Symposium on Relativistic Astrophysics (Chicago, IL, 1996)*, edited by A. Olinto (World Scientific, Singapore, 1997).

83. J. Peacock and S. Dodds, *Mon. Not. Roy. astron. Soc.* **267**, 1020 (1994).

84. A.R. Liddle et al., *Mon. Not. R. astron. Soc.* **282**, 281 (1996); *ibid* **281**, 531 (1995); J.S. Bagla, T. Padmanabhan and J.V. Narlikar, astro-ph/9511102; S. Cole et al, astro-ph/9702082.

85. S.D.M. White, G. Efstathiou, and C.S. Frenk, *Mon. Not. Roy. astron. Soc.* **262**, 1023 (1993).

86. R. Carlberg et al., *J. R. Astron. Soc. Canada* **88**, 39 (1994); P.T.P. Viana and A. Liddle, *Mon. Not. R. astron. Soc.* **281**, 323 (1996); J.R. Bond and S. Myers, *Astrophys. J. Supp.* **103**, 63. (1996); V.R. Eke, S. Cole and C.S. Frenk, astro-ph/9601088; U.-L.Pen, astro-ph/9610147.

87. K. Lanzetta, A.M. Wolfe, and D.A. Turnshek, *Astrophys. J.* **440**, 435 (1995); L.J. Storrie-Lombardi, R.G. McMahon, M.J. Irwin, and C. Hazard, *Mon. Not. R. astron. Soc.*, 282, 1330, astro-ph/9503089 (1996).

88. J.P. Ostriker, *Ann. Rev. Astron. Astrophys.* **31**, 689 (1993).

89. A. Liddle and D. Lyth, *Phys. Repts.* **231**, 1 (1993).

90. M. White and J. Silk, *Phys. Rev. Lett.* **77** 4704 (1996).

91. See e.g., W. Freedman, astro-ph/9612024; A. Riess, R.P. Krishner, and W. Press, *Astrophys. J.* **438**, L17 (1995); R. Giovanelli et al., *Astrophys. J.* **477**, 1, astro-ph/9612072.

92. M. Bolte and C.J. Hogan, *Nature* **376**, 399 (1995).

93. B. Chaboyer et al, *Science* **271**, 957 (1996).

94. J. Cowan, F. Thieleman, and J. Truran, *Ann. Rev. Astron. Astrophys.* **29**, 447 (1991).

95. S.D.M. White et al, *Nature* **366**, 429 (1993).

96. U.G. Briel et al, *Astron. Astrophys.* **259**, L31 (1992).

97. D.A. White and A.C. Fabian, *Mon. Not. R. astron. Soc.* **273**, 72 (1995).

98. A. Evrard, astro-ph/9701148.

99. E. Turner, *Astrophys. J.* **365**, L43 (1990).

100. C.S. Kochanek, *Astrophys. J.* **466**, 638 (1996).

101. L. Krauss and M.S. Turner, *Gen. Rel. Grav.* **27**, 1137 (1995); J.P. Ostriker and P.J. Steinhardt, *Nature* **377**, 600 (1995); J.S. Bagla, T. Padmanabhan and J.V. Narlikar, Comm. Astrophy. 18, 275, astro-ph/9511102; A.R. Liddle et al., *Mon. Not. R. astron. Soc.* **282**, 281 (1996).

102. J. Bartlett et al., *Science* **267**, 980 (1995);

103. C.H. Lineweaver, D. Barbosa, A. Blanchard, and J. Bartlett, *Ast. & Astrophysics*, 322, 365, astro-ph/9612146, (1997)

104. See e.g., E.L. Turner, R. Cen, and J.P. Ostriker, *Astron. J.* **103**, 1427 (1992); X. Shi, L.M. Widrow, and L.J. Dursi, *Mon. Not. R. astron. Soc.* **281**, 565 (1996); X. Shi and M.S. Turner, astro-ph/9705029.

105. T. Kundic et al., *Astrophys. J.*, **482**, 68, astro-ph/9610162; P.L. Schechter et al, *Astrophys. J.*, **475**, 85, astro-ph/9611051; W.L. Holzapfel et al, astro-ph/9702224; S.T. Meyers et al, astro-ph/9703123.

106. See http://www-sdss.fnal.gov:8000/.

107. See http://www.ast.cam.ac.uk/AAO/2df.

108. See e.g., J.E. Ellis, *Physics World*, July 1994, p. 31.

109. See e.g., L. Knox, *Phys. Rev. D* **52**, 4307 (1995); G. Jungman, M. Kamionkowski, A. Kosowsky, and D. Spergel, *Phys. Rev. D* **54**, 1332 (1996); J.R. Bond, G. Efstathiou, and M. Tegmark, astro-ph/9702100; M. Zaldarriaga, D.Spergel and U. Seljak, astro-ph/9702157.

110. S. Dodelson, E. Gates, and A.S. Stebbins, *Astrophys. J.* **467**, 10 (1996).

PRIMORDIAL CHEMISTRY
AND COSMIC BACKGROUND RADIATION

M. SIGNORE

Ecole Normale Superieure, Laboratoire de Radioastronomie
24, rue Lhomond 75231 Paris cedex 05 (France)

P. ENCRENAZ, R. MAOLI

Observatoire de Paris, D.E.M.I.R.M.
61, avenue de l'Observatoire 75014 Paris (France)

B. MELCHIORRI, F. MELCHIORRI

Universita di Roma-La Sapienza, Dipartimento di Fisica
2, piazzale Aldo Moro 0185 Roma (Italy)

D. PUY

International School for Advanced Studies
2, via Beirut 34013 Trieste (Italy)

Observatoire de Paris-Meudon, D.A.E.C.
5, place Jules Janssen 92190 Meudon (France)

1. Introduction

Some of the most important questions about the interaction of the primordial molecules with the Cosmic Background Radiation (CBR) are briefly discussed. Key issues include i) cosmological nucleosynthesis, ii) primordial chemistry, iii) the main molecular radiative processes. Searches for some primordial molecular signals are also presented.

C. H. Lineweaver et al. (eds.), The Cosmic Microwave Background, 345–364.
© *1997 Kluwer Academic Publishers.*

2. A Brief Overview of Cosmological Nucleosynthesis

2.1. PHYSICAL CONDITIONS OF STANDARD BIG BANG NUCLEOSYNTHESIS (SBBN)

The Universe has passed through a hot phase with $T > 10^{12}$K during which, it was homogeneous and isotropic and its components were in thermal equilibrium (see Pierre Salati's contribution to this volume). As long as the protons and the neutrons were in thermal equilibrium, which was maintained by the reactions:

$$n + e^+ \longleftrightarrow p + \bar{\nu}$$

$$p + e^- \longleftrightarrow n + \nu$$

$$n \longleftrightarrow p + e^- + \nu$$

the neutron-proton ratio was given by the relation:

$$\frac{n_n}{n_p} \sim exp(-\Delta_m c^2/k_B T)$$

where $\Delta mc^2 = |m_n - m_p|c^2 \sim 1.29$ MeV.

The reaction rate Γ was roughly equal to the expansion rate H at $t_\star \sim 1$ sec when $T_\star \sim 10^{10}$K and $(n_n/n_p)_\star \sim 1/6$. Then, the neutron-proton ratio left its *equilibrium curve*, i.e., decreased more slowly until the value $(n_n/n_p)_{nuc} \sim 1/7$ when nucleosynthesis effectively began at $t_{nuc} \sim 10^2$ sec. All the while, neutrons and protons were colliding to make deuterium:

$$n + p \to d + \gamma$$

But for $T > 10^9$ K ($t < 100$ sec) photons had enough energy to dissociate d. Thus, at early times, the reaction:

$$d + \gamma \to n + p$$

was faster than

$$n + d \to {}^3H + \gamma$$

$$p + d \to {}^3He + \gamma$$

and abundances of d, ${}^3He$, ${}^4He$ were small. When $T < 10^9$K ($t > 100$ sec), these laster reactions were faster and abundances of d, ${}^3H$, ${}^3He$ built up and these elements further reacted to form ${}^4He$:

$$n + {}^3He \to {}^4He$$

$$p + {}^3H \to {}^4He$$

$$d + d \rightarrow {}^4He.$$

Then, collisions of d, ${}^3H$ and ${}^3He$ on ${}^4He$ began producing heavier elements: ${}^6Li$, ${}^7Li$ and ${}^7Be$. The reactions of primary importance in the production of the light elements are shown in Figure 1. As the temperature dropped, the reactions slowed down relative to the expansion rate until the abundances froze out. At $t \sim 10^3$ sec, the epoch of BBN was over.

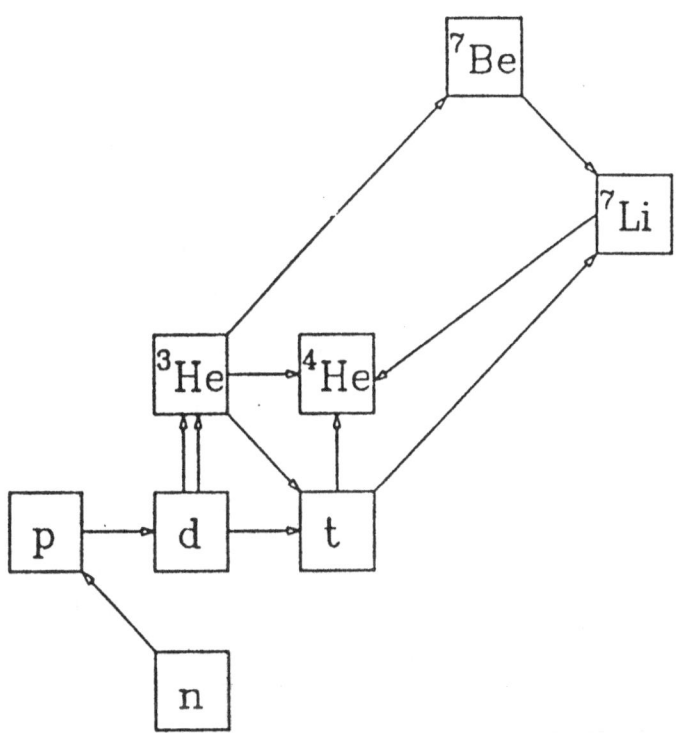

Figure 1. Network diagram of the main reactions in the SBBN model. Deuterium, the first *brick* of BBN, is a *passage obligé* of BBN.

2.2. PREDICTED AND INFERRED PRIMORDIAL ABUNDANCES OF SBBN

Cosmologists, from the reaction rates corresponding to the reactions of Figure 1, compute and predict the abundances of d, ${}^3He$, ${}^4He$ and ${}^7Li$ as a function of the ratio of nucleons to photons η and predict also very tiny abundances of heavier elements. η -or Ω_b, the density of baryons in units of the critical density- is the SBB free parameter. For a recent review on the SBBN model, see Schramm (1991), Smith et al (1993), Steigman (1995).

Astrophysicists deduce the primordial abundances of d and ^3He from observations of the solar wind and of meteorites, the primordial abundance of ^4He from observations of nearby HII-regions and dwarf galaxies, the primordial abundance of ^7Li from observations of old population II stars.

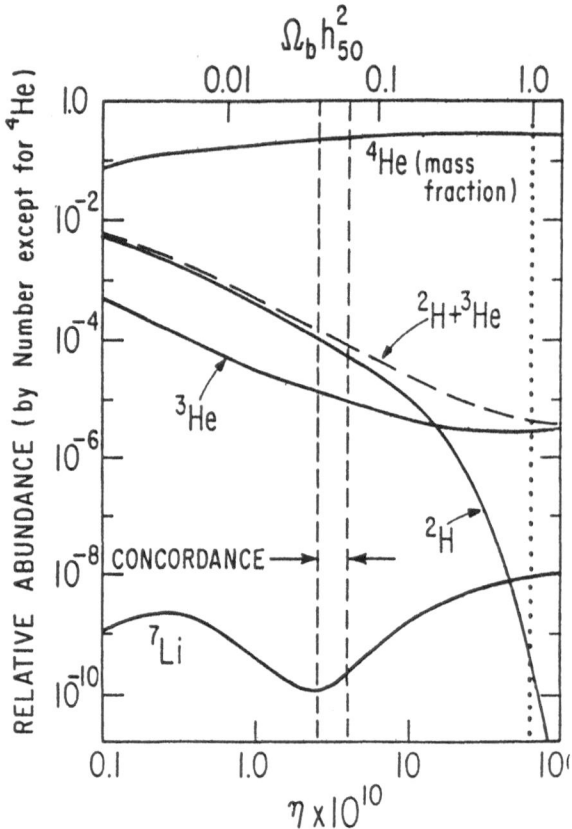

Figure 2. Light element abundances versus η for the ^4He mass fraction (Y_p) and $(d+^3He)/H$. $^7Li/H$. The vertical band shows the values of η for which predicted and inferred abundances are compatible (from Schramm 1991).

TABLE 1. Inferred primordial abundances

deuterium	deuterium + helium 3
$d/H \geq 1.8\,10^{-5}$	$(d+^3He)/H \leq 9\,10^{-5}$
helium 4	lithium
$0.21 < Y_p < 0.24$	$1.1\,10^{-10} < \,^7Li/H < 2.3\,10^{-10}$

The comparison of predicted abundances with the primordial abundances inferred from observational constraints on η or Ω_b (see Figure 2 and Table 1).

2.3. RECENT DEVELOPMENTS

2.3.1. *On the theoretical side*

Nuclear astrophysicists have recently focused on non standard models. One class of which postulates the existence of small scale inhomogeneities which might have been produced by the quark-hadron phase transition (if it was a first-order phase transition) at some $t \sim 10^{-5}$sec after the Big Bang. These inhomogeneities created zones of neutron-rich nucleosynthesis and could naturally produce, in particular, extra d and ^7Li:

$$d/H \sim 10^{-4} \qquad ^7Li/H \sim 10^{-9}$$

Moreover, due to these inhomogeneities short-lived nuclides were created (see below) which allowed for more reaction pathways to the heavier elements. Therefore some inhomogeneous BBN models (IBBN) produce estimates well in excess of those in SBBN not only for d and ^7Li but also for heavier nuclides from ^7Li to ^{11}B.

These results suggest that astronomers can potentially determine the level of inhomogeneity at some 10^{-5}sec after the Big Bang. Unfortunately these IBBN models contain more parameters, many unknown reaction rates (see below) and thus more uncertainties in their predictions. For a review of these IBBN models, see Malaney and Mathews (1993) and references therein and also Jedamzik et al (1994).

2.3.2. *On the experimental side*

As said above, the nuclei from ^7Li to ^{11}B have very recently come to be perceived as very important because their abundances have been put forward as possible tests of BBN theories. Moreover, all nuclides heavier than 11 a.m.u. (atomic mass unit) are funneled through ^{11}B on their nucleosynthesis paths through reactions such as:

$$^8Li + {}^4He \rightarrow {}^{11}B + n$$

In fact several reactions involving ^8Li, a nucleus with a half-life of 0.840 sec could be crucial: network calculations indicate that ^8Li is indeed pivotal to nucleosynthesis in IBBN models. Therefore, any reaction that either makes or destroys ^8Li seems to be very important (see Figure 3).

Recently, radioactive nuclear beam studies of the reactions of short-lived nuclides have already yielded some cross sections and reaction rates of reactions involved in Big Bang nucleosynthesis. In particular:

$$^8Li + {}^4He \rightarrow {}^{11}B + n,$$

but also:

$$^{8}Li + d \rightarrow {}^{7}Li + {}^{3}H$$

$$^{8}Li + d \rightarrow {}^{9}Be + n \ \text{etc...}$$

have been studied by several radiactive beam facilities (University of Notre Dame in USA, Riken in Japan). From the very preliminary results of these new experimental studies, it seems that differences in the abundance predictions of ^{7}Li, ^{9}Be, ^{11}B in SBBN and IBBN models may be large. For a review of the preliminary results of these experimental studies see Boyd (1993) and references therein and also Balbes et al (1995), Gu et al (1995).

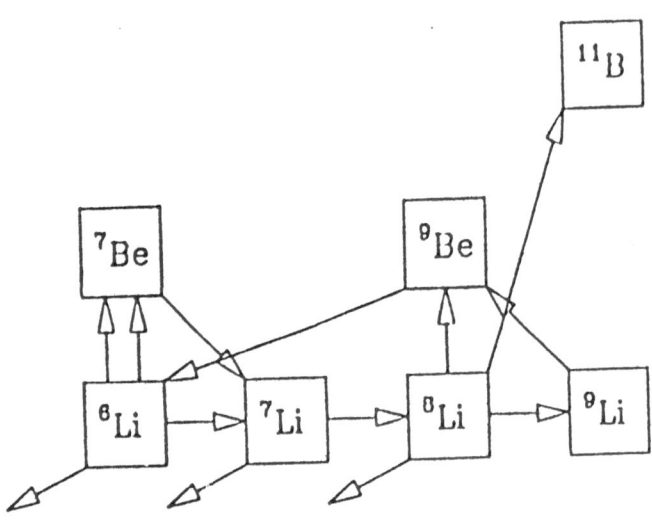

Figure 3. Reaction network in region of ^{8}Li

2.3.3. *On the observational side*

Some observations of metal deficient stars seem to indicate that ^{9}Be and ^{11}B may be products of Big Bang nucleosynthesis and therefore may be compatible with some IBBN models (Gilmore et al 1991, Ryan et al 1992, Duncan et al 1992). For a recent description of the lithium problem see Signore et al (1994) and references therein. More recently, observations of deuterium in Lyman-limit clouds, at high z, with the KECK-10 m telescope, determine a d/H ratio of the order of 10^{-4} which may reflect the primordial d/H ratio (see Songalia et al 1994, Carswell et al 1994, Rugers & Hogan 1996). But observational uncertainties include the possibility of interlopers that may lead to an overestimate of the d/H ratio (see in particular Tytler 1995).

2.3.4. *Summary*
For the following, let us only recall that in:

• Standard Big Bang Nucleosynthesis:

$$d/H \sim 10^{-5}, \ ^7Li/H \sim 10^{-10}$$

There are only traces of Be, B.

• Inhomogeneous Big Bang Nucleosynthesis:

$$d/H \sim 10^{-4}, \ ^7Li/H \sim 10^{-9}$$

The amounts of Be, B and even heavier elements are more important than in SBBN.

3. Evolution of Primordial Chemistry

In this section, we will discuss the evolution of primordial chemistry:

i) just after the short period in the expansion of the Universe referred to as the epoch of recombination
ii) later in the evolution of the Universe; during the gravitational collapse of protostructures.

3.1. MOLECULAR EVOLUTION AFTER RECOMBINATION

In standard Big Bang theory, chemistry took place around the epoch of recombination. At $z \sim 1000$, the chemical species essentially were:

$$H, \ H^+, \ D, \ D^+, \ He \text{ and } Li.$$

As the Universe expanded and cooled, different routes led to molecular formation; the possible chemical reactions are listed in Table 2. The main primordial molecules essentially were:

$$H_2, \ HD \text{ and } LiH.$$

TABLE 2. The reaction rates are $k=\gamma \left(\dfrac{T}{300}\right)^{\alpha} \exp(\dfrac{-\beta}{T})$, where γ is in $cm^3 s^{-1}$, β in Kelvin. The enthalpy ΔH is in eV.

	reactions	γ	α	β	ΔH
(1)	$H + e \longmapsto H^- + h\nu$	$3\ 10^{-16}$	1	0	
(2)	$H^- + H \longmapsto H_2 + e$	$1.5\ 10^{-9}$	0	0	3.71
(3)	$H^+ + H \longmapsto H_2^+ + h\nu$	$1.8\ 10^{-18}$	1.5	0	
(4)	$H_2^+ + H \longmapsto H_2 + H^+$	$6.4\ 10^{-10}$	0	0	1.81
(5)	$H_2 + h\nu \longmapsto H + H$	see Section 1			
(6)	$H^- + h\nu \longmapsto H + e$				
(7)	$H_2^+ + h\nu \longmapsto H^+ + H$				
(8)	$H^+ + H^- \longmapsto H + H$	$2.3\ 10^{-7}$	-0.5	0	12.84
(9)	$H^+ + H^- \longmapsto H_2^+ + e$	$8.83\ 10^{-15}$	-0.32	0	1.9
(10)	$H_2^+ + H^- \longmapsto H_2 + H$	$2.3\ 10^{-7}$	-0.5	0	14.66
(11)	$H^+ + e \longmapsto H + h\nu$	$3.5\ 10^{-12}$	-0.70	0	
(12)	$H_2^+ + e \longmapsto H + H$	$1.68\ 10^{-8}$	-0.29	0	10.95
(13)	$H^+ + D \longmapsto D^+ + H$	10^{-9}	0	41	-0.01
(14)	$H + D \longmapsto HD + h\nu$	10^{-25}	0	0	
(15)	$D^+ + H_2 \longmapsto H^+ + HD$	$1.7\ 10^{-9}$	0	0	0.06
(16)	$D^+ + H \longmapsto HD^+ + h\nu$	$1.8\ 10^{-18}$	1.5	0	
(17)	$H^+ + D \longmapsto HD^+ + h\nu$	$1.8\ 10^{-18}$	1.5	0	
(18)	$HD^+ + H \longmapsto H^+ + HD$	$6.4\ 10^{-10}$	0	0	1.85
(19)	$HD + h\nu \longmapsto H + D$				
(20)	$HD^+ + h\nu \longmapsto H^+ + D$				
(21)	$HD^+ + h\nu \longmapsto D^+ + H$				
(22)	$D^+ + e \longmapsto D + h\nu$	$3.5\ 10^{-12}$	-0.70	0	
(23)	$HD^+ + e \longmapsto D + H$	$1.39\ 10^{-8}$	-0.29	0	10.93
(24)	$D^+ + H \longmapsto H^+ + D$	10^{-9}	0	0	0.01
(25)	$H^+ + HD \longmapsto D^+ + H_2$	$1.7\ 10^{-10}$	0	462	-0.06
(26)	$H_2^+ + D \longmapsto D^+ + H_2$	$6.4\ 10^{-10}$	0	0	1.81
(27)	$D + H_2 \longmapsto HD + H$	$1.09\ 10^{-18}$	0	0	0.05
(28)	$Li + H \longmapsto LiH + h\nu$	10^{-17}	0	0	
(29)	$LiH + h\nu \longmapsto Li + H$				

Many authors have described in detail primordial molecular formation and evolution; in particular, Puy et al (1993) have simultaneously solved the coupled chemical equations corresponding to the reactions of Table 2 and the density and temperature evolution equations; let us recall their equations for:

-the co-moving particle density evolution:

$$\frac{1}{n}\cdot\frac{dn}{dt} = -3H_o(1+z)^{3/2} + \frac{1}{n}\cdot(\frac{dn}{dt})_{ch}$$

-the adiabatic expansion of radiation:

$$\frac{dT_{rad}}{dt} = -T_{rad}H_o(1+z)^{3/2}$$

-the thermal evolution of matter:

$$\frac{dT_{mat}}{dt} = -2T_{mat}H_o(1+z)^{3/2} + \frac{8\sigma_T a_{bb}}{3m_e c}(T_{rad} - T_{mat})x_e$$
$$+ 2\frac{(\Gamma_{mol} - \Lambda_{mol})}{3nk} + [\frac{2}{3}\frac{\Theta_{ch}}{nk} - \frac{T_{mat}}{n}\cdot(\frac{dn}{dt})_{ch}]$$

where Ω_0 is the density paramater, T_{rad} and T_{mat} are the temperatures of the radiation and of the matter respectively, H_o the Hubble constant, a_{bb} the black body constant, σ_T the Thomson cross-section, x_e the ionisation fraction, m_e the electron mass and c the speed of light. Γ_{mol} is the energy gain from photons per unit volume, Λ_{mol} the energy radiated per unit volume; $(\frac{dn}{dt})_{ch}$ is the variation of density due to chemistry; $\Theta_{ch} = \sum_i k_i n_i^X n_i^Y \Delta H_i$; k_i is the reaction rate of reaction i, n_i^X is the abundance of the species X in the reaction i of enthalpy ΔH_i given in Table 2.

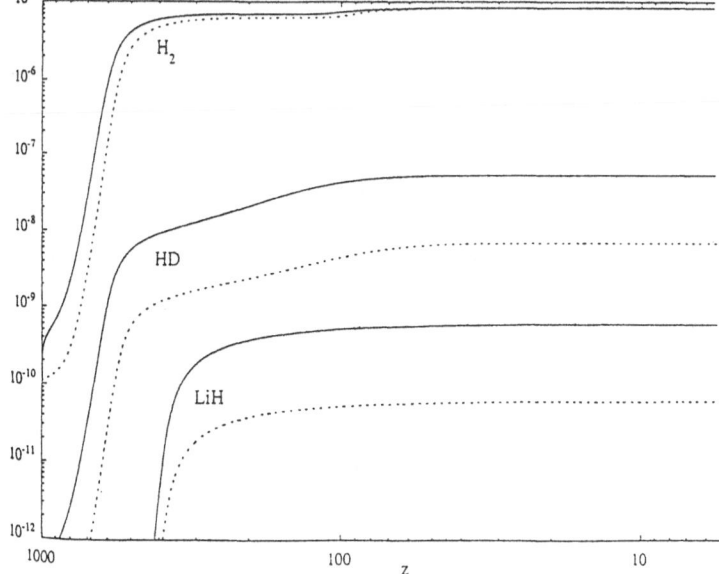

Figure 4. Fractional abundances of H_2, HD, LiH -versus redshift z- for model A (dotted line) SBBN; for model B (continuous line) IBBN, from Puy et al (1993).

Relative abundances of H_2, HD and LiH are shown on Figure 4 for two models: model A corresponds to initial conditions given by SBBN: $d/H \sim 10^{-5}$, $Li/H \sim 10^{-10}$; model B corresponds to initial conditions given by IBBN: $d/H \sim 10^{-4}$, $Li/H \sim 10^{-9}$.

3.2. MOLECULAR EVOLUTION DURING THE GRAVITATIONAL COLLAPSE OF PROTOSTRUCTURES

In the post-recombination epoch, most of the structure formation scenarios involve gravitational instability which leads to large primordial clouds which thereafter collapse. Because the protocloud temperature increased with contraction, a cooling mechanism was crucial to the first generation structure formation by lowering pressure opposing gravity, i.e., by allowing continued collapse of Jeans-unstable protoclouds.

Many authors have examined this problem (Hutchins 1976, Palla et al 1983, Silk 1983) introducing molecular coolants. Lahav (1986) elaborated a very simple description of the evolution of a protocloud with a three-phase model and with H_2 as the main cooling agent. More recently, Puy & Signore (1996), from this simple description, but with a more complete chemistry (primordial H_2, HD and LiH molecules) considered the three phases of the protoclouds:

i) a linear evolution which approximatively follows the expansion

ii) a *turn around* epoch when the protocloud reaches its maximum value

iii) a non-linear evolution of the collapse of the protocloud

• The overdense region is supposed to be initially spherical with an initial mass spectrum:

$$\frac{\delta\rho}{\rho} = (\frac{M}{M_\star(o)})^\alpha (1+z)^{-1}$$

where $M_\star(o) = 10^{15} M_\odot$, $\alpha = -1/3$.

• They adopted the IBBN abundances (Figure 4, model B), the molecular abundances calculated in the Section 3.1 (Puy et al 1993) as the initial conditions of the collapse phase, i.e., at the turn-around point:

$$1 + z_{ta} = (\frac{3\pi}{4})^{-2/3} \cdot (\frac{M}{M_\star(o)})^{-1/3}$$

$$r_{ta} = \frac{1.66\ 10^{20}}{1 + z_{ta}} (\frac{M}{\Omega_b h^2 M_\odot})^{1/3}\ \text{cm}$$

$$\rho_{ta} = (\frac{3\pi}{4})^2 1.88\ 10^{-29}\ \Omega_b h^2 (1 + z_{ta})^3\ \text{g/cm}^3$$

$$T_{ta} = \left(\frac{3\pi}{4}\right)^{4/3} T_{mat}(z_{ta})$$

where: Ω_b is the baryonic density in units of critical density; $h = H_o/100$ km/s/Mpc and $T_{mat}(z_{ta})$ is the temperature of the matter during the expansion of the Universe at the redshift z_{ta} which is also calculated in Section 3.1

• Let us recall the evolution equations for the collapse phase:

-the evolution of the density

$$\frac{dn}{dt} = -\frac{3n}{r}\cdot\frac{dr}{dt} + \left(\frac{dn}{dt}\right)_{ch}$$

-the evolution of the temperature of the matter

$$\frac{dT_{mat}}{dt} = -2\frac{T_{mat}}{r}\cdot\frac{dr}{dt} + \frac{2}{3nk}\Psi_{mol}$$

-the evolution of the cloud radius

$$\frac{d^2r}{dt^2} = \frac{5k}{\mu m_H}\cdot\frac{T_{mat}}{r} - \frac{GM}{r^2}$$

where M is the protocloud mass, m_H is the mass of the hydrogen atom, μ is the molecular weight and

$$\Psi_{mol} = \Gamma_{mol} - \Lambda_{mol} + \Theta_{mol}$$

where Γ_{mol}, Λ_{mol}, Θ_{ch} have the same meanings as in Section 3.1. Moreover, the dynamical equations are coupled with the chemical equations -corresponding to the chemical network (Table 2)- written for a gravitational collapse.

Figures 5 and 6 show the dynamical and chemical evolution of a 10^9 $M_\odot$ cloud and of a 10^{10} $M_\odot$ cloud which collapse at $z_{ta} \sim 55$ and at $z_{ta} \sim 25$, respectively. Let us only note that for this range of cloud masses, H_2 and LiH increase while HD decreases during the collapse phase of these protostructures. This point can be crucial for observations of lines expected to be emitted by these primordial molecules in collapsing protoclouds.

These observations may constrain protocloud evolution and, in that sense, primordial molecules could be seen as tracers of structure evolution. Moreover, if these molecules are LiH, observations of their lines may also constrain the primordial ^7Li abundance and may help to discriminate between SBBN and IBBN models and, therefore, may provide unique information on the physical conditions of the Universe at some 10^{-5} sec after the Big Bang.

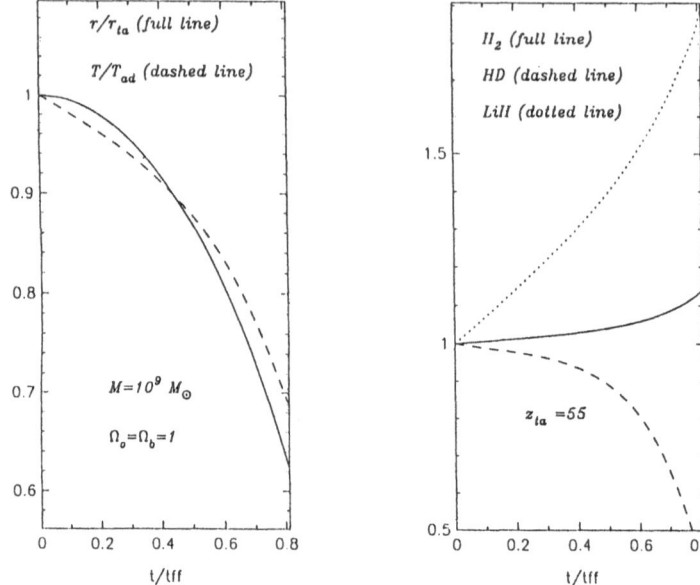

Figure 5. Evolution of radius and temperature of a 10^9 M$_\odot$ cloud (left); evolution of relative abundances from $z_{ta} \sim 55$ and $T_{ta} \sim 120$K (right) (from Puy & Signore 1996).

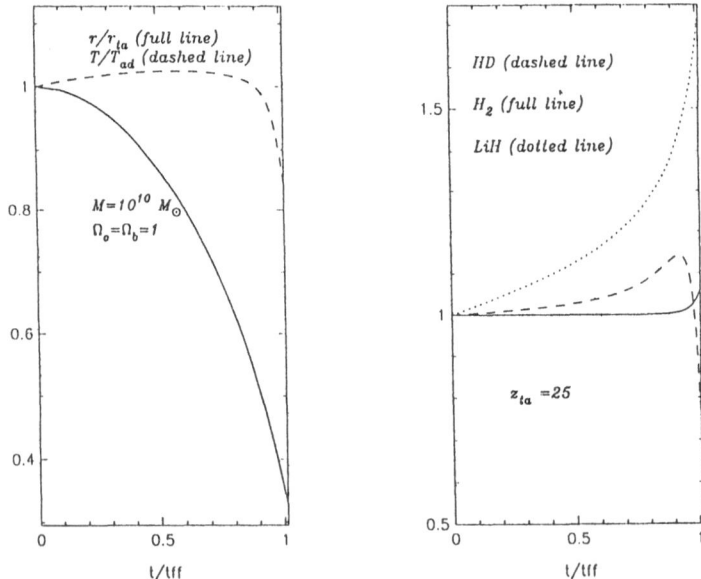

Figure 6. Evolution of a 10^{10} M$_\odot$ cloud from $z_{ta} \sim 25$ and $T_{ta} \sim 25$K (from Puy & Signore 1996).

3.3. NEW DEVELOPMENTS

Recently Dalgarno (private communication) pointed out the possibility that at the time of hydrogen recombination the lithium was still ionized because of its lower ionization potential and so its reactions would initiate a more complete lithium chemistry. Moreover, with the inclusion of a new quantal rate coefficient for the radiative association of lithium and hydrogen, Stancil et al (1996) calculated a fractional abundance of LiH which is smaller than in previous studies. Nevertheless, the situation in the context of gravitational collapse is better, in the sense that the three body reactions are possibly efficient and of course increase the abundance of LiH inside the collapse, which could provide an observational signature. This new field is in progress.

Sciama (1990) developed the idea that dark matter particles (neutrinos in particular) can decay with the emission of ionizing radiation. In this context, matter becomes fully ionized and so collisional reactions with ions are important. In this sense, the chemistry could also be strongly modified, opening the possibility of an observational signature.

But as we will see in the next section, the observability of the LiH lines -more generally of the effects of LiH molecules on the CBR- strongly depends on the final LiH abundance which is a function of lithium produced by BBN and of the efficiency of its conversion into LiH. Let us only recall that for: i) the primordial ^7Li: $^7Li/H \sim 10^{-10}$ in SBBN; $^7Li/H \sim 10^{-9}$ in IBBN; ii) the LiH conversion ratio $[LiH]/[Li]$ during the expansion is summarized in Table 3.

TABLE 3. ratio $[Lih]/[Li]$

$[LiH]/[Li]$	authors
0.01-0.1	Dubrovich 1977
0.002-0.1	Lepp & Shull 1984
0.6	Puy et al 1993
1	Khersonski & Lipovka 1993
0.001-0.002	Palla, Galli & Silk 1995
2-3 10^{-7}	Stancil, Lepp & Dalgarno 1996

4. Primordial Molecules and Cosmic Background Radiation

4.1. MAIN MOLECULAR RADIATIVE PROCESSES

The main molecular radiative processes are: emission, absorption, elastic scattering between CBR photons and primordial molecules and luminescence.

• Emission and absorption

Puy et al (1993) recently analysed in detail the first two processes. Taking into account collisional excitation and the radiative de-excitation of rotational levels, they introduced and calculated a molecular cooling function Λ_{mol} and therefore estimated the emission of photons which can add to the CBR; considering radiative excitation and collisional de-excitation of rotational levels, they defined a molecular heating function Γ_{mol} and also estimated the absorption of CBR photons. Then, they compared the expected fluxes of each molecule to the observed CBR and concluded that the diffuse radiative background due to radiative transitions of the primordial molecules has a maximum far below the CBR.

• **Scattering of CBR photons and primordial molecules**

During an elastic scattering between CBR photons and primordial molecules, a photon is absorbed and reemitted at the same frequency but not in the same direction. This process could be negligible because of the low abundance of primordial molecules. La (1989), Dubrovich (1993) and more recently Maoli et al (1994, 1996), de Bernardis et al (1994) and Signore et al (1996) have analysed in detail the resonant scattering between CBR photons and primordial molecules. They have shown that this effect -which could not alter the CBR spectrum- could alter the primary spatial distribution of the CBR, i.e., the Cosmic Background Anisotropies (CBA).

4.2. CONDITIONS FOR BLURRING OF PRIMARY CBA

For a given molecular line $i - j$, a CBR photon arriving from the last scattering surface (LS) could be resonantly scattered if, for the frequencies, we have the relation:

$$\nu_{ij} = \left(\frac{1+z}{1+z_{LS}} \right) . \nu_{LS}$$

Thus we estimate the resonant scattering depth

$$\tau_\nu = \int \sum_j \sigma_j n_{spec} dl \ , \ \sigma_j = \frac{\lambda^3 A_j}{4e} . \frac{\nu}{\Delta \nu_D}$$

where σ_j is the resonant scattering cross-section, n_{spec} the *relative abundance* of the rotational level j and dl a cosmological differential distance; $\Delta \nu_D$ is the Doppler broadening and A_j the Einstein coefficients.

Maoli et al (1994), adopting the following three values for LiH abundances:

$$[LiH] = 2 \times 10^{-12}, \ 5 \times 10^{-10}, \ 2 \times 10^{-9}$$

assuming LiH formation at $z_f \sim 400$ and LiH destruction at $z_d \sim 5$ and 70 calculated the optical depth τ (see Figure 7).

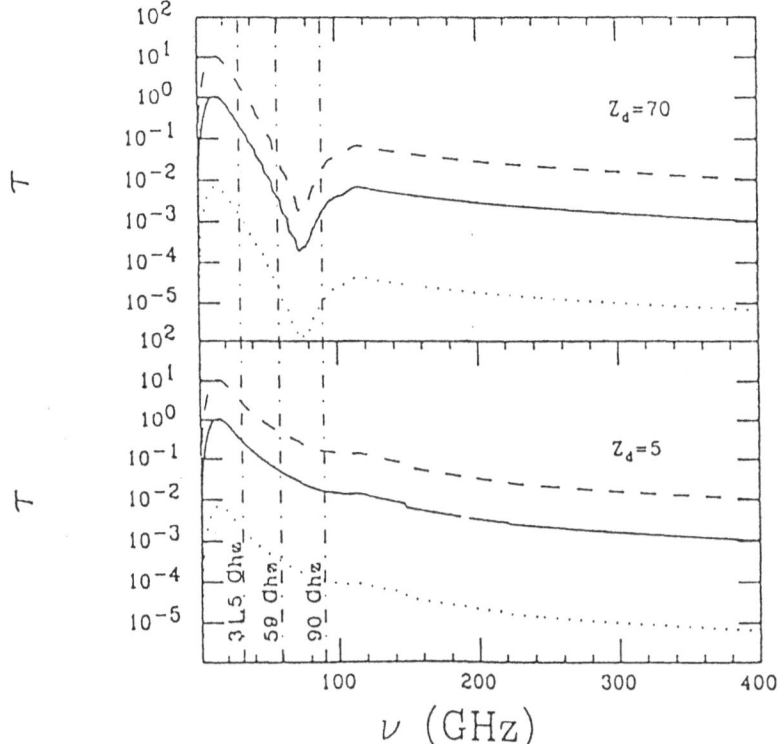

Figure 7. Total optical depth due to all molecules. (only *LiH* is important). $z_f \sim 400$; $z_d \sim 70$ (top); $z_d \sim 5$ (bottom). Three *LiH* abundances: $2\,10^{-12}$, $5\,10^{-10}$, $2\,10^{-9}$ (from bottom to top). The scattering is efficient ($\tau > 1$) at low frequencies ($\nu < 60$ GHz) for the highest *LiH* abundances (from Maoli et al 1994).

They found $\tau > 1$ for $[LiH] > 2\,10^{-9}$ and $\nu < 60$ GHz.
Let us also stress that a given power spectrum of primary CBR anisotropies could be affected through resonant scattering, by an amount $exp(-\tau_\nu)$ and on angular scales θ, if and only if

$$\tau_\nu \geq 1 \text{ and } \theta < \theta_H$$

where θ_H is the angular diameter of the horizon at the redshift where the scattering is effective (see Figure 8).
Therefore, with this crude lithium chemistry, one may conclude that, in the *radio region* ($\nu < 60$ GHz), primary CBA, at intermediate angular scale, could be erased by resonant molecular scattering.
In the *millimetric region*, this effect of resonant molecular scattering is negligible. Let us remark that this effect is very similar to the blurring of CBA due to a secondary ionization. But in the molecular case, it is strongly frequency dependent.

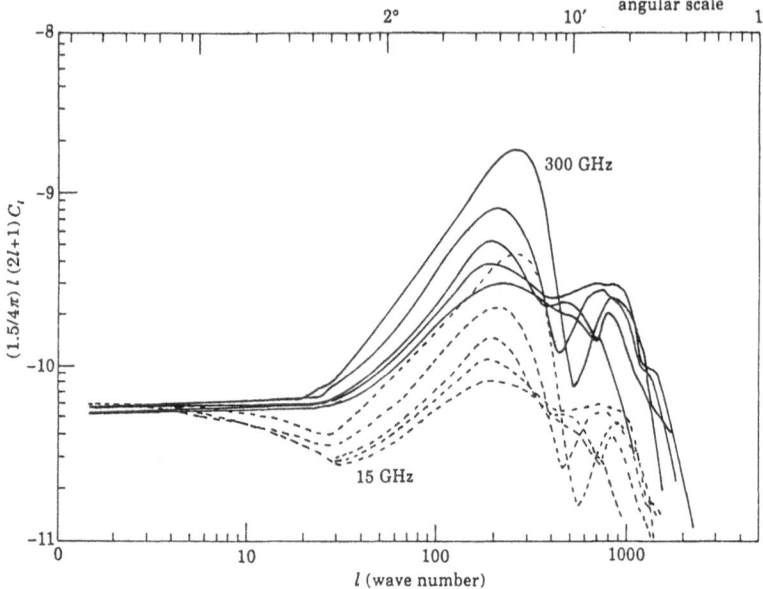

Figure 8. Power spectrum of CBR anisotropies in CDM theories, for different values of $\Omega_b = 0.5, 0.2\,0.05, 0.03, 0.01$ (from top to bottom):
-scattering negligible at high frequencies (solid curves)
-scattering efficient at low frequencies (dashed curves) at small angular scales (i.e., large $l \sim 1/\theta$).

The COBRAS/SAMBA mission -which will use tuned radioreceivers (HEMTs) over the frequency range 30-125 GHz (COBRAS) and cooled bolometers over the range 100-300 GHz (SAMBA)- will provide a near all-sky map of the CBA in eight channels over the frequency range 30-800 GHz with a peak sensitivity of $\frac{\Delta T}{T} \sim 10^{-6}$ per pixel in the frequency range 50 to 300 GHz. Therefore the COBRAS/SAMBA maps of the CBA will be able to address the following issues:

-possible effects of primordial molecules on CBA
-discrimination between effects of reionization and of primordial molecules on primary CBA
-primordial lithium abundance and possible discrimination between SBBN and IBBN models.

4.3. CONDITIONS FOR CREATION OF SECONDARY CBA

If a protocloud is moving relatively to the Hubble flow, the resonant scattering of the CBR photons by diatomic molecules could give raise to a molecular signal (Dubrovich 1993, Maoli et al 1994, de Bernardis et al

1994). The intensity of the effect is given by:

$$\frac{\Delta I}{I_{CBR}} = (3 - \alpha_\nu).\frac{v_{pec}}{c}(1 - e^{\tau_\nu})$$

where α_ν is the spectral index of the photon distribution:

$$\alpha_\nu = \frac{\nu}{I}.\frac{dI}{d\nu}$$

and τ_ν is the optical depth of the protocloud at the observational frequency, v_{pec} is the component along the line of sight of the peculiar velocity of the protocloud and c the velocity of light. Therefore, just as primary CBA could be erased, secondary anisotropies could be expected by molecular resonant scattering. They could arise from protoclouds which evolved from perturbations present at the last-scattering surface.

In a very recent paper, Maoli et al (1996) estimate the possible molecular signals expected during the three stages of evolution of the very simple protocloud model with always the same network of chemical reactions. They have done the calculations, in the framework of a cold dark matter scenario and in a more general case, of the intensity $\frac{\Delta I}{I_{CBR}}$ and of the line width $\frac{\Delta \nu}{\nu}$ of the expected signal during the linear stage, the turn-around epoch and the non-linear collapse of a protocloud evolution.

Secondary anisotropies may be tested at small angular scales: a search for them at angular scales of galaxies by means of the IRAM radio telescope, at 10 arcseconds of angluar resolution, has been undertaken. A first attempt to search for LiH lines in the linear stage of a high redshift ($z \sim 200$) protocloud has already been carried out (de Bernardis et al 1993). In their recent paper, Maoli et al (1996), studied the observability of a molecular signal from a primordial cloud by means of a radiotelescope and showed that, for standard observational conditions:

$$10^{-5} < \frac{\Delta \nu}{\nu} < 10^{-3}$$

$$\frac{\Delta I}{I_{CBR}} > 10^{-4}$$

the beginning of the non-linear collapse phase just after the turn around is the best one for a detection of the first two LiH rotational lines which are observable for the frequency range (see Figure 9)

$$30 < \nu < 250 \text{ GHz}$$

with an angular scale range

$$7'' < \theta < 20''$$

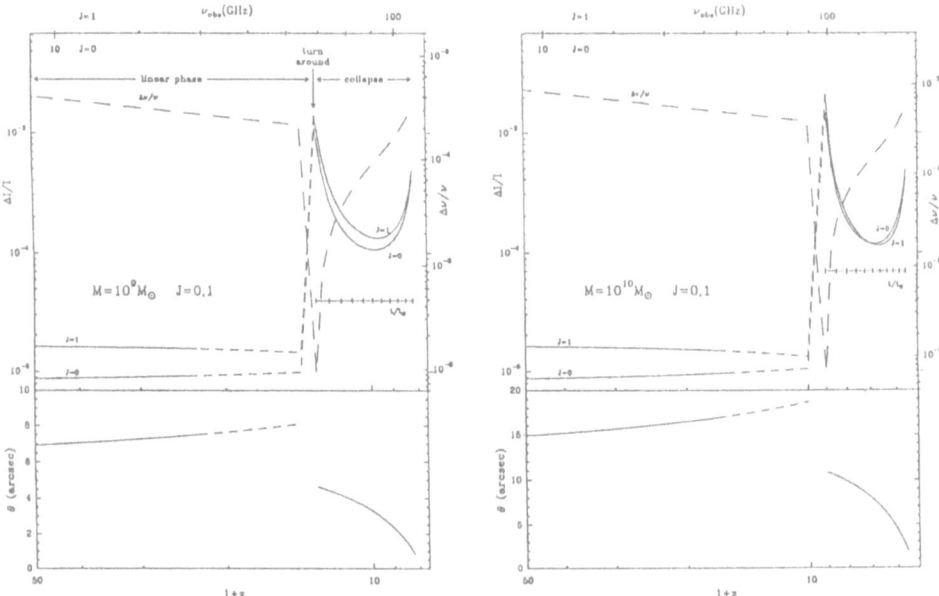

Figure 9. Top: intensity and line width for the $j = 0, 1$ lines of primordial clouds of 10^9 $M_\odot$ (left) and 10^{10} $M_\odot$ (right) in the CDM scenario. The upper scale is the observational frequency for the two lines. Bottom: angular dimension of the cloud (from Maoli et al 1996)

The IRAM 30m telescope, with its $10''$ of angular resolution and its frequency channels, is well suited for this kind of observations.

5. Conclusions

The effects of LiH molecules on the CBR anisotropies (erasing of primary CBA and creation of secondary ones) strongly depend on the final LiH abundance which is a function of lithium produced by primordial nucleosynthesis and of the efficiency of its conversion into LiH. Let us emphasize, once again, that the primordial lithium abundance and the percentage of lithium converted to LiH are both quite uncertain. These uncertainties reflect the uncertainties of nuclear and chemical reactions.

Let us remark that just as radioactive nuclear beam studies of reactions of short-lived nuclides may help to solve puzzles of primordial nucleosynthesis, CBR anisotropy observations may help to solve puzzles of primordial chemistry. Observations at the IRAM 30m telescope and above all, maps made by the COBRAS/SAMBA mission (the major space mission of the next decade dedicated to the CBR) should be excellent tools for discovering and analyzing anisotropies from resonant scattering of primordial molecules with CBR photons. Moreover, accurate measurements of anisotropies at intermediate and small angular scales would provide information on the abundances of LiH and Li. Any decrease of $\Delta T/T$ with wavelength could suggest the presence of resonant molecular scattering. Any upper limits to the difference in amplitude between infrared anisotropies (SAMBA) and radio anisotropies (COBRAS), at the same angular scale, should be used to constrain LiH and Li productions.

Finally, let us notice that all the above searches can be seen as a propaedeutic to the search of primordial molecules, which is one of the scientific objectives of the cornerstone FIRST (Far InfraRed Space Telescope) of the HORIZON 2000 plan of ESA.

AKNOWLEDGMENTS

This work has been supported by the EEC network CHRX-CT 920079 on *the CMB radiation measurements and interpretations*. Part of the work of R.M. has been supported by the European Space Agency. Part of the work of D.P. has been supported by the European Community (HCMP-ERBCHRTCT94 1163).

References

1. Balbes M.J. (1995) Cross sections and reactions rates of $d+{}^8Li$ reactions involved in Big Bang nucleosynthesis, *Nuclear Physics*, **A584**, pp 315-334
2. Boyd R.N. (1993) Recent experimental advances in primordial nucleosynthesis, *Physics Reports* **227**, pp 57-63
3. Carswell R.F. et al (1994) Is there deuterium in the $z = 3.32$ complex in the spectrum of 0014-813 ?, *Mon. Not. R. Astron. Soc.* **268**, pp L1-L4
4. COBRAS/SAMBA (1996) Report on the phase A study, *ESA-SCI* **96,3**
5. de Bernardis P. et al (1993) Search for LiH lines at high redshift, *Astron & Astroph.* **269**, pp 1-6
6. de Bernardis P. et al (1994) Microwave Background Anisotropies: Future plans, in *Present and future of the Cosmic Microwave Background* Springer Verlag Publishers, pp 188-207
7. Dubrovich V.K. (1977) Molecules of cosmological origin, *Sov. Astron. Lett.* **3**, pp 128-129
8. Dubrovich V.K. (1993) Blurring of spatial microwave background fluctuations by molecular line scattering, *Sov. Astron. Lett.* **19(I)**, pp 53-54
9. Duncan D.K., Lambert D.L., Lemke M. (1992) The abundance of boron in three halo stars, *Astrophys. Journ.* **401**, pp 584-595
10. Gilmore G., Edvardsson B., Nissen P.E. (1991) First detection of beryllium in a very metal poor star: a test of the standard Big Bang model, *Astrophys. Journ.* **378**, pp 17-21
11. Gu X. et al (1995) The ${}^8Li(\alpha n){}^{11}B$ reaction and primordial nucleosynthesis, *Phys. Lett.* **B343**, pp 31-35
12. Hutchins J.B. (1976) The thermal effects of H_2 molecules in rotating and collapsing spheroidal gas clouds, *Astrophys. Journ.* **205**, pp 103-121
13. Jedamzik K. et al (1994) Enhanced heavy-element formation in baryon-inhomogeneous big bang models, *Astrophys. Journ.* **422**, pp 423-429

14. Khersonski V.K. and Lipovka A.A. (1993) Free-bound transitions in LiH molecules, *Astrofiz. Issled* **36**, pp 88-106
15. La D. (1989) The existence of limiting wavelength for small-scale temperature fluctuations, *Astrophys. Journ.* **341**, pp 575-578
16. Lahav O. (1986) Cooling of populations III objects in a pressure supported collapse, **Mon. Not. R. Astr. Soc. 220**, pp 259-269
17. Lepp S. & Shull J.M. (1984) Molecules in the early Universe, *Astrophys. Journ.* **280**, pp 465-469
18. Malaney R.A. & Mathews G.J. (1993) Probing the early Universe: nucleosynthesis beyond the standard Big Bang, *Physics Reports* **229,N4**, pp 145-219
19. Maoli R., Melchiorri F. & Tosti D. (1994) Molecules in the postrecombination Universe and microwave background anisotropies, *Astrophys. Journ.* **423**, pp 372-381
20. Maoli R. et al (1996) Molecular signals from primordial clouds at high redshift, *Astrophys. Journ.* **457**, pp 1-12
21. Palla F., Salpeter E.E. & Stahler S.W. (1983) Primordial star formation: the role of molecular hydrogen, *Astrophys. Journ.* **271**, pp 632-641
22. Palla F., Galli D. & Silk J. (1995) Deuterium in the Universe, *Astrophys. Journ.* **451**, pp 44-50
23. Puy D. et al (1993) Formation of primordial molecules and therm al balance in the early Universe, *Astron. & Astrophys.* **267**, 337-346
24. Puy D. & Signore M. (1996) Primordial molecules in the early cloud formation, *Astron. & Astrophys.* **305**, pp 371-378
25. Rugers M., Hogan C.J. (1996) Confirmation of high deuterium abundance in quasar absorbers, *Astrophys. Journ.* **459**, pp L1-L4
26. Ryan S.G. et al (1992) Evolution of beryllium abundance in the Galactic halo, *Astrophys. Journ.* **388**, pp 184-189
27. Schramm D.N. (1993) The first three minutes: 1990 version, in *After the First Three Minutes* American Institute of Physics, pp 12-40
28. Sciama D.W. (1990) Dark matter and the ionization of hydrogen throughout the Universe*Comments Astrophysics* **15**, pp 71-86
29. Signore M. et al (1994) the lithium problem with IRAM, OSSE and INTEGRAL, *Astrophys. Journ. Sup* **92**, pp 535-537
30. Signore M. et al (1996) Primordial molecules and Cosmic Background Radiation Anisotropies, *Astrophys. Letters and Commun.*, in press
31. Silk J. (1983) The first stars, *Mon. Not. R. Astr. Soc.* **205**, pp 705-718
32. Smith M.S., Kawano L.H. & Malaney R.A. (1993) Experimental, computational and observational analysis of primordial nucleosynthesis, *Astrophys. Journ. Sup.* **85**, pp 219-247
33. Songalia A. et al (1994) Deuterium abundance and background radiation temperature in high redshift clouds, *Nature* **368**, pp 599-604
34. Stancil P.C., Lepp S. & Dalgarno A. (1996) the lithium chemistry in the early Universe, *Astrophys. Journ.* **458**, pp 401-406
35. Steigman G. (1995) Big Bang nucleosynthesis, *Nuclear Physics B* **37**C, pp 68-73
36. Tytler et al (1995) Ionisation and abundances of intergalactic gas, in *QSO absorption lines* Springer Verlag Publishers, pp 289-298

THE COSMIC BACKGROUND RADIATION AND ELEMENTARY PARTICLES

PIERRE SALATI

Laboratoire de Physique Théorique ENSLAPP,
B.P. 110, 74941 Annecy-le-Vieux Cedex, France
Université de Savoie, B.P. 1104, 73011 Chambéry Cedex, France
Institut universitaire de France

Abstract. In the early Universe, particles decouple from thermal equilibrium. Their fossil distributions, processed during the big bang, may still be present in the intergalactic medium. If these particles decay, energy may be released in the primordial plasma, generating spectral distortions of the cosmic background radiation. These cosmic imprints may incidentally be today precious signatures of the primeval epoch. In Section 1, the properties of what Gamow called the *Ylem* are reviewed. In that primordial mixture, matter is completely dissociated and is in equilibrium with the radiation. In Section 2, the decoupling of light and heavy neutrinos is presented as a generic example of the primeval behaviour of elementary particles. The fossil distribution of neutrinos is discussed in Section 3. Then, the delicate interplay between radiation and matter is investigated in Section 4. When thermal contact is established between photons and electrons , the radiation spectrum relaxes towards a Bose-Einstein distribution as a consequence of inverse Compton scatterings. The spectrum then achieves a Planck distribution as bremsstrahlung and double Compton emissions proceed. In Section 5, a variety of time scales relevant to our analysis are presented. The μ and y spectral distortions are discussed. Constraints on the amount of energy which elementary particles may release in the cosmic background radiation are deduced.

1. The Primordial Plasma

At early times, the Universe is filled up with an extremely dense and hot gas. Matter does not exist in an organized state, as it appears today. On the

C. H. Lineweaver et al. (eds.), The Cosmic Microwave Background, 365–407.
© *1997 Kluwer Academic Publishers.*

contrary, matter is completely dissociated during the Big Bang. Particles which are now fleetingly and with difficulty created in huge accelerators freely pervade the space at that epoch. Because the temperature and the density are so large, particles are steadily annihilating with each other and are created back from other species.

1.1. A FEW REMINDERS ABOUT THERMODYNAMICS

Let us consider a generic population of particles which we will denote by A. Because spectral distortions of the cosmic background radiation (CBR) occur after a time $\sim 10^4$ seconds, we will be here mostly concerned with photons, neutrinos as well as with electrons and their antipartners, positrons. When dealing with a fourth generation neutrino or with supersymmetry, we will also be interested in much heavier particles and the properties of the surrounding plasma will be quite relevant.

Collisions are so numerous that the various species are thermalized with each other, reaching a common temperature T. Moreover, the asymmetry between matter and antimatter may be neglected in first approximation so that the density of particles A is identical to the density of their antiparticles $\bar{A}$. The chemical potential μ_A of the population A is therefore the same as $\mu_{\bar{A}}$, the chemical potential relevant to the antiparticles $\bar{A}$. Furthermore, annihilation and creation reactions such as

$$A + \bar{A} \rightleftharpoons e^- + e^+ , \tag{1.1}$$

or

$$3\gamma \rightleftharpoons e^- + e^+ \rightleftharpoons 2\gamma , \tag{1.2}$$

are so fast that the corresponding chemical equilibria are reached, with

$$\mu_A + \mu_{\bar{A}} = \mu_{e^-} + \mu_{e^+} = 2\mu_\gamma = 3\mu_\gamma . \tag{1.3}$$

The chemical potential of photons is inferred to be zero. Their spectrum is therefore a pure black-body one and is given by the Planck distribution. The chemical potential μ_A turns out to be the opposite of $\mu_{\bar{A}}$. That relation and the condition that there is no asymmetry between matter and antimatter both lead to the fact that all the chemical potentials of the various species vanish. The density of particles A is therefore given by the Bose-Einstein or the Fermi-Dirac distribution functions. It may be expressed as a sum over the momentum $\vec{p}$ of the relevant statistical distribution function

$$n_A = \int \frac{d^3\vec{p}}{h^3} g_A \left\{ e^{E/kT} - \epsilon \right\}^{-1} , \tag{1.4}$$

where $\epsilon = 1$ when the spin is an integer and the particle a boson, whereas the case $\epsilon = -1$ corresponds to a fermion with an half-integer spin. The energy E is related to the momentum $\vec{p}$ and the mass M of the species A by

$$E^2 = M^2 + p^2 . \qquad (1.5)$$

The speed of light c is measured in units where it is equal to unity. The temperature of the plasma is T and h stands for the Planck constant. The number of different spin states is denoted by g_A. The spin degeneracy is 1 for spin 0 particles such as the neutral pion π^0. It is 2 for fermions with a spin $1/2$ such as the electron e^-, the positron e^+ or the neutrino ν. For the photon γ, g_A corresponds to the transverse helicity states and is equal to 2. If, in addition, one works in a system of units where the reduced Planck constant $\hbar = h / 2\pi$ and the Boltzmann constant k are both equal to unity, expression (1.4) simplifies into

$$n_A = \frac{g_A}{2\pi^2} T^3 \int_0^\infty x^2 \, dx \, \{e^y - \epsilon\}^{-1} , \qquad (1.6)$$

where $x = p/T$ and $y = E/T$. To get back to the usual system of units (CGS or MKSA), a multiplicative factor needs to be introduced. That factor is built from the constants $\hbar$, k and c. In the case of the number density, K^3 must be translated into cm^{-3}.

Exercise 1 *Show that expression (1.6) needs to be multiplied by*

$$\left(\frac{k}{\hbar c}\right)^3 = 83.22 \text{ K}^{-3} \text{ cm}^{-3} . \qquad (1.7)$$

In the regime of high temperatures, the mass M becomes negligible with respect to T and particles are ultra-relativistic, with statistical velocities approaching the velocity c of light. The variables x and y are basically equal so that the density simplifies into

$$n_A = \frac{g_A}{\pi^2} T^3 \zeta(3) \begin{cases} 1 & \text{(Boson)} \\ 3/4 & \text{(Fermion)} \end{cases} , \qquad (1.8)$$

where $\zeta(3) = 1.20205$. In that regime, remarkably enough, the ratio n_A / T^3 is constant. For photons, it is approximately equal to 20 K^{-3} cm^{-3}. For neutrinos with two spin states, it is ~ 15 K^{-3} cm^{-3}. In the regime where the temperature is small with respect to the mass M, the gas becomes non-relativistic. In that limit, the ratio $a = M/T$ is much larger than unity and the density simplifies into :

$$n_A = g_A T^3 e^{-a} \left(\frac{a}{2\pi}\right)^{3/2} . \qquad (1.9)$$

The energy density ρ and the pressure P of a population of particles A are respectively given by the integral over the phase space of the energy

E and of the product $\vec{p} \cdot \vec{v}/3$ corresponding to each quantum state of propagation :

$$\rho_A = \int \frac{d^3\vec{p}}{h^3} g_A \left\{ \frac{E}{e^{E/kT} - \epsilon} \right\} , \qquad (1.10)$$

and

$$P_A = \int \frac{d^3\vec{p}}{h^3} g_A \left\{ \frac{pv/3}{e^{E/kT} - \epsilon} \right\} . \qquad (1.11)$$

We assume here that the primordial plasma behaves as a perfect gas. The thermal energy well exceeds the potential energy resulting from long range interactions. That property is generally true for most of the species. It is not valid however in the case of the quark/hadron phase transition where strong interactions lead to the confinement of quarks inside particles such as protons, neutrons or pions. In a system of units where $\hbar = k = c = 1$, the energy density merely reduces to the integral

$$\rho_A = \frac{g_A}{2\pi^2} T^4 \int_0^\infty x^2 \, dx \left(\frac{y}{e^y - \epsilon} \right) . \qquad (1.12)$$

That expression needs to be multiplied by the quantity

$$\left(\frac{k^4}{\hbar^3 c^3} \right) \simeq 1.15 \times 10^{-21} \text{ J cm}^{-3} \text{ K}^{-4} , \qquad (1.13)$$

in order to restore proper units. In the non-relativistic regime of low temperatures for which T is much smaller than the mass M, the energy density ρ is negligible. It actually vanishes exponentially as e^{-a}. However, in the ultra-relativistic regime where M is negligible with respect to the temperature T, expression (1.12) simplifies.

Exercise 2 *Show that the energy density of ultra-relativisitic bosons or fermions (with vanishing or negligible mass) is given by :*

$$\rho_A = \frac{\pi^2}{15} \frac{g_A}{2} T^4 \left\{ \begin{array}{ll} 1 & \text{(Boson)} \\ 7/8 & \text{(Fermion)} \end{array} \right. . \qquad (1.14)$$

You may use the relation

$$\int_0^\infty dx \left(\frac{x^n}{e^x - \epsilon} \right) = \Gamma(n+1) \, \zeta(n+1) \left\{ \begin{array}{ll} 1 & \text{if } \epsilon = 1 \\ 1 - \dfrac{1}{2^n} & \text{if } \epsilon = -1 \end{array} \right. . \qquad (1.15)$$

The function $\zeta(s)$ is defined by the expansion

$$\zeta(s) = \sum_{n=1}^\infty \frac{1}{n^s} , \qquad (1.16)$$

and $\Gamma(s)$ denotes the Euler Gamma function.

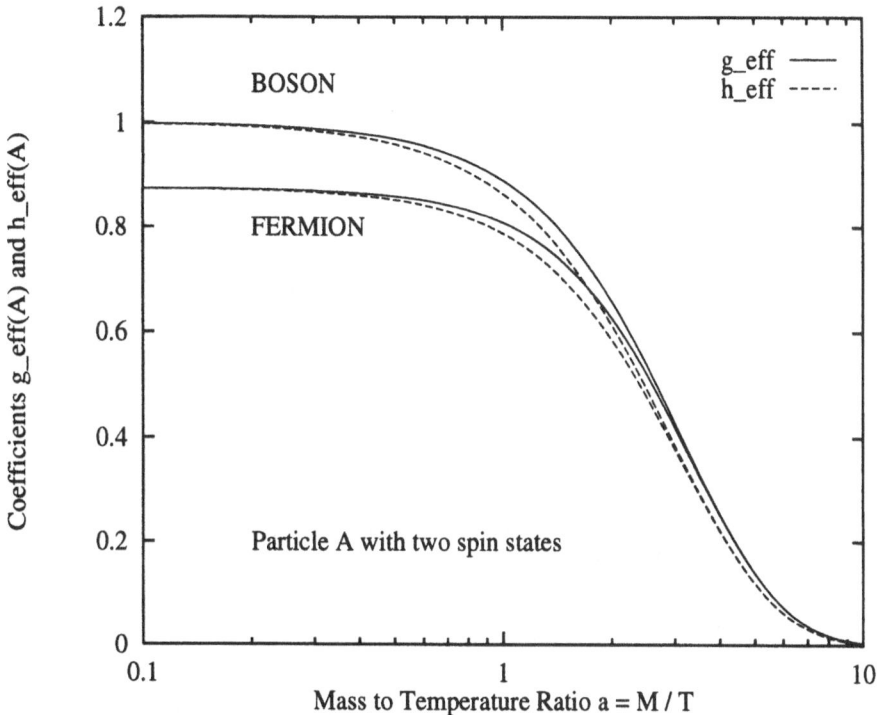

Figure 1. This figure features a population of generic particles A with 2 spin states, hence $g_A = 2$. The energy and entropy densities are expressed as a function of those for a photon gas with the same temperature. The effective number of degrees of freedom g_{eff} relative to the energy is plotted as a solid line for both the boson (upper curve) and fermion (lower curve) cases. The dashed line stands for the entropic number of degrees of freedom h_{eff}. The energy and entropy densities decrease as the mass to temperature ratio $a = M/T$ increases.

Therefore, the energy density of a photon gas may be expressed by relation

$$\rho_\gamma = a_\gamma T^4 \ . \tag{1.17}$$

The coefficient a_γ is given by

$$a_\gamma = \frac{\pi^2}{15} \left(\frac{k^4}{\hbar^3 c^3} \right) = 7.57 \times 10^{-16} \ \text{J m}^{-3} \ \text{K}^{-4} \ , \tag{1.18}$$

and is related to the Stefan-Boltzmann constant by $\sigma_S = a_\gamma c / 4$. It is convenient to express the energy density of the primordial plasma in units where the energy density of a photon gas with same temperature is unity. The population of species A is simply described by the effective number of degrees of freedom

$$g_{\text{eff}}(A) = \frac{\rho_A}{\rho_\gamma} \ . \tag{1.19}$$

Figure 1 features the case of a particle A with two spin states $(g_A = 2)$. In the ultra-relativistic regime, the ratio $a = M/T$ is extremely small and tends towards 0. The coefficient g_{eff} tends towards 1 for bosons and towards 7/8 for fermions. When the temperature decreases with respect to the mass M, the ratio a increases and the energy density diminishes. The evolution of g_{eff} as a function of the mass-to-temperature ratio M/T is featured by the two solid curves of Figure 1. In the non-relativistic regime $(a \gg 1)$, bosons and fermions behave identically in so far as both the Bose-Einstein and the Fermi-Dirac regimes reduce to the Maxwell-Boltzmann statistics.

The entropy of the primordial plasma can be computed while imposing a vanishing chemical potential for all the species. The first law of thermodynamics leads to the relation

$$T \sigma_A = \rho_A + P_A \ , \tag{1.20}$$

where σ_A denotes the entropy density of the gas of particles A. In the system of units where $\hbar = k = c = 1$, the entropy density may be expressed by the dimensionless integral :

$$\sigma_A = \frac{g_A}{2\pi^2} T^3 \int_0^\infty x^2 \, dx \left\{ y + \frac{x^2}{3y} \right\} \{e^y - \epsilon\}^{-1} \ . \tag{1.21}$$

In the ultra-relativistic regime, the entropy and energy densities may be simply related

$$\sigma_A = \frac{4}{3} \frac{\rho_A}{T} = \frac{4\pi^2}{45} \frac{g_A}{2} T^3 \left\{ \begin{array}{ll} 1 & \text{(Boson)} \\ 7/8 & \text{(Fermion)} \end{array} \right. \ . \tag{1.22}$$

The entropy of a population of particles may also be expressed as a function of the entropy σ_γ relative to a photon gas with the same temperature. The corresponding entropy coefficient is defined by

$$h_{\text{eff}}(A) = \frac{\sigma_A}{\sigma_\gamma} \ . \tag{1.23}$$

The behaviour of h_{eff} is featured in Figure 1 by the two dashed curves. At high temperature, the coefficients g_{eff} and h_{eff} are equal. They become different in the non-relativistic regime where bosons and fermions behave identically.

The energy density and the entropy density of the primordial plasma are given by the corresponding coefficients $g_{\text{eff}}(T)$ and $h_{\text{eff}}(T)$. The latter take into account all the different species which exist at the epoch of interest

$$g_{\text{eff}}(T) = \frac{\rho(T)}{\rho_\gamma(T)} \quad \text{and} \quad h_{\text{eff}}(T) = \frac{\sigma(T)}{\sigma_\gamma(T)} \ . \tag{1.24}$$

The only particles to significantly contribute to the energy and entropy densities are ultra-relativistic, with a mass smaller than the overall temperature. If those densities are expressed as a function of those for a photon gas with the same temperature, the coefficients $g_{\text{eff}}(T)$ and $h_{\text{eff}}(T)$ may be written as a sum over the spin states of the bosons and of the fermions which are ultra-relativistic at temperature T :

$$g_{\text{eff}}(T) \simeq h_{\text{eff}}(T) \simeq \sum_{M_B < T} \frac{g_B}{2} + \sum_{M_F < T} \frac{7\,g_F}{16} \ . \tag{1.25}$$

Exercise 3 *Show that for a temperature of 1 MeV, the energy density of the primordial plasma is given by $g_{\text{eff}} = 43/8$. Three families of light neutrinos , each with two helicity states, will be assumed here.*

In Figure 2, the coefficients relative to the energy $g_{\text{eff}}(T)$ and to the entropy $h_{\text{eff}}(T)$ have been computed exactly by taking directly into account the integrals (1.12) and (1.21). The solid curve features the evolution of g_{eff} while the temperature drops from 10 GeV down to 10 MeV. The dashed line refers to the number h_{eff} of entropic degrees of freedom. The dotted curve stands for the approximation (1.25). Its behaviour is discontinuous because as soon as the temperature becomes smaller than the mass of a given species, the latter stops immediately to be taken into account in the sum (1.25). Remember that the actual contribution of a particle to the overall energy density smoothly vanishes as the temperature drops. Note that, in any case, the approximation (1.25) is fairly satisfactory.

1.2. THE EXPANSION OF THE UNIVERSE

The distribution of galaxies and clusters of galaxies is approximately homogeneous on large scales, for domains with typical size in excess of ~ 100 Mpc. Large scale structures are also distributed isotropically, without much variation with respect to the direction towards which telescopes point. This isotropy is confirmed by the lack of strong inhomogeneities in the CBR. Because space is homogeneous and isotropic, it is maximally symmetric and may be described by the Robertson and Walker metric

$$d\tau^2 = c^2 dt^2 - R(t)^2 \left\{ \frac{dr^2}{1 - kr^2} + r^2 d\theta^2 + r^2 sin^2\theta \, d\phi^2 \right\} \ . \tag{1.26}$$

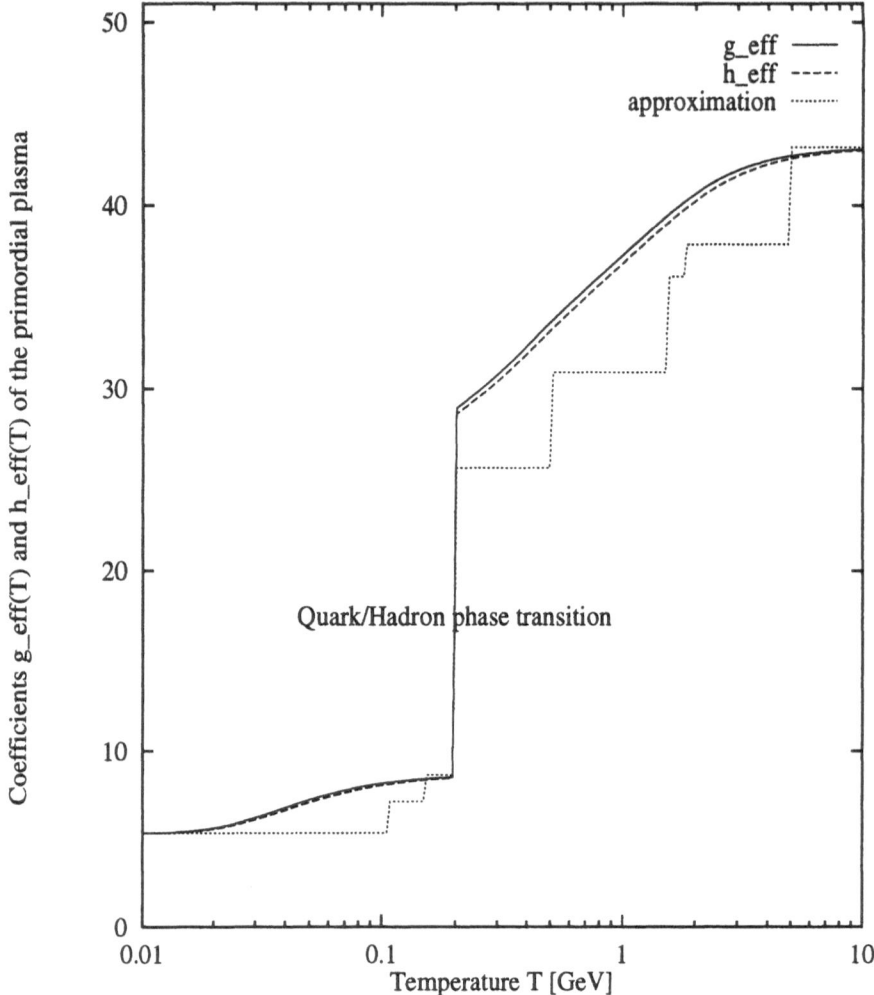

Figure 2. The effective number of degrees of freedom g_{eff} (solid) et h_{eff} (dashed) respectively describe the energy and entropy densities of the primordial plasma. The densities are expressed in units of those for a photon gas. These coefficients decrease when the temperature T drops from 10 GeV down to 10 MeV. When the Universe cools down, more and more species vanish. The dotted curve stands for the approximation (1.25) discussed in the text.

If the parameter k vanishes, the Minkowski metric of a 3D flat Euclidian space is recovered with the usual spherical coordinate system (r,θ,ϕ). If k is positive, the 3D physical space is spherical and finite. An explorer moving straight ahead in any direction would come back to the point of departure. If k is negative, space is infinite and hyperbolic, with a negative curvature. The equations of general relativity relate the time evolution of the scale factor $R(t)$ to the energy density ρ of the plasma which fills up the Universe :

$$H^2 = \left(\frac{\dot{R}}{R}\right)^2 = \frac{8\pi G}{3}\rho - \frac{k}{R^2} . \tag{1.27}$$

The time derivative of the scale factor $R(t)$ is denoted by $\dot{R}$, G is Newton's constant of gravitation and the parameter $H(t)$ describes the expansion rate of space at time t. Today, the two terms on the right-hand side of the equation of expansion (1.27) are of the same order of magnitude. However, in the past, the curvature term behaves like $1/R^2$ whilst the energy term increases much more rapidly when the scale factor R decreases. It varies like $1/R^3$ as long as the energy density is dominated by non-relativistic matter. At very early times, when the Universe is filled up with pure radiation, it varies more steeply like $1/R^4$. During primordial nucleosynthesis, for instance, the temperature of the primordial plasma is approximately 3×10^9 Kelvins. The scale factor $R(t)$ is a billion times smaller than today so that expression (1.27) is completely dominated by the energy density term :

$$H^2 = \frac{8\pi G}{3}\rho . \tag{1.28}$$

Another important consequence of general relativity is the conservation of the energy-momentum tensor related to the matter content of the Universe. This property leads to the conservation of the entropy of the primordial plasma. That entropy remains constant during the expansion of the Universe. Let us consider a volume that follows that dilation and expands with the expanding space. Such a volume is called a **covolume** because its comoving coordinates r, θ et ϕ are just constant. For the sake of simplicity, we will consider here a volume equal to R^3. The entropy of the gas contained in that cubic covolume may be expressed as :

$$S = \sigma(T) R^3 , \tag{1.29}$$

and does not vary in time. Taking into account relation (1.24), the entropy S may be directly expressed as a function of the temperature T and of the scale factor $R(t)$:

$$S = h_{\text{eff}}(T) \frac{4\pi^2}{45} T^3 R^3 . \tag{1.30}$$

In first approximation, the variations of the coefficient h_{eff} with the temperature may be neglected. Actually in Figure 2, h_{eff} varies only by a factor of 8 while the temperature decreases by three orders of magnitude, from 10 GeV down to 10 MeV. The conservation of the entropy of the primordial gas implies that the product of the temperature by the scale factor does not vary in time :

$$T \times R(t) = \text{Constant} . \qquad (1.31)$$

By expressing the energy density of the primordial plasma as a function of the energy density of a photon gas with the same temperature, relation (1.28) leads to a differential equation that describes the time evolution of the temperature

$$-\frac{\dot{T}}{T} = \frac{\dot{R}}{R} = \left(\frac{8\pi G}{3} g_{\text{eff}}(T) \frac{\pi^2}{15} T^4 \right)^{1/2} . \qquad (1.32)$$

Exercise 4 *Show that expression (1.32) may be written as*

$$-\frac{\dot{T}}{T^3} = \left(\frac{8\pi^3 G}{45} g_{\text{eff}}(T) \right)^{1/2} . \qquad (1.33)$$

If the variations of the coefficient g_{eff} with respect to the temperature are neglected, show that the age of the Universe, i.e., the time t since the creation of the Universe and the Big Bang, is related to the temperature T by :

$$\frac{1}{T^2} \simeq \left(\frac{32\pi^3 G}{45} g_{\text{eff}}(T) \right)^{1/2} t . \qquad (1.34)$$

You need to take, as the initial condition of the problem, an infinite temperature at time $t = 0$.

In order to evaluate numerically the previous expression, the Planck mass needs to be defined. That mass, or energy, is associated to Newton constant of gravitation G. The latter is approximately equal to $6.67 \times 10^{-11} \text{ m}^3 \text{ kg}^{-1} \text{ s}^{-2}$. It may be multiplied by the appropriate cocktail of fundamental constants $\hbar$ and c in order to derive a quantity whose dimension is an energy. That Planck mass is defined as :

$$M_P = \sqrt{\frac{\hbar c^5}{G}} \simeq 1.96 \times 10^9 \text{ J} \simeq 1.22 \times 10^{22} \text{ MeV} . \qquad (1.35)$$

Because an inverse MeV corresponds to the duration $\hbar/1 \text{ MeV}$, i.e., numerically to 6.6×10^{-22} seconds, relation (1.34) can be translated into :

$$t \simeq \frac{1.7 \text{ second}}{\sqrt{g_{\text{eff}}(T)}} \left(\frac{1 \text{ MeV}}{T} \right)^2 . \qquad (1.36)$$

During primordial nucleosynthesis, the temperature decreases from 1 MeV down to 0.1 MeV, i.e., from 10 to 1 billion Kelvins. The age of the Universe lies in the range between one second and three minutes. Energy release can alter the spectrum of the CBR after a time of order 10^4 seconds, for a temperature less than 10^8 Kelvins.

2. The Primordial Behaviour of Elementary Particles

We would like to investigate now the generic behaviour of elementary particles in the early Universe. As we shall see, species decouple from thermodynamical equilibrium. Their statistical distribution gets frozen because they stop colliding with each other. It is actually possible to compute the features of that quenching and to predict the present abundance of their remnants, provided those are stable. As an example, the CBR is the fossil radiation of the primordial photon gas which decoupled from baryons at recombination.

2.1. THE THERMAL DECOUPLING

Thermalization is achieved by the numerous collisions of A's with the other species. By numerous, we mean that the collision rate exceeds the expansion rate so that the temperature T_A has plenty of time to relax towards the temperature of the other populations. To illustrate this idea, we present the thermal behaviour of neutrinos . The rate of their collisions with electrons and positrons is given by

$$\Gamma_C = < \sigma_C V > n_e \simeq \left(G_F^2 T^2 \right) \times \left(\frac{3}{\pi^2} \zeta(3)\, T^3 \right) \simeq 4.9 \times 10^{-23} \text{MeV} \left(\frac{T}{1\,\text{MeV}} \right)^5 ,$$
$$(2.1)$$

where the Fermi constant is $G_F = 1.16 \times 10^{-11}$ MeV^{-2}. As discussed in the previous section, the curvature of space may be neglected at early times so that the Hubble parameter may be expressed as

$$H = \left\{ \frac{8\pi}{3} \frac{\rho}{M_P^2} \right\}^{1/2} = \left\{ \frac{8\pi^3}{45} g_{\text{eff}} \right\}^{1/2} \frac{T^2}{M_P} \simeq 4.5 \times 10^{-22}\,\text{MeV} \left(\frac{T}{1\,\text{MeV}} \right)^2 .$$
$$(2.2)$$

The Planck mass is $M_P = 1.22 \times 10^{22}$ MeV while the energy density ρ has been expressed in units of the photon energy density $\rho_\gamma = \pi^2 T^4 / 15$, hence an effective number of degrees of freedom $g_{\text{eff}} = 43/8$. The collision rate can be compared now with the expansion rate

$$\frac{\Gamma_C}{H} \simeq \left(\frac{T}{2\,\text{MeV}} \right)^3 ,$$
$$(2.3)$$

and we readily infer a decoupling temperature of ~ 2 MeV below which collisions are so rare that neutrinos no longer see their surroundings and become a fossil radiation.

2.2. ANNIHILATION AND THE CHEMICAL FREEZE-OUT

If neutrinos have a mass M larger than the decoupling temperature of $\sim$ 2 MeV, they may substantially annihilate before they freeze out. As an illustration, we focus here on the generic case of heavy neutrinos with mass $M = 2$ GeV.

At high temperature, for $T > 2$ GeV, heavy neutrinos are in chemical equilibrium. They steadily annihilate into $f\bar{f}$ pairs while the reverse process is also very active. The annihilation/production reaction

$$A + \overline{A} \rightleftharpoons f + \overline{f} \tag{2.4}$$

is in equilibrium and the A's density relaxes towards its equilibrium value (1.4). Below 2 GeV, A and $\overline{A}$ annihilate. As long as the chemical reaction (2.4) is in equilibrium, the A's density is given by relation (1.4). As the temperature decreases, the A's are significantly depleted by annihilation and their density drops down. The antiparticles $\overline{A}$ with which A's annihilate also become rare. At $T \simeq 100$ MeV, the density n_A is so low, particles and antiparticles are so much depleted that reaction (2.4) ceases to be in equilibrium. The probability for an A to encounter an antipartner becomes less than unity per typical expansion time. Annihilations stop under the combined action of the dilution due to the expansion of space, and of the severe depletion of the A species which has occurred between $T = 2$ GeV and $T = 100$ MeV. Below 100 MeV, annihilations are inhibited and the codensity f_A remains constant. Around $T \simeq 2$ MeV, heavy neutrinos stop colliding with the other species and become mere fossils of the early stages of the Universe. If they are stable, they pervade the intergalactic medium until the present epoch, and may even contribute a significant fraction to the closure mass.

The density n_A evolves according to the differential equation

$$\frac{dn_A}{dt} = -3Hn_A - \langle \sigma_{\rm an}v \rangle n_A^2 + \langle \sigma_{\rm an}v \rangle n_A^{0\,2} , \tag{2.5}$$

where the first expression on the right-hand side refers to the dilution resulting from the expansion. The second term accounts for the A annihilations while the last expression describes the retro-creation of $A\overline{A}$ pairs from light fermions and assumes detailed balance. The density n_A^0 corresponds to thermodynamical equilibrium. Its various expressions (1.4), (1.6), (1.8) and (1.9) have already been given above in Section 1.1. In particular, in the non-relativistic regime, that density may be expressed as

$$n_A^0 = g_A\, T^3\, e^{-a} \left(\frac{a}{2\,\pi}\right)^{3/2} , \tag{2.6}$$

where a denotes the mass to temperature ratio M/T. In terms of the co-density $f_A = n_A/T^3$, the evolution equation simplifies into

$$\frac{df_A}{dt} + (<\sigma_{an}v> n_A)\, f_A = <\sigma_{an}v> T^3 f_A^{0\,2} \ . \qquad (2.7)$$

In order to solve the differential equation (2.7), two typical time scales may be defined.

1) When equilibrium is reached, $i.e.$, when $f_A = f_A^0$, the time derivative $df_A/dt \simeq 0$. The characteristic time scale of the relaxation of f_A towards its kinetic equilibrium value f_A^0 is merely related to the annihilation rate

$$\tau_{rel}^{-1} = <\sigma_{an}v> n_A \ . \qquad (2.8)$$

2) The time scale of the variations of the equilibrium f_A^0 itself may be expressed as

$$\tau_{eq}^{-1} = -\frac{d}{dt}\mathrm{Log}\left\{f_A^{0\,2}T^3\right\} = 2aH \ , \qquad (2.9)$$

in the non-relativistic regime.

As is clear in Figures 3 and 4, two stages may be distinguished for the decoupling :

• At high temperature, as long as $\tau_{rel} < \tau_{eq}$, f_A has plenty of time to relax towards the equilibrium f_A^0 which evolves at a much slower pace. As a consequence, the annihilation reaction (2.4) is in chemical equilibrium so that $f_A = f_A^0$. In Figure 3, the solid line f_A cannot be distinguished from the dashed curve f_A^0. As is clear from Figure 4, as T decreases and $a = M/T$ increases, relaxation becomes progressively less efficient until the ratio τ_{rel}/τ_{eq} eventually reaches 1 where the decoupling from equilibrium occurs.

• Below the freeze-out temperature T_F, the relaxation time τ_{rel} exceeds τ_{eq}. Whilst f_A^0 drops down and vanishes, f_A still decreases a little bit. The freeze-out point may be derived from the equality $\tau_{rel} = \tau_{eq}$ which, for heavy neutrinos whose annihilation cross-section is

$$\sigma_{an}v\,(A\overline{A} \to f\overline{f}) = \frac{G_F^2}{2\pi} M^2 N_A \ , \qquad (2.10)$$

translates into

$$\sqrt{a_F}\,e^{a_F} = \frac{3\sqrt{5}}{(2\pi)^4}G_F^2\,M^3\,M_P\,N_A\,g_{\mathrm{eff}}^{-1/2} \simeq 0.7{\times}10^7\left(\frac{M}{1\,\mathrm{GeV}}\right)^3\frac{N_A}{\sqrt{g_{\mathrm{eff}}}} \ . \qquad (2.11)$$

The previous relation corresponds to a non-relativistic decoupling. Actually a_F is an order of magnitude larger than unity. For a 2 GeV neutrino, the

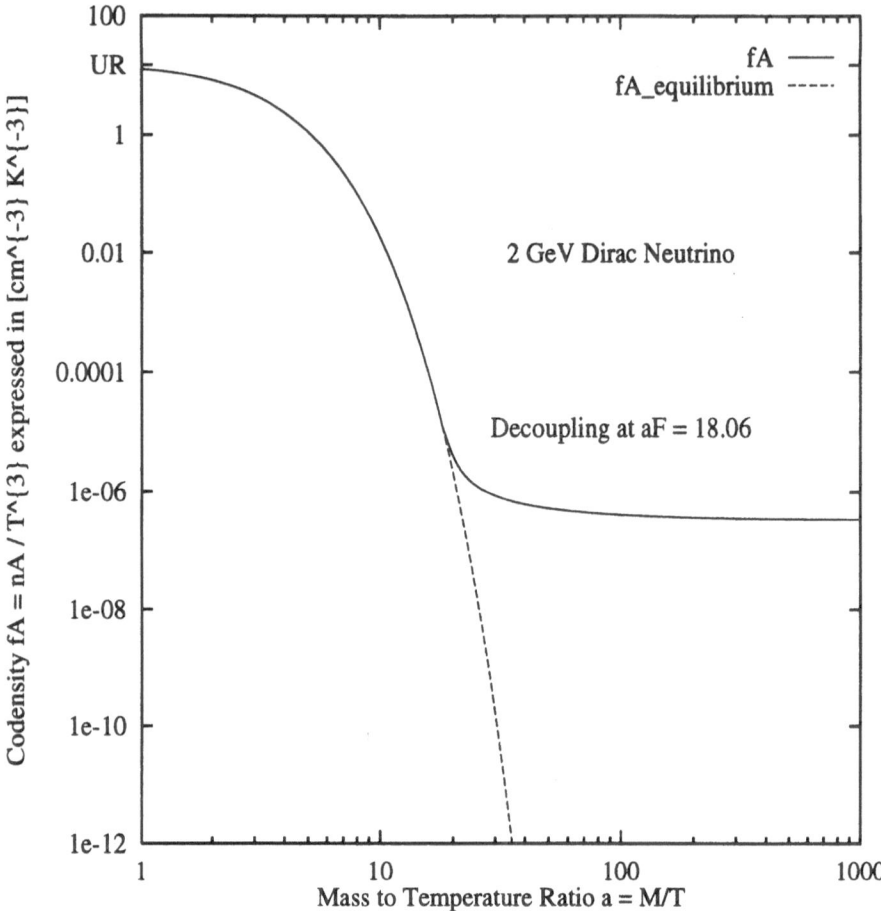

Figure 3. The codensity $f_A = n_A/T^3$ (solid) is presented as a function of the ratio $a = M/T$, for a 2 GeV neutrino. At large temperatures, it approaches its ultra-relativistic value of ~ 15 K^{-3} cm^{-3} (UR label on the vertical axis). The codensity follows the thermodynamical equilibrium f_A^0 (dashed) down to the critical point where decoupling occurs. In this example, the freeze-out temperature corresponds to $a_F \sim 18$. Then, the codensity is unable to follow f_A^0 which rapidly decreases. After the decoupling, f_A smoothly reaches its present value.

effective number of annihilation channels is $N_A \sim 14$, whilst the effective number of degrees of freedom is $g_{\text{eff}} = 57/8$ at a temperature $T_F \sim 100$ MeV. We infer from the previous equation a freeze-out point at $a_F \simeq 18.06$. In Figure 4, the freeze-out is indicated by the arrow labelled *Decoupling* at $a = a_F$. Note that as soon as the ratio τ_{rel}/τ_{eq} exceeds unity, the codensity f_A (solid line of Figure 3) decouples from the equilibrium value f_A^0 (dashed curve). From decoupling until now, f_A has been slightly decreasing

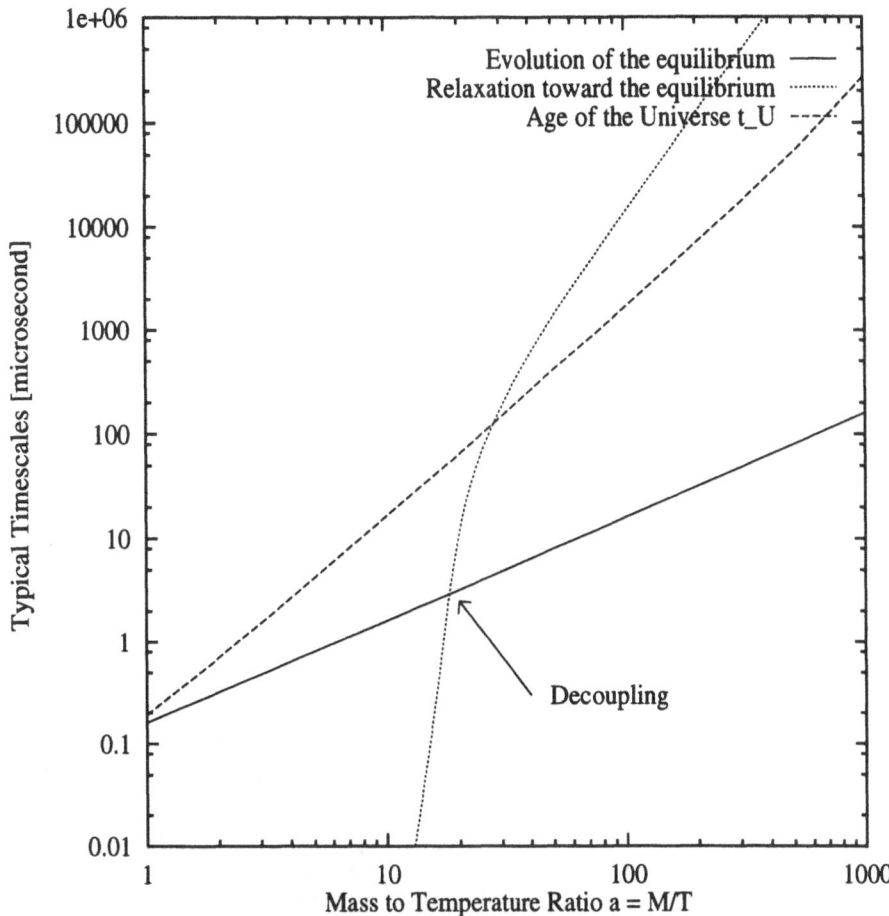

Figure 4. The typical timescale τ_{rel} with which the annihilation reaction relaxes towards its chemical equilibrium is plotted (dotted curve) as a function of the ratio $a = M/T$. The chemical equilibrium itself evolves with the timescale τ_{eq} featured by the solid line. The age t_U of the Universe corresponds to the dashed curve.

according to the relation

$$\frac{df_A}{dt} = - <\sigma_{an}v> n_A f_A \qquad (2.12)$$

which, in terms of the parameter $x = 1/a$, simplifies into

$$\frac{df_A}{f_A{}^2} = \sqrt{\frac{45}{8\pi^3}} \, g_{eff}{}^{-1/2} \, <\sigma_{an}v> M \, M_P \, dx \ . \qquad (2.13)$$

Integration of that differential equation from $x_F = 1/a_F$ where the codensity decouples from equilibrium, *i.e.*, where $f_A(a_F) = f_A^0(a_F) = f_F$, until the present epoch $x = 0$, where f_A reaches its asymptotic value f_A^{asy}, yields

$$f_A^{asy} = \frac{f_F}{1 + 2a_F} \simeq \sqrt{\frac{8\pi^3}{45}} \, g_{\text{eff}}^{1/2} \, (< \sigma_{\text{an}} v > M \, M_P \, x_F)^{-1} \, . \quad (2.14)$$

At present, the codensity of a stable 2 GeV neutrino is $f_A^{asy} \sim 4 \times 10^{-9}$ which translates into $\sim 3.3 \times 10^{-7}$ cm^{-3} K^{-3}. Note finally the dependence of f_A^{asy} on the annihilation cross section : the larger $< \sigma_{\text{an}} v >$, the lower the relic abundance f_A^{asy}.

3. Heavy Neutrinos versus LEP

The recent results [1] of the Large Electron Positron collider (LEP) at CERN, Geneva, as well as the COBE measurements [2] of the CBR spectrum have changed the cosmological rôle of heavy neutrinos .

3.1. THE RELIC DENSITY OF HEAVY NEUTRINOS

Slightly after the thermal freeze-out of neutrinos at a temperature ~ 2 MeV, electrons and positrons annihilate. Radiation is reheated and, as a consequence, the photon to neutrino temperature ratio T_γ/T_ν increases by a factor $(11/4)^{1/3} \simeq 1.4$. We infer a neutrino temperature $T_\nu^0 \simeq 2$ K at present. The careful reader may feel uncomfortable with a neutrino temperature which does not make any *thermodynamical* sense below 2 MeV where, indeed, T_ν is more a typical scale factor than a temperature per se, and varies as R^{-1}. Depending on the mass of the neutrino, two regimes may be discussed in order to derive the relic abundance ρ_ν [3].

• For $M < 2$ MeV, neutrinos decouple before they may annihilate. The resulting fossil population is ultra-relativistic at decoupling so that the present neutrino density is merely given by expression (1.8). If neutrinos are stable, we infer a relic abundance

$$\rho_\nu = \frac{3\zeta(3)}{2\pi^2} T_\nu^{0\,3} M \simeq 110 \text{ keV cm}^{-3} \left(\frac{M}{1 \text{ keV}}\right) , \quad (3.1)$$

to be contrasted with the closure density

$$\rho_C \simeq 2 \times 10^{-29} \, h^2 \text{ g cm}^{-3} \simeq 11.25 \, h^2 \text{ keV cm}^{-3} , \quad (3.2)$$

which corresponds to a *flat* Friedmann-Lemaître cosmology. If the mass density exceeds ρ_C, the Universe is spherical and will recontract in the future while, in the opposite case, its geometry is hyperbolic and the expansion will last for ever. The Hubble constant h is expressed in units of

100 km/s/Mpc. The relic density ρ_ν may be conveniently expressed in units of the closure density

$$\Omega_\nu h^2 \simeq \frac{M}{100\,\text{eV}}\,. \tag{3.3}$$

The larger M, the larger the relic abundance $\Omega_\nu h^2$ which, as a matter of fact, cannot reasonably exceed ~ 1 under the penalty of overclosing the Universe, whose age should be larger, at least, than the age of our galactic disk, i.e., ~ 9 billion years. We readily infer a cosmological upper bound on the mass of light stable neutrinos of ~ 100 eV.

• If $M > 2$ MeV, neutrinos may significantly annihilate before reaction (2.4) gets quenched. The freeze-out temperature below which annihilations stop as a result of the severe depletion of the neutrino density may be derived from equation (2.11) with appropriate values for N_A and g_{eff}. As may be guessed from (2.11), the larger M, the larger $a_F = M/T_F$. The relic density of heavy neutrinos may be expressed as

$$f_A^{asy} \simeq 9 \times 10^{-9}\, \frac{g_{\text{eff}}^{1/2}}{N_A}\, a_F \left(\frac{1\,\text{GeV}}{M}\right)^3 , \tag{3.4}$$

which translates into the ratio

$$\Omega_\nu h^2 \simeq 0.98 \left(\frac{g_{\text{eff}}^{1/2}\, a_F}{N_A}\right) \left(\frac{1\,\text{GeV}}{M}\right)^2 , \tag{3.5}$$

to which both neutrinos and antineutrinos contribute. For a neutrino mass $M \sim 2$ GeV, we may take $g_{\text{eff}} = 57/8$ and $N_A = 14$ so that we infer a fossil abundance

$$\Omega_\nu h^2 \sim \left(\frac{1.8\,\text{GeV}}{M}\right)^2 , \tag{3.6}$$

in fairly good agreement with Figure 5 where the annihilation cross sections of heavy *Majorana* (solid curve) and *Dirac* (dashed line) neutrinos have been carefully computed. As is obvious in relation (2.14), the relic abundance $\Omega_\chi h^2$ of any stable species χ is inversely proportional to the associated annihilation cross section. As a consequence of expression (2.10), the fossil density $\Omega_\nu h^2$ decreases approximately as M^{-2} as soon as the neutrino mass exceeds 2 MeV (see both curves of Figure 5). The neutrino relic abundance $\Omega_\nu h^2$ is peaked around $M \sim 2$ MeV and is fairly well described by expression (3.3) for $M < 2$ MeV whilst, above 2 MeV, relation (3.6) is valid.

Since Majorana neutrinos are their own antipartners, their annihilation is suppressed with respect to the Dirac case. As a result, the relic abundance $\Omega_\nu h^2$ of Majorana species is slightly larger than for Dirac particles. Provided a *stable* heavy Dirac neutrino has a mass in the range $\sim 2 - 10$ GeV, it may

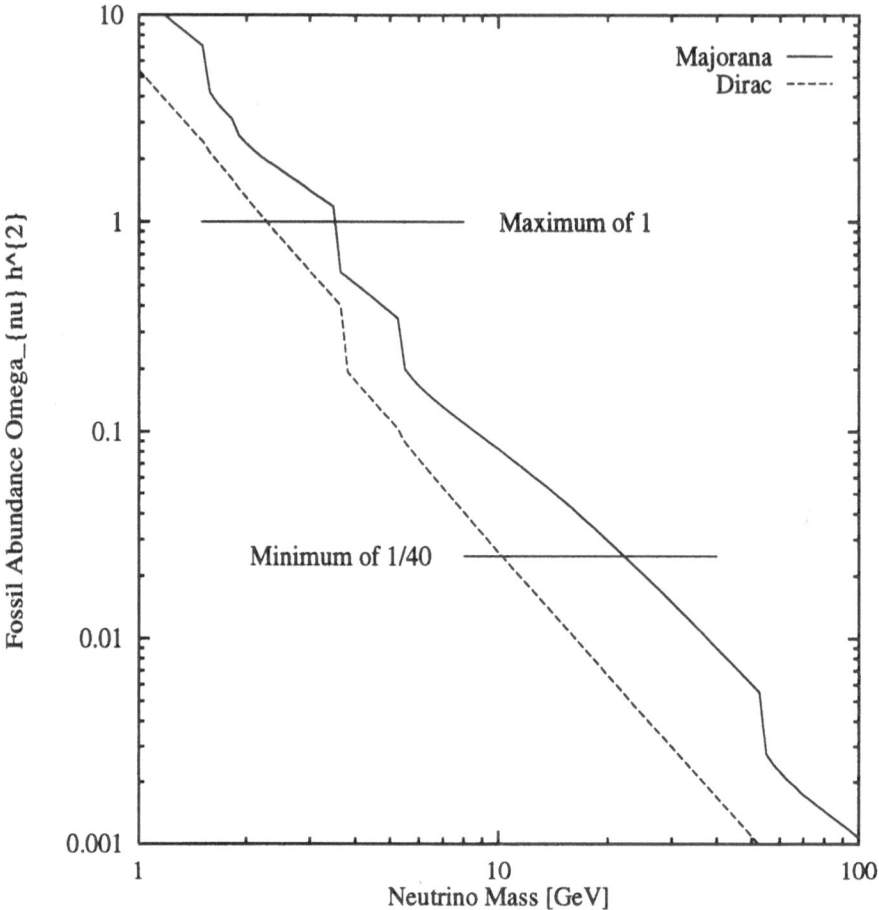

Figure 5. The abundance $\Omega_\nu h^2$ of heavy remnant neutrinos is plotted as a function of their mass M for Majorana (solid curve) and Dirac (dashed line) fermions. The parameter Ω_ν is defined as the ratio of the density ρ_ν to the critical density ρ_C. The Hubble constant h is expressed in units of 100 km/s/Mpc. The fossil density $\Omega_\nu h^2$ approximately decreases as M^{-2}. Below ~ 0.025, it is so low that it cannot account for the astronomical missing mass, even around galaxies.

contribute a significant fraction to the closure mass and may be a plausible candidate to the cosmological dark matter *. For a Majorana neutrino, the relevant mass range is between 3 and 20 GeV.

*See however Section 3.2 where the LEP results are discussed. It is shown that neutrinos cannot explain the missing mass.

383

3.2. LEP, HEAVY NEUTRINOS AND DARK MATTER

Recent experiments at LEP [1] have observed the reaction

$$e^+ e^- \to Z^0 \to f\bar{f} \ , \tag{3.7}$$

where an electron and a positron annihilate into a pair of light fermions. As the center of mass energy $\sqrt{s}$ of the electron-positron pair varies, the cross section of the previous process exhibits a characteristic Breit-Wigner resonance, with a peak at $\sqrt{s} = M_{Z^0}$ which corresponds to the production of the Z^0 boson. The width of this lineshape is merely the total decay rate of the Z^0 boson. From the direct observation of the Z^0 decays into quarks (the constituents of nucleons) and charged leptons (there are three charged leptons : the electron, the muon and the tau), and from the determination of the total decay rate, the invisible width, *i.e.*, the decay rate into invisible neutrinos , has been for the first time measured to be ~ 0.49 GeV. Note that the Z^0 decay rate into a neutrino-antineutrino pair is given by

$$\Gamma_{Z^0} = \frac{G_F}{12\sqrt{2}\pi} M_{Z^0}^3 \ \kappa(\beta) \simeq 165 \, \text{MeV} \ \kappa(\beta) \ , \tag{3.8}$$

where β denotes the velocity of the final neutrinos in the rest frame of the decaying Z^0 boson. The kinematic function $\kappa(\beta) = \beta^3$ for Majorana neutrinos whilst, for Dirac species, $\kappa(\beta) = \beta(\beta^2 + 3)/4$. For massless particles, $\beta = 1$ and $\kappa = 1$, so that the decay width is a mere 165 MeV. The invisible width corresponds to three light neutrino species which were already known to exist in association with the above mentioned charged leptons. A fourth family heavy neutrino may exist but should not contribute to the Z^0 decay width, *i.e.*, it should not be produced in Z^0 decays. We readily infer the lower bound $M > M_{Z^0}/2 \sim 45$ GeV, so that should such a species exist and be stable, its remnants would not contribute significantly to the closure mass density of the Universe as may be readily inferred from Figure 5. A heavy neutrino cannot be a good dark matter candidate as was incidentally advocated before the LEP results. LEP has shown that the conventionnal model of the electro-weak interactions is distressingly valid.

4. The Interplay between Matter and Radiation

As a result of the linearity of Maxwell's equations, electromagnetic waves do not interact with each other. Photons achieve thermal equilibrium because they merely collide with free electrons which, therefore, act as the seeds of the radiation thermalization. The Thomson-Compton cross section may be expressed as a function of the classical radius of the electron

$$r_0 = e^2/4\pi m_e c^2 = 2.818 \times 10^{-13} \, \text{cm}$$

$$\sigma_T = \frac{8\pi}{3} r_o^2 \simeq 6.67 \times 10^{-25} \, \text{cm}^2 \ . \tag{4.1}$$

Note that when an incident photon with energy $h\nu << m_e c^2$ collides with an electron at rest, the angular distribution of the scattered photon favours the forward and backward directions

$$\frac{1}{\sigma_T} \frac{d\sigma_T}{d\Omega} = \frac{3}{16\pi} \left(1 + cos^2\theta\right) \ , \tag{4.2}$$

where θ denotes the angle between the initial and final photon momenta. Energy is transferred from the photon to the electron

$$\nu - \nu' \simeq \frac{h\nu^2}{m_e c^2} \left(1 - cos\theta\right) \ . \tag{4.3}$$

As soon as recombination occurs and electrons neutralize with protons into hydrogen atoms, the CBR drops out of equilibrium and its spectrum gets quenched.

This section is devoted to the intricate interplay between the radiation and the plasma of free electrons . Our aim is to derive the Kompaneets equation [4] which accounts for the relaxation of the photon distribution $\eta(\nu)$ towards a Bose-Einstein spectrum in thermal equilibrium at temperature $T_\gamma = T_e$. We focus on non-relativistic electrons and consider photons with energy $h\nu << m_e c^2$.

4.1. A HEURISTIC APPROACH

To commence, we assume a hot electron gas. Photon energies $h\nu$ are small with respect to the electron temperature T_e. This condition is met for instance in the Rayleigh-Jeans region of the spectrum. We specially focus on an electron which, in the cosmological frame $\mathcal{R}_C$, has velocity $\vec{\beta} = \beta \vec{U}$, with $\beta << 1$. When a photon with initial momentum $(h\nu_i/c) \, \vec{U}$ collides with that electron, its frequency undergoes, on average, a Doppler shift.

In the rest frame $\mathcal{R}_e$ of the electron, the frequency of the incident photon is

$$\nu_i' = \sqrt{\frac{1-\beta}{1+\beta}} \, \nu_i \ . \tag{4.4}$$

Notice that this equation is valid for a photon moving in the same direction as the electron. Since $h\nu_i' << m_e c^2$, relation (4.3) implies that the frequency ν' is not changed by the scattering in the rest frame $\mathcal{R}_e$. In addition, the final momentum may be crudely assumed, with equal probabilities, either aligned with, or opposite to the incident direction $\vec{U}$. In the first case, the scattered photon points towards the same direction as the electron, and its frequency, in $\mathcal{R}_C$, is not modified. In the opposite situation, the scattered photon and the electron are back to back so that, in $\mathcal{R}_C$, the frequency ν_f is red-shifted with respect to the initial value ν_i. In the cosmological frame $\mathcal{R}_C$, the situation may therefore be summarized by

$$\nu_f(\rightarrow) = \nu_i(\rightarrow) \quad \text{while,} \quad \nu_f(\leftarrow) = \left(\frac{1-\beta}{1+\beta}\right) \nu_i(\rightarrow) . \qquad (4.5)$$

If the photon scatters from the forward direction, the same reasoning applies and we readily infer a blue-shift of the frequency

$$\nu_f(\leftarrow) = \nu_i(\leftarrow) \quad \text{while,} \quad \nu_f(\rightarrow) = \left(\frac{1+\beta}{1-\beta}\right) \nu_i(\leftarrow) . \qquad (4.6)$$

Photons with incident momenta along the same direction as the electron undergo an average frequency shift

$$< \nu_f > \text{(after collision)} \simeq \left(2 + \frac{1-\beta}{1+\beta} + \frac{1+\beta}{1-\beta}\right) \left(\frac{\nu_i}{4}\right) , \qquad (4.7)$$

which, once expanded up to second order in the electron velocity β, leads to

$$\frac{\Delta\nu}{\nu} \simeq \beta^2 . \qquad (4.8)$$

Photons with incident momenta transverse to the electron velocity are also, on average, scattered perpendicularly so that their frequencies are not shifted. That possibility has two chances out of three of occurring in each collision.

Since every photon undergoes $\sigma_T c n_e$ collisions each second, a third of which only are not sterile and lead to the frequency shift (4.8), we infer the average increase

$$< \Delta\nu > \simeq \frac{1}{3} (\sigma_T c n_e) < \beta^2 > \nu \Delta t , \qquad (4.9)$$

during the time interval Δt. Since electrons , whose density is denoted by n_e, behave as a perfect gas, the mean value $< \beta^2 >$ may be expressed as a function of T_e and m_e. The parameter y is defined as

$$dy = \frac{d\nu}{\nu} = (\sigma_T c n_e) \left(\frac{kT_e}{m_e c^2}\right) dt = \frac{dt}{t_C} , \qquad (4.10)$$

where t_C is the typical time scale for the inverse Comptonization of the photons. As is clear from expressions (4.9) and (4.10), thermal contact

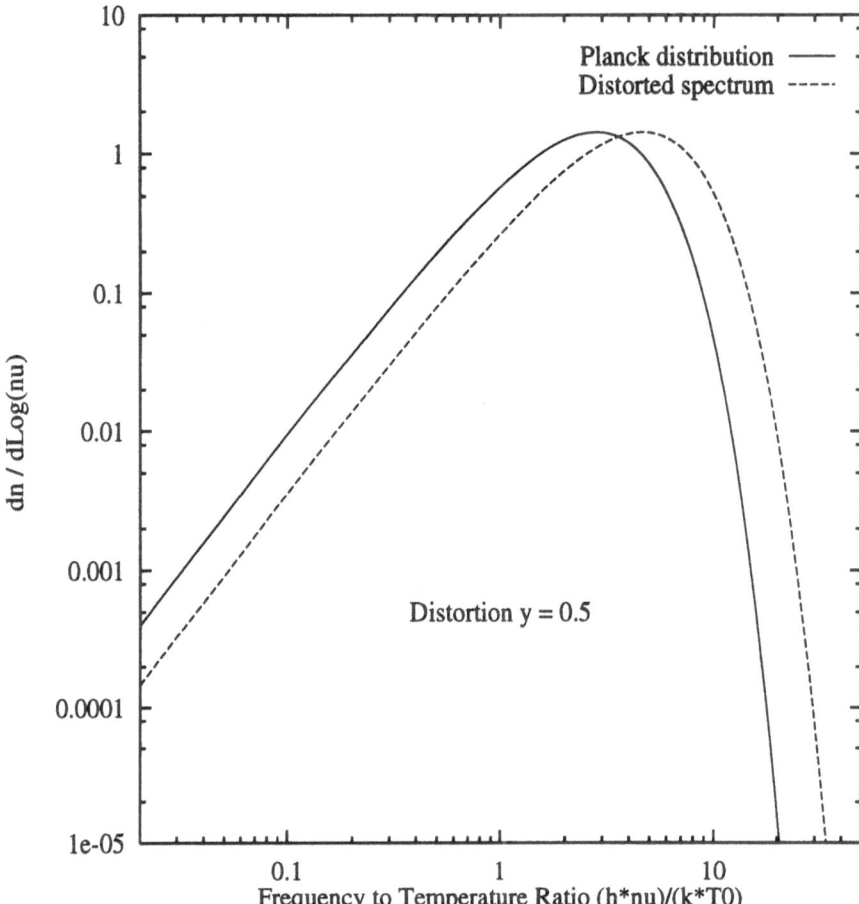

Figure 6. A Planck distribution (solid curve) experiences a $y = 0.5$ spectral distortion which results into a shift of the spectrum (dashed curve) towards higher frequencies.

between a hot electron gas and the radiation leads to the so-called y spectral distortion [5]

$$\nu(t) = \nu(0)\, e^y \;, \tag{4.11}$$

where the parameter y evolves according to

$$y = \int \frac{dt}{t_C} \;. \tag{4.12}$$

Since the frequency of the distorted spectrum is rescaled upwards by the factor e^y, the number of photons per logarithmic interval $dLog\nu$ is constant as is clear in Figure 6. The solid line stands for a pure black-body spectrum

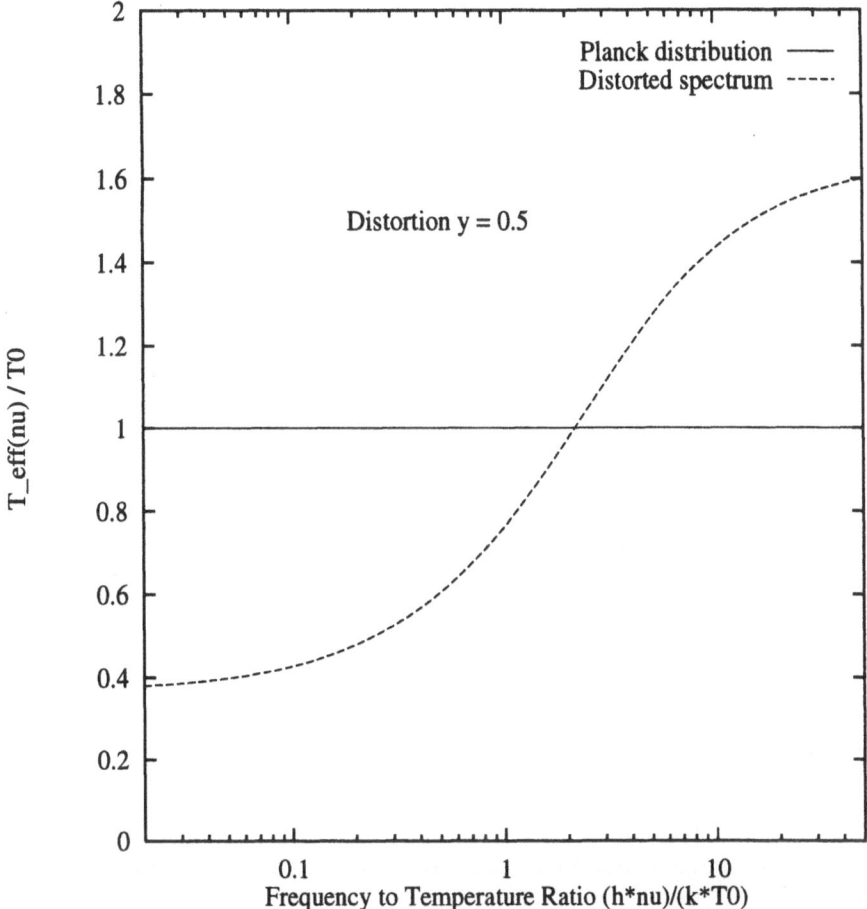

Figure 7. In the Rayleigh-Jeans region, the effective black-body temperature $T_{eff}(\nu)$ of the distorted spectrum (dashed curve) is shifted downwards by an e-fold with respect to the initial temperature T_0 (solid line).

at temperature T_0

$$\frac{dn}{dLog\nu} = \frac{\nu^3}{e^{h\nu/kT_0} - 1} \,, \tag{4.13}$$

while the dashed curve exhibits a $y = 0.5$ distortion and is shifted towards the high energy region of the diagram. Figure 7 displays the effective black-body temperature T_{eff}. For the pure Planck spectrum (solid line), the temperature has the constant value $T_{eff} = T_0$. Since photons collide with hot electrons and gain energy, high frequencies are populated at the expense of low energies. If on the one hand, the distorted spectrum (dashed curve) exhibits an excess in the Wien region, on the other hand, the Rayleigh-

Jeans part is deficient. Note that T_{eff} is indeed shifted downwards with respect to the pure black-body curve. For low frequencies, the spectrum merely reduces to $\propto \nu^2 T$, so that the effective Rayleigh-Jeans temperature decreases by a factor e^{-2y}

$$T_{RJ}(\nu) = e^{-2y} T_0 \; . \tag{4.14}$$

Since in our example $y = 0.5$, the Rayleigh-Jeans temperature is shifted downwards by an e-fold, as is clear in Figure 7.

We have analysed the spectral distortion generated by the thermal contact between hot electrons and low energy photons with $h\nu \ll kT_e$. We have neglected the frequency shift which photons undergo in the electron rest frame in which, moreover, the radiation is *not* isotropic. We have also disregarded stimulated emission which may be important during Thomson collisions. Notice, however, that the y-distortion effect is typical of the Rayleigh-Jeans region of the spectrum for which the condition $h\nu \ll kT_e$ is always fulfilled.

4.2. THE ENERGY TRANSFER BETWEEN ELECTRONS AND PHOTONS

We first assume a pure black-body radiation with temperature T_0. In the cosmological frame $\mathcal{R}_C$, the radiation is isotropic. However, in the rest frame $\mathcal{R}_e$ of an electron moving with velocity $\vec{\beta}$, photons from the forward direction are blue-shifted, while those from the backward direction are red-shifted. In $\mathcal{R}_C$, the photon distribution is

$$\eta(\nu) = \frac{1}{e^{h\nu/kT_0} - 1} \; . \tag{4.15}$$

Since that distribution is a Lorentz invariant, it may be expressed as

$$\eta(\nu', \theta) = \frac{1}{e^{h\nu'/kT_\theta} - 1} \tag{4.16}$$

in the electron rest frame $\mathcal{R}_e$. The angle between the electron and photon momenta is denoted by θ, and ν' stands for the frequency. We readily infer an anisotropy of the temperature distribution in $\mathcal{R}_e$

$$T_\theta \simeq (1 - \beta cos\theta) T_0 \; , \tag{4.17}$$

for non-relativistic $\beta \ll 1$ electrons . In $\mathcal{R}_e$, photons with initial direction θ transfer to the electron an average momentum $(h\nu'/c) cos\theta$ per collision. Since there are many more photons coming from the forward ($\theta = \pi$) than from the backward ($\theta = 0$) directions, the net transfer of momentum results into the drag force

$$F_{drag} = \int \left(\frac{h\nu'}{c} \cos\theta \right) (\sigma_T c) \left(\eta(\nu', \theta) \frac{2\nu'^2 d\nu' d\Omega}{c^3} \right) . \qquad (4.18)$$

Once frequencies have been integrated out, that expression simplifies

$$F_{drag} = \left(\frac{8\pi^5}{15} \frac{k^4}{h^3 c^3} \right) T_0^4 \sigma_T \int \left(\frac{T_\theta}{T_0} \right)^4 \cos\theta \left(\frac{d\Omega}{4\pi} \right) , \qquad (4.19)$$

and the photon energy density ρ_γ may be factored out. The drag force is therefore a mere function of the electron velocity β and of the energy density ρ_γ

$$\vec{F}_{drag} = -\frac{4}{3} \sigma_T \rho_\gamma \vec{\beta} . \qquad (4.20)$$

Radiation acts as a viscous medium with respect to the electrons which experience a damping of their propagation. That friction results into the heating of the photon gas at the expense of the electrons , whose kinetic energy is transferred to the radiation with the rate

$$\epsilon_{e \to \gamma} = < F_{drag} \beta c > n_e , \qquad (4.21)$$

per unit volume. For non-relativistic free electrons , that energy transfer simplifies finally into

$$\epsilon_{e \to \gamma} = 4 \rho_\gamma (\sigma_T c n_e) \left(\frac{kT_e}{m_e c^2} \right) = 4 \rho_\gamma \frac{dy}{dt} . \qquad (4.22)$$

The careful reader will have noticed that stimulated emission is absent from our discussion. However, we have not neglected that important effect. In $\mathcal{R}_e$, if a photon with momentum $\vec{p_1}$ scatters into the state $\vec{p_2}$, it gives to the electron a momentum proportional to $\eta_1 (1 + \eta_2) (\vec{p_1} - \vec{p_2})$. The reverse reaction is possible and results in the momentum transfer $\eta_2 (1 + \eta_1) (\vec{p_2} - \vec{p_1})$. The stimulated emission factors vectorially cancel because $d\sigma_T/d\Omega$ depends only on the angle $(\vec{p_1}, \vec{p_2})$ between the photon momenta. In addition, since $h\nu \ll m_e c^2$, frequencies in the initial and final states are identical and $\nu_1 = \nu_2$. For a hot electron gas, i.e., $T_\gamma \ll T_e$, the energy flow from radiation to electrons is negligible. The energy density ρ_γ evolves therefore as

$$d\rho_\gamma \simeq \epsilon_{e \to \gamma} dt = 4 \rho_\gamma dy , \qquad (4.23)$$

and exponentially increases with the y parameter

$$\rho_\gamma(t) = \rho_\gamma(0) e^{4y} . \qquad (4.24)$$

As a matter of fact, we could have readily guessed that result. Since $T_\gamma \ll T_e$, a y-distortion obtains and frequencies are shifted according to the exponential growth (4.11). Remember that ρ_γ is an integral over $\nu^3 \, d\nu$, and is proportional to the fourth power of the average frequency which increases as e^y.

In order to derive the energy transfer $\epsilon_{\gamma \to e}$ from the radiation to the electrons, we will assume the latter at rest in the cosmological frame $\mathcal{R}_C$. When a photon with energy $h\nu$ collides with an electron, it loses an average energy

$$h\Delta\nu \simeq \left(\frac{h\nu}{m_e c^2}\right) h\nu \; , \tag{4.25}$$

where the factor $cos\theta$ of relation (4.3) has been integrated out. That energy transfer is second order in β since, if we grossly assume $T_\gamma \sim T_e$, the fraction $(h\nu/m_e c^2) \sim (kT_\gamma/m_e c^2)$ and is of order $(kT_e/m_e c^2) \sim \beta^2$. Stimulated emission does not cancel now and, after integration over the photon distribution $\eta(\nu)$, the rate $\epsilon_{\gamma \to e}$ may be expressed as

$$\epsilon_{\gamma \to e} = \sigma_T c \, n_e \int \frac{8\pi\nu^2}{c^3} \frac{(h\nu)^2}{m_e c^2} \eta(\nu) \left[1 + \eta(\nu)\right] d\nu \; . \tag{4.26}$$

4.3. THE KOMPANEETS EQUATION

A proper derivation of the Kompaneets equation [4] may be found, for instance, in [6] and is fairly lengthy. We will obtain that equation quite crudely, but rapidly. Frequencies may be conveniently expressed in units of the electron temperature so that we define

$$x = \frac{h\nu}{kT_e} \; . \tag{4.27}$$

The energy density of the photon bath is

$$\rho_\gamma = 8\pi \frac{(kT_e)^4}{(hc)^3} \int x^3 \, \eta(x) \, dx \; , \tag{4.28}$$

and varies according to the energy balance between radiation and electrons

$$\frac{d\rho_\gamma}{dt} = \epsilon_{e \to \gamma} - \epsilon_{\gamma \to e} = 8\pi \frac{(kT_e)^4}{(hc)^3} \int x^3 \, dx \, (4 - x - x\eta) \frac{\eta}{t_C} \; . \tag{4.29}$$

The variation of ρ_γ is directly related to the time derivative of the photon distribution so that, after integration by parts of the various integrals, the Kompaneets equation yields

$$t_C \left.\frac{\partial\eta}{\partial t}\right|_K = \frac{1}{x^2} \frac{\partial}{\partial x} \left\{ x^4 \left(\frac{\partial\eta}{\partial x} + \eta + \eta^2\right) \right\} \; , \tag{4.30}$$

where $\partial\eta/\partial t$ is expressed as a function of the first and second derivatives of the photon distribution η with respect to the frequency x.

As an illustration, we derive directly from the Kompaneets equation the time evolution of the Rayleigh-Jeans temperature, *i.e.*, relation (4.14). The distribution of low-frequency photons at temperature T_γ simplifies into

$$\eta(x) \simeq \frac{kT_\gamma}{h\nu} = \frac{\alpha(t)}{x} \quad , \tag{4.31}$$

where the time-dependent ratio T_γ/T_e is denoted by α, and is assumed to be very small with respect to 1. In the Rayleigh-Jeans region of the spectrum, the ordering $x << \alpha(t) << 1$ obtains, and we may neglect η and η^2 with respect to $\partial\eta/\partial x$ in equation (4.30). The evolution of α is readily inferred to be

$$t_C \frac{d\alpha}{dt} = -2\,\alpha(t) \quad , \tag{4.32}$$

and leads to relation (4.14) with an exponential decrease of the Rayleigh-Jeans temperature.

Note finally that, under the action of Thomson-Compton collisions, a photon spectrum evolves until it reaches the equilibrium configuration for which

$$\frac{\partial\eta}{\partial x} + \eta + \eta^2 = 0 \quad . \tag{4.33}$$

The previous relation is satisfied for a Bose-Einstein spectrum

$$\eta_{BE}(x) = \frac{1}{e^{(x+\mu)} - 1} \quad , \tag{4.34}$$

where the constant of integration μ of the differential equation (4.33) may be translated into the chemical potential $-(\mu \times kT_e)$ of the Bose-Einstein distribution.

4.4. THE RELAXATION OF A BOSE-EINSTEIN DISTRIBUTION TOWARDS A PLANCK SPECTRUM

When heat is injected into radiation, the average energy per photon increases and a spectral distortion develops. After a time $\sim t_C$, Thomson collisions redistribute the additional energy, and the spectrum relaxes towards a Bose-Einstein distribution with chemical potential $\mu > 0$ and temperature $T_\gamma = T_e$. Thomson collisions conserve the number of photons. If they operate alone, the spectrum (4.34) does not evolve towards the pure black-body shape

$$\eta_{Pl}(x) = \frac{1}{e^x - 1} \quad . \tag{4.35}$$

In fact, as a result of the combined action of bremsstrahlung and double Compton reactions, for which the number of photons may increase, μ relaxes towards 0.

The bremsstrahlung process.

The bremsstrahlung reaction $ep \to ep\gamma$ replenishes the radiation with additional photons. If an electron is accelerated by an ion of charge Z_i, the corresponding electric dipole moment $\vec{d}$ of the electron-ion pair radiates a power

$$P = \frac{2}{3c^3} \frac{\partial^2 d}{\partial t^2} \; , \tag{4.36}$$

so that the total energy emitted at frequency $\nu = \omega/2\pi$ during the passage of the electron near the ion is

$$\frac{dW}{d\omega} = \frac{8\pi}{3c^3} \omega^4 \, \hat{d}^2(\omega) \; . \tag{4.37}$$

Note that $\hat{d}$ stands for the Fourier transform of the dipole moment

$$\omega^2 \hat{d}(\omega) = \frac{Z_i e^3}{\pi m_e vb} \; , \tag{4.38}$$

where an undeflected electron trajectory with velocity v and parameter of impact b, has been assumed. Each electron radiates an energy d^2W during dt and per frequency interval $d\omega$

$$\frac{d^2W}{dt\,d\omega} = \frac{16}{3c^3} \frac{Z_i^2 e^6}{m_e^2} \frac{n_i}{v} \int_{b_{min}}^{b_{max}} \frac{db}{b} \; , \tag{4.39}$$

where n_i is the density of ions. Above $b_{max} \sim v/\omega$, the period associated with the frequency ω is so small with respect to the duration of the passage of the electron in the vicinity of the ion, that the Fourier transform of $\vec{d}$ is completely chopped out and vanishes. On the other hand, if b is too small, deflection may be important, and quantum effects come into play as soon as b gets smaller than the de Broglie wavelength $h/m_e v$ of the electron. Integration over those electrons whose kinetic energy exceeds $\hbar\omega$ leads to the bremsstrahlung evolution equation [7]

$$t_{\gamma e} x^3 \left. \frac{\partial \eta}{\partial t} \right|_B = Q e^{-x} g(x) \left[1 - \eta(x,t)(e^x - 1) \right] \; , \tag{4.40}$$

with the Q factor of the form

$$Q = \frac{\alpha}{(2\pi)^{7/2}} \left(\frac{hc}{kT_e} \right)^3 \left(\frac{m_e c^2}{kT_e} \right)^{1/2} \sum n_i Z_i^2 \; . \tag{4.41}$$

The Gaunt factor $g(x)$ may be expressed as

$$g(x) = Log\left(\frac{2.2}{x}\right) \qquad x \leq 1$$

$$\text{while,} \qquad (4.42)$$

$$g(x) = \frac{Log(2.2)}{\sqrt{x}} \qquad x \geq 1 \ .$$

The typical collision time of photons with electrons is denoted by $t_{\gamma e} = (n_e c \sigma_T)^{-1}$. Note that the bremsstrahlung production and absorption processes are in balance as soon as a Planck distribution obtains.

The Double Compton reaction.

The differential cross section of the *double* Compton reaction $e\,\gamma(\nu') \rightarrow e\,\gamma(\nu_1)\,\gamma(\nu_2)$ is of order the fine structure constant of electromagnetism $\alpha \sim 1/137$ with respect to the Thomson-Compton cross section

$$\frac{d\sigma_{DC}}{d\nu_2} = \frac{4\alpha}{3\pi} \left(\frac{h\nu'}{m_e c^2}\right)^2 (1 - cos\theta_1) \frac{1}{\nu_2} d\sigma_T \ . \qquad (4.43)$$

One of the final photons has approximately the same energy as its progenitor, *i.e.*, $\nu_1 \simeq \nu'$. When integrated over the photon and electron distributions, the evolution of the radiation spectrum owing to double Compton emission and absorption takes the form [7]

$$t_{\gamma e} x^3 \left.\frac{\partial \eta}{\partial t}\right|_{DC} = \frac{4\alpha}{3\pi} \left(\frac{kT_e}{m_e c^2}\right)^2 \mathcal{I}(t) \left[1 - \eta(x,t)(e^x - 1)\right] \ . \qquad (4.44)$$

The integral $\mathcal{I}(t)$ over the field distribution of photons is the driving term for the double Compton process

$$\mathcal{I}(t) = \int x'^4 \eta(x',t) \left[1 + \eta(x',t)\right] dx' \ , \qquad (4.45)$$

where $x' = h\nu'/kT_e \simeq h\nu_1/kT_e$. In the absence of photons, double Compton would be completely ineffective.

Suppose that at some initial time $t = 0$, a pure black-body radiation is thermally coupled to a hot plasma of ions and electrons at temperature $T_e > T_\gamma$. The photon distribution $\eta(x,t)$ evolves under the combined action of Thomson collisions, and bremsstrahlung and double Compton reactions

$$\left.\frac{\partial \eta}{\partial t}\right|_{tot} = \left.\frac{\partial \eta}{\partial t}\right|_K + \left.\frac{\partial \eta}{\partial t}\right|_B + \left.\frac{\partial \eta}{\partial t}\right|_{DC} \ , \qquad (4.46)$$

where $x = h\nu/kT_e$. This equation must be integrated numerically. The main features of the evolution of $\eta(x,t)$ may nevertheless be sketched. Comptonization stops as soon as a Bose-Einstein spectrum is established, while bremsstrahlung and double Compton drive the photon distribution towards a Planck spectrum.

On the one hand, Comptonization occurs on the time scale t_C, on the other hand, bremsstrahlung and double Compton absorptions are respectively associated with the time scales

$$t_B^{abs} = \frac{x^2}{Q\,g(x)}\,t_{\gamma e} \tag{4.47}$$

and

$$t_{DC}^{abs} = \frac{3\pi}{4\alpha}\left(\frac{m_e\,c^2}{kT_e}\right)^2 \mathcal{I}^{-1}(t)\,x^2\,t_{\gamma e}\;, \tag{4.48}$$

in the low energy part of the spectrum. Provided its frequency is lower than the critical value

$$x_B = \left\{g(x_B)\frac{Q}{8}\left(\frac{m_e\,c^2}{kT_e}\right)\right\}^{1/2}, \tag{4.49}$$

for which $t_B^{abs}(x_B) = t_C/8$, a low energy photon is more likely to be absorbed by inverse bremsstrahlung than to migrate towards higher frequencies. Below x_B, bremsstrahlung is therefore more efficient than electron scattering, and a Planck spectrum readily obtains. For the double Compton process, that critical frequency is

$$x_{DC} = \left\{\frac{\alpha}{6\pi}\left(\frac{kT_e}{m_e\,c^2}\right)\mathcal{I}(t)\right\}^{1/2}. \tag{4.50}$$

As shown in [7], double Compton dominates bremsstrahlung whenever

$$\frac{n_\gamma}{n_e} \geq 0.1\left(\frac{m_e\,c^2}{kT_e}\right)^{5/2}, \tag{4.51}$$

i.e., for a hot and diluted gas. For a colder and denser medium, bremsstrahlung takes over.

5. CBR Spectral Distortions and Particle Physics

5.1. THE VARIOUS TIME SCALES

• As discussed above, the typical collision time of photons with free electrons is $t_{\gamma e} = (n_e c\sigma_T)^{-1}$. When recombination proceeds, the electron density n_e drops down while $t_{\gamma e}$ increases.
• The redistribution of energy among photons, described by the Kompaneets equation (4.30), is due to the inverse Compton scatterings of low

frequency photons by energetic electrons , and is associated with the typical time scale

$$t_C \simeq \left(\frac{m_e c^2}{k T_e} \right) t_{\gamma e} \ . \tag{5.1}$$

• The evolution of the electron temperature is due to the energy balance between matter and radiation

$$3 n_e k \frac{dT_e}{dt} = \epsilon_{\gamma \to e} - \epsilon_{e \to \gamma} \ . \tag{5.2}$$

Note that protons contribute an additional factor $3 k n_e / 2$ to the heat capacity of the plasma. The electron temperature evolves according to the differential equation

$$\frac{dT_e}{dt} + \frac{4 \sigma T \rho_\gamma}{3 m_e c} \left(T_e - T_e^{equi} \right) = 0 \ , \tag{5.3}$$

and relaxes towards its kinetic equilibrium value

$$T_e^{equi} = \int \frac{8 \pi \nu^2}{c^3} \frac{(h\nu)^2}{4 k \rho_\gamma} \eta(\nu) \left[1 + \eta(\nu) \right] d\nu \ , \tag{5.4}$$

with the typical time scale

$$t_{e\gamma} = \frac{3 m_e c}{4 \sigma T \rho_\gamma} \ . \tag{5.5}$$

• As shown by Lightman [7], the typical time scale for which bremsstrahlung may replenish a deficient photon population can be expressed as

$$t_B \simeq \frac{2 \zeta(3)}{Q} g^{-2}(x_B) t_{\gamma e} \ , \tag{5.6}$$

while the restoration of a Planck distribution by double Compton emission occurs on the time scale

$$t_{DC} \simeq \frac{\pi}{8 \alpha} \left(\frac{m_e c^2}{k T_e} \right)^2 t_{\gamma e} \tag{5.7}$$

In Figure 8, those various time scales are displayed as a function of the red-shift $1 + z = T_\gamma(t)/2.375$ K, and are compared with the corresponding age $t_U(z)$ of the Universe. The expansion equation has been integrated numerically with three light neutrino species, radiation and a baryon density of $n_B/n_\gamma = 3 \times 10^{-10}$, in agreement with the primordial nucleosynthesis estimates.

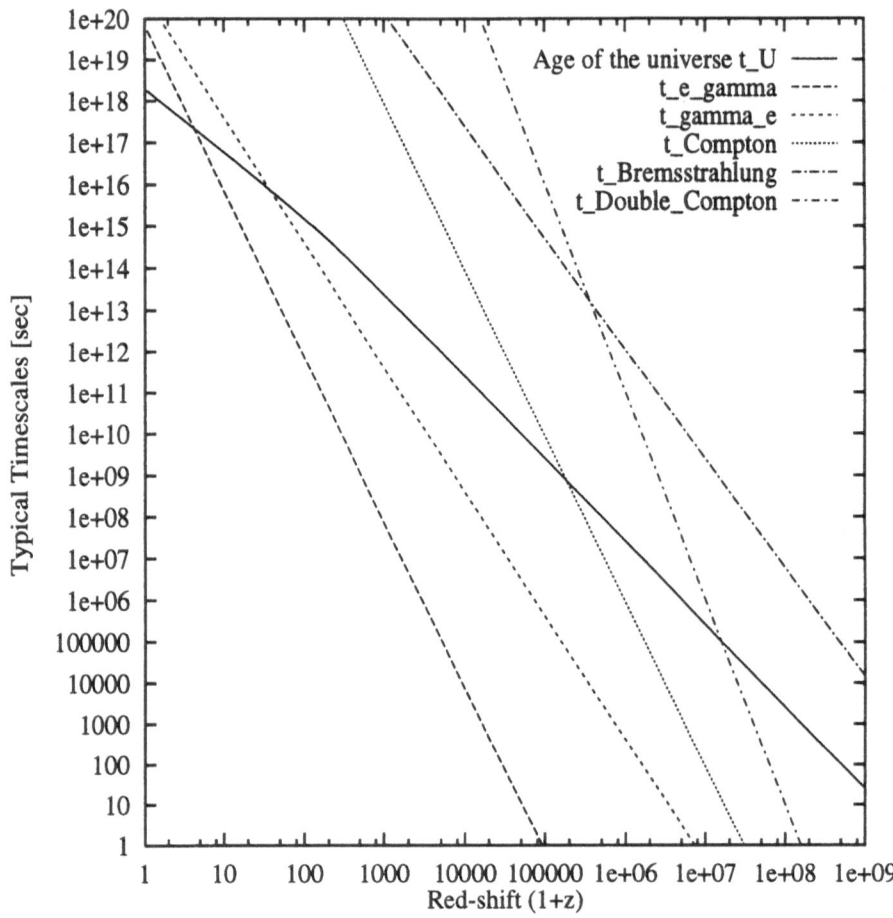

Figure 8. The age of the Universe t_U (solid) is presented as a function of the red-shift (1 + z) and is compared to the relevant time scales $t_{e\gamma}$ (long dashed), $t_{\gamma e}$ (short dashed), t_C (dotted), t_B (dotted – long dashed) and finally t_{DC} (dotted – short dashed). Complete ionization of the baryonic gas has been assumed here at any given time. This is not valid after recombination, for a red-shift $z \leq 10^3$.

5.2. THE Y AND μ DISTORTIONS

In Figure 8, the solid line stands for the age t_U of the Universe. Note that $t_{e\gamma}$ (long dashed curve) is always much smaller than all the other relevant time scales, so that the electron temperature T_e relaxes rapidly towards its equilibrium value at any given time and for any given photon distribution.

Photons stop to collide with electrons after recombination. However, for a red-shift $z \geq 10^3$, $t_{\gamma e}$ (short dashed curve) is smaller than the typical expansion time t_U. However, thermalization of the photon bath is only effective above a red-shift $z_{BE} \sim 2 \times 10^5$ below which the inverse Comptonization time scale t_C (dotted curve) exceeds the expansion time t_U. The red-shift z_{BE} corresponds to an age of $\sim 10^9$ sec.

For the low baryon abundance which we have adopted, double Compton emission is more efficient than bremsstrahlung. Indeed, the time scale t_B (dotted – long dashed curve) always exceeds t_U. On the other hand, double Compton processes restore the Planck form of the radiation spectrum above the critical red-shift $z_{Pl} \sim 1 - 2 \times 10^7$, at which t_{DC} (dotted – short dashed curve) crosses t_U. The red-shift z_{Pl} is associated with an age of $\sim 10^5$ sec.

We may therefore sketch the scenario of the spectral distortions of the CBR. Four possibilities may occur, depending on when the energy is released into the primordial plasma.

1) $z \geq z_{Pl}$

This case corresponds to the situation where the age of the Universe t_U exceeds both t_C and t_{DC}. Double Compton emission acts efficiently and any spectral distortion is erased during an expansion time. No matter how large the energy release is, the microwave background is not affected.

2) $z_{BE} \leq z \leq z_{Pl}$

The age of the Universe t_U is still larger than the Comptonization time scale t_C so that thermalization of the photon gas ensues. However, the production of additional photons is frozen because the double Compton time scale t_{DC} exceeds t_U. Any heat release generates a μ-distortion. The radiation spectrum relaxes towards a Bose-Einstein distribution faster than the Universe expands ($t_C \leq t_U$). However, a pure black-body radiation cannot be achieved since double Compton emission is slower than expansion ($t_{DC} \geq t_U$). The photon bath is still in thermal equilibrium, but with an average photon energy larger than for a Planck spectrum, which translates into a non-zero chemical potential μ. In units where $\hbar = k = c = 1$, the energy ρ_{BE}, number n_{BE} and entropy σ_{BE} densities may be expressed as

$$\rho_{BE} = \frac{T_e^4}{\pi^2} I_3(\mu), \quad n_{BE} = \frac{T_e^3}{\pi^2} I_2(\mu) \quad \text{and} \quad \sigma_{BE} = \frac{T_e^3}{\pi^2} \left\{ \frac{4}{3} I_3(\mu) + \mu I_2(\mu) \right\},$$
(5.8)

where the integral $I_n(\mu)$ is defined as

$$I_n(\mu) = \int_0^\infty x^n \left(e^{x+\mu} - 1 \right)^{-1} dx .$$
(5.9)

If $\Delta\rho_\gamma$ (or Δn_γ) denotes the increase of the energy (or number) density due to the heat release, the plasma temperature T_e and chemical potential μ evolve according to

$$\Delta n_\gamma = \left(\frac{\partial n_{BE}}{\partial T_e} \right) dT_e + \left(\frac{\partial n_{BE}}{\partial \mu} \right) d\mu$$

$$\text{while} \hspace{6cm} (5.10)$$

$$\frac{\Delta\rho_\gamma}{T_e} = \left(\frac{\partial\sigma_{BE}}{\partial T_e}\right) dT_e + \left(\frac{\partial\sigma_{BE}}{\partial\mu}\right) d\mu \ .$$

Note that

$$\frac{dI_n}{d\mu} = -n\, I_{n-1}(\mu) \ . \hspace{4cm} (5.11)$$

Since in the practical cases of interest μ is small – COBE [2] sets the stringent constraint $\mu \leq 9 \times 10^{-5}$ (95% CL) – the previous integrals may be evaluated at $\mu = 0$ and simplify into $I_1(0) = \pi^2/6$, $I_2(0) = 2\zeta(3)$ and $I_3(0) = \pi^4/15$.

The electron temperature T_e relaxes towards the radiation temperature T_γ much faster than expansion. In general, high energy photons are injected into the plasma so that $\Delta n_\gamma \ll n_\gamma$. This leads to a unique relation between T_e and μ, and we may therefore express the variation of the chemical potential parameter μ as a function of the relative increase of the photon energy

$$\frac{\Delta\rho_\gamma}{\rho_\gamma} = \left\{\left(\frac{2\pi^2}{9\zeta(3)}\right) - \left(\frac{90\zeta(3)}{\pi^4}\right)\right\} d\mu \simeq 0.714\, d\mu \ . \hspace{1cm} (5.12)$$

3) $z_{rec} \leq z \leq z_{BE}$

Both t_{DC} and t_C exceed the expansion time t_U. Photons may still collide with electrons so that a y-distortion ensues, which corresponds to the small value $y \sim t_U/t_C$. If energy is released in the primordial plasma, electrons and protons are heated up to the temperature $T_e \gg T_\gamma$. As a result of rare inverse Compton reactions, the Rayleigh-Jeans region is slightly depleted by a factor e^{-2y} with respect to the rest of the spectrum. Remember that COBE [2] sets the bound $y \leq 1.5 \times 10^{-5}$.

4) $z \leq z_{rec}$

Below $z_{rec} \sim 10^3$, free electrons combine with protons. Figure 8 has assumed a constant ionization fraction $x_e = 1$. In fact, as soon as recombination drives the population of free electrons to extinction, $t_{\gamma e}$ exceeds t_U and the primordial radiation decouples from the neutralized matter. The microwave spectrum becomes frozen and its features are fossilized until now. If photons are released during that period, the CBR spectrum is directly distorted. The CBR is therefore of crucial importance as regards the thermal history of the Universe. It provides us with a unique probe of its recent evolution.

5.3. CONSTRAINTS FROM THE μ-DISTORTION

If energy is injected in the primordial plasma, a μ-distortion may develop. The relevant parameter is the energy η_Q released per CBR photon. That quantity is just the ratio of the total energy Q_{inj} received by each unit volume of the photon gas over the photon density n_γ :

$$\eta_Q = \frac{Q_{inj}}{n_\gamma} . \tag{5.13}$$

We are concerned here with decaying particles whose lifetime is denoted by τ. Observational limits on the maximal value of the CBR chemical potential μ may be translated into an exclusion diagram in the plane (τ, η_Q).

Let us consider the case of an unstable species χ which releases a total energy $\eta_Q = 1$ eV per CBR photon. That particle has a lifetime $\tau = 10^8$ seconds and a mass M_χ. In Figure 9, the age of the Universe t_U (solid curve) is presented as a function of the red-shift $1 + z$. The thermalization timescale t_C (dashed) and the double-Compton timescale t_{DC} (dotted) are also featured. The Hubble constant H_0 is 50 km/s/Mpc today while the baryon to photon ratio is fixed at $\eta_{10} = 3$. As long as the age of the Universe t_U exceeds both t_C and t_{DC}, a Planck spectrum readily obtains. When t_{DC} becomes larger than t_U, the chemical potential μ cannot relax any longer towards 0. It increases according to the relation :

$$d\mu = 1.4 \frac{d\rho_\gamma}{\rho_\gamma} , \tag{5.14}$$

where ρ_γ stands for the energy density of the photon gas while $d\rho_\gamma$ is the energy released by the decays during the time dt. That energy is related to the lifetime τ and to the total amount Q_{inj} of heat which the radiation receives per unit volume

$$d\rho_\gamma = Q_{inj} \, e^{-t/\tau} \frac{dt}{\tau} . \tag{5.15}$$

The energy Q_{inj} depends in turn on the mass M_χ of the unstable particle, on its density n_χ and on the branching ratio B_γ of the channel $\chi \to X + \gamma$:

$$Q_{inj} = B_\gamma \left(\frac{M_\chi}{2} \right) n_\chi . \tag{5.16}$$

As featured in Figure 10, when the particle of our example starts to decay, the Comptonization time scale t_C is still less than the age of the Universe while the double-Compton creation of additional CBR photons is ineffective. A μ-distortion therefore develops until t_C exceeds t_U as shown by the solid curve. When t_C exceeds t_U, the chemical potential μ gets frozen. In our case, the final value $\sim 4 \times 10^{-3}$, well exceeds the observational constraint [2]

$$\mu \leq 9 \times 10^{-5} \quad (95\% \text{CL}) . \tag{5.17}$$

400

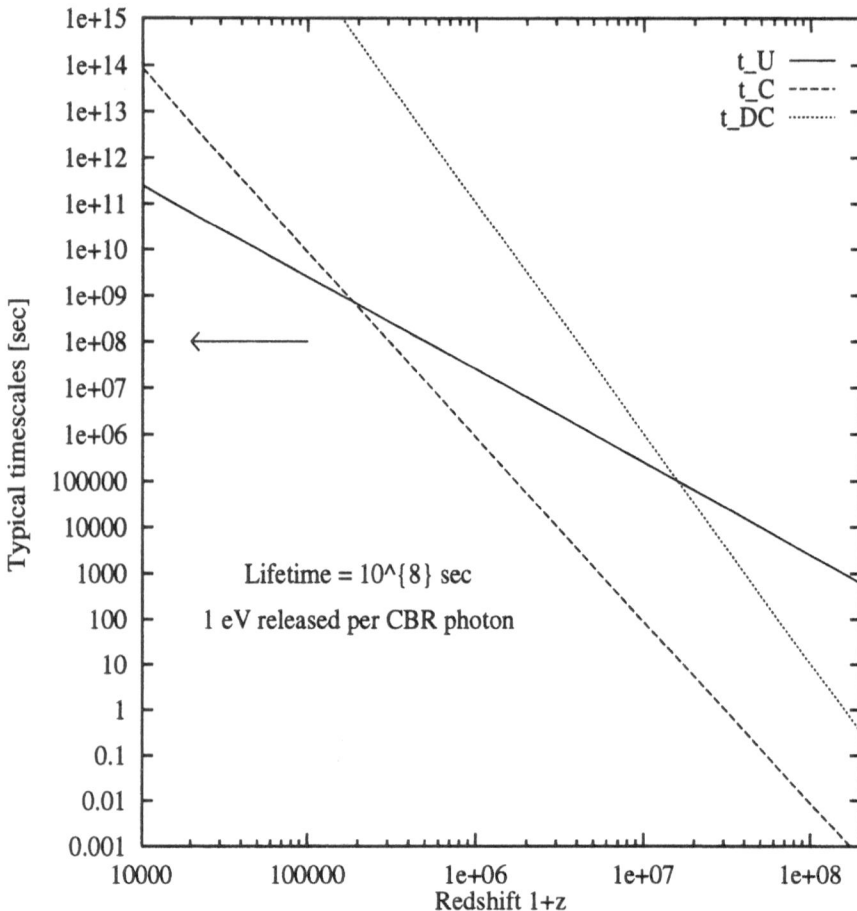

Figure 9. The age of the Universe t_U (solid curve) is presented as a function of the red-shift $1 + z$. The thermalization timescale t_C (dashed) and the double-Compton timescale t_{DC} (dotted) are also featured. The Hubble constant H_0 is 50 km/s/Mpc while the baryon abundance is fixed at $\eta_{10} = 3$.

In Figure 11, the maximal energy η_Q released in the early Universe (per CBR photon) is plotted as a function of the decay lifetime τ. The solid line stands for the constraint $\mu < 9 \times 10^{-5}$ while the dashed curve is an order of magnitude more stringent, with $\mu < 10^{-5}$.

5.4. CONSTRAINTS FROM THE Y-DISTORTION

Let us consider the reaction

$$\chi \to X + \gamma \ , \tag{5.18}$$

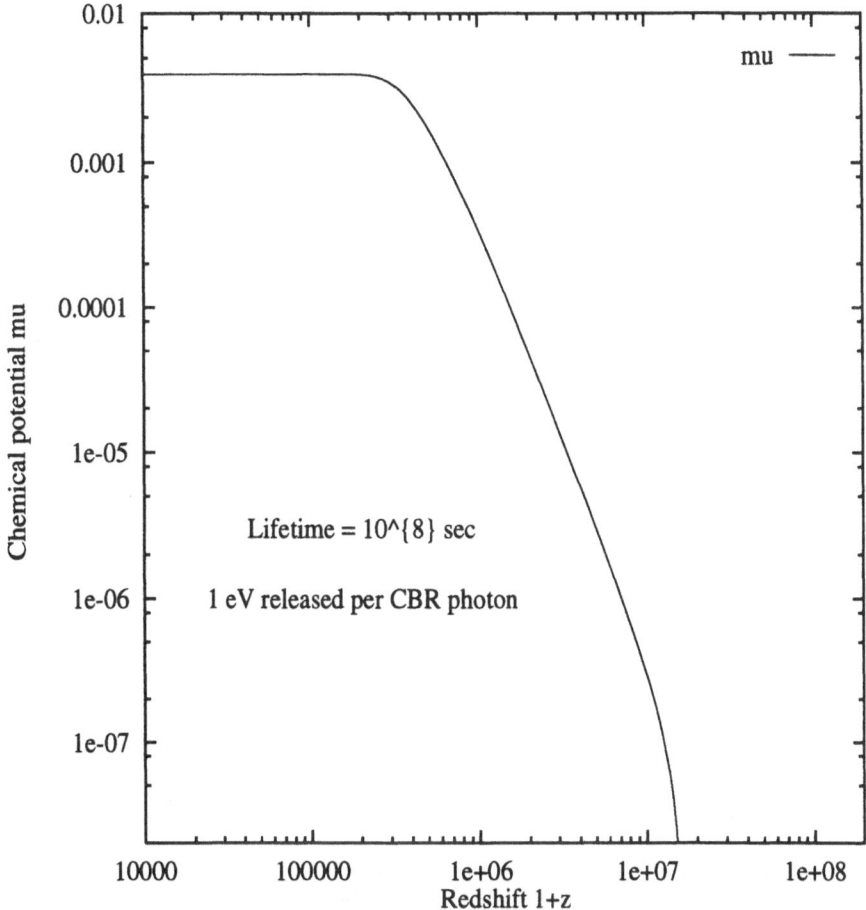

Figure 10. As long as the age of the Universe t_U exceeds both t_C and t_{DC}, a Planck spectrum readily obtains. When t_{DC} becomes larger than t_U, the chemical potential μ cannot relax any longer towards 0. It increases as the Universe expands as shown by the solid curve. Then t_C exceeds t_U and μ gets frozen.

where X is a decay product that does not interact through electromagnetism. In the case of the y-distortion, three parameters are relevant :
- the energy ϵ_γ of the decay photon,
- the number F_γ of decay photons released per initial CBR photon,
- and finally, the lifetime τ of the unstable species.

The three next diagrams feature the case of an $\epsilon_\gamma = 1$ keV decaying particle with a lifetime $\tau = 10^{11}$ seconds. The number of decay photons released per CBR photon has been set equal to $F_\gamma = 10^{-4}$ so that the total energy released per CBR photon is 0.1 eV. In Figure 12, the age of the Universe t_U (solid curve) and the thermalization timescale t_C (dashed) are

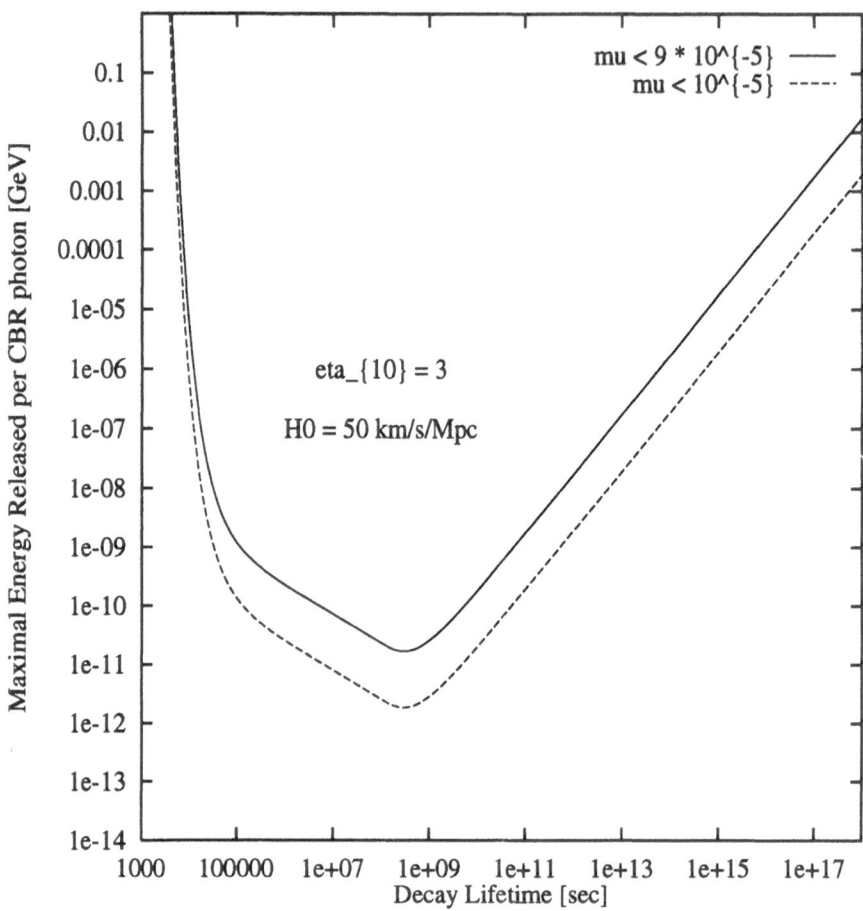

Figure 11. The maximal energy η_Q released in the early Universe (per CBR photon) is plotted as a function of the decay lifetime τ. The solid line stands for the constraint $\mu < 9 \times 10^{-5}$ while the dashed curve is an order of magnitude more stringent, with $\mu < 10^{-5}$.

featured as a function of the red-shift $1 + z$. The Hubble constant H_0 has been set equal to 50 km/s/Mpc. The baryon abundance is fixed at $\eta_{10} = 3$. The two vertical arrows define an interval for the age of the Universe during which a y-distortion may develop. Actually, the decay photons are no longer thermalized with the CBR so that they directly contribute to the spectral distortion

$$\delta\eta(x) = 2\zeta(3) \frac{F_\gamma}{x^3} \frac{e^{-t/\tau}}{H(t)\tau} \, , \tag{5.19}$$

where

$$x = \epsilon_\gamma / T_\gamma \, , \tag{5.20}$$

and where $H(t)$ is the expansion rate at time t. That spectral distortion is presented in Figure 13 in the case of our example. Today, it sits in the far infrared region of the spectrum. It should be buried in the infrared emission of the dust and of the cold molecular clouds which pervade the disc of our galaxy. Remember also that our example leads to a y-distortion more than two orders of magnitude larger than the current observational limits.

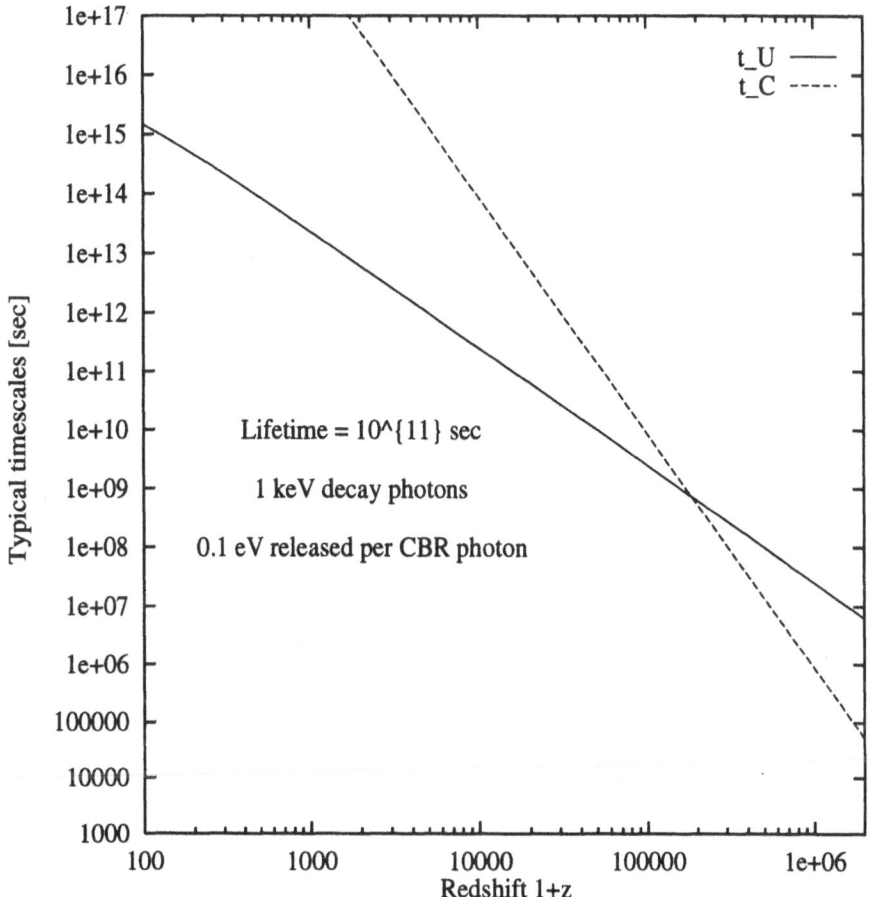

Figure 12. The age of the Universe t_U (solid curve) and the thermalization timescale t_C (dashed) are featured as a function of the red-shift $1 + z$. The Hubble constant H_0 has been set equal to 50 km/s/Mpc. The baryon abundance is fixed at $\eta_0 = 3$.

The spectral distortion of the CBR generates an increase of the electron temperature which, according to relation (5.4), is given by :

$$\frac{T_e - T_\gamma}{T_\gamma} = \frac{15}{\pi^4} \int \left\{ \frac{x}{4} \left(2\eta_0 + 1 \right) - 1 \right\} \delta\eta \, x^3 \, dx \ , \qquad (5.21)$$

404

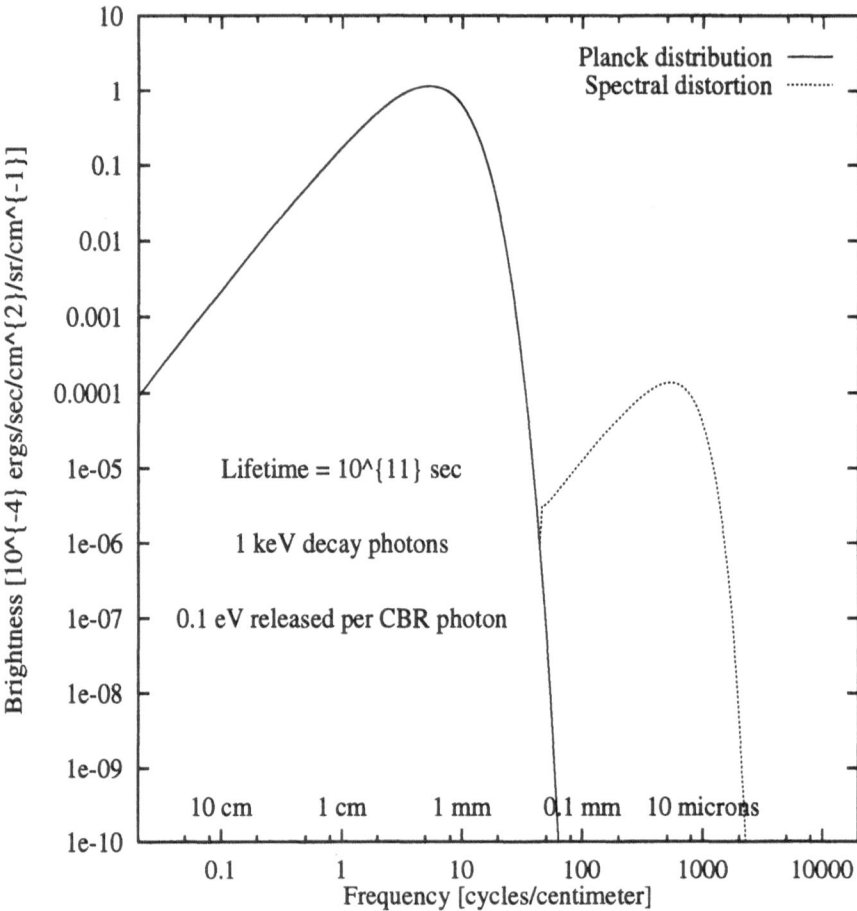

Figure 13. On top of the y-distortion, the decay photons directly generate a significant spectral distortion in the Wien part of the CBR spectrum. The brightness of the latter is plotted at present as a function of the frequency.

where $\eta_0 = (e^x - 1)^{-1}$ is the Planck distribution. The dashed curve of Figure 14 features the evolution of the ratio $(T_e - T_\gamma)/T_\gamma$. Because the electron temperature increases, a y-distortion may be generated. Remember that the thermalization timescale t_C must exceed the age of the Universe t_U. The plasma must also be ionised. The increase of the y-parameter follows the equation :

$$y = \int \left\{ \frac{T_e - T_\gamma}{T_e} \right\} \frac{dt}{t_C} \ . \tag{5.22}$$

As soon as recombination takes place, for a red-shift of $\sim 10^3$, y stops increasing and the solid curve of Figure 14 reaches a plateau in the low

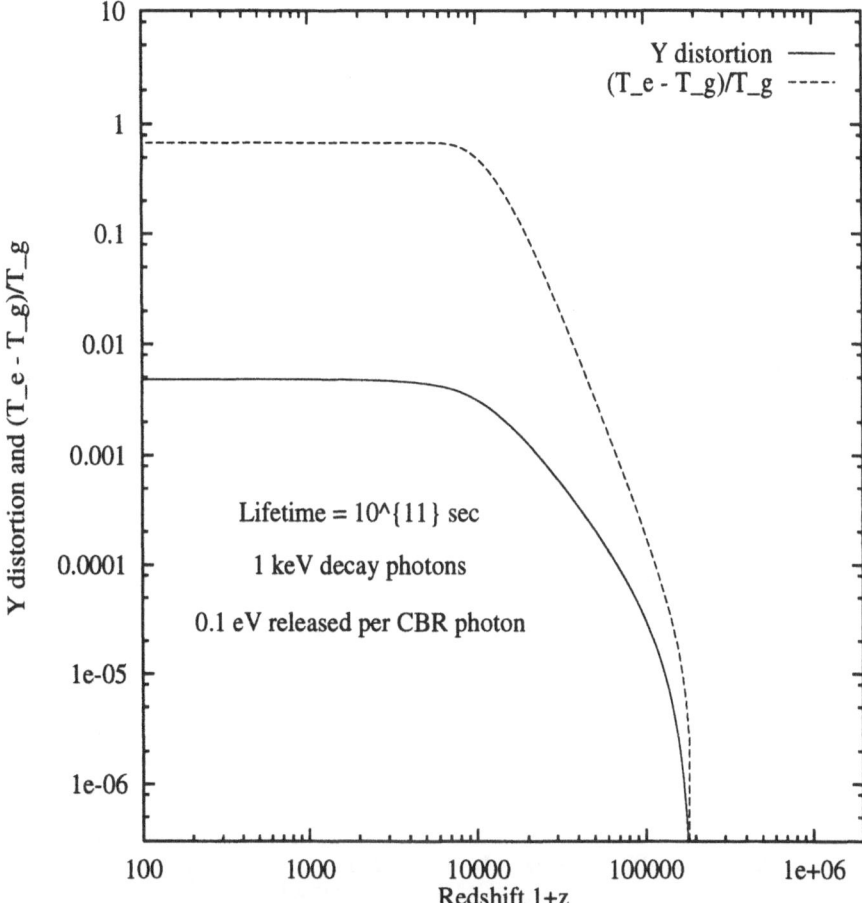

Figure 14. A y-distortion (solid line) may be generated when the thermalization timescale t_C exceeds the age of the Universe t_U. The plasma must also be ionised. As soon as recombination takes place, at a red-shift of $\sim 10^3$, y stops increasing. The electron temperature T_e slightly exceeds the photon temperature T_γ. The dashed curve features the evolution of the ratio $(T_e - T_\gamma)/T_\gamma$.

red-shift region of the diagram.

Analysis of the spectrum of the CBR yields the observational constraint [2]

$$y \leq 1.5 \times 10^{-5} \quad (95\%\mathrm{CL}) \ . \tag{5.23}$$

That limit may be translated into an exclusion plot in the $(\tau, \eta_Q = F_\gamma \epsilon_\gamma)$ plane of Figure 15. The maximal energy η_Q released in the early Universe (per CBR photon) is actually presented as a function of the decay lifetime τ for two values of ϵ_γ. The solid line stands for decay photons with an energy of 1 keV. For the dotted case, that energy is 10 keV.

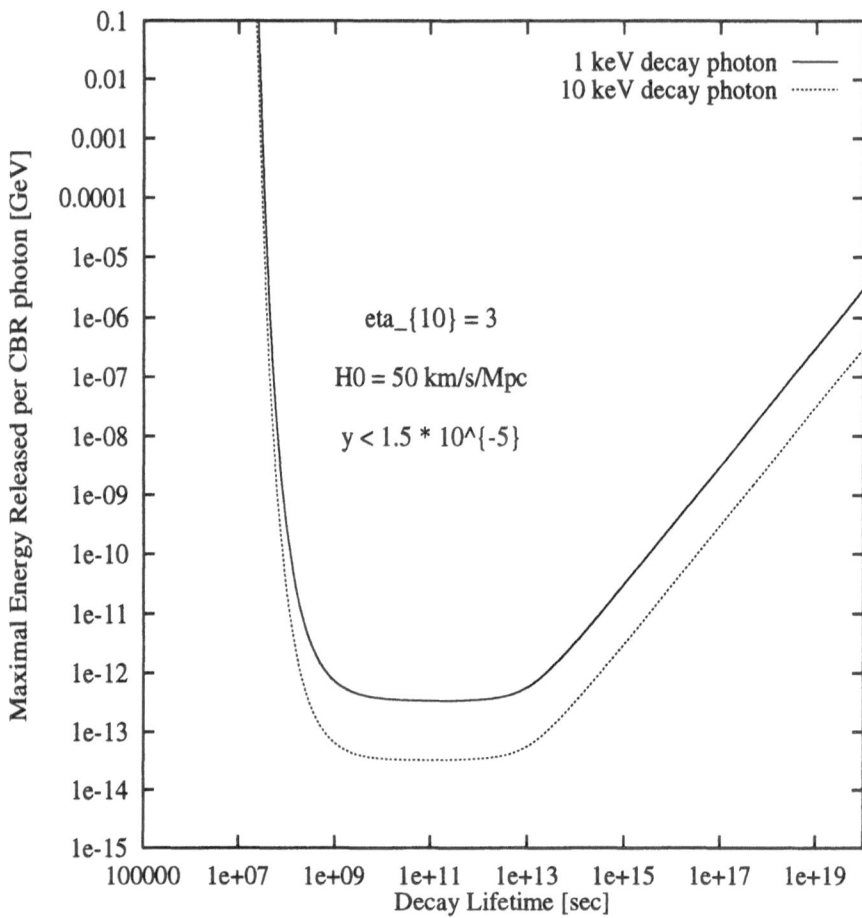

Figure 15. The maximal energy η_Q released in the early Universe (per CBR photon) is plotted as a function of the decay lifetime τ. This bound is derived from the requirement that the y-distortion cannot exceed 1.5×10^{-5}. The solid line stands for decay photons with an energy of 1 keV. In the dotted case, that energy is 10 keV.

References

1. L3 Collaboration, Adeva, B. *et al.* (1989) *Phys. Lett.* **B231**, 509;
   ALEPH Collaboration, Decamp, D. *et al.* (1989) *Phys. Lett.* **B231**, 519;
   OPAL Collaboration, Akrawy, M. Z. *et al.* (1989) *Phys. Lett.* **B231**, 530;
   DELPHI Collaboration, Aarnio, P. *et al.* (1989) *Phys. Lett.* **B231**, 539;
   see also Fernandez, E. (1990) 'Proceedings of the Neutrino 90 Conference', 11-15 June 1990, CERN, Geneva, Switzerland and references therein.
2. Mather, J. C. *et al.* (1990) *Astrophys. J.* **354**, L37;
   Mather, J. C. *et al.* (1994) *Astrophys. J.* **420**, 439;
   Fixsen, D. *et al.* (1994) *Astrophys. J.* **420**, 445;
   Fixsen, D. *et al.* (1996) preprint astro-ph/9605054, *Astrophys. J.* in the press;
   see also Smoot, G. F., Scott, D. (1996) preprint astro-ph/9603157.

3. Kolb, E. W., Turner, M. S. (1990) *The Early Universe*, Frontiers In Physics Series, Addison-Wesley.
4. Kompaneets, A. S. (1957) *Soviet Phys. – JETP* **4**, 730.
5. Sunyaev, R. A., Zel'dovich, Ya. B. (1970) *Astrophys. Space Sci.* **7**, 3;
   Sunyaev, R. A., Zel'dovich, Ya. B. (1970) *Astrophys. Space Sci.* **9**, 368;
   Sunyaev, R. A., Zel'dovich, Ya. B. (1972) *Astron. Astrophys.* **20**, 189;
   Sunyaev, R. A., Zel'dovich, Ya. B. (1980) 'Microwave background radiation as a probe of the contemporary structure and history of the universe', *Ann. Rev. Astron. Astrophys.* **18**, 537-560 and references therein.
6. Peebles, P. J. E. (1971) 'Physical Cosmology'.
7. Lightman, A. P. (1981) 'Double Compton emission in radiation dominated thermal plasmas', *Astrophys. J.* **244**, 392-405.

POLARIZATION OF THE MICROWAVE BACKGROUND: EXPERIMENTS

FRANCESCO MELCHIORRI
Consorzio Interuniversitario di Fisica Spaziale CIFS, Rome Group, University of Rome La Sapienza

CHIARA MONTECCHIO AND GIAMPAOLO PISANO
Dept. of Phys. University of Rome La Sapienza

AND

BIANCA MELCHIORRI-OLIVO
Istituto di Fisica dell'Atmosfera del CNR, Rome

Abstract.
We review the available results on Cosmic Background Polarization (CBP): we discuss the advantages and disadvantages of ground based observations of CBP versus satellite observations. Atmospheric noise can be reduced to a very small amount by an appropriate design of a polarimeter. Various polarimeters are discussed and the spurious poalarization of optical components is described. Finally we present a new type of polarimeter with spurious polarization smaller than 10^{-7} and capable of measuring CBP at angular scales of 5-10 arcminutes from ground.

1. Introduction: Cosmic Background Polarization (CBP) versus Cosmic Background Anisotropy (CBA).

We define as Cosmic Background Polarization, or CBP the quantity

$$P = \frac{I_p(\lambda)}{BB(\lambda, 2.73K)}$$

Where $I_p(\lambda)$ is the linear polarized brigthness measured at wavelength λ and $BB(\lambda, 2.73K)$ is the brigthness of a blackbody at a given temperature and the same wavelength. Figure 1 summarizes the results obtained in searching for CBP; details are given in Table I.

C. H. Lineweaver et al. (eds.), The Cosmic Microwave Background, 409–418.

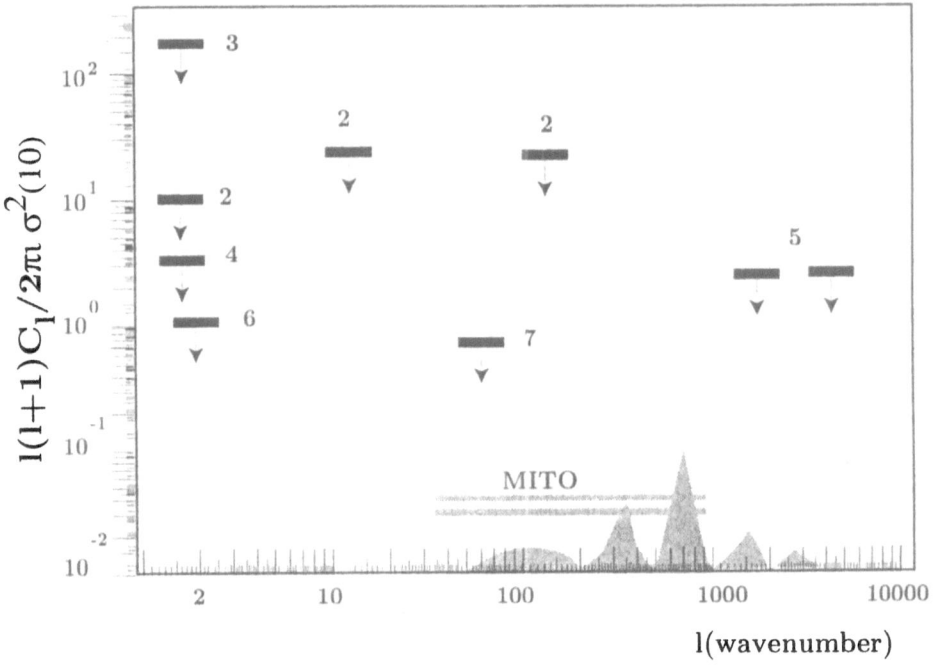

Figure 1. Available upper limits on CBP: we have also indicated the theoretical values (see the paper by Melchiorri and Vittorio this volume) and the expected sensitivity of the Italian MITO Polarimeter

The situation is similar to that of CBA in the eighties: upper limits only are available at a level of about one order of magnitude greater than the theoretical expectations (two orders in power). Generally speaking, CBR, if isotropic cannot be polarized. Simple theoretical considerations about the Rayleigh scattering prove (see the paper by Melchiorri and Vittorio in this book) that the quadrupole component of each anisotropy can be polarized through the scattering with free electrons. Since the anisotropies may be polarized, (not the isotropic flux) and the degree of polarization cannot exceed the unity, each anisotropy measurement represents a stringent upper limit for the total polarization of CBR, as previously defined. For instance, COBE-DMR results of 30 ± 1 μK at an angular scale of 10 degrees indicates that CBR polarization at this angular scale cannot exceed 10^{-5}.

TABLE 1. Available Upper Limits to CBP

Author	Angular Scale	Frequency	Upper limit to lin. Pol(circ. Pol)
1- Penzias,Wilson(1965)	10'	4.3 GHz	0.1
2- Caderni et al.(1978)	0.5-90 deg	100-600 GHz	10^{-4}
3- Nanos (1979)	15 deg	9.4 GHz	6×10^{-4}
4- Lubin, Smoot(1981)	45-90 deg	33 GHz	6×10^{-5} (6×10^{-4})
5- Partridge et al.(1988)	18-160 arcsec.	5 GHz	5×10^{-5} (5×10^{-5})
6- COBE-DMR(1992)	10 deg	53-90 GHz	10^{-5}
7- Wollack et al.(1996)	1 deg.	26-36 GHz	9×10^{-6}

Usually CBP has been searched as a byproduct of CBA measurements with the exception of the experiment by Caderni et al. (1978), which was dedicated to CBP search alone. This experiment is still of some interest because it made a survey of the galactic plane in the wavelength range 300 - 1500 μm thereby providing an upper limit of 10^{-3} for the polarization of dust emission in the far infrared at angular scales 0.5 $\Rightarrow$ 90 degrees. On the basis of this result we compare in Figure 2 the foregrounds for CBP with those relative to CBA: although this is a very qualitative picture, one is led to the conclusion that the optimum frequency for observations of CBP is shifted towards higher frequencies with respect to the case of CBA, due to the presence of significant polarization in synchrotron radiation and the lack of significant polarization in free-free and dust emission.

There is a widely diffused opinion that CBA observations must be performed from space in order to get a sensitivity of a few microkelvin per pixel (see, for instance the various ESA and NASA proposals, like the recent FIRE, MAP and SAMBA-COBRAS or the previous CIRBS, COS P, AELITA etc). To answer this question in a quantitative way, B. Melchiorri et al (1996) have studied the limitations arising from atmospheric fluctuations in ground based observations. They have shown that atmospheric fluctuations cannot be corrected better than by a factor of 20-30 through multifrequency observations and this fact alone limits CBA observations to very special sites, like in Antarctica and/or at high altitudes. In the case of CBP this situation is simplified and improved by the fact that observations can be carried out at *the same* frequency but with ortogonal polarization planes.

Non polarized fluctuations of atmospheric emission can be removed within the accuracy of zero of the two-bolometer bridge: usually, the quality of this is tested by chopping all the atmospheric emission at the focal plane

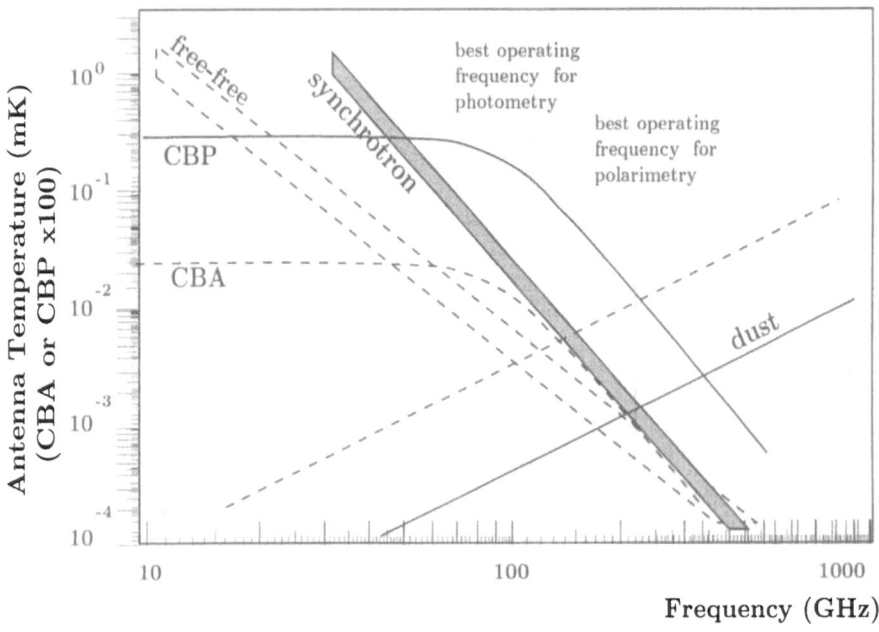

Figure 2. Limitations posed by diffuse galactic radiations to CBA observations (dotted lines) and to CBP observations (continuous lines multiplied by 100): note that the range of frequencies useful for CBP observations is around 200-400 GHz, while that for CBA is around 50-150 GHz

of the telescope. Preliminary tests have shown that the drift of the zero may be limited to less than $10\mu K/hour$; therefore, atmospheric fluctuations up to several mK can be removed at a level of 10% of expected CBP.

In order to achieve this accuracy it is crucial to reduce the spurious instrumental polarization to below 10^{-7} and this turned out to be the main constraint for ground based observations.

We may conclude that CBP observations are possible from ground at a level limited by the detector intrinsic noise (roughly 1% of anisotropy) and there is less compelling evidence for satellite experiments, once a polarimeter is realized with a spurious polarization smaller than 10^{-7}.

2. Instrumental Spurious Polarization

We have performed a systematic analysis of all the sources of spurious polarization in a polarimeter attached to the focal plane of a Cassegrain telescope.

To make our analysis a bit more practical we refer in the following the project of a polarimeter to be attached at the focal plane of the MITO telescope (MITO= Millimeter and Infrared Testagrigia Observatory) Table II provides some useful information on the MITO telescope.

TABLE 2. Main Parameters of MITO Telescope

item	characteristics
Primary Diameter	2600 mm
Primary Illuminated Diameter	2000 mm
f-number Primary	0.48
f-number Illuminated Primary	0.62
Secondary Diameter	410 mm
Photometer f-number	4.07
Actual Magnification	6.56
Actual Focal Length	8151 mm
Distance primary vertex-focal plane	461 mm
Primary-Secondary distance	1018.7 mm
Secondary vertex-rotation axis distance	228 mm
Focal plane scale	25 arcsec/mm
Detector dimensions	10-40 mm
FWHM	4.2-16.6 arcmin
Largest modulation in sky	63 arcmin.
Largest secondary tilt	2.5 deg.
angular resolution at 1mm	1.6 arcmin
throughput	0.02-1.52 $cm^2\ sr$

First of all we have considered the polarization introduced in an ideal telescope (all the optical elements are correctly aligned) when a source is travelling in front of it. The optical system consists of the MITO telescope + a detector with a circular dimension of 40 mm. The worst case is that of the largest possible angular acceptance (about 17 arcminutes in our case): the behaviour of spurious polarization is shown in Figure 3

. In the same figure we have plotted the spurious polarization in the case of the secondary being tilted by 2.5 degrees, corresponding to a sky modulation of about 17 arcminutes. An unpolarized source would produce a signal in the polarimeter corresponding to a fraction between 10^{-5} and 10^{-7} of the unpolarized power, since the spurious polarization is higher at the edge of the field of view. If we are interested in detecting CBP at the level of 1 μK, we require atmospheric fluctuations to be less than 100 mK, in order not to be confused with CBP. However, one should note that

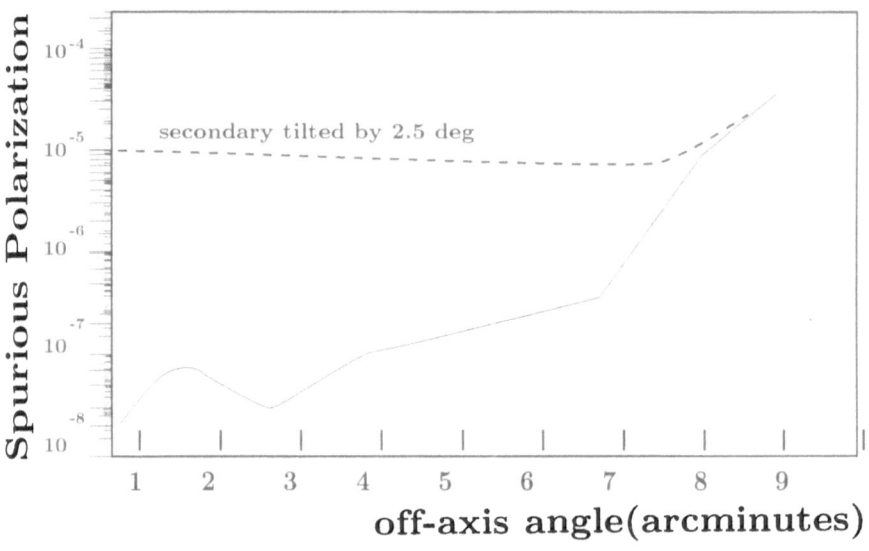

Figure 3. Spurious polarization in a Cassegrain telescope, like MITO, versus the off-axis distance of a non-polarized source. The request to stay below 10^{-7} is fulfilled for a field of view of 5 arcminutes.

only a small fraction (less than 50%) of the atmospheric fluctuation at the angular scale of 17 arcminutes is spuriously polarized at the level of 10^{-5}. Therefore, even 200 mK of r.m.s. fluctuations of the atmosphere could be tolerated. However, atmospheric uniform emission spuriously polarized by a factor 10^{-5} would produce a systematic offset of about 3mK and a careful choice of the phase reference and a very stable electronic gain is needed to get such an offset stable within $1\mu K$ for several hours of integration. This fact suggests limiting the angular aperture of the system to less than ± 5 arcminutes.

As a second step we have analysed the spurious polarization introduced by the Winston cones, which are the optical elements usually employed both to condense the radiation and to realize a sharp angular response [8,9]. Figure 4 illustrates the spurious polarization for a gold plated cone at 4 K.

One should note that cones introduce a significant depolarization; this means that they cannot be used as field-optics elements (double-cones are

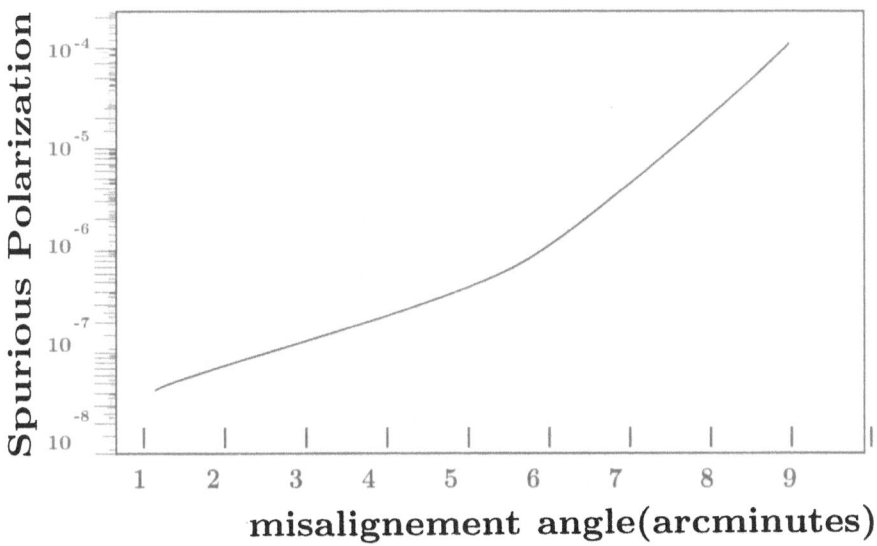

Figure 4. Spurious polarization due to a double-cone system at 4.2 K misaligned with respect to the optical axis by an angular amount given in abscissa

usually employed in far infrared photometers to transfer the radiation from the entrance of the cryostat to the detector box). Figure 5 and Table III summarize the results of our analysis.

TABLE 3. Sources of Spurious Polarization

cause	spurious linear polarization
Misalignement mirrors of 10'	1.4×10^{-8}
As above+cones in axis	10^{-7}
Misalagniment of mirrors and cones of 10'	2×10^{-7}
Source at 5' from the axis	10^{-7}
Depolarization by cones	0.72

416

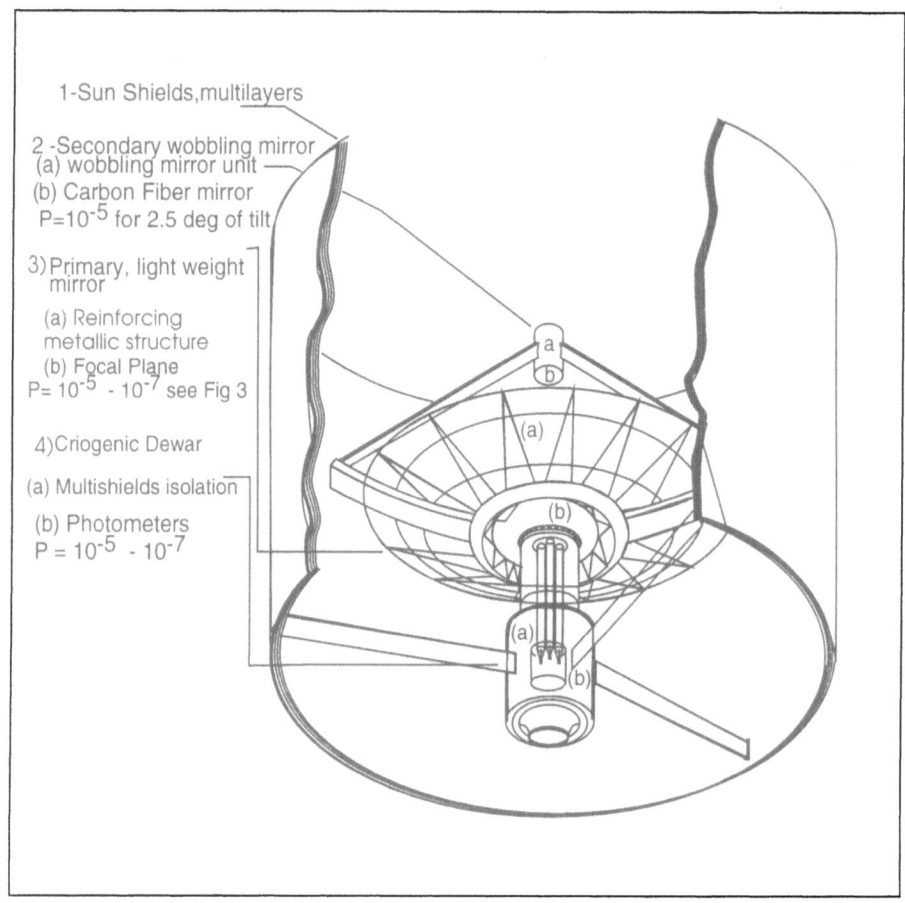

1-Sun Shields,multilayers

2 -Secondary wobbling mirror
(a) wobbling mirror unit
(b) Carbon Fiber mirror
P=10^{-5} for 2.5 deg of tilt

3)Primary, light weight
mirror

(a) Reinforcing
metallic structure
(b) Focal Plane
P= 10^{-5} - 10^{-7} see Fig 3

4)Criogenic Dewar

(a) Multishields isolation

(b) Photometers
P = 10^{-5} - 10^{-7}

Figure 5. Sources of spurious polarization in the MITO Cassegrain configuration

3. Far Infrared Polarimeters

The simplest polarimeter consists of a rotating grid polarizer (GP) and a detector, as in the case of the Caderni et al. (1978) experiment. In this case all the incoming radiation is polarized by the GP and any spurious polarization after the GP will produce a signal sinchronous with CBP. This was the main reason for the large spurious polarization of the Caderni et al. (1978) experiment. Moreover, the rotating element will emit and reflect radiation thereby producing a signal at half the frequency of the polarized radiation.

Hildebrand (1986) has suggested using a K-mirror to rotate both the image and the polarization plane: a K-mirror has a small but nevertheless not negligible residual polarization. The most appropriate solution is shown in Figure 6 and consists of a rotating double Fresnel romb.

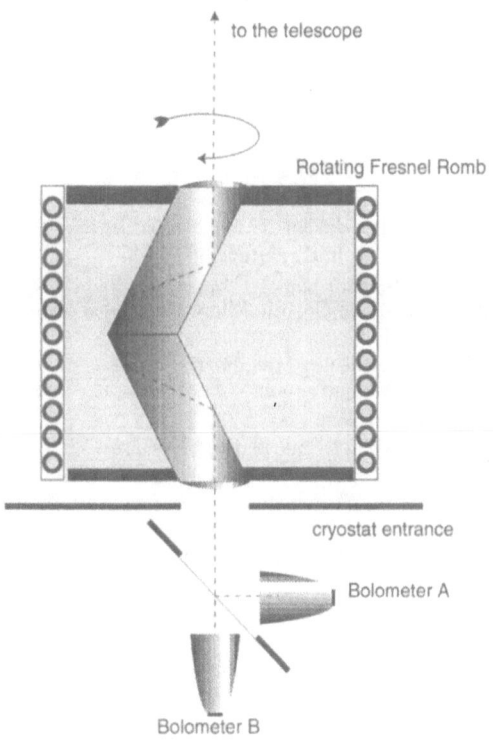

Figure 6. Sketch of the optical layout of the MITO Polarimeter: a double rotating Fresnel romb rotates the polarization plane without rotating the image. The two bolometers A and B are finely tuned to give the same signal for unpolarized atmospheric fluctuations: the differential output is therefore proportional to the polarized component alone

Pisano (1996) has proposed realizing two field lenses at the two faces of the romb in order to optimize the optical efficiency. The gains of the two bolometers are adjusted in such a way as to minimize the fluctuations of the unpolarized radiation. A polarizing interferometer can substitute the polarizing beam-splitter, thereby producing the spectrum of the polarized component alone.

4. Acknowledgements

Our heartfelt thanks to the organizers of the School for the logistical support and the warm hospitality. This work was funded by the 40% MURST Italian program " Polarizzazione del Fondo Cosmico" .

References

1. Caderni,N., Fabbri, R., Melchiorri, B., Melchiorri, F., Natale, V. (1978) Polarization of the Microwave Background Radiation: an Infrared Survey of the Sky, *Phys. Rev. D*, **17**, 1908
2. Nanos, G.P. (1979) Polarization of the Blackbody Radiation at 3.2 Centimeters, *Ap.J.*,**232**, 341
3. Lubin, P.M., Smoot G.F. (1981) Polarization of the Cosmic Background Radiation,*Ap.J.*,**245**,1
4. Lubin, P., Malese, P., Smoot, G.F. (1983) Linear and Circular Polarization of the Cosmic Background Radiation, *Ap.J.*,**273**,L51
5. Partridge, R.B., Nowakowski, H.M., Martin, H.M.(1988) Linear Polarized Fluctuations in the Cosmic Microwave Background, *Nature*,**331**, 146
6. Wollack, E.J., Jarosik, C.B., Netterfield, L.A., Page, L.A., Wilkinson, D.T. (1993) A Measurement of the Anisotropy in the Cosmic Microwave Background Radiation at Degree Angular Scale, *Ap.J.*, **419**, L49
7. Melchiorri, B., De Petris, M., Guarini, G., Melchiorri F., Signore, M. (1996) " Limitations to the Accuracy of CBA Measurements: Atmospheric Fluctuations, *Ap. J.* accepted for publication (November 1996)
8. Chipman, R.A. (1992) The Mechanics of Polarization Ray Tracing, *Proc. SPIE*,**1746**, 62
9. Welford, W.T., Winston, R. (1978), *The Optics of Nonimaging Concentrators*, Academic Press Inc. London
10. Hildebrand, R.H. (1986) Focal Plane Optics in Far-Infrared and Submillimeter Astronomy, *Opt. Eng.*,**25**, 323
11. Auton J.P. (1967) Infrared Transmission Polarizers by Photolithography, *Appl. Opt.*,**6**, 1023
12. Dall'Oglio, G., Melchiorri, B., Melchiorri, F., Natale, V., Aiello, S., Mencaraglia, F. (1972) Astronomical Polarimetry in the Far Infrared,*Planet, Stars and Nebulae Studied with Photopolarimetry*, Gehrels ed.
13. Pisano, G. (1996) Progettazione di un Polarimetro per Misure di Fondo Cosmico nel lontano IR, *Thesis* Rome University

POLARIZATION OF THE MICROWAVE BACKGROUND: THEORETICAL FRAMEWORK

ALESSANDRO MELCHIORRI AND NICOLA VITTORIO
Dipartimento di Fisica
Università "Tor Vergata", Roma, Italy

Abstract.
We present a brief review of the polarization properties of the cosmic microwave background in dark matter models for structure formation. Quite independently of the model parameters, the polarization level is expected to be $\sim 10\%$ of the anisotropy signal at angular scales $\leq 1°$. Detections of polarization at larger angular scales would provide strong evidence in favour of an early reionization of the intergalactic medium.

1. Introduction: Some Historical Remarks

Most of the early theoretical work on the polarization of the cosmic microwave background (CMB) has focused, after Rees' pioneering work [1], on anisotropic cosmological models [2,3,4,5,6,7,8,9]. The degree of Faraday rotation expected in the presence of a universal magnetic field and the use of polarization measurements to constrain the amplitude of such a field were also considered [14,33]. More recently, it has been shown that even in isotropic cosmological models the anisotropic component of the CMB is polarized [10]. Detailed numerical predictions have been made for dark matter dominated models with adiabatic fluctuations [see e.g. 11,12,13], with and without an early reionization of the intergalactic medium [15].These calculations have shown that the level of polarization can be 10 percent of the anisotropy signal. After the COBE/DMR result [16], new attention has been dedicated to the tensor modes of metric fluctuations, which also produce anisotropy on large angular scales [17,18,19,20,21,22,23]. The polarization due to a background of primordial gravitational waves has been widely discussed [24,25,26,27,28]. For describing the statistics of the polarization field the polarization - anisotropy correlation function has also

419

C. H. Lineweaver et al. (eds.), The Cosmic Microwave Background, 419–440.

been introduced [29,30], while other authors [31,32] have shown that ne-
glecting polarization yields a theoretical overestimate of the anisotropy at
small angular scale.

From the experimental side, in spite of a continuous increment in the
experimental sensitivity, no polarization was found and only upper limits
were given [34,35,36,37], with the best upper limit to date of $\sim 25\mu K$ from
the Saskatoon experiment [38]. As we show in Section 5, the level of CMB
polarization expected is in most of the models at least a factor 10 below this
limit, so is not clear if the present sensitivity of the CMB experiments is
sufficient to detect polarization. However, in view of forthcoming high sensitivity new experiments, it is of interest to discuss the general properties
of the polarization pattern and its dependence on the various cosmological
parameters. So, the aim of this work is to review the basic steps behind the
theoretical calculation of CMB polarization. The plan of the paper is as follows. In Section 2 we briefly review the definition of the Stokes parameters
and their variations in a Thomson scattering. In Section 3 and 4 we write
the set of equations necessary to describe anisotropy and polarization of the
CMB. In Section 5 we review some of the results obtained by numerically
integrating those equations. Finally, in Section 6, we summarize the main
findings.

2. An Elementary Description of the Polarization of Light

For an elliptically polarized wave, the components of the electric field along
two orthogonal directions, ξ and τ say, can be written as :

$$\begin{cases} E_\xi = E_\xi^0 \sin(\omega t - \epsilon_1) \\ E_\tau = E_\tau^0 \sin(\omega t - \epsilon_2) \end{cases} \tag{1}$$

where $E_{\xi,\tau}^0$ and $\epsilon_{1,2}$ are constants. The polarization of the radiation field is
conveniently described in terms of the Stokes parameters :

$$\begin{cases} I = E_\xi^{0^2} + E_\tau^{0^2} \equiv I_\xi + I_\tau \\ Q = E_\xi^{0^2} - E_\tau^{0^2} \equiv I_\xi - I_\tau \\ U = 2E_\xi^0 E_\tau^0 \cos[\epsilon_1 - \epsilon_2] \\ V = 2E_\xi^0 E_\tau^0 \sin[\epsilon_1 - \epsilon_2] \end{cases} \tag{2}$$

The parameter I is proportional to the intensity of the wave (we omit
the proportionality factor) V measures the ratio of the principal axes of
the polarization ellipse while Q or U measures the orientation of the ellipse
relative to the ξ axes. In general $I^2 \geq Q^2 + U^2 + V^2$, the equality holding
for an elliptically polarized wave.

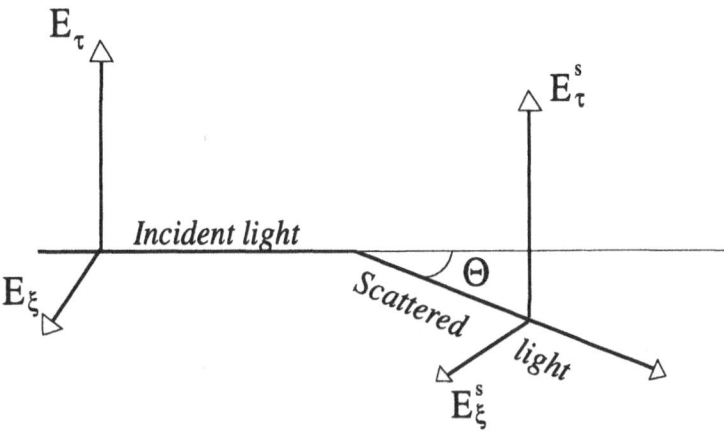

Figure 1. Thomson scattering of a photon by an electron.

A clockwise rotation by an angle Ξ of the $\xi - \tau$ axes in the polarization plane leaves unchanged the I and V parameters and it is equivalent to applying the operator:

$$\hat{\mathbf{L}}(\Xi) = \begin{pmatrix} \cos^2 \Xi & \sin^2 \Xi & \frac{\sin 2\Xi}{2} & 0 \\ \sin^2 \Xi & \cos^2 \Xi & -\frac{\sin 2\Xi}{2} & 0 \\ -\sin 2\Xi & \sin 2\Xi & \cos 2\Xi & 0 \\ 0 & 0 & 0 & 1 \end{pmatrix} \tag{3}$$

to the vector $\vec{I} \equiv (I_\xi, I_\tau, U, V)$.

For Thomson scattering, the light scattered in a direction making an angle Θ with the direction of incidence (see Figure 1) is

$$\begin{cases} E_\xi^s = \sqrt{\frac{3}{2}\sigma_T} E_\xi^0 \cos \Theta \sin(\omega t - \epsilon_1) \\ E_\tau^s = \sqrt{\frac{3}{2}\sigma_T} E_\tau^0 \sin(\omega t - \epsilon_2) \end{cases} \tag{4}$$

In analogy with Equation (2) the Stokes parameters of the scattered light are:

$$\begin{cases} I_\xi^s = (3/2)\sigma_T I_\xi \cos^2 \Theta \\ I_\tau^s = (3/2)\sigma_T I_\tau \\ U^s = (3/2)\sigma_T U \cos \Theta \\ V^s = (3/2)\sigma_T V \cos \Theta \end{cases} \tag{5}$$

or, in matrix form, $\vec{I}^s = \sigma_T \hat{R} \times \vec{I}$, where

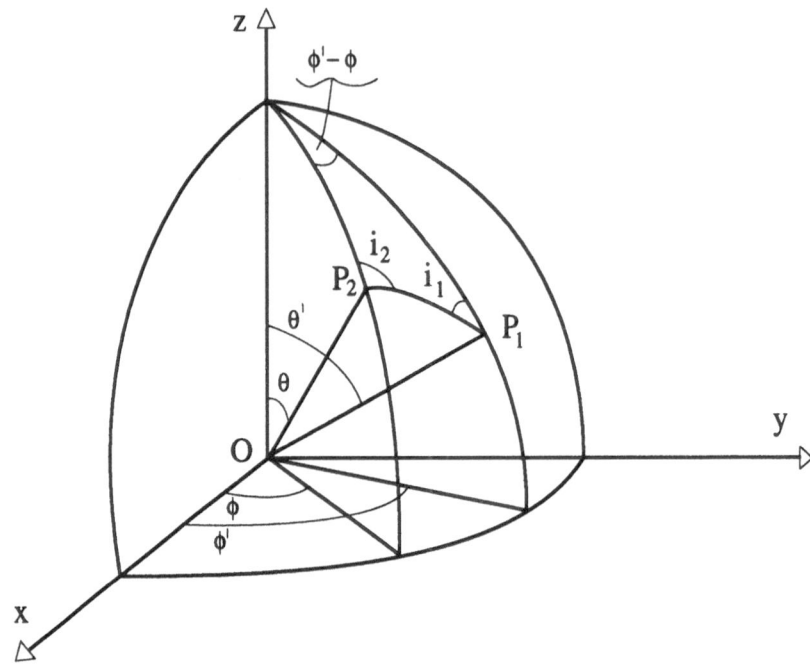

Figure 2. Coordinate system needed to describe Thomson scattering in the lab frame.

$$\hat{R} = \frac{3}{2} \begin{pmatrix} \cos^2 \Theta & 0 & 0 & 0 \\ 0 & 1 & 0 & 0 \\ 0 & 0 & \cos \Theta & 0 \\ 0 & 0 & 0 & \cos \Theta \end{pmatrix} \tag{6}$$

In order to study the variations of Stokes parameters in the lab frame (see Figure 2) we have to :

1- apply the transformation $\hat{L}(-i_1)$ to $\vec{I}$, where i_1 is the angle between the meridian and the scattering planes. In this way we obtain the Stokes parameters of the incident light in the frame of Figure 1.

2- apply $\hat{R}$ to these parameters in order to obtain the Stokes parameters of the scattered light, again in the frame of Figure 1.

3- apply the transformation $\hat{L}(\pi - i_2)$, where i_2 represents the angle between the plane OP_2Z and OP_1P_2. In this way we are back to the lab frame.

Thus, the radiation scattered in the (θ, ϕ) direction, relative to the lab frame, can be written as [39]:

$$\vec{I}^s(\theta, \phi) = \frac{1}{4\pi} \int_{4\pi} [\hat{P}(\theta, \phi; \theta', \phi') \times \vec{I}(\theta', \phi')] d\Omega' \tag{7}$$

where

$$\hat{P} = \hat{Q} \times [\hat{P}^0(\mu, \mu') + \sqrt{(1 - \mu^2)}\sqrt{(1 - \mu'^2)}\hat{P}^1(\mu, \phi, \mu', \phi') + \hat{P}^2(\mu, \phi, \mu', \phi')] \tag{8}$$

$$\hat{Q} = \begin{pmatrix} 1 & 0 & 0 & 0 \\ 0 & 1 & 0 & 0 \\ 0 & 0 & 2 & 0 \\ 0 & 0 & 0 & 2 \end{pmatrix}, \tag{9}$$

$$\hat{P}^0 = \frac{3}{4} \begin{pmatrix} 2(1 - \mu^2)(1 - \mu'^2) + \mu^2\mu'^2 & \mu^2 & 0 & 0 \\ \mu'^2 & 1 & 0 & 0 \\ 0 & 0 & 0 & 0 \\ 0 & 0 & 0 & \mu\mu' \end{pmatrix}, \tag{10}$$

$$\hat{P}^1 = \frac{3}{4} \begin{pmatrix} 4\mu\mu' \cos(\phi - \phi') & 0 & 2\mu \sin(\phi' - \phi) & 0 \\ 0 & 0 & 0 & 0 \\ -2\mu' \sin(\phi - \phi') & 0 & \cos(\phi - \phi') & 0 \\ 0 & 0 & 0 & \cos(\phi - \phi') \end{pmatrix}, \tag{11}$$

$$\hat{P}^2 = \frac{3}{4} \begin{pmatrix} \mu^2\mu'^2 \cos 2(\phi' - \phi) & -\mu^2 \cos 2(\phi' - \phi) & \mu^2\mu' \sin 2(\phi' - \phi) & 0 \\ -\mu'^2 \cos 2(\phi' - \phi) & \cos 2(\phi' - \phi) & -\mu' \sin 2(\phi' - \phi) & 0 \\ -\mu'^2\mu \sin 2(\phi' - \phi) & \mu \sin 2(\phi' - \phi) & \mu'\mu \cos 2(\phi' - \phi) & 0 \\ 0 & 0 & 0 & 0 \end{pmatrix}, \tag{12}$$

and μ and μ' are defined as $\cos\theta$ and $\cos\theta'$, respectively.

3. The Boltzmann Transfer Equation for Polarized Light

In order to study anisotropy and polarization of the CMB we need to write down the transfer equation for the Stokes parameters. We restrict ourselves to isotropic universes where, to zeroth order, anisotropy and polarization vanish. The perturbations to the Stokes parameters and to the other relevant quantities (see Section 3.3) are written in the synchronous gauge formalism. Following Peebles [40,41] we introduce the fractional fluctuations of the Stokes parameters as follows:

$$\begin{pmatrix} I \\ Q \\ U \\ V \end{pmatrix} = \frac{\rho_\gamma(t)}{4\pi} \begin{pmatrix} 1+\iota \\ q \\ u \\ v \end{pmatrix}, \tag{13}$$

where (ι,q,u,v) are functions of the observer's position $\vec{x}$, of the line of sight observation $\hat{\gamma} \equiv (\gamma_1, \gamma_2, \gamma_3)$ and of the cosmic time t. To first order the transfer equations become

$$\frac{\partial}{\partial t} \begin{pmatrix} \iota \\ q \\ u \\ v \end{pmatrix} + \frac{\gamma_\alpha}{a} \frac{\partial}{\partial x^\alpha} \begin{pmatrix} \iota \\ q \\ u \\ v \end{pmatrix} + \begin{pmatrix} y \\ 0 \\ 0 \\ 0 \end{pmatrix} = \sigma_T n_e \left[\begin{pmatrix} \iota^s \\ q^s \\ u^s \\ v^s \end{pmatrix} - \begin{pmatrix} \iota \\ q \\ u \\ v \end{pmatrix} \right], \tag{14}$$

where $y = -2\dot{h}_{\alpha\beta}\gamma_\alpha\gamma_\beta$ is the term containing the linear perturbation to the metric tensor, (ι^s, q^s, u^s, v^s) are evaluated in the comoving frame and refer to the radiation scattered in the $\vec{\gamma}$ direction, and a is the scale factor. In order to avoid spatial dependence it is convenient to work in Fourier space. We choose for each k mode a reference system with the z axis parallel to $\vec{k}$, in order to achieve an azimuthal symmetry.

3.1. SCALAR MODES

For scalar modes, the only non-vanishing components of the perturbed metric tensor are the diagonal ones : h_{11}, h_{22}, h_{33} ($h_{00} \equiv 0$ because of the chosen gauge). Thus, the gravitational term in Equation (14) has the form: $y = (1 - 3\mu^2)\dot{h}_{33} - (1 - \mu^2)\dot{h}$, and each Fourier mode is independent of the azimuthal angle ϕ. After integrating over this angle in Equation (7) it can be proved that $\hat{P}^1$ and $\hat{P}^2$ give no contribution. Therefore we can assume $U = 0$ in this case. Also the equation for V is decoupled from the others: if V vanishes at the beginning, it also vanishes afterwards. Therefore only the perturbations ι and q of the I and Q parameters are of interest. Their evolution is described by the following transfer equation [10,11,12,13]:

$$\frac{\partial}{\partial t} \begin{pmatrix} \iota \\ q \end{pmatrix} + \frac{ik\mu}{a} \begin{pmatrix} \iota \\ q \end{pmatrix} - \begin{pmatrix} y \\ 0 \end{pmatrix} = \sigma_T n_e \left(\int_{-1}^1 \hat{M}_S(\mu,\mu') \begin{pmatrix} \iota' \\ q' \end{pmatrix} d\mu' - \begin{pmatrix} \iota + 4\mu v_b \\ q \end{pmatrix} \right) \tag{15}$$

where $y = (1 - 3\mu^2)\dot{h}_{33} - (1 - \mu^2)\dot{h}$ is the term taking into account the effects of gravitational potential, and where the 2×2 matrix $\hat{M}_S$ is composed of the first two rows and columns of the matrix $\hat{P}^0$ in the (I, Q, U, V) basis:

$$\hat{M}_S(\mu,\mu') = \frac{3}{16}\begin{pmatrix} 3 - \mu'^2 - \mu^2 + 3\mu^2\mu'^2 & 1 - \mu'^2 - 3\mu^2(1-\mu'^2) \\ 1 - 3\mu'^2 - \mu^2 + 3\mu^2\mu'^2 & 3 - 3\mu'^2 - 3\mu^2(1-\mu'^2) \end{pmatrix}.$$
(16)

The angular dependence in Equation (15) can be eliminated by expanding ι and q in Legendre polynomials :

$$\iota = \sum_\ell (\sigma_{2\ell}^k(t) P_{2\ell}(\mu) + i\sigma_{2\ell+1}^k(t) P_{2\ell+1}(\mu))$$
(17)

$$q = \sum_\ell (\eta_{2\ell}^k(t) P_{2\ell}(\mu) + i\eta_{2\ell+1}^k(t) P_{2\ell+1}(\mu)).$$
(18)

Because of the orthogonality of the Legendre polynomials, Equation (15) becomes:

$$\begin{cases} \frac{\partial \iota}{\partial t} + \frac{ik\mu}{a}\iota - y_S = \sigma_T n_e \left(\sigma_0 - 4\mu v_b - \iota + P_2(\mu)1/2(\sigma_2/5 - \eta_0 + \eta_2/5) \right) \\ \frac{\partial q}{\partial t} + \frac{ik\mu}{a}q = \sigma_T n_e \left(-q + 1/2(1 - P_2(\mu))(-\sigma_2/5 + \eta_0 - \eta_2/5) \right) \end{cases}$$
(19)

These equations are coupled together through the quadrupole term, *i.e.*, the radiation must have a quadrupole anisotropy to get polarized.

3.2. TENSOR MODES

For tensor perturbations the only non-vanishing components of the perturbed metric tensor are $h_{11} = h_{22} = h_+$ and $h_{12} = h_{21} = h_\times$ where the two values h_+, $h_\times$ refer to the two polarization states of the gravitational waves . The equation of transfer has the following form [24] :

$$\frac{\partial}{\partial t}\begin{pmatrix} \iota \\ q \\ u \end{pmatrix} + \frac{ik\mu}{a}\begin{pmatrix} \iota \\ q \\ u \end{pmatrix} - \begin{pmatrix} y \\ 0 \\ 0 \end{pmatrix} = \sigma_T n_e \left(\int_\Omega \hat{M}_T(\mu,\phi;\mu',\phi') \begin{pmatrix} \iota' \\ q' \\ u' \end{pmatrix} \frac{d\Omega'}{4\pi} - \begin{pmatrix} \iota \\ q \\ u \end{pmatrix} \right)$$
(20)

where now $y = -\dot{h}_+(1 - \mu^2)\cos(2\phi) + \dot{h}_\times(1 - \mu^2)\sin(2\phi)$, and the 3×3 matrix $\hat{M}_T$ is composed of the first three rows and columns of the matrix $\hat{P}^2$ in the (I, Q, U, V) basis:

$$\hat{M}_T = \frac{3}{8}\begin{pmatrix} K_-(\mu)K_-(\mu')\cos\Delta_\phi & -K_-(\mu)K_+(\mu')\cos\Delta_\phi & -2\mu' K_-(\mu)\sin\Delta_\phi \\ K_+(\mu)K_-(\mu')\cos\Delta_\phi & -K_+(\mu)K_+(\mu')\cos\Delta_\phi & 2\mu' K_+(\mu)\sin\Delta_\phi \\ \mu K_-(\mu')\sin\Delta_\phi & -\mu K_+(\mu')\sin\Delta_\phi & 2\mu\mu'\cos\Delta_\phi \end{pmatrix}$$
(21)

with $K_\pm(\mu) = 1 \pm \mu^2$, $\Delta_\phi = 2(\phi' - \phi)$. The particular form of the metric tensor makes ι still dependent on ϕ in spite of the choice of a special reference system. However, this dependence is not too cumbersome. In fact, it is possible to introduce a change of variables [24] to eliminate the dependence on the azimuthal angle. The new quantities, $\tilde{I}$, $\tilde{Q}$ and $\tilde{U}$ are related to the old ones by the following relation:

$$
\begin{cases}
I(\mu, \phi) = \tilde{I}_+(\mu)(1 - \mu^2) \cos 2\phi + \tilde{I}_\times(\mu)(1 - \mu^2) \sin 2\phi \\
Q(\mu, \phi) = \tilde{Q}_+(\mu)(1 + \mu^2) \cos 2\phi + \tilde{Q}_\times(\mu)(1 + \mu^2) \sin 2\phi \\
U(\mu, \phi) = -\tilde{U}_+ 2\mu \sin 2\phi + \tilde{U}_\times 2\mu \cos 2\phi
\end{cases}
\tag{22}
$$

It is easy to prove that with these new variables only $\hat{P}^2$ provides a non-vanishing contribution to the integral of Equation (7) over ϕ. This is why we have considered only this term in Equation (21). Also, as the Boltzmann equation becomes independent of ϕ, we can still develop fluctuations of $\tilde{I}, \tilde{Q}$ and $\tilde{U}$ in Legendre polynomials. Thus, Equation (20) becomes:

$$
\begin{cases}
\dot{\tilde{\iota}}_\epsilon + \frac{ik\mu}{a}\tilde{\iota}_\epsilon - 2\dot{h}_\epsilon = -\sigma_T n_e(\tilde{\iota}_\epsilon + \Psi) \\
\dot{\tilde{q}}_\epsilon + \frac{ik\mu}{a}\tilde{q}_\epsilon = -\sigma_T n_e(\tilde{q}_\epsilon - \Psi) \\
\tilde{q}_\epsilon + \tilde{u}_\epsilon = 0
\end{cases}
\tag{23}
$$

where

$$
\Psi = 3/5\tilde{\eta}_{\epsilon,0} + 6/35\tilde{\eta}_{\epsilon,2} + 1/210\tilde{\eta}_{\epsilon,4} - 1/10\tilde{\sigma}_{\epsilon,0} + 1/35\tilde{\sigma}_{\epsilon,2} - 1/210\tilde{\sigma}_{\epsilon,a} \tag{24}
$$

and ϵ identifies either the $+$ or the $\times$ polarization state of the gravitational wave.

3.3. NUMERICAL CALCULATIONS

We restrict ourselves to a universe composed of baryons, cold dark matter, photons and three families of massless neutrinos. The equations describing anisotropy and polarization of the CMB have been written above. In Fourier space, the equations describing fractional fluctuations in the remaining cosmic components are [42,43,44]:

$$
\frac{\partial \iota_\nu}{\partial t} + i\frac{k\mu}{a}\iota_\nu = y \tag{25}
$$

$$
\ddot{h} + 2\frac{\dot{a}}{a}\dot{h} = 8\pi G \left(\rho_B \delta_B + \rho_{CDM}\delta_{CDM} + 2\rho_\gamma \delta_\gamma + 2\rho_\nu \delta_\nu\right) \tag{26}
$$

$$
\dot{h}_{33} - \dot{h} = \frac{16\pi Ga}{k}\left(\rho_B v + \rho_\gamma f_\gamma + \rho_\nu f_\nu\right) \tag{27}
$$

$$\dot{\delta}_B = \frac{\dot{h}}{2} - i\frac{kv}{a} \qquad (28)$$

$$\dot{v}_b + H(t)v_b = \sigma_T n_e \left(f_\gamma - \frac{4}{3}v_b\right) \qquad (29)$$

$$\dot{\delta}_{CDM} = \frac{\dot{h}}{2} \qquad (30)$$

and

$$\ddot{h}_{+,\times} + 3\frac{\dot{a}}{a}\dot{h}_{+,\times} + \frac{k^2}{a^2}h_{+,\times} = 0 \qquad (31)$$

for scalar and tensor fluctuations, respectively.

Eq.(25) describes the fluctuations in the massless neutrinos . We follow this component only when the perturbation proper wavelength is larger than one tenth of the horizon. Afterwards, free streaming rapidly damps fluctuations in this hot component.

The time evolution of the baryon and CDM density contrasts and of the baryon peculiar velocity are described by Eqs.(28), (30) and (29) respectively. The system for the scalar fluctuations is closed by Eqs.(26) and (27) describing the field equations for the trace and the $3-3$ component of the metric perturbation tensor, while Eq.(31) is all we need to describe the evolution of the metric perturbations for tensor fluctuations. We numerically integrate the previous equations from redshift $z = 10^7$ up to the present. The abundance of free electrons , n_e, is evaluated following a standard recombination scheme [45,46] for H and ^4He, taken in the ratio $77 : 33$. In the following we also consider the possibility that the Universe reionized instantaneously at redshift $z_{rh} << 1000$, and remained completely reionized up to the present.

4. Computing the Correlation Function for Anisotropy and Polarization

Under the assumption of gaussian initial conditions , the statistical properties of the CMB anisotropy and polarization patterns are fully described in terms of their correlation functions. The stochastic anisotropic component of the CMB is conveniently expanded in spherical harmonics: $\delta T(\hat{\gamma})/T_0 = \sum_{lm} a_{lm} Y_m^l(\hat{\gamma})$. The coefficients a_{lm} are random gaussian variables with zero mean and rotationally invariant variances, $C_\ell \equiv \langle| a_{lm} |^2\rangle$. The mean (over the ensemble) correlation function of the anisotropy pattern has the standard expression:

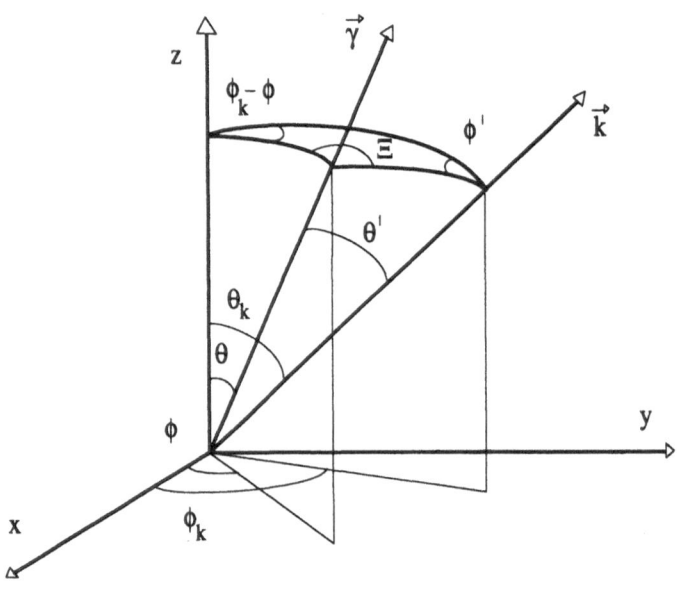

Figure 3. Laboratory reference system.

$$\langle\frac{\delta T(\vec{\gamma}_1)}{T_0}\frac{\delta T(\vec{\gamma}_2)}{T_0}\rangle = \frac{1}{4\pi}\sum_{\ell}(2\ell+1)C_{\ell}P_{\ell}(\cos\theta) \tag{32}$$

where $\cos\theta = \vec{\gamma}_1 \cdot \vec{\gamma}_2$, and

$$C_{\ell} = \frac{A_S}{8\pi}\int_0^{\infty}\frac{\mid\sigma_{\ell}(k)\mid^2}{(2\ell+1)^2}k^{n_S+2}dk. \tag{33}$$

Here the primordial power spectrum of scalar fluctuations is assumed to have the standard form $P(k) = A_S k^{n_S}$. In the case of tensor fluctuations, the change of variables needed to achieve rotationally symmetry [see Equation (22)] must be taken into account. Then, the correlation function of the CMB anisotropy induced by tensor modes reads:

$$\langle\frac{\delta T(\vec{\gamma}_1)}{T_0}\frac{\delta T(\vec{\gamma}_2)}{T_0}\rangle = \frac{A_T}{128\pi^3}\int\sum_{\ell_1\ell_2}\Pi_{\ell_1,\ell_2}(k,\vec{\gamma}'_1,\vec{\gamma}'_2)P_{\ell_1}(\mu'_1)P_{\ell_2}(\mu'_2)k^{n_T-3}d^3k \tag{34}$$

where

$$\Pi_{\ell 1,\ell 2} = K_-(\mu'_1)K_-(\mu'_2)[\tilde{\sigma}^\times_{\ell 1}\tilde{\sigma}^{\times *}_{\ell 2}\cos(2\phi'_1)\cos(2\phi'_2)+\tilde{\sigma}^+_{\ell 1}\tilde{\sigma}^{+*}_{\ell 2}\sin(2\phi'_1)\sin(2\phi'_2)] \tag{35}$$

where the power spectrum of metric fluctuations due to tensor modes is assumed to be $\tilde{P}(k) = A_T k^{n_T-3}$. Assuming $\tilde{\sigma}^+_\ell = \tilde{\sigma}^\times_\ell$ and making some algebraic manipulations yields:

$$\langle\frac{\delta T(\vec{\gamma}_1)}{T_0}\frac{\delta T(\vec{\gamma}_2)}{T_0}\rangle = \frac{A_T}{128\pi^3}\int\sum_{\ell 1\ell 2}\Upsilon(\vec{\gamma}'_1,\vec{\gamma}'_2)\tilde{\sigma}^+_{\ell 1}\tilde{\sigma}^{+*}_{\ell 2}P_{\ell 1}(\mu'_1)P_{\ell 2}(\mu'_2)k^{n_T-3}d^3k \tag{36}$$

where

$$\Upsilon = [2(\vec{\gamma}_1\cdot\vec{\gamma}_2 - \mu'_1\mu'_2)^2 - K_-(\mu'_1)K_-(\mu'_2)] \tag{37}$$

and, finally [47,48,49],

$$\langle\frac{\delta T(\vec{\gamma}_1)}{T_0}\frac{\delta T(\vec{\gamma}_2)}{T_0}\rangle = \frac{1}{4\pi}\sum_\ell(2\ell+1)\tilde{C}_\ell P_\ell(\cos\theta) \tag{38}$$

with

$$\tilde{C}_\ell = \frac{A_T}{8\pi}\frac{(\ell+2)!}{(\ell-2)!}\int_0^\infty\frac{|\tilde{\Sigma}_\ell(k)|^2}{(2\ell+1)^2}k^{n_T-1}dk \tag{39}$$

and

$$\tilde{\Sigma}_\ell(k) = \frac{\tilde{\sigma}^+_{\ell-2}}{(2\ell-1)(2\ell-3)} - 2\frac{\tilde{\sigma}^+_\ell}{(2\ell-1)(2\ell+3)} + \frac{\tilde{\sigma}^+_{\ell+2}}{(2\ell+5)(2\ell+3)}. \tag{40}$$

As discussed in Section 2, the Q and U Stokes parameters vary because of a clockwise rotation Ξ of the reference system in the polarization plane :

$$\begin{cases} Q_\Xi = Q\cos(2\Xi) + U\sin(2\Xi) \\ U_\Xi = -Q\sin(2\Xi) + U\cos(2\Xi) \end{cases} \tag{41}$$

So, the perturbations to Q and U in the lab frame are related with those in $\vec{k}$ space by the following relation:

$$\frac{Q(\vec{x},\theta,\phi)}{T_0} = \frac{1}{32\pi^3}\int q(\vec{k})e^{i\vec{k}\vec{x}}\cos[2\Xi(\vec{k})]d^3k \tag{42}$$

$$\frac{U(\vec{x},\theta,\phi)}{T_0} = \frac{-1}{32\pi^3}\int q(\vec{k})e^{i\vec{k}\vec{x}}\sin[2\Xi(\vec{k})]d^3k \tag{43}$$

The correlation function for the Q parameter for scalar modes can be written as follows:

$$\langle\frac{Q(\vec{\gamma}_1)}{T_0}\frac{Q(\vec{\gamma}_2)}{T_0}\rangle = \frac{A_S}{128\pi^3}\int\sum_{\ell_1,\ell_2}\eta_{\ell 1}^*\eta_{\ell 2}\cos(2\Xi_1)\cos(2\Xi_2)P_{\ell 1}(\mu_1')P_{\ell 2}(\mu_2')k^{ns}d^3k$$

(44)

The correlation function for U has a similar expression with $\cos 2\Xi \rightarrow \sin 2\Xi$. Let us identify the line of sight $\vec{\gamma}_1$ with the z-axis of the lab frame. In this case, $(\theta_1, \phi_1) = (0,0)$, $\theta_1' = \theta_k$ and $\phi_1' = -\phi_k$ (see Figure 3). In the small angle approximation, $\cos(2\Xi_1) \sim \cos(2\Xi_2) \sim \cos(2\phi_k)$, and Equation (44) yields :

$$\langle\frac{Q(\vec{z})}{T_0}\frac{Q(\vec{\gamma}_2)}{T_0}\rangle = A(\theta) + B(\theta, \phi),$$

(45)

where

$$A(\theta) = \frac{A_S}{256\pi^3}\int\sum_{\ell 1 \ell 2}\eta_{\ell 2}(k)\eta_{\ell 1}^*(k)P_{\ell 1}(\mu'_k)P_{\ell 2}(\mu'_2)k^{ns}d^3k$$

(46)

and

$$B(\theta, \phi) = \frac{A_S}{256\pi^3}\int\sum_{\ell 1 \ell 2}\eta_{\ell 2}(k)\eta_{\ell 1}^*(k)P_{\ell 1}(\mu'_k)P_{\ell 2}(\mu'_2)\cos(4\phi_k)k^{ns}d^3k.$$

(47)

The first term has the standard expression :

$$A(\theta) = \frac{1}{4\pi}\sum_{\ell}(2\ell + 1)C_\ell^Q P_\ell(\cos\theta)$$

(48)

with

$$C_\ell^Q = \frac{A_S}{16\pi}\int_0^\infty\frac{|\eta_\ell(k)|^2}{(2\ell+1)^2}k^{ns+2}dk.$$

(49)

For the second term, we can develop $P_{\ell 1}(\cos\theta_k)$ in associated Legendre polynomials with $m = 4$:

$$P_{\ell 1}(\cos(\theta_k)) = \sum_{\ell' \geq 4}^{\infty}\alpha_{\ell 1,\ell'}P_{\ell'}^4(\cos(\theta_k)),$$

(50)

where

$$\alpha_{\ell 1\ell'} = \frac{2\ell'+1}{2}\frac{(\ell'-4)!}{(\ell'+4)!}A_{\ell 1,\ell'}$$

(51)

and $A_{\ell 1,\ell'} = \int_{-1}^{1}P_{\ell 1}P_{\ell'}^4 d(\cos\theta_k)$ has the following values :

$$
\begin{cases}
A_{\ell,\ell} = \frac{2}{2\ell+1}\frac{\ell!}{(\ell-4)!} & \ell' \equiv \ell \\
A_{\ell,\ell'} = 8(\ell'^2 + \ell' - 3(\ell^2 + \ell + 2)) & \ell' = \ell+2, \ell+4, ..., \ell+2n \\
A_{\ell,\ell'} = 0 & \ell' = \ell+1, \ell+3, ..., \ell+2n+1 \\
A_{\ell,\ell'} = 0 & \ell' < \ell
\end{cases}
\tag{52}
$$

From the definition of the spherical harmonics it also follows:

$$
P_{\ell'}^4(\cos\theta_k) \cdot \cos(4\phi_k) = \frac{1}{2}\sqrt{\frac{4\pi}{2\ell'+1}\frac{(\ell'+4)!}{(\ell'-4)!}}(Y_{\ell'}^4(\vec{k}) + Y_{\ell'}^{-4}(\vec{k})).
\tag{53}
$$

Now, execute the following steps:

1. Insert Equation (50) into Equation (47)
2. Develop the product between the Legendre polynomial and the cosine with Equation (53)
3. Integrate over $d\Omega_k$
4. Use Equation (53) again to transform the spherical harmonics in P_ℓ^4
5. Develop the P_ℓ^4 in P_ℓ.

At the end, in the small angle approximation, it is possible to write [29,50,71]:

$$
\langle \frac{Q(\vec{z})}{T_0} \frac{Q(\vec{\gamma}_2)}{T_0} \rangle = \frac{1}{4\pi}\sum_\ell (2\ell+1)(C_\ell^Q + \cos(4\phi)B_\ell)P_\ell(\cos\theta)
\tag{54}
$$

where the two terms C_ℓ^Q and B_ℓ are defined, for scalar perturbations, as follows:

$$
C_\ell^Q = \frac{A_S}{16\pi}\int \frac{|\eta_\ell|^2}{(2\ell+1)^2}k^{n_S+2}dk
\tag{55}
$$

$$
B_\ell = \frac{A_S}{64\pi}\sum_{\ell_1\ell_2}\frac{(\ell_2-4)!}{(\ell_2+4)!}A_{\ell,\ell_2}A_{\ell_1,\ell_2}\int \eta_{\ell_1}^*\eta_{\ell_2}k^{n_S+2}dk.
\tag{56}
$$

For tensor fluctuations, the calculation is similar. The final result is [30,50,71]:

$$
\langle \frac{Q(\vec{z})}{T_0} \frac{Q(\vec{\gamma}_2)}{T_0} \rangle = \frac{1}{4\pi}\sum_\ell (2\ell+1)(\tilde{C}_\ell^Q + \cos(4\phi)\tilde{B}_\ell)P_\ell(\cos\theta)
\tag{57}
$$

where

$$
\tilde{C}_\ell^Q = \frac{A_T}{16\pi}\int \frac{(|T_\ell|^2 + 4|R_\ell|^2)k^{n_T-1}dk}{(2\ell+1)^2},
\tag{58}
$$

$$\tilde{B}_\ell = \frac{A_T}{64\pi} \sum_{\ell 1 \ell 2} \frac{(\ell_2 - 4)!}{(\ell_2 + 4)!} A_{\ell,\ell 2} A_{\ell 1,\ell 2} \int (T_{\ell 1}{}^* T_{\ell 2} + 4 R_{\ell 1}{}^* R_{\ell 2}) k^{n_T - 1} dk, \quad (59)$$

$$R_\ell = \frac{\ell + 1}{2\ell + 3} \tilde{\eta}_{\ell+1}^+ + \frac{\ell}{2\ell - 1} \tilde{\eta}_{\ell-1}^+, \quad (60)$$

$$T_\ell = \frac{(\ell + 2)(\ell + 1)}{(2\ell + 3)(2\ell + 5)} \tilde{\eta}_{\ell+2}^+ + 2 \frac{6\ell^3 + 9\ell^2 - \ell - 2}{(2\ell + 3)(2\ell - 1)(2\ell + 1)} \tilde{\eta}_\ell^+ + \frac{\ell(\ell - 1)}{(2\ell - 1)(2\ell - 3)} \tilde{\eta}_{\ell-2}^+, \quad (61)$$

and $A_{\ell 1,\ell 2} = \int P_{\ell 1}(x) P_{\ell 2}{}^4(x) dx$.

An interesting case is the correlation function between CMB anisotropy and polarization. For scalar fluctuations and in the small angle approximation, the result is [29,50,71]:

$$\langle \frac{\delta T(\vec{z})}{T_0} \frac{Q(\vec{\gamma}_2)}{T_0} \rangle = \frac{1}{4\pi} \sum_\ell (2\ell + 1) C_\ell^{QT} \cos(2\phi) P_\ell^2(\cos\theta) \quad (62)$$

where

$$C_\ell^{QT} = \frac{A_S}{16\pi} \sum_{\ell 1} \frac{(\ell - 2)!}{(\ell + 2)!} B_{\ell 1,\ell} \int \frac{\sigma_{\ell 1}{}^* \eta_\ell}{(2\ell + 1)} k^{n_S + 2} dk \quad (63)$$

and the integral $B_{\ell 1,\ell 2} = \int P_{\ell 1}(x) P_{\ell 2}{}^2(x) dx$ has the values :

$$\begin{cases} B_{\ell,\ell} = -\frac{2}{2\ell+1} \frac{\ell!}{(\ell-2)!} & \ell' = \ell \\ B_{\ell,\ell'} = 4 & \ell' = \ell + 2, \ell + 4, ..., \ell + 2n \\ B_{\ell,\ell'} = 0 & \ell' = \ell + 1, \ell + 3, ..., \ell + 2n + 1 \\ B_{\ell,\ell'} = 0 & \ell' < \ell \end{cases} \quad (64)$$

5. Numerical Results and Discussion

With the formalism developed in the previous sections, we are now able to make theoretical predictions for CMB anisotropy and polarization. As stated before, we restrict ourselves to the cold dark matter scenario. This is not quite enough to completely define the model, as we have to deal with quite a number of parameters. We have to fix : i) the total density parameter, Ω_0, and the cosmological constant, Λ; ii) the baryonic abundance, Ω_b; iii) the Hubble constant; iv) the primordial spectral index for spectral fluctuations; v) the relative amplitude of scalar and tensor fluctuations; vi) the spectral index for tensor fluctuations; vii) the thermal history of the Universe; viii) the overall amplitude for scalar fluctuations.

The baryonic abundance is quite severly restricted by primordial nucle-osynthesis. We consider the fiducial value of $\Omega_b = 0.05 \pm 0.02$ as representative of the possible uncertainty in this parameter. We remind that changes in Ω_b yield variations in the pressure of the photon-baryon fluid before recombination, and then variations in the amplitude of the first acoustic peak in the anisotropy power spectrum.

For flat models the value of the Hubble constant is usually taken to be $H_0 = 50 \ km \ s^{-1}/Mpc$ for age considerations, even if the estimated globular cluster age allows for slightly different values: $40 \ km \ s^{-1}/Mpc < H_0 < 65 \ km \ s^{-1}/Mpc$ [63]. Small variations in the Hubble constant yield huge variations in the radiation power spectrum, modifying both the amplitudes and positions of the acoustic peaks .

The primordial index for scalar fluctuations, n_S, is usually taken to be unity, as inflation suggests. However, in more general inflationary scenarios, n_S can be either smaller or larger than unity [64]. This modifies the relative power between the anisotropy on small and large angular scale. Power-law inflationary models predict $n_S < 1$, but also a background of gravitational waves . A standard prediction is that the ratio of the quadrupoles induced by scalar and tensor fluctuations is:

$$\frac{\tilde{C}_2}{C_2} = -7 n_T \sim 7(1 - n_s), \tag{65}$$

which relates the amplitudes and shapes of primordial power spectra for scalar and tensor fluctuations, respectively [18,49,64].

The thermal history of the Universe, in its standard form, assumes recombination of the primordial plasma at redshift ~ 1000. However, both the Gunn-Peterson test [65] and the enriched composition of the intracluster medium [74,75] suggest the possibility of a considerable energy release during the early stages of galaxy formation and evolution.

Finally, the amplitude of fluctuations is still unknown from first principles, and it is fixed in order to match the observed rms temperature fluctuations ($29 \pm 1 \mu K$) of the COBE/DMR anisotropy maps [53].

In Figure 4 we show theoretical predictions for CMB anisotropy and polarization of a standard cold dark matter model with $\Omega_0 = 1$, $\Omega_b = 0.05$, $H_0 = 50 \ km \ s^{-1}/Mpc$, $n_S = 1$ and standard recombination. The anisotropy power spectrum for scalar fluctuations has a flat behavior at low ℓ's, where the Sachs-Wolfe effect [66] dominates, and a structure of peaks at higher ℓ's, defined by the acoustic oscillations in the photon-baryon fluid experienced before recombination by fluctuations smaller then the acoustic horizon. The damping at high ℓ's is due to the finite thickness of the last-scattering surface [67]. The first peak at $\ell \sim 200$ corresponds to fluctuations that entered the horizon at recombination, which subtend an angle $\sim 2^{\circ} h^{-1}$.

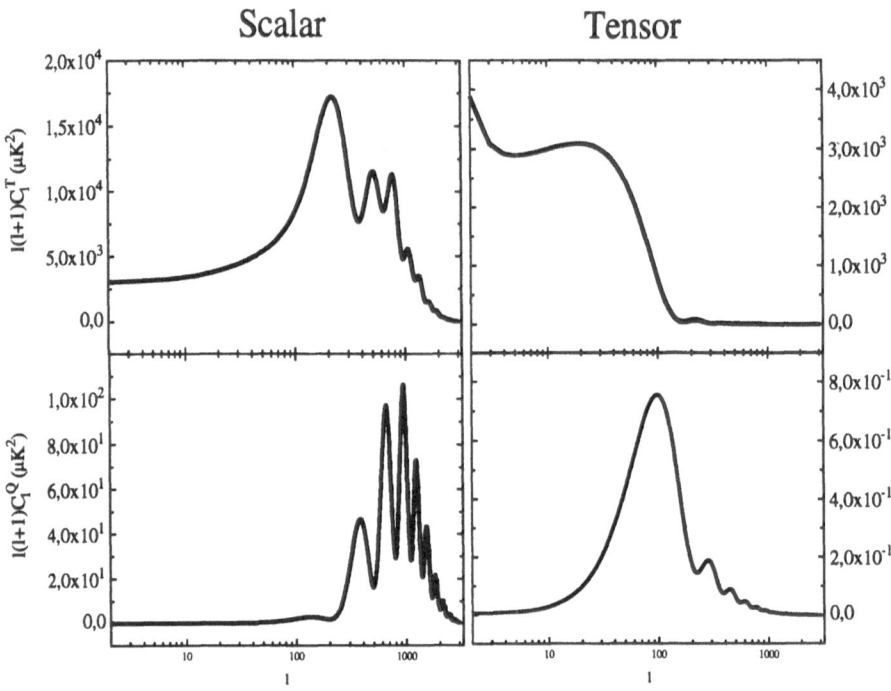

Figure 4. Anisotropy (top) and polarization (bottom) power spectrum for scalar (left) and tensor (right) fluctuations. Each model is normalized to COBE $\sigma(10°) = 29\,\mu K$.

The polarization power spectrum has instead power only at $\ell > 200$, *i.e.*, on scales $< 2°h^{-1}$. The case of pure tensor fluctuations is shown in Figure 4 only for didactic purposes. In this case, the anisotropy spectrum has power only at $\ell < 200$ and the polarization spectrum shows a prominent peak at $\ell \sim 100$. Note that the polarization spectrum has an amplitude lower than the anisotropy spectrum by a factor $\sim 10^2$ and $\sim 10^4$ for scalar and tensor fluctuations, respectively.

Real experiments are sensitive to a limited region of the power spectrum, because of the antenna beam and modulation techniques. For anisotropy experiments, this effect can be parameterized by a window function, W_ℓ, so that the variance of temperature fluctuations detected by an experiment can be written as:

$$\left\langle \left(\frac{\delta T(\vec{\gamma})}{T_0}\right)^2 \right\rangle = \frac{1}{4\pi} \sum_\ell (2\ell + 1) C_\ell W_\ell \qquad (66)$$

A similar expression holds for the variance of the fluctuations of the Q parameter:

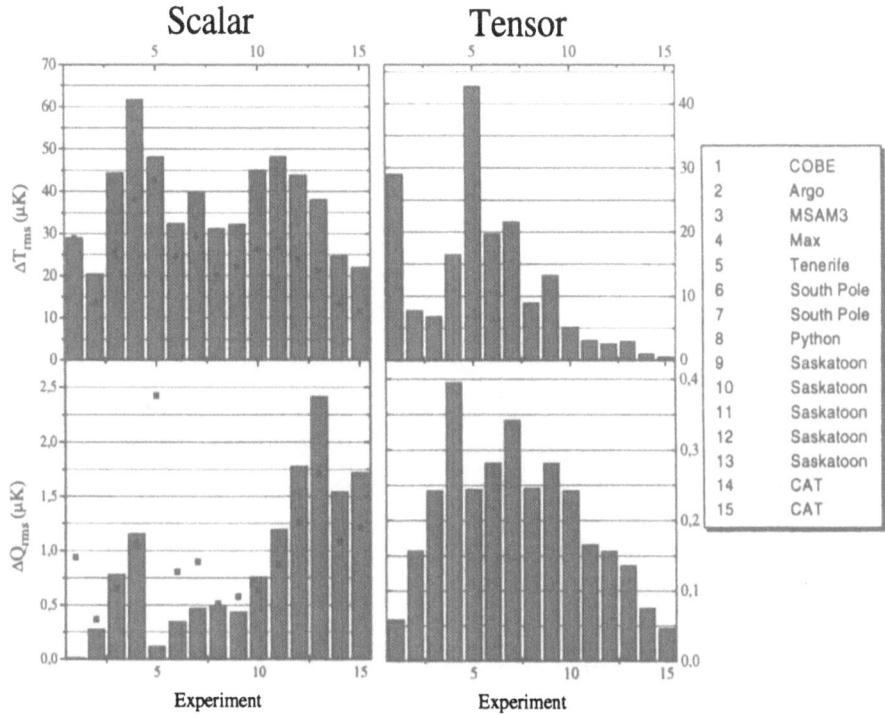

Figure 5. R.m.s. values for anisotropies (top) and polarization (bottom) scheduled for scalar (left) and tensor (right) models for different experiments [53...61]. The tensor model has $n_T = 0$ and the scalar model has $n_S = 1$. Each model is normalized to COBE $\sigma(10°) = 29\mu K$. The dot of the left side represents the values for a scalar reionized model at $z_r \sim 70$.

$$\left\langle \left(\frac{Q(\vec{\gamma})}{T_0} \right)^2 \right\rangle = \frac{1}{4\pi} \sum_\ell (2\ell + 1) C_\ell^Q W_\ell. \qquad (67)$$

In fact, it can be proved that the azimuthal contribution [see Equations (54) and (57)] to the Q variance vanishes. In Equation (67) we use the same anisotropy window functions , in order to give an order of magnitude estimate of the level of measurable polarization at different angular scales.

In Figure 5 we plot the expected rms values for CMB anisotropy and polarization using 15 different window functions corresponding to 9 different anisotropy experiments. The level of polarization from scalar modes is below the current experimental sensitivity, even for small scale experiments such as Saskatoon or CAT that are sensitive to multipole $\ell \sim 400$ where the polarization has the first two peaks. The rms level from pure tensor modes, even for the MAX experiment that seems to have the best

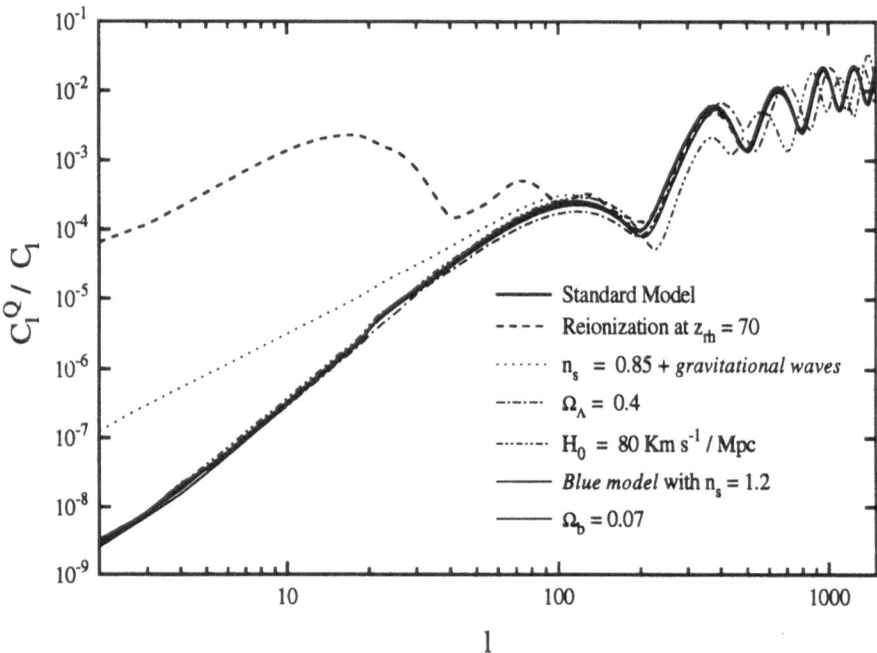

Figure 6. Dependence of polarization on cosmological parameters.

window function, is below 0.5 μK, so the separation between scalar and tensor fluctuations does not seem to be at hand with polarization measurements [25,26,27,28,73]. This can be done by combining anisotropy measurements at both large (where tensor modes could contribute) and small (where tensor modes do not contribute) angular scales [49,72,73]. An accurate mapping of the anisotropy pattern with both high sensitivity and high angular resolution will be provided by planned, dedicated space missions such as COBRAS/SAMBA [68] and MAP [52]. At the moment the bulk of degree-scale detections, combined with the COBE/DMR and Tenerife experiments, seems to suggest a spectral index for scalar fluctuations $n_S \geq 1$ [69] and a negligible contribution of tensor modes.

As mentioned above, there are several free parameters, each with its own uncertainty, which define a theoretical model. So, it is interesting to explore the sensitivity of the polarization level relative to the anisotropy one. To show this, we plot in Figure 6 the quantity C_ℓ^Q/C_ℓ as a function of ℓ, for different choices of the model parameters. Generally speaking the effects of the variation of these parameters on anisotropy and polarization are similar: both quantities tend to decrease with increasing H_0 and tend to

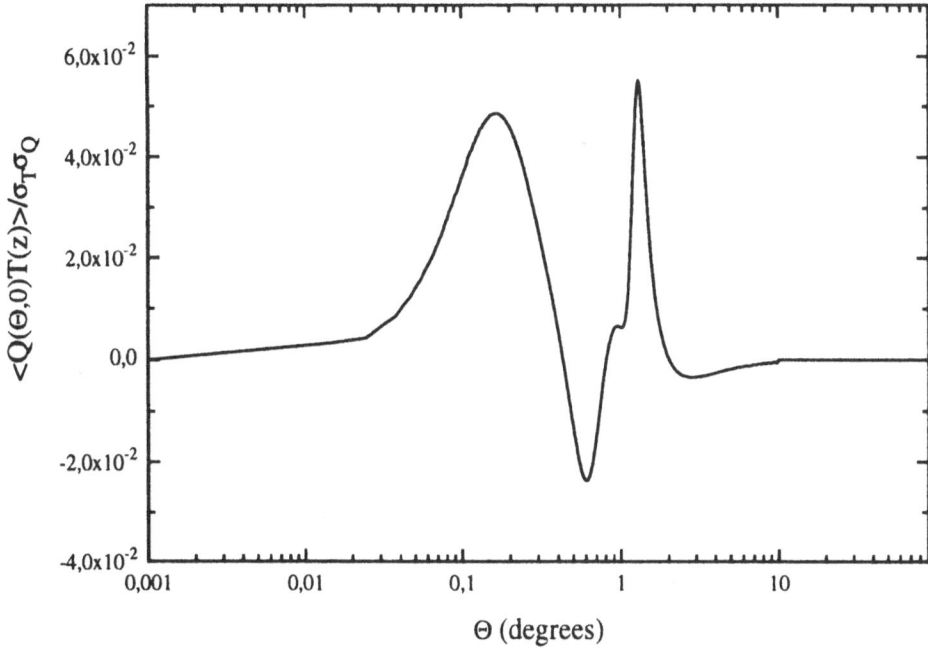

Figure 7. Polarization - Anisotropy correlation function.

increase when a cosmological constant $\Omega_\Lambda = 1 - \Omega_0$ is taken into account. In particular, varying n_S or Ω_b yields basically no variations on the C_ℓ^Q/C_ℓ ratio. Moreover, decreasing the spectral index and adding gravitational waves increases the large scale polarization, but, as we have shown, not enough to pass the threshold of present day detector sensitivity. So, even taking into account reasonable uncertainties in the parameters, it seems that only with a huge increment in the experimental sensitivity (see the accompanying paper by F. Melchiorri et al. in this volume) and/or a space mission [52,70,71] a robust detection of the polarization spectrum over a wide range of ℓ's would be possible. Coulson et al. [29] suggested searching for a correlation between the temperature in one direction and the polarization in a circle at distance Θ from that direction. The shape of the correlation function (62) measurable in this way is shown in Figure 7. As we can see the cross correlation is positive on scales $> 1^o$, negative on scales between 0.5^o and 1^0 and positive again on scales $< 0.5^o$. According to [52] the future MAP satellite would have the capability to measure the expected amplitude of this signal.

The final item to be investigated is the dependence of CMB anisotropy and polarization on the thermal history of the Universe. A reionization at $z \leq 100$ produces a new, later and thicker last scattering surface. The effect of such a new last scattering surface is to smooth the anisotropy on small angular scales and to leave unchanged the level of anisotropy on large angular scales. The effect of reionization on polarization is to reduce the polarization on small scales but to increase the polarization level at large angular scales. This is shown again in Figure 6. Thus, possible detection of polarization between 1° and 10°, say, would be evidence for an early re-heating of the intergalactic medium.

6. Conclusions

Numerical solutions of the Boltzmann transport equation show that a certain degree of polarization must be present as a consequence of the primordial fluctuations responsible for structure formation. The level of polarization depends strongly on the angular scale, much more than in the case of anisotropy, quite independently of the choice of the model parameters. At angular scales larger than one degree we do not expect detectable polarization unless the Universe was reionized at early times $z \geq 40$. At small angular scales the polarization may reach the $5 - 10\%$ of the anisotropy. However, a polarization of a few percent at angular scales of 1-5 degrees could be explained only with reionization: a search for polarization at these scale is therefore important in the study of the thermal history of the Universe. Also, it seems hard to disentangle tensor from scalar perturbations through measurements of polarization, due to the tenuity of the signal.

7. Acknowledgments

We would like to thank Paolo de Bernardis for comments and suggestions. AM thanks Arthur Kosowsky for helpful discussions. This work has been supported by MURST.

References

1. Rees M. J., (1968) Polarization and Spectrum of the Primeval Radiation in an Anisotropic Universe, *Astr. J. Lett.*, **153**, L1
2. Caderni N., Fabbri R., Melchiorri B., Melchiorri F., Natale V., (1978) Polarization of the Microwave Background Radiation I: Theory, *Phys. Rev. D*, **17**
3. Dautcourt G., Rose K., (1978), *Astron. Nachrichten*, **299**, 13
4. Basko M.M., Polnarev A.G., (1980), *Sov. Astron.*, **24**, 268
5. Negroponte, Silk J., (1980), *Phys. Rev. Lett.*, **44**, 1433
6. Tolman B.W., Matzner R.A., (1984), *Proc. R. Soc. Lond.*, **391**, A392
7. Fabbri R. Milaneschi E., (1985) , *A&A*, **151**, 714
8. Milaneschi E. Fabbri R., (1986) , *A&A*, **162**, 6
9. Fabbri R., Tamburrano, (1987) , *A&A*, **179**, 11F
10. Kaiser N., (1983), *Mont. No. R. Astr. Soc.*, **101**, 1169

11. Bond J. R., Efstathiou G., (1984), *Ap. J.*, **285**, L45
12. Bond J. R., Efstathiou G., (1987), *Mon. No. R. Astr. Soc.*, **226**, 655
13. Milaneschi E., Valdarnini R., (1986), *A&A*, **162**, 5-12
14. Ceccarelli C., Dall'Oglio G., de Bernardis P., Masi S., Melchiorri B., Melchiorri F., Moreno G., Pietranera L., Pucacco G., (1982), The Polarization of the Cosmic Background and the Universal Magnetic Field, *The Birth of the Universe* Eds. J. Audouze, J. Tran Thanh Van, Edition Frontieres, 191
15. Efstathiou G., (1988), *Large Scale Motions in the Universe: a vatican study week. Edited by Vera C. Rubin George V. Coyne*, Princeton series in physics
16. Smoot G. et al., (1992), *Ap. J.*, **396**, L1
17. Krauss L.M., White M., (1992), *Phys. Rev. Lett.*, **69**, 869-872
18. Davis R.L. et al., (1992), *Phys. Rev. Lett.*, **69**, 1856
19. Liddle A.R., Lyth D., (1992), *Phys. Lett.*, **B 291**, 391
20. Lidsey J.E. Coles P., (1992), *Mon. Not. R. Astr. Soc.*, **358**, 57
21. Salopek D., (1992), *Phys. Rev. Lett.*, **69**, 3602
22. Lucchin F., Matarrese S., Mollerach S., (1992), *Ap. J.*, **401**, L49
23. Sourdeep T. et al., (1992), *Phys. Lett.*, **A 7**, 3541
24. Polnarev, A.G., (1985), *Sov. Astr.*, **29**, 607
25. Crittenden R., Davis R.L., Steinhardt P.J., (1993), *Ap. J.*, **417**, L13
26. Frewin R.A., Polnarev A.G., Coles P. , (1994), *Mon. No. R. Astr. Soc.*, **266**, L21
27. Sazhin M.V., Benitez N., (1995), *Astro. Lett. and Communications*, **32**, 105
28. Ng K.L., Ng K.W., (1995), *Ap. J.*, **445**, 521
29. Coulson D., Crittenden R., Turok N.G. , (1994), *Phys. Rev. Lett.*, **73**, 2390
30. Crittenden R., Coulson D., Turok, N.G. , (1995), *Phys. Rev. D*, **52**, 5402
31. Hu W., Scott D., Sugiyama N., White M., (1995), *Phys. Rev. D*, **52**, 5498
32. Chun-Pei M., Bertschinger E., (1995), *Apj*, **455**, 7
33. Kosowsky A., Loeb A. , (1996), *Apj*, in press
34. Penzias A.A., Wilson R.W., (1965) ,*Ap.J.*, **142**, 419
35. Nanos G.P., (1979) , *Ap. J.*, **232**, 341
36. Caderni N., Fabbri R., Melchiorri B., Melchiorri F., Natale V., (1978) Polarization of the Microwave Background Radiation I: an Infrared Survey of the Sky, *Phys. Rev. D*, **17**
37. Lubin P., Smoot G., (1981) , *Ap. J.*, **273**, L51
38. Wollack et al, (1993), *Ap. J.*, **419**, L49
39. Chandrasekhar, S., (1960), *Radiation Transfer*, Ed. Dover
40. Peebles P.J.E., Yu J.T., (1970), *Ap. J.*, **162**, 815
41. Peebles P.J.E., (1980), The large scale structure of the Universe.
42. Peebles P.J.E, (1981), *Ap. J.*, **243**, L119
43. Peebles P.J.E., (1982), *Ap. J.*, **258**, 415
44. Peebles P.J.E., (1982), *Ap. J.*, **263**, L1
45. Peebles P.J.E., (1968), *Ap. J.*, **153**, 1
46. Jones B.J., Wyse R., (1985), *A & A*, **149**, 144-150
47. Fabbri R., Pollock M., (1983), *Phys. Letters*, **125B**, 445
48. Abbot L.F., Wise M., (1984), *Nucl. Phys.*, **B244**, 541
49. Crittenden R., Bond J.R., Davis R.L., Efstathiou G., Steinhardt P., (1993), *Phys. Rev Lett.*, **71**, 324c
50. Kosowsky A., (1996), *Annals of Physics*, **246**, 49
51. Steinhardt P., (1995), *Astro-ph/9502024*
52. Bennett C.L. et al., (1995), MAP MIDEX mission proposal.
53. Bennett C.L. et al., (1996), *Ap. J.*, **464**, L1
54. Hancock S. et al, (1994), *Nature*, **367**, 333
55. Gundersen J.O. et al., (1993), *Ap. J.*, **413**, L1
56. Dragovan M. et al., (1993), *Ap. J.*, **427**, L67
57. de Bernardis P. et al., (1994), *Ap. J.*, **422**, L33
58. Tanaka S.T. et al., (1996), *Ap. J.*, 468, L81

59. Cheng E.S. et al., (1994), *Ap. J.*, **422**, L40
60. Netterfield et al., (1997), *Ap. J.*, **474**, 47, *Astro-ph/9601197*
61. Scott P.F. et al., (1996), *Ap. J.*, **461**, L1
62. Jackson J.D., (1962), *Classical electrodynamics*, John Wiley, NY
63. Kolb E.W. Turner M.S., (1993), *The Early Universe*, Addison Wesley, CA
64. Kolb E.W. Vadas S.L., (1994), *Phys. Rev. D*, **50**, 4, 2479
65. Gunn J.E. Peterson B.A., (1965), *Ap. J.*, **142**, 1633
66. Sachs R.K. Wolfe A.M., (1967), *Ap. J.*, **147**, 73
67. Silk J., (1968), *Ap. J.*, *151*, 459
68. Bersanelli M. et al., (1996), *COBRAS/SAMBA proposal*, D/SCI(96)3
69. de Bernardis P., Balbi A., de Gasperis G., Melchiorri A., Vittorio N., (1997), *Ap. J.*, *480*, 1, *Astro-ph/9609154*.
70. Kamionkowski M., Kosowsky A., Stebbins A., (1996), *Astro-ph/9609132*
71. Seljak U., (1996), *Astro-ph/9608131*
72. Knox L. Turner M.S., (1994), *Phys. Rev. Lett.*, **73**
73. Turner M.S., (1996), *Astro-ph/9607066*
74. Mushotzky R.F., (1984), *Physica Scripta*, **T7**, 157
75. Edge A.C., Stewart G.C., (1991), *Mont. No. R. Astr. Soc.*, **252**, 414

INDEX

446